At a Glance

To James Lafayette German III, MD
New York

Physician—Human Biologist—Musician,

Mentor and Friend

Color Atlas of Genetics

Eberhard Passarge, MD
Professor of Human Genetics
Former Director
Institute of Human Genetics
University Hospital Essen
Essen, Germany

Third edition, revised and updated

With 202 color plates prepared
by Jürgen Wirth

Thieme
Stuttgart · New York

Library of Congress Cataloging-in-Publication Data

Passarge, Eberhard
[Taschenatlas der Genetik. English]
Color Atlas of genetics/Eberhard Passarge; with 202 color plates prepared by Jürgen Wirth. – 3rd ed., rev. and updated.
p.; cm.
Includes bibliographical references and index.
ISBN-13: 978-3-13-100363-8
(GTV: alk. paper)
ISBN-10: 3-13-100363-4 (GTV: alk. paper)
ISBN-13: 978-1-58890-336-5
(TNY: alk. paper)
ISBN-10: 1-58890-336-2 (TNY: alk. paper)
1. Genetics–Atlases. 2. Medical genetics–Atlases. I. Tile.
[DNLM: 1. Genetics, Medical–Atlases, 2. Genetics, Medical–Handbooks, QZ 17 P286t 2006a]
QH436.P3713 2006
576.5022'2–dc22
2006023813

1st German edition 1994
1st English edition 1995
1st French edition 1995
1st Japanese edition 1996
1st Chinese edition 1998
1st Italian edition 1999
1st Turkish edition 2000
2nd English edition 2001
2nd French edition 2003
2nd German edition 2004
1st Polish edition 2004
1st Portuguese edition 2004
1st Spanish edition 2004
1st Greek edition 2005

© 2007 Georg Thieme Verlag KG
Rüdigerstraße 14, D-70469 Stuttgart, Germany
http://www.thieme.de
Thieme New York, 333 Seventh Avenue, New York, NY 10001 USA
http://www.thieme.com

Color plates prepared by Jürgen Wirth, Professor of Visual Communication, Dreieich, Germany

Typesetting by Druckhaus Götz GmbH, D-71636 Ludwigsburg
Printed and bound in India by Replika Press Pvt. Ltd.

ISBN 10: 3-13-100363-4 (GTV)
ISBN 13: 978-3-13-100363-8 (GTV)
ISBN 10: 1-58890-336-2 (TNY)
ISBN 13: 978-1-58890-336-5 (TNY)
1 2 3 4 5 6

Preface

The aim of this book is to give an account of the scientific field of genetics based on visual displays of selected concepts and related facts. Additional information is presented in the introduction, with a chronological list of important discoveries and advances in the history of genetics, in an appendix with supplementary data in tables, in an extensive glossary explaining genetic terms, and in references, including websites for further in-depth studies. This book is written for two kinds of readers: for students of biology and medicine, as an introductory overview, and for their mentors, as a teaching aid. Other interested individuals will also be able to gain information about current developments and achievements in this rapidly growing field.

Gerhardus Kremer (1512–1594), the mathematician and cartographer known as Mercator, first used the term atlas in 1594 for a book containing a collection of 107 maps. The frontispiece shows a figure of the Titan Atlas holding the globe on his shoulders. When the book was published a year after Kremer's death, many regions were still unmapped. Genetic maps are a leitmotif in genetics and a recurrent theme in this book. Establishing genetic maps is an activity not unlike mapping new, unknown territories 500 years ago.

This third edition has been extensively rewritten, updated, and expanded. Every sentence and illustration was visited and many changed to improve clarity. The general structure of the previous editions, which have appeared in 11 languages, has been maintained:
Part I, *Fundamentals*; Part II, *Genomics*; Part III, *Genetics and Medicine*.

Each color plate is accompanied by an explanatory text on the opposite page. Each double page constitutes a small, self-contained chapter. The limited space necessitates a concentration on the most important threads of information at the expense of related details not included. Therefore, this book is a supplement to, rather than a substitute for, classic textbooks.

New topics in this third edition, represented by new plates, include overviews of the taxonomy of living organisms ("tree of life"), cell communication, signaling and metabolic pathways, epigenetic modifications, apoptosis (programmed cell death), RNA interference, studies in genomics, origins of cancer, principles of gene therapy, and other topics.

A single-author book of this size cannot provide all the details on which specialized scientific knowledge is based. However, it can present an individual perspective suitable as an introduction. This hopefully will stimulate further interest. I have selected many topics to emphasize the intersection of theoretical fundamentals and the medical applications of genetics. Diseases are included as examples representing genetic principles, but without the many details required in practice.

Throughout the book I have emphasized the importance of evolution in understanding genetics. As noted by the great geneticist Theodosius Dobzhansky, "Nothing in biology makes sense except in the light of evolution." Indeed, genetics and the science of evolution are intimately connected. For the many young readers naturally interested in the future, I have included a historical perspective. Whenever possible and appropriate, I have referred to the first description of a discovery. This is a reminder that the platform of knowledge today rests on previous advances.

All color plates were prepared for publication by Jürgen Wirth, Professor of Visual Communication at the Faculty of Design, University of Applied Sciences, Darmstadt, Germany 1986–2005. He created all the illustrations from computer drawings, hand sketches, or photographs assembled for each plate by the author. I am deeply indebted to Professor Jürgen Wirth for the most pleasant cooperation. His most skillful work is a fundament of this book. I thank my wife, Mary Fetter Passarge, MD, for her careful editing of the manuscript and for her numerous helpful suggestions. At Thieme International, Stuttgart, I was guided and supported by Stephan Konnry. I also wish to thank Stefanie Langner and Elisabeth Kurz of the Production Department for the pleasant cooperation.

Eberhard Passarge

Acknowledgements

In preparing this third edition many colleagues from different countries again kindly provided illustrations, valuable comments, or useful information. I am grateful to them and to anyone who suggests possible improvements for future editions.

I wish to express my gratitude to Alireza Baradaran (Mashhad, Iran), John Barranger (Pittsburgh), Claus R. Bartram (Heidelberg), Laura Carrel (Hershey, Pennsylvania), Thomas Cremer (München), Nicole M. Cutright (Creighton, Pennsylvania), Andreas Gal (Hamburg), Robin Edison (NIH, Bethesda, Maryland), Evan E. Eichler (Seattle), Wolfgang Engel (Göttingen), Gebhard Flatz (Bonn, formerly Hannover), James L. German (New York), Dorothea Haas (Heidelberg), Cornelia Hardt (Essen), Reiner Johannisson (Lübeck), Richard I. Kelley (Baltimore), Kiyoshi Kita (Tokyo), Christian Kubisch (Köln), Nicole McNeil and Thomas Ried (NIH, Bethesda, Maryland), Roger Miesfeld (Tucson, Arizona), Clemens Müller-Reible (Würzburg), Maximilian Muenke (NIH, Bethesda, Maryland), Stefan Mundlos (Berlin), Shigezuku Nagata (Osaka), Daniel Nigro (Long Beach City College, California), Alfred Pühler (Bielefeld), Helga Rehder (Marburg), André Reis (Erlangen), David L. Rimoin (Los Angeles), Michael Roggendorf (Essen), Hans Hilger Ropers (Berlin), Gerd Scherer (Freiburg), Axel Schneider (Essen), Evelin Schröck (Dresden), Eric Schulze-Bahr (Münster), Peter Steinbach (Ulm), Gesa Schwanitz and Heredith Schüler (Bonn), Michael Speicher (Graz, formerly München), Manfred Stuhrmann-Spangenberg (Hannover), Gerd Utermann (Innsbruck), Thomas Voit (Essen), Michael Weis (Cleveland), Johannes Zschocke (Heidelberg).

In addition, the following colleagues at our Department of Human Genetics, Universitäsklinikum Essen, made helpful suggestions: Karin Buiting, Hermann-Josef Lüdecke, Bernhard Horsthemke, Dietmar Lohmann, Beate Albrecht, Michael Zeschnigk, Stefan Böhringer, Dagmar Wieczorek, and Sven Fischer. In secretarial matters I was supported by Liselotte Freimann-Gansert and Astrid Maria Noll. Figures were provided by Beate Albrecht, Karin Buiting, Gabriele Gillessen-Kaesbach (now Lübeck), Bernhard Horsthemke, Elke Jürgens, and Dietmar Lohmann.

About the Author

The author is a medical scientist in human genetics at the Medical Faculty of the University of Duisburg–Essen, Germany. He graduated from the University of Freiburg in 1960 with an MD degree and received training in different fields of medicine in Hamburg, Germany, and Worcester, Massachusetts/USA, between 1961 and 1963, in part with a stipend from the Ventnor Foundation. During a residency in pediatrics at the University of Cincinnati, Children's Medical Center, he worked in human genetics as a student of *Josef Warkany* from 1963–1966 before working as a research fellow in human genetics with *James German* at the Cornell Medical Center New York from 1966–1968. Thereafter he established cytogenetics and clinical genetics at the Department of Human Genetics, University of Hamburg (1968–1976). In 1976 he became Founding Chairman of the Department of Human Genetics, University of Essen, Germany. He retired from the chair in 2001, but remains active in teaching human genetics. The author's field of research covers the genetics and clinical delineation of hereditary disorders, in particular Hirschsprung disease and Bloom syndrome, and associated congenital malformations, and includes chromosomal and molecular studies documented in more than 230 peer-reviewed research articles and in textbooks. He is former President of the German Society of Human Genetics (1990–1996), Secretary-General of the European Society of Human Genetics (1989–1992), and a member of various scientific societies in Europe and the USA. The practice of medical genetics and teaching of human genetics are of particular interest to the author. He received the Hufeland Prize in 1978 and the Mendel Medal of the Czechoslovakian Biological Society in 1986. He is an honorary member of the Czechoslovakian Society for Medical Genetics and the Purkyne Society Prague, corresponding honorary member of the Romanian Academy of Medical Sciences, and corresponding member of the American College of Medical Genetics. He served as Vice Rector of the University of Essen from 1983–1988, as Chairman of the Ethics Committee Medical Faculty Essen from 1981–2001, and on the editorial board of several scientific journals in human genetics.

Table of Contents

Introduction

Reasons for Studying Genetics

Genetics is defined in dictionaries as the science that deals with heredity and variation in organisms, including the genetic features and constitution of a single organism, species, or group, and with the mechanisms by which they are effected (*Encyclopaedia Britannica* 15th edition, 1995; *Collins English Dictionary*, 5th edition 2001). New investigative methods and observations, especially during the last 50 years, have moved genetics into the mainstream of biology and medicine. Genetics is relevant to virtually all fields of medicine and biological disciplines, anthropology, biochemistry, physiology, psychology, ecology, and other fields of the sciences. As both a theoretical and an experimental science, it has broad practical applications in understanding and control of genetic diseases and in agriculture. Knowledge of basic genetic principles and their medical application is an essential part of medical education today.

The determination of the nearly complete sequence of the building blocks encoding the genetic information of man in 2004 marked an unprecedented scientific milestone in biology. The Human Genome Project, an international organization of several countries, reported this major achievement just 50 years after the structure of DNA, the molecule that encodes genetic information, was elucidated (IHGSC, 2004). Although much work remains before we know how the molecules of life interact and produce living organisms, through genetics we now have a good foundation for understanding the living world from a biological perspective.

Each of the approximately ten trillion (10^{13}) cells of an adult human contains a program with life-sustaining information in its nucleus (except red blood cells, which do not have a nucleus). This information is hereditary, transmitted from one cell to its descendent cells, and from one generation to the next. About 200 different types of cells carry out the complex molecular transactions required for life.

Genetic information allows organisms to convert atmospheric oxygen and ingested food into energy production, it regulates the synthesis and transport of biologically important molecules, protects against unwarranted invaders, such as bacteria, fungi, and viruses by means of an elaborate immune defense system,

and maintains the shape and mobility of bones, muscles, and skin. Genetically determined functions of the sensory organs enable us to see, to hear, to taste, to feel heat, cold, and pain, to communicate by speech, to support brain function with the ability to learn from experience, and to integrate the environmental input into cognate behavior and social interaction. Reproduction and detoxification of exogenous molecules likewise are under genetic control. Yet, the human brain is endowed with the ability to take free decisions in daily life and developing plans for the future.

The living world consists of two types of cells, the smallest membrane-bound units capable of independent reproduction: *prokaryotic* cells without a nucleus, represented by bacteria, and *eukaryotic* cells with a nucleus and complex internal structures, which make up higher organisms. Genetic information is transferred from one cell to both daughter cells at each cell division and from one generation to the next through specialized cells, the *germ cells*, *oocytes*, and *spermatozoa*.

The integrity of the genetic program must be maintained without compromise, yet it must be adaptable to long-term changes in the environment. Errors in maintaining and transmitting genetic information occur frequently in all living systems despite the existence of complex systems for damage recognition and repair.

Biological processes are mediated by biochemical reactions performed by biomolecules, called *proteins*. Each protein is made up of dozens to several hundreds of amino acids arranged in a linear sequence that is specific for its function. Subsequently, it assumes a specific three-dimensional structure, often in combination with other polypeptides. Only this latter feature allows biological function. Genetic information is the blueprint for producing the proteins in a given cell. Most cells do not produce all possible proteins, but a selection, depending on the type of cell. The instructions are encoded in discrete units, the *genes*.

Each of the 20 amino acids used by living organisms recognizes a code of three specific chemical structures. These are the nucleotide bases of a large molecule, DNA (deoxyribonucleic acid). DNA is a read-only memory device of a genetic information system, called the *genetic code*. In contrast to the binary system of strings of ones and zeros used in computers ("bits," which are then combined into "bytes," which are eight binary digits long), the genetic code in the living world uses a quaternary system of four nucleotide bases with chemical names having the initial letters A, C, G, and T (see Part I, Fundamentals). The quaternary code used in living cells uses three building blocks, called a *triplet codon*. This genetic code is universal and is used by all living cells, including plants and viruses. A gene is a unit of genetic information. It is equivalent to a single sentence in a text. Thus, genetic information is highly analogous to a linear text and is amenable to being stored in computers.

Genes

Depending on the organizational complexity of an organism, its number of genes ranges from about 5000 in bacteria, 6241 in yeast, 13,601 in the fruit fly *Drosophila melanogaster*, and 18,424 in a nematode to about 22,000 in humans and other mammals (which is much less than assumed a few years ago). The minimal number of genes required to sustain independent cellular life is surprisingly small; it takes about 250–400 for a prokaryote. Since many proteins are involved in related functions of the same pathway, they and their corresponding genes can be grouped into families of related function. It is estimated that the human genes form about 1000 gene families. The entirety of genes and DNA in each cell of an organism is called the *genome*. By analogy, the entirety of proteins of an organism is called the *proteome*. The corresponding fields of study are termed *genomics* and *proteomics*, respectively.

Genes are located on chromosomes. Chromosomes are individual, complex structures located in the cell nucleus, consisting of DNA and special proteins. Chromosomes come in pairs of homologous chromosomes, one derived from the mother, and one from the father. Man has 23 pairs, consisting of chromosomes 1–22 and an X and a Y chromosome in males or two X chromosomes in females. The number and size of chromosomes in different organisms vary, but the total amount of DNA and the total number of genes are the same for a particular species. Genes are arranged in linear order along each chromosome. Each gene has a defined position, called a *gene locus*. In higher organisms, genes are structured into contiguous sections of coding and noncoding sequences, called *exons*

(coding) and *introns* (noncoding), respectively. Genes in multicellular organisms vary with respect to size (ranging from a few thousand to over a million nucleotide base pairs), the number and size of exons, and regulatory DNA sequences. The latter determine the state of activity of a gene, called *gene expression*. Most genes in differentiated, specialized cells are permanently turned off. Remarkably, more than 90% of the 3 billion (3×10^9) base pairs of DNA in higher organisms do not carry known coding information (see Part II, Genomics).

The linear text of information contained in the coding sequences of DNA in a gene cannot be read directly. Rather, its total sequence is first transcribed into a structurally related molecule with a corresponding sequence of codons. This molecule is called RNA (*ribonucleic acid*). RNA is processed by removing the noncoding sections (*introns*). The coding sections (*exons*) are spliced together into the final template, called messenger RNA (mRNA). This serves as a template to arrange the amino acids in the sequence specified by the genetic code. This process is called *translation*.

Genes and Evolution

In *The Origin of Species*, Charles Darwin wrote in 1859 at the end of chapter IX, On the Imperfection of the Geological Record: "…I look at the natural geological record, as a history of the world imperfectly kept, and written in a changing dialect; of this history we possess the last volume alone, relating only to two or three countries. Of this volume, only here and there a short chapter has been preserved; and of each page, only here and there a few lines." Advances in genetics and new findings of hominid remains have provided new insights into the process of evolution.

Genes with comparable functions in different organisms share structural features. Occasionally they are nearly identical. This is the result of evolution. Living organisms are related to each other by their origin from a common ancestor. Cellular life was established about 3.5 billion years ago when land masses first appeared. Genes required for fundamental functions are similar or almost identical across a wide variety of organisms, e.g., in bacteria, yeast, insects, worms, vertebrates, mammals, and even plants. They control vital functions such as the cell cycle, DNA repair, or in embryonic development and differentiation. Similar or identical genes present in different organisms are referred to as *conserved in evolution*.

Genes evolve within the context of the genome of which they are a part. Evolution does not proceed by accumulation of mutations. Most mutations are detrimental to function and usually do not improve an organism's chance of surviving. Rather, during the course of evolution existing genes are duplicated or parts of genes reshuffled and brought together in a new combination. The duplication event can involve an entire genome, a whole chromosome or a part of it, or a single gene or group of genes. All these events have been documented in the evolution of vertebrates. The human genome contains multiple sites that were duplicated during evolution (see Part II. Genomics).

Humans, *Homo sapiens*, are the only living species within the family of *Hominidae*. All data available are consistent with the assumption that today's humans originated in Africa about 100 000 to 300 000 years ago, spread out over the earth, and populated all continents. Owing to regional adaptation to climatic and other conditions, and favored by geographic isolation, different ethnic groups evolved. Human populations living in different geographic regions differ in the color of the skin, eyes, and hair. This is often mistakenly used to define human races. However, genetic data do not support the existence of human races. Genetic differences exist mainly between individuals regardless of their ethnic origin. In a study of DNA variation from 12 populations living on five continents of the world, 93–95% of differences were between individuals; only 3–5% were between the populations (Rosenberg et al., 2002). Observable differences are literally superficial and do not form a genetic basis for distinguishing races. Genetically, *Homo sapiens* is one rather homogeneous species of recent origin. As a result of evolutionary history, humans are well adapted to live peacefully in relatively small groups with a similar cultural and linguistic history. However, humans have not yet adapted to global conditions. They tend to react with hostility to groups with a different cultural background in spite of negligible genetic differences.

Changes in Genes: Mutations

In 1901, H. De Vries recognized that genes can change the contents of their information. For this new observation, he introduced the term *mutation*. The systematic analysis of mutations contributed greatly to the developing science of genetics. In 1927, H. J. Muller determined the spontaneous mutation rate in *Drosophila* and demonstrated that mutations can be induced by roentgen rays. C. Auerbach and J. M. Robson in 1941 and, independently, F. Oehlkers in 1943 observed that certain chemical substances also could induce mutations. However, it remained unclear what a mutation actually was, since the physical basis for the transfer of genetic information was not known.

Genes of fundamental importance do not tolerate changes (mutations) that compromise function. As a result, deleterious mutations do not accumulate in any substantial number. All living organisms have elaborate cellular systems that can recognize and eliminate faults in the integrity of DNA and genes (*DNA repair*). Mechanisms exist to sacrifice a cell by programmed cell death (*apoptosis*) if the defect cannot be successfully repaired.

Early Genetics Between 1900 and 1910

In 1906, the English biologist William Bateson (1861–1926) proposed the term *genetics* for the new biological field devoted to investigating the rules governing heredity and variation. Bateson referred to heredity and variation when comparing the differences and similarities, respectively, of genealogically related organisms. Heredity and variation represent two views of the same phenomenon. Bateson clearly recognized the significance of the Mendelian rules, which had been rediscovered in 1900 by Correns, Tschermak, and De Vries.

The Mendelian rules are named after the Augustinian monk Gregor Mendel (1822–1884), who conducted crossbreeding experiments on garden peas in his monastery garden in Brünn (Brno, Czech Republic) in 1865. Mendel recognized that heredity is based on individual factors that are independent of each other. These factors are transmitted from one plant generation to the next in a predictable pattern, each factor being responsible for an observable trait. The trait one can observe is the *phenotype*. The underlying genetic information is the *genotype*.

Johann Gregor Mendel

cessary for normal development. This clearly indicated that the individual chromosomes possess different qualities.

Genetics became an independent scientific field in 1910 when Thomas H. Morgan introduced the fruit fly *(Drosophila melanogaster)* for systematic genetic studies at Columbia University in New York. Subsequent systematic genetic studies on *Drosophila* showed that genes are arranged on chromosomes in sequential order. Morgan summarized this in 1915 in the *chromosome theory of inheritance.*

The English mathematician G.H. Hardy and the German physician W. Weinberg independently recognized in 1908 that Mendelian inheritance accounts for certain regularities in the genetic structure of populations. Their work contributed to the successful introduction of genetic concepts into plant and animal breeding. Although genetics was well established as a biological field by the end of the second decade of the last century, knowledge of the physical and chemical nature of genes was sorely lacking. Structure and function remained unknown.

However, the fundamental importance of Mendel's conclusions was not recognized until 1900. The term *gene* for this type of a heritable factor was introduced in 1909 by the Danish biologist Wilhelm Johannsen (1857–1927). Beginning in 1901, Mendelian inheritance was systematically analyzed in animals, plants, and also in man. Some human diseases were recognized as having a hereditary cause. A form of brachydactyly (type A1, McKusick number MIM 112500) observed in a large Pennsylvania sibship by W. C. Farabee (PhD thesis, Harvard University, 1902) was the first condition in man to be described as being transmitted by autosomal dominant inheritance (Haws and McKusick, 1963).

Chromosomes were observed in dividing cells (in mitosis by Flemming in 1879; in meiosis by Strasburger in 1888). Waldeyer coined the term chromosome in 1888. Before 1902, the existence of a functional relationship between genes and chromosomes was not suspected. Early genetics was not based on chemistry or cytology. An exception is the prescient work of Theodor Boveri (1862–1915), who recognized the genetic individuality of chromosomes in 1902. He wrote that not a particular number but a certain combination of chromosomes is ne-

Thomas Hunt Morgan

Genetic Individuality

In 1902, Archibald Garrod (1857–1936), later Regius Professor of Medicine at Oxford University, demonstrated that four congenital metabolic diseases (albinism, alkaptonuria, cystinuria, and pentosuria) are transmitted by autosomal recessive inheritance. He called these *inborn errors of metabolism* (1909). Garrod was also the first to recognize that subtle biochemical differences among individuals result from individual genetic differences. In 1931, he published a prescient monograph entitled *The Inborn Factors in Disease*. He suggested that small genetic differences might contribute to the causes of diseases. Garrod, together with W. Bateson, introduced genetic concepts into medicine in the early years of genetics between 1902 and 1909. In late 1901, Garrod and Bateson began an extensive correspondence about the genetics of alkaptonuria and the significance of consanguinity, which Garrod had observed among the parents of affected individuals. Garrod clearly developed the idea of human biochemical individuality. In a letter to Bateson on 11 January 1902, Garrod wrote, "I have for some time been collecting information as to specific and individual differences of metabolism, which seems to me a little explored but promising field in relation to natural selection, and I believe that no two individuals are exactly alike chemically any more than structurally." (Bearn, 1993) However, Garrod's concept of the genetic individuality of man was not recognized at the time. One reason may have been that the structure and function of genes was totally unknown, in spite of early fundamental discoveries. Today we recognize that individual susceptibility to disease is an important factor in its causes (see Childs, 1999).

The sequence of DNA is not constant but differs between unrelated individuals within a group of organisms (a species). These individual differences occur about once in 1000 base pairs of human DNA between individuals (*single nucleotide polymorphism*, SNP). They occur in noncoding regions. Individual genetic differences in the efficiency of metabolic pathways are thought to predispose to diseases that result from the interaction of many genes, often in combination with particular environmental influences. They may also protect one individual from an illness to which another is prone. Such

Archibald Garrod

individual genetic differences are targets for individual therapies with specifically designed pharmaceutical substances aimed at high efficacy and a low risk of side effects. This is investigated in the field of *pharmacogenetics*.

A Misconception in Genetics: Eugenics

Eugenics, a term coined by Francis Galton in 1882, is the study of improvement of humans by genetic means. Such proposals date back to ancient times. Many countries between about 1900 and 1935 adopted policies and laws which were assumed to lead to the erroneous goals of eugenics. It was believed that the "white race" was superior to others, but proponents did not realize that genetically defined human races do not exist. Eugenists believed that sterilizing individuals with diseases thought to be hereditary would improve human society. By 1935, sterilization laws had been passed in Denmark, Norway, Sweden, Germany, and Switzerland, as well as in 27 states of the United States. Individuals with mental impairment of variable degree, epilepsy, criminals, and homosexuals were prime targets. Although in most cases the stated purpose was eugenic, sterilizations were carried out for social rather than genetic reasons.

The complete lack of knowledge of the structure and function of genes probably contributed to the eugenic misconceptions, which assumed that "bad genes" could be eliminated from human populations. However, the disorders targeted are either not hereditary or have a complex genetic background. Sterilization simply will not reduce the frequency of genes contributing to mental retardation and other disorders. In Nazi Germany, eugenics was used as a pretext for widespread discrimination and the murder of millions of innocent human beings claimed to be "worthless" (see Müller-Hill, 1988; Vogel and Motulsky, 1997; Strong, 2003). All reasons based on genetics are totally invalid. Modern genetics has shown that the ill-conceived eugenic approach to attempt to eliminate human genetic disease is impossible. Thus, incomplete genetic knowledge was applied to human individuals at a time when nothing was known about the structure of genes. Indeed, up to 1949 no fundamental advances in genetics had been obtained by studies in humans. Quite the opposite holds true today. It is evident that genetically determined diseases cannot be eradicated. Society has to adjust to their occurrence. No one is free from a genetic burden. Every individual carries about five or six potentially harmful changes in the genome which might manifest as a genetic disease in a child.

The Rise of Modern Genetics Between 1940 and 1953

With the demonstration in the fungus *Neurospora crassa* that one gene is responsible for the formation of one enzyme ("one gene, one enzyme," Beadle and Tatum in 1941), the close relationship of genetics and biochemistry became apparent. This is in agreement with Garrod's concept of inborn errors of metabolism. Systematic studies in microorganisms led to other important advances in the 1940s. Bacterial genetics began in 1943 when Salvador E. Luria and Max Delbrück discovered mutations in bacteria. Other important advances were genetic recombination demonstrated in bacteria by Lederberg and Tatum in 1946, and in viruses by Delbrück and Bailey in 1947; as well as spontaneous mutations observed in bacterial viruses, the bacteriophages, by Hershey in 1947. The study of genetic phenomena in microorganisms turned out to be as significant for the further development of genetics as the analysis of *Drosophila* had been 35 years earlier (see Cairns et al., 1978). A very influential, small book entitled *What is Life?* by the physicist E. Schrödinger (1944) postulated a molecular basis for genes. From then on, the elucidation of the molecular biology of the gene became a central theme in genetics.

Max Delbrück and Salvador E. Luria at Cold Spring Harbor (Photograph by Karl Maramorosch, from Judson, 1996)

Genetics and DNA

A major advance was the discovery by Avery, MacLeod, and McCarty at the Rockefeller Institute in New York in 1944 that a chemically relatively simple, long-chained nucleic acid (deoxyribonucleic acid, DNA) carries genetic information in bacteria (for historical reviews see Dubos, 1976; McCarty, 1985). Many years earlier in 1928, F. Griffith had observed that permanent (genetic) changes could be induced in pneumococcal bacteria by a cell-free extract derived from other strains of pneumococci (the *transforming principle*). Avery and his co-workers showed that DNA was this transforming principle. In 1952, Hershey and Chase proved that DNA alone carries genetic information and excluded other molecules. With this discovery, the question of the structure of DNA took center stage in biology.

This question was resolved most elegantly by James D. Watson, a 24-year-old American on a scholarship in Europe, and Francis H. Crick, a 36-year-old English physicist, at the Cavendish Laboratory of the University of Cambridge. On 25 April 1953, they proposed in a short article of one page in the journal *Nature* the structure of DNA as a double helix (Watson and Crick, 1953).

Oswald T. Avery

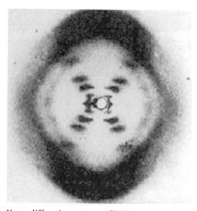

X-ray diffraction pattern of DNA
(Franklin & Gosling, 1953)

Although it was not immediately recognized as such, this discovery is the cornerstone of modern genetics in the 20th century. The novel features of this structure were derived from careful model building based on the X-ray diffraction pattern (see figure) and data provided by colleagues, mainly Maurice Wilkins and Rosalind Franklin. Franklin argued against a helical structure and announced (with R. Gosling) "… with great regret … the death of D.N.A. Helix (crystalline) on Friday 18th July, 1952. A memorial service will be held …." (Judson, 1996; Wilkins, 2003). An earlier basis for recognizing the importance of DNA was the discovery by E. Chargaff in 1950 that of the four nucleotide bases guanine was present in the same quantity as cytosine, and adenine in the same quantity as thymine. However, this was not taken to be the result of pairing (Wilkins, 2003).

The structure of DNA as a double helix with the nucleotide bases inside explains two fundamental genetic mechanisms: storage of genetic information in a linear, readable pattern and replication of genetic information to ensure its accurate transmission from generation to generation. The DNA double helix consists of two complementary chains of alternating sugar (deoxyribose) and monophosphate molecules, oriented in opposite directions. Inside the helical molecule are paired nucleotide bases. Each pair consists of a pyrimidine and a purine, either cytosine (C) and guanine (G) or thymine (T) and adenine (A). The crucial feature is that the base pairs are inside the molecule, not outside. That the authors fully recognized the significance for genetics of the novel structure is apparent from the closing statement of their article, in which they state, "It has not escaped our notice that the specific pairing we have postulated immediately suggests a possible copying mechanism for the genetic material." Vivid, albeit different, accounts of their discovery have been given by the authors (Watson, 1968; Crick, 1988) and by Wilkins (2003).

The elucidation of the structure of DNA is regarded as the beginning of a new era of molecular biology and genetics. The description of DNA as a double-helix structure led directly to an understanding of the possible structure of genetic information. When F. Sanger determined the sequence of amino acids of insulin in 1955, he provided the first proof of the primary structure of a protein. This supported the notion that the sequence of amino acids in proteins could correspond to the sequential character of DNA. However, since DNA is located in the cell nucleus and protein synthesis occurs in the cytoplasm, DNA cannot act directly. Rather, it is first

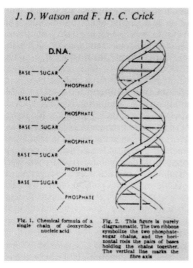

DNA structure 1953

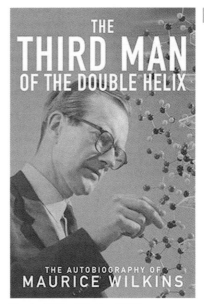

Maurice Wilkins (Maddox, 2002)

transcribed into a chemically similar messenger molecule, called *messenger ribonucleic acid* (mRNA) when it was discovered by Crick, Barnett, Brenner, and Watts-Tobin in 1961. mRNA, with a corresponding nucleotide sequence, is transported into the cytoplasm. Here it serves as a template for the amino acid sequence encoded in DNA. The genetic code for the synthesis of proteins from DNA and mRNA was determined in the years 1963–1966 by Nirenberg, Mathaei, Ochoa, Benzer, Khorana, and others. Detailed accounts of these developments have been presented by several authors (see Chargaff, 1978; Judson, 1996; Stent, 1981; Watson and Tooze, 1981; Crick, 1988; Watson, 2000; Wilkins, 2003).

Watson and Crick in 1953
(Photograph by Anthony Barrington Brown, Nature 421: 417, 2003)

Rosalind Franklin

With the structure of DNA known, the nature of the gene could be redefined in molecular terms. In 1955, Seymour Benzer provided the first genetic fine structure. He established a map of contiguous deletions of a region (rII) of the bacteriophage T4. He found that mutations fell into two functional groups, A and B. Mutants belonging to different groups could complement each other (eliminate the effects of the deletion); those belonging to the same group could not. This work showed that the linear array of genes on chromosomes also applied to the molecule of DNA. This defined the gene in terms of function and added an accurate molecular size estimate for the components of a gene.

New Methods in the Development of Genetics after 1953

From the beginning, genetics has been a field strongly influenced by the development of new experimental methods. In the 1950s and 1960s, the groundwork was laid for *biochemical genetics* and *immunogenetics*. Relatively simple but reliable procedures for separating complex molecules by different forms of electrophoresis, methods of synthesizing DNA in vitro (Kornberg in 1956), and other approaches were applied to genetics. The introduction of cell culture methods was of particular importance for the genetic analysis of humans. G. Pontecorvo introduced the genetic analysis of cultured eukaryotic cells (*somatic cell genetics*) in 1958. The study of mammalian genetics, with increasing significance for studying human genes, was facilitated by methods of fusing cells in culture (*cell hybridization*; T. Puck, G. Barski, B. Ephrussi in 1961) and the development of a cell culture medium for selecting certain mutants in cultured cells (*HAT medium*; Littlefield in 1964). The genetic approach that had been so successful in bacteria and viruses could now be applied in higher organisms, thus avoiding the obstacles of a long generation time and breeding experiments. A hereditary metabolic defect in man (galactosemia) was demonstrated for the first time in cultured human cells in 1961 (R. S. Krooth). The correct number of chromosomes in man was determined in 1956 (Tjio and Levan; Ford and Hamerton). Lymphocyte cultures were introduced for chromosomal analysis (Hungerford and co-workers in 1960). The replication pattern of human chromosomes

was described (German in 1962). These and other developments paved the way for a new field, *human genetics*. Since the late 1970s, this field has taken root in all areas of genetic studies, in particular molecular genetics.

Molecular Genetics

The discovery of reverse transcriptase, independently by H. Temin and D. Baltimore in 1970, upset a central dogma in genetics that the flow of genetic information is in one direction only, from DNA to RNA and from RNA to a protein as the gene product. *Reverse transcriptase* is an enzyme complex in RNA viruses (*retroviruses*) which transcribes RNA into DNA. This is not only an important biological finding, but this enzyme can be used to obtain *complementary* DNA (cDNA) that corresponds to the coding regions of an active gene. This allows one to analyze a gene directly without knowledge of its gene product. Enzymes cleaving DNA at specific sites, called *restriction endonucleases* or, simply, *restriction enzymes*, were discovered in bacteria by W. Arber in 1969, and by D. Nathans and H. O. Smith in 1971. They can be used to cleave DNA into fragments of reproducible and defined size, thus allowing easy recognition of an area to be studied. DNA fragments of different origin can be joined and their properties analyzed. Methods of probing for genes, producing multiple copies of DNA fragments (polymerase chain reaction, PCR, see part I), and sequencing the nucleotide bases of DNA were developed between 1977 and 1985 (see Part I, Polymerase chain reaction and DNA sequencing). All these methods are collectively referred to as *recombinant DNA technology*.

In 1977, recombinant DNA analysis led to a completely new and unexpected finding about the structure of genes in higher organisms. Genes are not continuous segments of coding DNA, but are interrupted by noncoding segments. The size and pattern of coding DNA segments, called *exons*, and of the noncoding segments, called *introns* (two new terms introduced by W. Gilbert in 1978) are characteristic for each gene. This is known as the *exon/intron structure* of eukaryotic genes. Modern molecular genetics allows the determination of the chromosomal location of a gene and the analysis of its structure without prior knowledge of the gene product. The extensive homologies of genes that regulate embryological development in different organisms and the similarities of genome structures have removed the boundaries in genetic analysis that formerly existed between different organisms (e.g., *Drosophila* genetics, mammalian genetics, yeast genetics, bacterial genetics). Genetics has become a broad, unifying discipline in biology, medicine, and evolutionary research.

Human Genetics

Human genetics deals with all human genes, normal and abnormal. However, it is not limited to humans, but applies knowledge and uses methods relating to many other organisms. These are mainly other mammals, vertebrates, yeast, fruit fly, and microorganisms. Arguably, human genetics was inaugurated when *The American Society of Human Genetics* and the first journal of human genetics, the *American Journal of Human Genetics*, were established in 1949. In addition, the first textbook of human genetics appeared in 1949, Curt Stern's *Principles of Human Genetics*.

The medical applications of human genetics contribute to the understanding of the underlying cause of a disease. This leads to improved precision in diagnosis. The concept of disease in human genetics differs from that in medicine. In medicine, diseases are usually classified according to organ systems, age, and gender. In human genetics, diseases are classified according to gene loci, genes, types of mutations (molecular pathology). Some genetic diseases result from rearrangements in different genes, or different rearrangements in one and the same gene may result in clinically different diseases. These diseases belong into different medical specialties, although the underlying genetic fault is the same. Without genetic knowledge, the common basis would go unrecognized.

The causes of diseases are not viewed as random processes, but rather as the consequences of individual attributes of a person's genome and its encounter with the environment, as first proposed in A. Garrod's *Inborn Factors in Disease* in 1931. Depending on the family history and the type of disease, it is possible to obtain diagnostic information about a disease that will manifest in the future. Not only the affected individual, the patient, but also other, unaffected family members, seek information about their own risk for a disease or the risk for a disease in their offspring. Thus, a family approach is the rule in the medical application of human genetics. The concept of disease in human genetics is widened beyond the patient and the borders of medical specialties. Thus, it provides a unifying basis for the understanding of diseases.

Two important discoveries in 1949 relate to a human disease that still poses a public health problem in tropical parts of the world. J.V. Neel showed that sickle cell anemia is inherited as an autosomal recessive trait. Pauling, Itano, Singer, and Wells demonstrated that a defined alteration in normal hemoglobin was the cause. This is the first example of a human molecular disease. The first biochemical basis of a human disease was demonstrated in liver tissue by Cori & Cori in 1952. It was an enzyme defect, glucose-6-phosphatase deficiency, in glycogen storage disease type I, also called von Gierke disease.

In 1959, the first chromosomal aberrations were discovered in three clinically well-known human disorders: trisomy 21 in Down syndrome by J. Lejeune, M. Gautier, R. Turpin; monosomy X (45,X) in Turner syndrome by Ford and co-workers; and an extra X chromosome (47,XXY) in Klinefelter syndrome by Jacobs & Strong. Subsequently, other numerical chromosome aberrations were shown to cause recognizable diseases in man: trisomy 13 and trisomy 18, by Patau and co-workers and Edwards and co-workers in 1960, respectively. The loss of a specific region (a deletion) of a chromosome was shown to be associated with a recognizable pattern of severe developmental defects by Lejeune and co-workers, 1963; Wolf, 1964; and Hirschhorn in 1964). The Philadelphia chromosome, a characteristic structural alteration of a chromosome in bone marrow cells of patients with chronic myelogenous leukemia, which was discovered by Nowell and Hungerford in 1962, showed a connection to the origins of cancer. The central role of the Y chromosome in establishing gender in mammals became apparent when it was realized that individuals without a Y chromosome are female and individuals with a Y chromosome are male, irrespective of the number of X chromosomes present. These observations further promoted interest in a new subspecialty, *human cytogenetics.*

Since the early 1960s, new insights into mechanisms in genetics in general have been obtained, often for the first time by studies in man. Analysis of genetically determined diseases in man has provided new knowledge about the normal function of genes in other organisms as well. Today, more is known about the general genetics of man than about that of any other species. Numerous subspecialties of human genetics have arisen, such as *biochemical genetics, immunogenetics, somatic cell genetics, cytogenetics, clinical genetics, population genetics, teratology, mutational studies*, and others. The development of human genetics has been well summarized by Vogel and Motulsky (1997), and McKusick (1992).

More than 3000 defined human genetic diseases are known to be due to a mutation at a single gene locus. These are monogenic diseases inherited according to a Mendelian mode of inheritance. About 1900 monogenic diseases have been recognized at the molecular level. Their manifestations differ widely with respect to the age of onset and organ systems involved. This reflects the wide spectrum of genetic information contained in the genes involved. Many monogenic diseases are pleiotropic, i.e., they affect more than one organ system. Monogenic diseases have been catalogued in *Mendelian Inheritance of Man* (McKusick, 1998). This rich source of indispensable information is available online (OMIM at www.ncbi.nlm.nih.gov/Omim). This synopsis, begun by V. A. McKusick in Baltimore in 1966, has established the systematic basis of human diseases and the genes involved. Throughout this book, the MIM catalog number is provided for every disease mentioned.

The enormous progress since about 1975 in clarifying the genetic etiology of human diseases has mainly been achieved by molecular methods, thereby providing insights into the structure and function of normal genes. The foundation of several new scientific journals dealing with human genetics since 1965 documents this: *American Journal of Medical Genetics, European Journal of Human Genetics, (Humangenetik, after 1976 Human Genetics), Clinical Genetics, Human Molecular Genetics, Journal of Medical Genetics, Genetics in Medicine, Annales de Génétique (now European Journal of Medical Genetics), Cytogenetics and Cell Genetics (now Chromosome Research)*, Prenatal Diagnosis, Clinical Dysmorphology, *Community Genetics, Genetic Counseling*, and others.

In recent years, a new area has been attracting attention: *epigenetics*. This refers to genetic mechanisms that influence the phenotype without altering the DNA sequence (see the section on Epigenetic Modifications in Part I).

Genetics in Medicine

A disease is genetically determined if it is mainly or exclusively caused by disorders in the genetic program of cells and tissues. However, most disease processes result from environmental influences interacting with the individual genetic makeup of the affected individual. These are multigenic or multifactorial diseases. They include many relatively common chronic diseases, e.g., high blood pressure, hyperlipidemia, diabetes mellitus, gout, psychiatric disorders, and certain congenital malformations. Another common category is cancer, a large, heterogeneous group of nonhereditary genetic disorders resulting from mutations in

Table 1. Categories and frequency of genetically determined diseases

Category of disease	Frequency per 1000 individuals
Monogenic diseases total	5–17
Autosomal dominant	2–10
Autosomal recessive	2–5
X-chromosomal	1–2
Chromosome aberrations	5–7
Multifactorial disorders	70–90
Somatic mutations (cancer)	200–250
Congenital malformations	20–25
Total	300–400

somatic cells. Chromosomal aberrations are also an important category. Thus, all medical specialties need to incorporate the genetic foundations of their discipline.

As a rule, the genetic origin of a disease cannot be recognized by familial aggregation. Instead, the diagnosis must be based on clinical features and laboratory data. Owing to new mutations and small family size in developed countries, genetic disorders usually do not affect more than one member of a family. About 90% occur isolated within a family. Since genetic disorders affect all organ systems and age groups, and frequently go unrecognized, their contribution to the causes of human diseases appears smaller than it actually is. Genetically determined diseases are not a marginal group, but make up a substantial proportion of diseases. More than one-third of all pediatric hospital admissions are for diseases and developmental disorders that, at least in part, are caused by genetic factors (Weatherall, 1991). The total estimated frequency of genetically determined diseases of different categories in the general population is about 3–5% (see Table 1).

The large number of individually rare genetically determined diseases and the overlap of diseases with similar clinical manifestations but different etiology cause additional diagnostic difficulties. This principle of genetic or etiological heterogeneity has to be taken into account when a diagnosis is made, to avoid false conclusions about the genetic risk.

The Dynamic Genome

Between 1950 und 1953, remarkable papers appeared entitled "The origin and behavior of mutable loci in maize" (McClintock, 1950), "Chromosome organization and genic expression" (McClintock, 1951), and "Introduction of instability at selected loci in maize" (McClintock, 1953). Here the author, Barbara McClintock of Cold Spring Harbor Laboratory, describes genetic changes in Indian corn plants (maize) and their effect on the phenotype induced by a mutation in a gene that is not located at the site of the mutation. Surprisingly, such a gene can exert a type of remote control. In subsequent work, McClintock described the special properties of this group of genes, which she called *controlling genetic elements*. Different controlling elements could be distinguished according to their effects on other genes and the mutations

caused. However, her work received little interest at the time (see Fox Keller 1983; Fedoroff and Botstein 1992). Thirty years later, at her 1983 Nobel Prize lecture (McClintock, 1984), things had changed. Today we know that genomes are not rigid and static structures. Rather, genomes are flexible and dynamic. They contain parts that can move from one location to another, called *mobile genetic elements* or *transposons*. This lends the genome flexibility to adapt to changing environmental conditions during the course of evolution. Although the precision of the genetic information depends on stability, complete stability would also mean static persistence. This would be detrimental to the development of new forms of life. Genomes are subject to alterations, as life requires a balance between the old and the new.

Genomics

The term *genomics* was introduced in 1987 by V.A. McKusick and F.H. Ruddle to define the new field. Genomics refers to the scientific study of the structure and function of genomes of different species of organisms. The genome of an animal, plant, or microorganism contains all biological information required for life and reproduction. It comprises the entire nucleotide sequence, all genes, their structure and function, their chromosomal localization, chromosome-associated proteins, and the architecture of the nucleus. Genomics integrates genetics, molecular biology, and cell biology. The scientific goals of genomics are manifold and all aimed at the entire genome of an organism: sequencing of the nucleotide bases of an organism, in particular all genes and gene-related sequences; analysis of all molecules involved in transcription and translation, and their regulation (the *transcriptome*); analysis of all proteins that a cell or an organism is able to produce (the *proteome*); identification of all genes and functional analysis (*functional genomics*); to establish genomic maps with regard to the evolution of genomes (*comparative genomics*); and assembly, storage, and management of data (*bioinformatics*).

The Human Genome Project

A new dimension was introduced into biomedical research by the Human Genome Project (HGP) and related programs in many other organisms (see Part II, Genomics; Lander and

Weinberg, 2000). The HGP is an international organization which represents several countries under the leadership of centers in the USA and UK. The main goal of the HGP was to determine the entire sequence of the 3 billion nucleotide pairs in the DNA of the human genome and to find all the genes within it. This daunting task began in 1990. It is comparable to deciphering each individual 1-mm-wide letter along a text strip 3000 km long. A first draft of a sequenced human genome covering about 90% of the genome was announced in June 2000 (IHGSC, 2001; Venter et al., 2001). The complete DNA sequence of man was published in 2004 (IHGSC, 2004). As of May 2006, all human chromosomes have been sequenced (see www.nature.com).

Ethical and Societal Aspects

From its start the HGP devoted attention and resources to ethical, legal, and social issues (the ELSI program). This is an important part of the HGP, in view of the far-reaching consequences of the current and expected knowledge about human genes and the genome. Here only a few areas can be mentioned. Among these are questions of validity and confidentiality of genetic data, of how to decide about a genetic test before the first manifestation of a disease (pre-symptomatic genetic testing), or whether to test for the presence or absence of a disease-causing mutation in an individual before any signs of the disease can be expected (predictive genetic testing). How does one determine whether a genetic test is in the best interests of the individual? Does she or he benefit from the information, or could it result in discrimination? How are the consequences defined? How is (genetic) counseling done and informed consent obtained? The use of embryonic stem cells is another area that concerns the public. Careful consideration of benefits and risks in the public domain will aid in reaching rational and balanced decisions. The decision on whether perform a genetic test has to take into account a person's view on an individual basis, and be obtained after proper counseling about the purpose, validity, and reliability, and the possible consequences of the test result. The application of genetic methods in the diagnosis of diseases can greatly augment the physician's resources in patient care and family counseling, but only if the information generated is used in the best interests of the individual involved, informed consent is obtained, and confidentiality of data is assured.

Education

Although genetic principles are quite straightforward, genetics is opposed by some and misunderstood by many. Scientists should seize any opportunity to inform the public about the goals of genetics and genomics and the principal methods employed. Genetics should be highly visible at the elementary and high school levels. Human genetics should be emphasized in teaching in medical schools.

References

Bearn AG: Archibald Garrod and the Individuality of Man. Oxford University Press, Oxford, 1993.

Cairns J, Stent GS, Watson JD, eds: Phage and the Origins of Molecular Biology. Cold Spring Harbor Laboratory Press, New York, 1978.

Childs B: Genetic Medicine. A Logic of Disease. Johns Hopkins University Press, Baltimore, 1999.

Crick F: What Mad Pursuit: A Personal View of Scientific Discovery. Basic Books, New York, 1988.

Dubos RJ: The Professor, the Institute, and DNA: O.T. Avery, his Life and Scientific Achievements. Rockefeller University Press, New York, 1976.

Dunn LC: A Short History of Genetics. McGraw-Hill, New York, 1965.

Fedoroff N, Botstein D, eds: The Dynamic Genome: Barbara McClintock's Ideas in the Century of Genetics. Cold Spring Harbor Laboratory Press, New York, 1992.

Fox Keller EA: A Feeling for the Organism: the Life and Work of Barbara McClintock. W.H. Freeman, New York, 1983.

Franklin RE, Gosling RG: Molecular configuration in sodium thymonucleate. Nature 171: 740–741, 1953.

Garrod AE: The Inborn Factors in Disease: an Essay. Clarendon Press, Oxford, 1931.

Haws DV, McKusick VA: Farabee's brachydactyly kindred revisited. Bull Johns Hopkins Hosp 113: 20–30, 1963.

IHGSC (International Human Genome Sequencing Consortium): Initial sequencing and analysis of the human geneome. Nature 409: 286–921, 2001.

IHGSC (International Human Genome Sequencing Consortium): Finishing the euchromatic sequence of the human genome. Nature 431: 931–945, 2004 (see Nature Web Focus: The Human Genome (www.nature.com/nature/focus/humangenome/index.html).

Judson HF: The Eighth Day of Creation. Makers of the Revolution in Biology, expanded edition. Cold Spring Harbor Laboratory Press, New York, 1996.

Lander ES, Weinberg RA: Genomics. Journey to the center of biology. Pathways of discovery. Science 287: 1777–1782, 2000.

McCarty M: The Transforming Principle. W.W. Norton, New York, 1985.

McClintock, B. The origin and behavior of mutable loci in maize. Proc Natl Acad Sci USA 36: 344–355, 1950.

McClintock B: Chromosome organization and genic expression. Cold Spring Harb Symp Quant Biol 16: 13–47, 1951.

McClintock B: Induction of instability at selected loci in maize. Genetics 38: 579–599, 1953.

McClintock B: The significance of responses of the genome to challenge. Science 226: 792–801, 1984.

McKusick VA: Presidential Address. Eighth International Congress of Human Genetics: The last 35 years, the present and the future. Am J Hum Genet 50: 663–670, 1992.

McKusick VA: Mendelian Inheritance in Man: A Catalog of Human Genes and Genetic Disorders, 12th ed. Johns Hopkins University Press, Baltimore, 1998 (online version available at http://www.ncbi.nlm.nih.gov/Omim/).

Müller-Hill B: Murderous Science. Oxford University Press, Oxford, 1988.

Rosenberg NA, Pritchard JK, Weber JL, et al: Genetic structure of human populations. Science 298: 2381–2385, 2002.

Schrödinger, E.: What Is Life? The Physical Aspect of the Living Cell. Penguin Books, New York, 1944.

Stent GS, ed.: James D. Watson. The Double Helix: A Personal Account of the Discovery of the Structure of DNA. Weidenfeld & Nicolson, London, 1981.

Stern C: Principles of Human Genetics. WH Freeman, San Francisco, 1949.

Strong C: Eugenics. In: Cooper DV, ed., Encyclopedia of the Human Genome. Vol. 2: 335–340, Nature Publishing Group, London, 2003.

Sturtevant AH: A History of Genetics. Harper & Row, New York, 1965.

Venter JC, Adams, MD, Myers EW et al.: The sequence of the human genome. Science 291: 1304–1351, 2001.

Vogel F, Motulsky AG: Human Genetics: Problems and Approaches, 3rd ed. Springer Verlag, Heidelberg, 1997.

Watson JD: The Double Helix. A Personal Account of the Discovery of the Structure of DNA. Atheneum, New York, 1968.

Watson JD: A Passion for DNA. Genes, Genomes, and Society. Cold Spring Harbor Laboratory Press, New York, 2000.

Watson JD, Crick FHC: A structure for deoxyribonucleic acid. Nature 171: 737, 1953.

Watson JD, Tooze J: The DNA Story: a documentary history of gene cloning. WH Freeman, San Francisco, 1981.

Weatherall DJ: The New Genetics and Clinical Practice, 3rd ed. Oxford Univ. Press, Oxford, 1991.

Wilkins M: The Third Man of the Double Helix. Oxford University Press, Oxford, 2003.

Selected Introductory Reading

Aase JM: Diagnostic Dysmorphology. Plenum Medical Book Company, New York, 1990.

Alberts B, Johnson A, Lewis, J, Raff M, Roberts K, Walter P: Molecular Biology of the Cell. 4th ed. Garland Publishing Co, New York, 2002.

Bateson W: Mendel's Principles of Heredity. Univ. of Cambridge Press, Cambridge, 1913.

Brown TA: Genomes, 2nd ed. Bios Scientific Publishers, Oxford, 2002.

Chargaff E: Heraclitean Fire: Sketches from a Life before Nature. Rockefeller University Press, New York, 1978.

Clarke AJ, ed.: The Genetic Testing of Children. Bios Scientific Publishers, Oxford, 1998.

Dobzhansky T: Genetics of the Evolutionary Process. Columbia University Press, New York, 1970.

Epstein CJ, Erickson, RP, Wynshaw-Boris, eds: Inborn Errors of Development. The Molecular Basis of Clinical Disorders of Morphogenesis. Oxford University Press, Oxford, 2004.

Gilbert SF: Developmental Biology. 7th ed., Sinauer, Sunderland , Massachussetts, 2003.

Gilbert-Barness E, Barness L: Metabolic Diseases. Foundations of Clinical Management, Genetics and Pathology. Eaton Publishing, Natick, MA 01760 USA, 2000.

Griffith AJF, Suzuki DT, Miller JH, Lewontin RC, Gelbart WM: An Introduction to Genetic Analysis. 7th ed. W.H. Freeman & Co., New York, 2000.

Harper PS: Practical Genetic Counselling. 6th ed., Edward Arnold, London, 2004.

Harper PS, Clarke AJ: Genetics, Society, and Clinical Practice. Bios Scientific Publishers, Oxford, 1997.

Horaitis R, Scriver CR, Cotton RGH: Mutation databases: Overview and catalogues, pp. 113–125. In: CR Scriver et al, eds: The Metabolic and Molecular Bases of Inherited Disease. 8th ed. McGraw-Hill, New York, 2001.

Jobling MA, Hurles M, Tyler-Smith C: Human Evolutionary Genetics. Origins, Peoples, and Disease. Garland Science, New York, 2004.

Jameson JL ed.: Principles of Molecular Medicine. Humana Press, Totowa, New Jersey, 1998.

Jones KL: Smith's Recognizable Patterns of Human Malformation. 6th ed. W.B. Saunders, Philadelphia, 2006.

Jorde LB, Carey JC, White RL, Bamshad MJ: Medical Genetics. 2nd ed. C.V. Mosby, St. Louis, 2001.

Kasper DL et al: Harrison's Principles of Internal Medicine. 16th ed. (with online access). McGraw-Hill, New York, 2005.

King R, Rotter J, Motulsky AG, eds: The Genetic Basis of Common Disorders. 2nd ed. Oxford University Press, Oxford, 2002.

King RC, Stansfield WD: A Dictionary of Genetics, 6th ed. Oxford University Press, Oxford, 2002.

Klein J, Takahata N: Where do we come from? The Molecular Evidence for Human Descent. Springer, Berlin, 2002.

Knippers, R.: Molekulare Genetik, 8. Aufl. Georg Thieme Verlag, Stuttgart–New York, 2005.

Koolman J, Roehm K-H: Color Atlas of Biochemistry. 2nd ed, Thieme, Stuttgart – New York, 2005.

Lodish H, Berk A, Matsudaira P, Kaiser CA, Krieger M, Scott MP, Zipursky SL, Darnell J: Molecular Cell Biology (with an animated CD-ROM). 5th ed. W.H. Freeman & Co., New York, 2004.

Macilwain C: World leaders heap praise on human genome landmark. Nature 405: 983–984, 2000.

Maddox B: Rosalind Franklin. Dark Lady of DNA. HarperCollins, London, 2002.

Miller OJ, Therman E: Human Chromosomes. 4th ed. Springer, New York, 2001.

Murphy EA, Chase GE: Principles of Genetic Counseling. Year Book Medical Publishers, Chicago, 1975.

Nussbaum RL, McInnes RR, Willard HF: Thompson & Thompson Genetics in Medicine, 6th ed. W. B. Saunders, Philadelphia, 2001.

Ohno S: Evolution by Gene Duplication. Springer Verlag, Heidelberg, 1970.

Passarge E: The human genome and disease, pp. 31–37. In: Molecular Nuclear Medicine. The Challenge of Genomics and Proteomics to Clinical Practice. L.E. Feinendegen et al, eds. Springer, Berlin-Heidelberg-New York, 2003.

Passarge E, Kohlhase J: Genetik, pp. 4–66. In: Klinische Pathophysiologie, 9. Auflage, W. Siegenthaler, H.E. Blum, eds., Thieme Verlag Stuttgart, 2006.

Pennisi E. Human genome. Finally, the book of life and instructions for navigating it. Science 288: 2304–2307, 2000.

Rimoin, DL, Connor JM, Pyeritz RE, Korf BR, eds.: Emery and Rimoin's Principles and Practice of Medical Genetics, 5th ed., Churchill-Livingstone, Edinburgh, 2006.

Stebbins GL: Darwin to DNA. Molecules to Humanity. W.H. Freeman, San Francisco, 1982.

Stent G, Calendar R: Molecular Genetics. An Introductory Narrative, 2nd ed. W.H. Freeman, San Francisco, 1978.

Stevenson RE, Hall JG, eds.: Human Malformations and Related Anomalies. 2nd ed. Oxford Univ. Press, Oxford, 2006.

Strachan T, Read AP: Human Molecular Genetics. 3rd ed. Garland Science, London, 2004.

Stryer L, Biochemistry. 4th ed. W. H. Freeman, New York, 2005.

Turnpenny PD, Ellard S: Emery's Elements of Medcal Genetics, 12th ed. Elsevier Churchill-Livingstone, Edinburgh-London-New York, 2005.

Vogelstein B, Kinzler KW, eds.: The Genetic Basis of Human Cancer. 2nd ed. McGraw-Hill, New York, 2002.

Watson JD, Baker TA, Bell SP, Gann A, Levine M, Losick R: Molecular Biology of the Gene. 5th ed. Pearson/ Benjamin Cummings and Cold Spring Harbor Laboratory Press, 2004.

Weinberg RA: The Biology of Cancer. Garland Science, New York, 2006.

Whitehouse HLK: Towards an Understanding of the Mechanism of Heredity, 3rd ed. Edward Arnold, London, 1973.

Selected Websites for Access to Genetic Information:

Online Mendelian Inheritance in Man, OMIM (TM). McKusick-Nathans Institute for Genetic Medicine. Johns Hopkins University (Baltimore, Maryland) and the National Center for Biotechnology Information, National Library of Medicine (Bethesda, Maryland), 2000, at World Wide Web URL: (http://www.ncbi.nlm.nih.gov/Omim/).

GeneClinics, a clinical information resource relating genetic testing to the diagnosis, management, and genetic counseling of individuals and families with specific inherited disorders: (http://www.geneclinics.com).

Information on Individual Human Chromosomes and Disease Loci: Chromosome Launchpad: (http://www.ornl.gov/hgmis/launchpad).

National Center of Biotechnology Information Genes and Disease Map: (http://www.ncbi. nlm.nih.gov/disease/).

Medline: (http://www.ncbi.nlm.nim.nih.gov/PubMed/).

MITOMAP: A human mitochondrial genome database: (http://www.gen.emory.edu/mitomap.html), Center for Molecular Medicine, Emory University, Atlanta, GA, USA, 2000.

Nature Web Focus: The Human Genome (www.nature.com/nature/focus/humangenome/index.html).

Important Advances that Contributed to the Development of Genetics

(This list represents a selection and should not be considered complete; apologies to all authors not included.)

1665 Cells described and named (*Robert Hooke*)

1827 Human egg cell described (*Karl Ernst von Baer*)

1839 Cells recognized as the basis of living organisms (*Schleiden, Schwann*)

1859 Concept and facts of evolution (*Charles Darwin*)

1865 Rules of inheritance by distinct "factors" acting dominantly or recessively (*Gregor Mendel*)

1869 "Nuclein," a new acidic, phosphorus-containing, long molecule (*F. Miescher*)

1874 Monozygotic and dizygotic twins distinguished (*C. Dareste*)

1876 "Nature and nurture" (*F. Galton*)

1879 Chromosomes in mitosis (*W. Flemming*)

1883 Quantitative aspects of heredity (*F. Galton*)

1888 Term "chromosome" (*W. Waldeyer*)

1889 Term "nucleic acid" (*R. Altmann*)

1892 Term "virus" (*R. Ivanowski*)

1897 Enzymes discovered (*E. Büchner*)

1900 Mendel's discovery recognized (*H. de Vries, E.Tschermak, K. Correns,* independently)
ABO blood group system (*Landsteiner*)

1902 Some diseases in man inherited according to Mendelian rules (*W. Bateson, A. Garrod*)
Sex chromosomes (*McClung*)
Chromosomes and Mendel's factors are related (*W. Sutton*)
Individuality of chromosomes (*T. Boveri*)

1906 Term "genetics" proposed (*W. Bateson*)

1908 Population genetics (*G.H. Hardy, W. Weinberg*)

1907 Amphibian spinal cord culture (*Harrison*)

1909 Inborn errors of metabolism (*A. Garrod*)
Terms *gene, genotype, phenotype* proposed (*W. Johannsen*)
Chiasma formation during meiosis (*Janssens*)
First inbred mouse strain DBA (*C. Little*)

1910 Beginning of *Drosophila* genetics (*T. H. Morgan*)
First *Drosophila* mutation (*white-eyed*)

1911 Sarcoma virus (*Peyton Rous*)

1912 Crossing-over (*Morgan and Cattell*)
Genetic linkage (*Morgan and Lynch*)
First genetic map (*A. H. Sturtevant*)

1913 First long-term cell culture (*A. Carrel*)
Nondisjunction (*C. B. Bridges*)

1915 Genes located on chromosomes (chromosomal theory of inheritance) (*Morgan, Sturtevant, Muller, Bridges*)
Bithorax mutant (*C.B. Bridges*)
First genetic linkage in vertebrates (*JBS Haldane, AD Sprunt, NM Haldane*)
Term "intersex" (*RB Goldschmidt*)

1917 Bacteriophage discovered (*F. d'Herelle*)

1922 Characteristic phenotypes of different trisomies in the plant *Datura stramonium* (*F. Blakeslee*)

1923 Chromosome translocation in *Drosophila* (*C.B. Bridges*)

1924 Blood group genetics (*Bernstein*)
Statistical analysis of genetic traits (*R.A. Fisher*)

1926 Enzymes are proteins (*J. Sumner*)

1927 Mutations induced by X-rays (*H. J. Muller*)
Genetic drift (*S. Wright*)

1928 Euchromatin/heterochromatin (*E. Heitz*)
Genetic transformation in bacteria (*F. Griffith*)

1933 Pedigree analysis (*Haldane, Hogben, Fisher, Lenz, Bernstein*)
Polytene chromosomes (*Heitz and Bauer, Painter*)

1935 First cytogenetic map in *Drosophila* (*C. B. Bridges*)

1937 Mouse H2 gene locus (*P. Gorer*)

1940 Polymorphism (*E. B. Ford*)
Rhesus blood groups (*Landsteiner and Wiener*)

1941 Evolution through gene duplication (*E. B. Lewis*)
Genetic control of enzymatic biochemical reactions (*Beadle and Tatum*)
Mutations induced by mustard gas (*C. Auerbach and J.M. Robson*)

1943 Mutations in bacteria (*S. E. Luria and M. Delbrück*)

1944 DNA as the material basis of genetic information (*Avery, MacLeod, McCarty*)
What is Life? The Physical Aspect of the Living Cell. An influential book (*E. Schrödinger*)

1946 Genetic recombination in bacteria (*Lederberg and Tatum*)

1947 Genetic recombination in viruses (*Delbrück and Bailey, Hershey*)

1949 Sickle cell anemia, a genetically determined molecular disease (*Neel, Pauling*)
Hemoglobin disorders prevalent in areas of malaria (*J. B. S. Haldane*)
X chromatin (*Barr and Bertram*)

1950 Defined relation of the four nucleotide bases (*E. Chargaff*)

1951 Mobile genetic elements in Indian corn, *Zea mays* (*B. McClintock*)
α-helix and β-sheet in proteins (*L. Pauling and R.B. Corey*)

1952 Genes consist of DNA (*Hershey and Chase*)
Plasmids (*Lederberg*)
Transduction by phages (*Zinder and Lederberg*)
First enzyme defect in man (*Cori and Cori*)
First linkage group in man (*Mohr*)
Colchicine and hypotonic treatment in chromosomal analysis (*Hsu and Pomerat*)
Exogenous factors as a cause of congenital malformations (*J. Warkany*)

1953 DNA structure (*Watson and Crick, Franklin, Wilkins*)

Conjugation in bacteria (*W. Hayes, L. L. Cavalli, J. and E. Lederberg, independently*)
Nonmendelian inheritance (*Ephrussi*)
Cell cycle (*Howard and Pelc*)
Dietary treatment of phenylketonuria (*Bickel*)

1954 DNA repair (*Muller*)
HLA system (*J. Dausset*)
Leukocyte drumsticks (*Davidson and Smith*)
Cells in Turner syndrome are X-chromatin negative (*P. Polani*)
Cholesterol biosynthesis (*K. Bloch*)

1955 First genetic map at the molecular level (*S. Benzer*)
First amino acid sequence of a protein, insulin (*F. Sanger*)
Lysosomes (*C. de Duve*)
Buccal smear (*Moore, Barr, Marberger*)
5-Bromouracil, an analogue of thymine, induces mutations in phages (*A. Pardee and R. Litman*)

1956 46 Chromosomes in man (*Tijo and Levan; Ford and Hamerton*)
Amino acid sequence of hemoglobin molecule (*V. Ingram*)
DNA synthesis in vitro (*S. Ochoa; A. Kornberg*)
Synaptonemal complex, the area of synapse in meiosis (*M.J. Moses; D. Fawcett*)
Genetic heterogeneity (*H. Harris, C.F. Fraser*)

1957 Genetic complementation (*Fincham*)
Genetic analysis of radiation effects in man (*Neel and Schull*)

1958 Semiconservative replication of DNA (*M. Meselson and F.W. Stahl*)
Somatic cell genetics (*G. Pontecorvo*)
Ribosomes (*Roberts, Dintzis*)
Cloning of single cells (*Sanford, Puck*)

1959 First chromosomal aberrations in man: trisomy 21 (*Lejeune, Gautier, Turpin*); Turner syndrome, 45,XO (*Jacobs and Strong*); Klinefelter syndrome: 47 XXY (*Ford*)
DNA polymerase (*A. Kornberg*)
Isoenzymes (*Vessel, Markert*)
Pharmacogenetics (*Motulsky, Vogel*)

1960   Phytohemagglutinin-stimulated lymphocyte cultures (*Nowell, Moorhead, Hungerford*)

1961   The genetic code is read in triplets (*Crick, Brenner, Barnett, Watts-Tobin*)
The genetic code determined (*Nirenberg, Mathaei, Ochoa*)
X-chromosome inactivation (*M. F. Lyon*, confirmed by *Beutler, Russell, Ohno*)
Gene regulation, concept of operon (*Jacob and Monod*)
Galactosemia in cell culture (*Krooth*)
Cell hybridization (*Barski, Ephrussi*)
Thalidomide embryopathy (*Lenz, McBride*)

1962   Philadelphia chromosome (*Nowell and Hungerford*)
Molecular characterization of immunoglobulins (*Edelman, Franklin*)
Identification of individual human chromosomes by ^{3}H-autoradiography (*J. German, O.J. Miller*)
Term "codon" for a triplet of (sequential) bases (*S. Brenner*)
Replicon (*Jacob* and *Brenner*)
Cell culture (*W. Szybalski* and E.K. Szybalska)
Xg, the first X-linked human blood group (*Mann, Race, Sanger*)
Screening for phenylketonuria (*Guthrie, Bickel*)

1963   Lysosomal storage diseases (*C. de Duve*)
First autosomal deletion syndrome (cri-du-chat syndrome) (*J. Lejeune*)

1964   Colinearity of gene and protein gene product (*C. Yanofsky*)
Excision repair (*Setlow*)
MLC test (*Bach and Hirschhorn, Bain and Lowenstein*)
Microlymphotoxicity test (*Terasaki and McClelland*)
Selective cell culture medium HAT (*J. Littlefield*)
Spontaneous chromosomal instability (*J. German, T.M. Schröder*)
Cell culture from amniotic fluid cells (*H. P. Klinger*)
Hereditary diseases studied in cell cultures (*Danes, Bearn, Krooth, Mellman*)
Population cytogenetics (*Court Brown*)
Fetal chromosomal aberrations in spontaneous abortions (*Carr, Benirschke*)

1965   Sequence of alanine transfer RNA from yeast (*R.W. Holley*)
Limited life span of cultured fibroblasts (*Hayflick, Moorhead*)
Crossing-over in somatic cells (*J. German*)
Cell fusion with Sendai virus (*H. Harris and J.F. Watkins*)

1966   Genetic code complete
Catalog of Mendelian phenotypes in man (*V.A. McKusick*)
Concept of epigenetics (*C.H. Waddington*)

1968   Restriction endonucleases (*H. O. Smith, Linn and Arber, Meselson and Yuan*)
Okazaki fragments in DNA synthesis (*R.T. Okazaki*)
HLA-D the strongest histocompatibility system (*Ceppellini, Amos*)
Repetitive DNA (*Britten and Kohne*)
Biochemical basis of the ABO blood group substances (*Watkins*)
DNA excision repair defect in xeroderma pigmentosum (*Cleaver*)
First assignment of an autosomal gene locus in man (*Donahue, McKusick*)
Synthesis of a gene in vitro (*H.G. Khorana*)
Neutral gene theory of molecular evolution (*M. Kimura*)

1970   Reverse transcriptase (*D. Baltimore, H. Temin*, independently)
Synteny, a new term to refer to all gene loci on the same chromosome (*Renwick*)
Enzyme defects in lysosomal storage diseases (*Neufeld, Dorfman*)
Individual chromosomal identification by specific banding stains (*Zech, Caspersson, Lubs, Drets and Shaw, Schnedl, Evans*)
Y-chromatin (*Pearson, Bobrow, Vosa*)
Thymus transplantation for immune deficiency (*van Bekkum*)

1971   Two-hit theory in retinoblastoma (*A.G. Knudson*)

1972   High average heterozygosity (*Harris and Hopkinson; Lewontin*)
Association of HLA antigens and diseases

1973   Receptor defects in the etiology of genetic defects, genetic hyperlipidemia

(*Brown, Goldstein, Motulsky*)
Demonstration of sister chromatid exchanges with BrdU (*S. A. Latt*)
Philadelphia chromosome as translocation (*J. D. Rowley*)

1974 Chromatin structure, nucleosome (*Kornberg, Olins and Olins*)
Dual recognition of foreign antigen and HLA antigen by T lymphocytes (*P. C. Doherty and R. M. Zinkernagel*)
Clone of a eukaryotic DNA segment mapped to a specific chromosome location (*D.S. Hogness*)

1975 Southern blot hybridization (*E. Southern*)
Monoclonal antibodies (*Köhler and Milstein*)
First protein-signal sequence identified (*G. Blobel*)
Model for promoter structure and function (*D. Pribnow*)
First transgenic mouse (*R. Jaenisch*)
Asilomar conference about recombinant DNA

1976 Overlapping genes in phage ΦX174 (*Barell, Air, Hutchinson*)
Loci for structural genes on each human chromosome known (*Baltimore Conference on Human Gene Mapping*)
First diagnosis using recombinant DNA technology (*W. Kan, M.S. Golbus, A. M. Dozy*)

1977 Genes contain coding and noncoding DNA segments (*R. J. Roberts, P. A. Sharp,* independently)
First recombinant DNA molecule that contains mammalian DNA
Methods to sequence DNA (*F. Sanger; Maxam and Gilbert*)
Sequence of phage ΦX174 (*F. Sanger*)
X-ray diffraction analysis of nucleosomes (*Finch and co-workers*)

1978 Terms *exon* and *intron* for coding and noncoding parts of eukaryotic genes (*W. Gilbert*)
β-Globulin gene structure (*Leder, Weissmann, Tilghman and others*)
Mechanisms of transposition in bacteria
Production of somatostatin with recombinant DNA
Introduction of "chromosome walking" to find genes

(*Brown, Goldstein, Motulsky*)
First genetic diagnosis using restriction enzymes (*Y.H. Kan and A.M. Dozy*)
DNA tandem repeats in telomeres (*E. H. Blackburn and J.G. Gall*)

1979 Small nuclear ribonuceleo-proteins ("snurps") (*M.R. Lerner and J.A. Steitz*)
Alternative genetic code in mitochondrial DNA (*B.G. Barell, A.T. Bankier, J. Drouin*)

1980 Restriction fragment length polymorphism for mapping (*D. Botstein and co-workers*)
Genes for embryonic development in *Drosophila* studied by mutational screen (*C. Nüsslein-Volhard and E. Wieschaus*)
First transgenic mice by injection of cloned DNA (*J. W. Gordon*)
Transformation of cultured mammalian cells by injection of DNA (*M. R. Capecchi*)
Structure of 16 S ribosomal RN (*C. Woese*)

1981 Sequence of a mitochondrial genome (*S. Anderson, S. G. Barrell, A. T. Bankier*)

1982 Tumor suppressor genes (*H. P. Klinger*)
Prions (proteinaceous infectious particles) as cause of central nervous system diseases (kuru, scrapie, Creutzfeldt-Jakob disease) (*S. B. Prusiner*)
Insulin made by recombinant DNA marketed (Eli Lilly)

1983 Cellular oncogenes (*H.E. Varmus and others*)
HIV virus (*L. Montagnier; R. Gallo*)
Molecular basis of chronic myelocytic leukemia (*C. R. Bartram, D. Bootsma and co-workers*)
First recombinant RNA molecule (*E.A. Miele, D.R. Mills, F.R. Kramer*)
Bithorax complex of *Drosophila* sequenced (*W. Bender*)

1984 Identification of the T-cell receptor (*Tonegawa*)
Homeobox (Hox) genes in Drosophila and mice (*W. McGinnis*)
Localization of the gene for Huntington disease (*Gusella*)
Description of *Helicobacter pylori* (*B. Marshall and R. Warren*)

1985 Polymerase chain reaction
(*K.B. Mullis, R.K. Saiki*)
Hypervariable DNA segments as
"genetic fingerprints" (*A. Jeffreys*)
Hemophilia A gene cloned (*J. Gietschier*)
Sequencing of the HIV-1 virus
Linkage analysis of the gene for cystic fibrosis (*H. Eiberg and others*)
Isolation of telomerase from *Tetrahymena* (*C. W. Greider and E.H. Blackburn*)
Isolation of a zinc finger protein from
Xenopus oocytes (*J.R. Miller,
A.D. McLachlin, A. Klug*)
Insertion of DNA by homologous recombination (*O. Smithies*)
Genomic imprinting in the mouse
(*B. Cattanach*)

1986 First cloning of human genes.
Human visual pigment genes characterized (*J. Nathans, D. Thomas, D.S. Hogness*)
RNA as catalytic enzyme (*T. Cech*)
First identification of a human gene
based on its chromosomal location
(positional cloning) (*B. Royer-Pokora and
co-workers*)

1987 Fine structure of an HLA molecule
(*Björkman, Strominger and co-workers*)
Cloning of the gene for Duchenne
muscular dystrophy (*L.M. Kunkel and
others*)
Knockout mouse (*M. Capecchi*)
A genetic map of the human genome
(*H. Donis-Keller and co-workers*)
Mitochondrial DNA and human evolution (*R.L. Cann, M. Stoneking,
A.C. Wilson*)

1988 Start of the Human Genome Project
Molecular structure of telomeres at the
ends of chromosomes (*E.H. Blackburn
and others*)
Mutations in human mitochondrial DNA
(*D.C. Wallace*)
Transposable DNA as rare cause of
hemophilia A (*H.H. Kazazian*)
Successful gene therapy in vitro

1989 Identification of the gene causing cystic
fibrosis (*L.-C. Tsui and others*)
Microdissection and cloning of a defined region of a human chromosome
(*Lüdecke, Senger, Claussen, Horsthemke*)

1990 Mutations in the *p53* gene as cause of
Li-Fraumeni syndrome (*D. Malkin*)
Mutations in the gene *wrinkled seed*
used by Mendel (*M.K. Bhattcharyya*)
A defective gene as cause of inherited
breast cancer (*Mary-Claire King*)

1991 Cloning of the gene for Duchenne
muscular dystrophy (*L.M. Kunkel and
others*)
Odorant receptor multigene family
(*Buck and Axel*)
Complete sequence of a yeast chromosome
Increasing use of microsatellites as polymorphic DNA markers

1992 Trinucleotide repeat expansion as a new
class of human pathogenic mutations
High density map of DNA markers on
human chromosomes
X chromosome inactivation center identified
p53 knockout mouse (*O. Smithies*)

1993 Gene for Huntington disease cloned
(*M.E. MacDonald*)
Developmental mutations in zebra fish
(*M.C. Mullins and C. Nüslein-Volhard*)

1994 First physical map of the human
genome in high resolution
Mutations in fibroblast growth factor
receptor genes as cause of achondroplasia and other human diseases
(*M. Muenke*)
Identification of genes
for hereditary breast cancer

1995 Cloning of the *BLM* (Bloom's syndrome)
gene (*N.A. Ellis, J. Groden, J. German and
co-workers*)
First genome sequence of a free living
bacterium, *Haemophilus influenzae*
(*R.D. Fleischmann, J.C. Venter and co-workers*)
Master gene of the vertebrate eye, *sey*
(small-eye) (*G.Halder, P. Callaerts,
W. J. Gehring*)
STS map of the human genome
(*T.J. Hudson and co-workers*)

1996 Yeast genome sequenced (*A. Goffean
and co-workers*)
Mouse genome map with more than
7000 markers (*E. S. Lander*)

1997   Sequence of *E. coli* (*F. R. Blattner and co-workers*), *Helicobacter pylori* (*J.F. Tomb*)
Neanderthal mitochondrial DNA sequences (*M. Krings, S. Pääbo and co-workers*)
Mammal ("Dolly, the sheep") cloned by transfer of an adult cell nucleus into an enucleated oocyte (*I. Wilmut*)

1998   RNA interference, RNAi (*A. Fire and co-workers*)
Nematode *C. elegans* genome sequenced
Human embryonic stem cells (*Thomson and Gearhart*)

1999   First human chromosome (22) sequenced
Ribosome crystal structure

2000   *Drosophila* genome sequenced (*M.D. Adams*)
First complete genome sequence of a plant pathogen (*Xylella fastidiosa*)
Arabidopsis thaliana, the first plant genome sequenced

2001   First draft of the complete sequence of the human genome (*F.H. Collins; J.C. Venter and co-workers*)

2002   Genome sequence of the mouse (*R.H. Waterston and co-workers*)
Sequence of the genome of rice, *Oryza sativa* (*J. Yu, S. A. Goff and co-workers*)
Sequence of the genomes of malaria parasite, *Plasmodium falciparum*, and its vector, *Anopheles gambiae*
Earliest hominid, *Sahelanthropos tchadiensis* (*M. Brunet*)

2003   International HapMap Project launched
Sequence of the human Y chromosome (*H. Skaletsky, D.C. Page and co-workers*)
Homo sapiens idaltù, the oldest anatomically modern man from pleistocene 154–160 years ago (*T.D. White and co-workers*)

2004   Genome sequence of the Brown Norway rat
A new small bodied hominin from Flores island, Indonesia (*P. Brown and co-workers*)

2005   Genome sequence of the chimpanzee (*R.H. Waterston, E.S. Lander, R.K. Watson and co-workers*)

1.58 million human single-nucleotide polymorphisms mapped (*D.A. Hinds, D.R. Cox and co-workers*)
Human haplotype map
Sequence of the human X chromosome (*M.T. Ross and co-workers*)
Inactivation profile of the human X chromosome (*L. Carrel and H.F. Willard*)

2006   All human chromosomes sequenced

References for the Chronology

In addition to personal notes, dates are based on the following main sources:

Dunn LC: A Short History of Genetics. McGraw-Hill, New York, 1965.

King RC, Stansfield WD: A Dictionary of Genetics, 6th ed. Oxford University Press, Oxford, 2002.

Lander ES, Weinberg RA: Genomics. A journey to the center of science. Science 287: 1777–1782, 2000.

McKusick VA: Presidential Address. Eighth International Congress of Human Genetics: The last 35 years, the present and the future. Am J Hum Genet 50: 663–670, 1992.

Stent GS, ed.: James D. Watson. The Double Helix: A Personal Account of the Discovery of the Structure of DNA. Weidenfeld & Nicolson, London, 1981.

Sturtevant AH: A History of Genetics. Harper & Row, New York, 1965

The New Encyclopaedia Britannica, 15th ed. Encyclopaedia Britannica, Chicago, 1995.

Vogel F, Motulsky AG: Human Genetics: Problems and Approaches, 3rd ed. Springer Verlag, Heidelberg, 1997.

Whitehouse HLK: Towards an Understanding of the Mechanism of Heredity, 3rd ed. Edward Arnold, London, 1973

Fundamentals

Taxonomy of Living Organisms: The Tree of Life

In his *Origin of Species,* Charles Darwin wrote (1859): "Probably all of organic beings which have ever lived on this Earth have descended from some primordial form." Thus, if all living organisms are derived from a common ancestor, in theory it should be possible to establish their relationship (taxonomy) based on the type and number of characteristics they share. This poses enormous difficulties, because data about previously living organisms are restricted to scanty records. But phylogenetic relationships can be based on anatomical features, proteins, DNA, or other molecules (phylogenomics, Delsuc et al., 2005). There is overall agreement that the earth is a little more than 4.5 billion years old and that early forms of life date back about 3.5 billion years.

A. The three domains of living organisms

The formal evolutionary hierarchy of groups of organisms proceeds from the largest to the smallest groups: domain – kingdom – phylum – order – class – family – genus – species. Living organisms are grouped according to the type of cells they consist of, either *prokaryotic* cells or *eukaryotic* cells. Prokaryotes have a simple internal architecture without a nucleus. Eukaryotes have a distinct internal structure with a nucleus containing the genetic material. A third group of living organisms was recognized in the late 1960s, the *Archaea* (also called archaebacteria). They differ from ordinary bacteria by their plasma membrane (isoprene ether lipids rather than fatty acid ester lipids) and lifestyle. They are assigned to two classes, *Crenarchaeota* and *Euryarcheota.*
Archaea can live without molecular oxygen at high temperatures (70°C–110°C, thermophiles) or at low temperatures (psychrophiles), in water with high concentrations of sodium chloride (halophiles) or sulfur (sulfothermophiles), in a highly alkaline environment (pH as high as 11.5, alkaliphiles) or in acid conditions with pH near zero (acidophiles) or a combination of such adverse conditions that would boil or dissolve ordinary bacteria. It is assumed that prokaryotes predate eukaryotes, and that two preexisting prokaryotes contributed their genomes to the first eukaryotic genome.

Eukaryotes consist of several kingdoms, including animals, fungi, plants, algae, protozoa, and others. The three domains have a presumed common progenitor, called the last universal common ancestor.

B. Phylogeny of metazoa (animals)

The phylogeny of metazoa differs, depending on whether it is based on the traditional interpretation or on molecular evidence as revealed mainly by rRNA sequence comparisons. Here a simplified version of the molecule-based interpretation is shown.

C. Mammalian phylogeny

Mammals arose about 100 million years ago in the late Mesozoic period of the Earth. The time scale is only approximate. Of the 4629 known mammalian species, 4356 are placentals, which fall into 12 orders. The first five placental orders according to their number of species are rodents (2015), followed by bats (925), insectivores (385), carnivores (271), and primates (233). (Figures modified from Klein & Takahata, 2001.)

References

Allers T, Mevarech M: Archaeal genetics – the third way. Nature Rev Genet 6: 58–74, 2005.
Delsuc F, Brinkmann H, Philippe H: Phylogenomics and the reconstruction of the tree of life. Nature Rev Genet 6: 361–375, 2005.
Delsuc F et al: Tunicates and not cephalochordates are the closest living relatives of vertebrates. Nature 439: 965–968, 2006.
Hazen RM: Genesis: the Scientific Quest for Life's Origins. Joseph Henry Press, 2005.
Klein, J, Takahata, N: Where do we Come from? The Molecular Evidence for Human Descent. Springer, Berlin-Heidelberg, 2001.
Murphy WJ et al: Molcular phylogenetics and the origins of placental mammals. Nature 409: 614–618, 2001.
Rivera MC, Lake MA: The ring of life provides evidence for a genome fusion origin of eukaryotes. Nature 431: 152–155, 2004.
Woese CR: Interpreting the universal phylogenetic tree. Proc Nat Acad Sci 97: 8392–8396, 2000.
Woese CR: On the evolution of cells. Proc Nat Acad Sci 99: 8742–8747, 2002.
Woese CR: A new biology for a new century. Microbiol & Mol Biol Rev 68: 173–186, 2004.

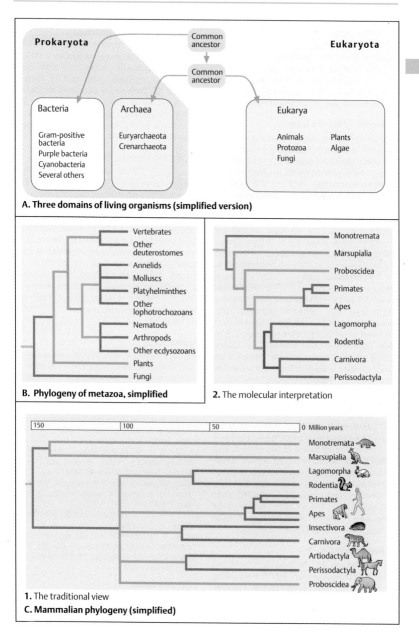

A. Three domains of living organisms (simplified version)

B. Phylogeny of metazoa, simplified

2. The molecular interpretation

1. The traditional view

C. Mammalian phylogeny (simplified)

Human Evolution

Humans are the only living species, *Homo sapiens*, within the family of *Hominidae*. All available data are consistent with the assumption that today's humans originated in Africa about 100 000–300 000 years ago, spread out over the earth, and populated all continents.

A. Hominid family tree

The last common ancestor of man and the chimpanzee lived about 6–7 million years ago (mya). The oldest identified hominid skeletal remains were found in Eastern Africa, in Chad (*Sahelanthropus tchadensis*) in 2002 (ca. 6–7 mya) and Kenyia (*Orronin tugensis*, ca. 5.8–6.1 mya). Fossils from 5 and 4 mya belong to the genus *Australopithecus*. A member of this group is *Ardipethicus ramidus* (ca. 4.5 mya). Bipedal gait developed early, about 4.5 to 4 mya. Several different species originated about 4.5 to 2 mya. The best known is *A. afarensis*, represented by the famous partial skeleton "Lucy" (3.2 mya), with signs of bipedalism. During the Pliocene epoch (5.3 to 1.6 mya) fundamental changes in morphology and behavior occurred, presumably to adapt to a change in habitat, from the forest to the plains: after early bipedalism, brain volume increased dramatically, accompanied by tool making and other complex behavior. Modern humans as they exist today date back about 30 000–40 000 years. They arrived on the five continents at different times.

B. Important hominid finds

The transition from *Homo erectus* to *Homo sapiens*, i.e., the origin of modern humans, likely occurred according to one of two models: (i) a multiregional model, assuming several transitions, at different times and locations, or (ii) an "out-of-Africa" model, proposing that the transition occurred recently (< 200 000 years ago), only once, in Africa. Genetic data favor the out-of-Africa model. (Figure adapted from Wehner & Gehring, 1995)

C. Neanderthals

Modern humans and Neanderthals coexisted about 30 000–40 000 years ago, but according to genetic data did not interbreed. Pairwise comparison of mitochondrial DNA (mtDNA, see p. 130) of humans, Neanderthals (DNA extracted from fossils), and chimpanzees indicates that Neanderthals did not contribute mitochondrial DNA to modern humans (**1**). At three locations about 2000 km apart (Feldhofer Cave, Neandertal; Mezmaiskaya Cave, northern Caucasus; Vindija, southern Balkans), mtDNA from Neanderthal specimens shows little diversity (3.5%) compared with that of modern humans (**2**). Preliminary data from Y-chromosomal sequences confirm the differences between Neanderthal and human DNA also in the Y chromosome (Dalton, 2006). (Figures adapted from Krings et al., 1997.)

D. A phylogenetic tree

Studies of the Y chromosome (inherited through fathers only) and mitochondrial DNA (inherited through mothers only) are consistent with the out-of-Africa hypothesis. Construction of a phylogenetic tree from the mtDNA of 147 modern humans of African, Asian, Australian, New Guinean, and European origin could be traced to an ancestral haplotype dating back about 200 000 years (Cann et al., 1987). Although this result (dubbed "mitochondrial Eve") remains controversial, the major conclusion that there is a recent African origin has been supported. (Figure adapted from Cann et al., 1987).

References

Cann RL, Stoneking M, Wilson AC: Mitochondrial DNA and human evolution. Nature 325: 31–36, 1987.

Caroll SB: Genetics and the making of *Homo sapiens*. Nature 422: 849–857, 2003.

Dalton R: Neanderthal DNA yields to genome foray. Nature 441: 260–261, 2006.

Denell R, Roebroeks W: An Asian perspective on early human dispersal from Africa. Nature 438:1099–1104, 2005.

Jobling MA, Hurles, M, Tyler-Smith C: Human Evolutionary Genetics. Origins, Peoples, and Disease. Garland Publishing, New York, 2004.

Klein J, Takahata N: Where do we come from? The Molecular Evidence for Human Descent. Springer, Berlin, 2002.

Krings M et al.: Neanderthal mtDNA diversity. Nature Genet. 26: 144–146, 2000.

Mellers P: Neanderthals and the modern human colonization of Europe. Nature 432: 461–465, 2004.

Wehner R, Gehring W: Zoologie, 23rd ed. Thieme Verlag, Stuttgart, 1995.

Online information:

Human evolution and fossils (www.archaeologyinfo.com).

(www.modernhumanorigins.com).

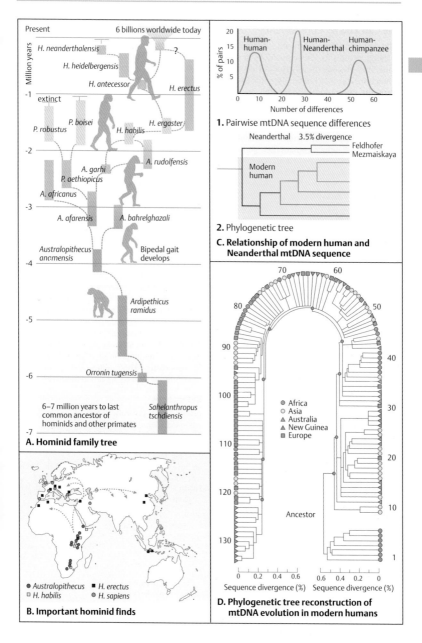

A. Hominid family tree

Present

H. neanderthalensis
H. heidelbergensis
H. antecessor
H. erectus
extinct
P. robustus
P. boisei
H. habilis
H. ergaster
A. rudolfensis
A. garhi
P. aethiopicus
A. africanus
A. afarensis
A. bahrelghazali
Australopithecus anamensis
Bipedal gait develops
Ardipethicus ramidus
Orronin tugensis
6–7 million years to last common ancestor of hominids and other primates
Sahelanthropus tschdiensis

6 billions worldwide today

Million years

1. Pairwise mtDNA sequence differences

Neanderthal   3.5% divergence
Feldhofer
Mezmaiskaya
Modern human

2. Phylogenetic tree

C. Relationship of modern human and Neanderthal mtDNA sequence

● Africa
○ Asia
▲ Australia
▲ New Guinea
■ Europe

Ancestor

Sequence divergence (%)   Sequence divergence (%)

D. Phylogenetic tree reconstruction of mtDNA evolution in modern humans

● Australopithecus   ■ H. erectus
□ H. habilis   ● H. sapiens

B. Important hominid finds

The Cell and Its Components

Cells are the smallest organized structural units of living organisms. Surrounded by a membrane, they are able to carry out a wide variety of functions during a limited life span. Each cell originates from another living cell, as postulated by R. Virchow in 1855 ("*omnis cellula e cellula*"). Two basic types of cells exist: *prokaryotic cells*, which carry their functional information in a circular genome without a nucleus, and *eukaryotic cells*, which contain their genome in individual chromosomes in a nucleus and have a well-organized internal structure. Robert Hooke introduced the word *cell* in 1665 for the tiny cavities in cork, which reminded him of the small rooms in which monks sleep. Cells were recognized as the "elementary particles of organisms," animal and plant, by Mathias Schleiden and Theodor Schwann in 1839. Today we understand many of the biological processes of cells at the molecular level.

A. Scheme of a prokaryotic cell

Prokaryotic cells (bacteria) are typically rod-shaped or spherical with few micrometers in diameter, without a nucleus or special internal structures. Within a cell wall consisting of a bi-layered cell membrane, bacteria contain on average 1000–5000 genes tightly packed in a circular molecule of DNA (p. 42). In addition, they usually contain small circular DNA molecules named *plasmids*. These replicated independently of the main chromosome and generally contain genes which confer antibiotic resistance (p. 94).

B. Scheme of a eukaryotic cell

A eukaryotic cell consists of cytoplasm and a nucleus. It is enclosed by a plasma membrane. The eukaryotic cell nucleus contains the genetic information. The cytoplasm contains a complex system of inner membranes that form discrete structures (organelles). These are the mitochondria (in which important energy-delivering chemical reactions take place), the endoplasmic reticulum (a series of membranes in which important molecules are formed), the Golgi apparatus (for transport functions), lysosomes, in which some proteins are broken down, and peroxisomes (formation or degradation of certain molecules). Animal cells (**1**) and plant cells (**2**) share several features, but differ in important structures. A plant cells contains chloroplasts for photosynthesis. Plant cells are surrounded by a rigid wall of cellulose and other polymeric molecules, and they contain vacuoles for water, ions, sugar, nitrogen-containing compounds, or waste products. Vacuoles are permeable to water but not to other substances enclosed within them.

C. Plasma membrane of the cell

Cells are surrounded by plasma membranes. These are water-resistant membranes composed of bipartite molecules of fatty acids. These molecules are phospholipids arranged in a double layer (bilayer). The plasma membrane contains numerous molecules that traverse the lipid bilayer once or many times to perform special functions. Cells communicate with each other by means of a broad repertoire of molecular signals. *Different types of membrane proteins* can be distinguished: (i) transmembrane proteins used as channels to transport molecules into or out of the cell, (ii) proteins connected with each other to provide stability, (iii) receptor molecules involved in signal transduction, and (iv) molecules with enzyme function to catalyze internal chemical reactions in response to an external signal, and (v) gap junctions in specialized cells forming pores between adjacent cells. Gap junction proteins are composed of connexins. They allow the passage of molecules as large as 1.2 nm in diameter.

References

Alberts B et al: Molecular Biology of the Cell, 5th ed. Garland Science, New York, 2002.

Alberts B et al: Essential Cell Biology. An Introduction to the Molecular Biology of the Cell. Garland Publishing, New York, 1998.

de Duve C: A Guided Tour of the Living Cell, 2 Vols. Scientific American Books Inc, New York, 1984.

Lodish H et al: Molecular Cell Biology, 5th ed. WH Freeman & Co, New York, 2005.

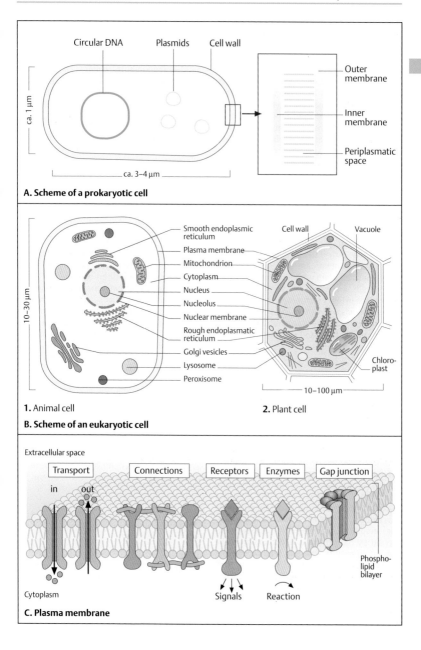

A. Scheme of a prokaryotic cell

Circular DNA   Plasmids   Cell wall

ca. 1 μm

ca. 3–4 μm

Outer membrane

Inner membrane

Periplasmatic space

B. Scheme of an eukaryotic cell

10–30 μm

Smooth endoplasmic reticulum
Plasma membrane
Mitochondrion
Cytoplasm
Nucleus
Nucleolus
Nuclear membrane
Rough endoplasmatic reticulum
Golgi vesicles
Lysosome
Peroxisome

Cell wall   Vacuole

Chloro-plast

10–100 μm

1. Animal cell

2. Plant cell

C. Plasma membrane

Extracellular space

Transport   Connections   Receptors   Enzymes   Gap junction

in   out

Cytoplasm

Signals   Reaction

Phospho-lipid bilayer

Some Types of Chemical Bonds

Chemical bonds form between molecules and allow building of complex structures. Each atom can establish chemical bonds with another in a defined way. Strong forces of attraction are present in covalent bonds, when two atoms share one pair of electrons. Weak forces of attraction occur in noncovalent bonds. They play a major role in many biomolecules such as carbohydrates, lipids, nucleic acids, and proteins. Four major types of noncovalent interactions are distinguished: hydrogen bonds, ionic interactions, van der Waals interactions, and hydrophobic effects.

Close to 99% of the weight of a living cell is composed of just four elements: carbon (C), hydrogen (H), nitrogen (N), and oxygen (O). Almost 50% of the atoms are hydrogen atoms; about 25% are carbon, and 25% oxygen. Apart from water (about 70% of the weight of the cell) almost all components are carbon compounds. Carbon, a small atom with four electrons in its outer shell, is the central chemical building block of the living world. It can form four strong covalent bonds with other atoms. But most importantly, carbon atoms can combine with each other to build chains and rings, and thus large complex molecules with specific biological properties.

A. Functional groups with hydrogen (H), oxygen (O), and carbon (C)

Four simple combinations of these atoms occur frequently in biologically important molecules: hydroxyl ($-OH$; alcohols), methyl ($-CH_3$), carboxyl ($-COOH$), and carbonyl ($C = O$; aldehydes and ketones) groups. They impart characteristic chemical properties to the molecules, including possibilities to form compounds.

B. Acids and esters

Many biological substances contain a carbon–oxygen bond with weak acidic or basic (alkaline) properties. The degree of acidity is expressed by the pH value, which indicates the concentration of H^+ ions in a solution, ranging from 10^{-1} mol/L (pH 1, strongly acidic) to 10^{-14} mol/L (pH 14, strongly alkaline). Pure water contains 10^{-7} moles H^+ per liter (pH 7.0). An ester is formed when an acid reacts with an alcohol. Esters are frequently found in lipids and phosphate compounds.

C. Carbon–nitrogen bonds (C—N)

C—N bonds occur in many biologically important molecules: in amino groups, amines, and amides, especially in proteins. Of paramount significance are the amino acids (see. p. 38), the building blocks of proteins. Proteins have specific roles in the functioning of an organism.

D. Phosphate compounds

Ionized phosphate compounds play an essential biological role. HPO_4^{2-} is a stable inorganic phosphate ion from ionized phosphoric acid. A phosphate ion and a free hydroxyl group can form a phosphate ester. Phosphate compounds play an important role in energy-rich molecules and numerous macromolecules because they can store energy.

E. Sulfur groups

Sulfur often joins biological molecules together, especially when two sulfhydryl groups ($-SH$) react to form a disulfide bridge ($-S-S-$). Sulfur is a component of two amino acids (cysteine and methionine) and of some polysaccharides and sugars. Disulfide bridges play an important role in many complex molecules, serving to stabilize and maintain particular three-dimensional structures.

References

Alberts B et al: Molecular Biology of the Cell, 4th ed. Garland Publishing Co, New York, 2002.

Koolman J, Roehm KH: Color Atlas of Biochemistry, 2nd ed. Thieme, Stuttgart – New York, 2005.

Lodish H et al: Molecular Cell Biology, 5th ed. WH Freeman, New York, 2004.

Pauling L: The Nature of the Chemical Bond. 3rd ed. Cornell University Press, Ithaca, New York, 1960.

Stryer L: Biochemistry, 4th ed. WH Freeman & Co, New York, 1995.

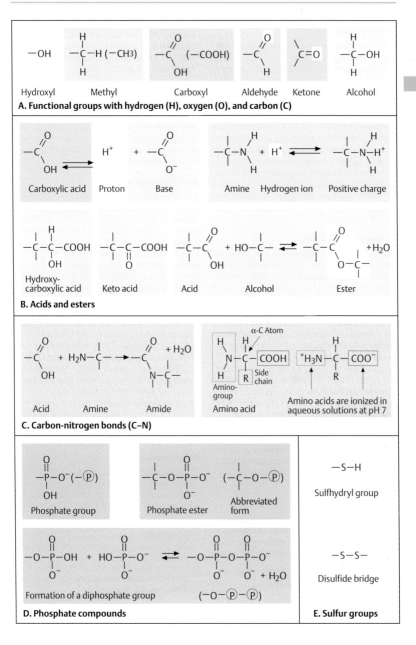

A. Functional groups with hydrogen (H), oxygen (O), and carbon (C)

Hydroxyl   Methyl   Carboxyl   Aldehyde   Ketone   Alcohol

B. Acids and esters

Carboxylic acid   Proton   Base   Amine   Hydrogen ion   Positive charge

Hydroxy-carboxylic acid   Keto acid   Acid   Alcohol   Ester

C. Carbon-nitrogen bonds (C–N)

Acid   Amine   Amide

α-C Atom   Amino-group   Side chain   Amino acid

Amino acids are ionized in aqueous solutions at pH 7

D. Phosphate compounds

Phosphate group   Phosphate ester   Abbreviated form

Formation of a diphosphate group

E. Sulfur groups

Sulfhydryl group

Disulfide bridge

Carbohydrates

Carbohydrates are carbonyl compounds (aldehydes, ketones) that occur widely in living organisms as part of biomolecules. Carbohydrates are one of the most important classes of biomolecules. Their main functions can be classified into three groups: (i) to deliver and store energy, (ii) to help provide the basic framework for DNA and RNA, the information-carrying molecules (see pp. 44), and (iii) to form structural elements of cell walls of bacteria and plants (polysaccharides). In addition, they form cell surface structures (receptors) used in conducting signals from cell to cell. Combined with numerous proteins and lipids, carbohydrates are important components of numerous internal cell structures.

A. Monosaccharides

Monosaccharides (simple sugars) are aldehydes ($-C = O$, $-H$) or ketones ($C = O$) with two or more hydroxy groups (general structural formula: CH_2O_n). The aldehyde or ketone group can react with one of the hydroxy groups to form a ring. This is the usual configuration of sugars that have five or six C atoms (pentoses and hexoses). The C atoms are numbered sequentially. The D- and the L-forms of sugars are mirror-image isomers of the same molecule.

The naturally occurring forms are the D-(dextro) forms. These further include β- and α-forms as stereoisomers. In the cyclic forms, the C atoms of sugars are not on a plane, but three-dimensionally take the shape of a chair or a boat. The β-D-glucopyranose configuration (glucose) is the energetically favored one, since all the axial positions are occupied by H atoms. The arrangement of the $-OH$ groups can differ, so that stereoisomers such as mannose or galactose are formed.

B. Disaccharides

These are compounds of two monosaccharides. The aldehyde or ketone group of one can bind to an α-hydroxy or a β-hydroxy group of the other. Sucrose and lactose are frequently occurring disaccharides.

C. Derivatives of sugars

When certain hydroxy groups are replaced by other groups, sugar derivatives are formed. These occur especially in polysaccharides. In a large group of genetically determined syndromes, complex polysaccharides cannot be degraded owing to reduced or absent enzyme function (mucopolysaccharidoses, mucolipidoses) (see p. 358).

D. Polysaccharides

Short (oligosaccharides) and long chains of sugars and sugar derivatives (polysaccharides) form essential structural elements of the cell. Complex oligosaccharides with bonds to proteins or lipids are part of cell surface structures, e.g., blood group antigens.

Medical relevance

Examples of human hereditary disorders in the metabolism of carbohydrates are:

Diabetes mellitus (MIM 125850): a heterogeneous group of disorders characterized by elevated levels of blood glucose, with complex clinical and genetic features (see p. 372).

Disorders of fructose metabolism: Benign fructosuria (MIM 229800), hereditary fructose intolerance with hypoglycemia and vomiting (MIM 229600), and hereditary fructose 1.6-bisphosphatase deficiency with hypoglycemia, apnea, lactic acidosis, and often with lethal outcome in newborn infants (MIM 229700).

Galactose metabolism: Inherited disorders with acute galactose toxicity and long-term effects (Galactosemia, MIM 230400; galactokinase deficiency, MIM 230200; Galactose epimerase deficiency, MIM 230350), and others.

Glycogen storage diseases: Eight types of disorders of glycogen metabolism that differ in clinical symptoms and the genes and enzymes involved (MIM 232200, 232210–232800).

References

Gilbert-Barness E, Barness L: Metabolic Diseases. Foundations of Clinical Management, Genetics, and Pathology. Eaton Publishing, Natick, MA, USA, 2000.

Koolman J, Roehm KH: Color Atlas of Biochemistry, 2nd ed. Thieme, Stuttgart – New York, 2005.

MIM—McKusick, VA: Mendelian Inheritance in Man, 12th ed., Johns Hopkins University Press Baltimore, 1998, available online at www.ncbi.nlm.nih.gov/ Omim.

Scriver CR, Beaudet AL, Sly WS, Valle D (eds): The Metabolic and Molecular Bases of Inherited Disease, 8th ed. McGraw-Hill, New York, 2001.

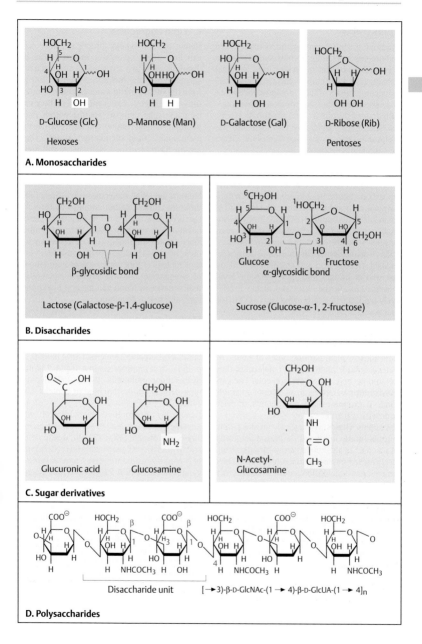

A. Monosaccharides

D-Glucose (Glc)   D-Mannose (Man)   D-Galactose (Gal)   D-Ribose (Rib)

Hexoses   Pentoses

B. Disaccharides

β-glycosidic bond

Lactose (Galactose-β-1.4-glucose)

α-glycosidic bond

Glucose   Fructose

Sucrose (Glucose-α-1, 2-fructose)

C. Sugar derivatives

Glucuronic acid   Glucosamine   N-Acetyl-Glucosamine

D. Polysaccharides

Disaccharide unit   [→3)-β-D-GlcNAc-(1 → 4)-β-D-GlcUA-(1 → 4]n

Lipids (Fats)

Lipids are essential components of membranes and precursors of biomolecules, such as steroid hormones and other molecules used in signal transduction. Lipids are important energy-carrying components of food (dosage-dependent). They also form important compounds with carbohydrates (glycolipids) and phosphate groups (phospholipids). Lipids are classified into hydrolyzable (able to undergo hydrolytic cleavage) and nonhydrolyzable.

A. Fatty acids

A fatty acid is composed of an unbranched hydrocarbon chain of 4–24 carbon atoms with a terminal carboxylic acid group. A fatty acid is polar, with a hydrophilic ($-COOH$) and a hydrophobic end ($-CH_3$). Saturated fatty acids without a double bond and unsaturated fatty acids with one or more double bonds are distinguished. Linoleic acid and arachidonic acid are essential in human nutrition. A double bond causes a kink in the chain and makes it relatively rigid. The free carboxyl group ($-COOH$) of a fatty acid is ionized ($-COO^-$).

B. Lipids

Fatty acids can combine with other groups of molecules to form various types of lipids. As water-insoluble (hydrophobic) molecules, they are soluble only in organic solvents. The carboxyl group can enter into an ester or an amide bond. Triglycerides are compounds of fatty acids with glycerol.

Glycolipids (lipids with sugar residues) and phospholipids (lipids with a phosphate group attached to an alcohol derivative) are the structural bases of important macromolecules. Their intracellular degradation requires the presence of numerous enzymes, disorders of which have a genetic basis and lead to diseases.

Sphingolipids are an important group of molecules in biological membranes. Here, sphingosine, instead of glycerol, is the fatty acid-binding molecule. Sphingomyelin and gangliosides contain sphingosine. Gangliosides make up 6% of the central nervous system lipids. They are degraded by a series of enzymes.

C. Lipid aggregates

Owing to their bipolar properties, fatty acids can form lipid aggregates in water. The hydrophilic ends are attracted to their aqueous surroundings; the hydrophobic ends protrude from the surface of the water and form a surface film. If completely under the surface, they may form a micelle, compact and dry within. Phospholipids and glycolipids can form two-layered membranes (lipid membrane bilayer). These are the basic structural elements of cell membranes.

D. Other lipids: steroids

Steroids are small molecules consisting of four different rings of carbon atoms. Cholesterol is the precursor of five major classes of steroid hormones: prostagens, glucocorticoids, mineralocorticoids, androgens, and estrogens. Each of these hormone classes is responsible for important biological functions such as maintenance of pregnancy, fat and protein metabolism, maintenance of blood volume and blood pressure, and sexual development.

Medical relevance

Several groups of disorders of lipoprotein and lipid metabolism exist. Important examples are familial hypercholesterolemia (MIM 143890, see p. 368), hyperlipoproteinemia (MIM 238600), dysbetalipoproteinemia (MIM 107741), and high-density lipoprotein binding protein (MIM 142695).

Genetically determined disorders of ganglioside catabolism lead to severe diseases, e.g., Tay–Sachs disease (MIM 272800) due to defective degradation of ganglioside GM2 (deficiency of β-N-acetylhexosaminidase), several types of gangliosidoses (MIM 230500, 305650), Sandhoff disease (MIM 268800), and others.

References

Gilbert-Barness E, Barness L: Metabolic Diseases. Foundations of Clinical Management, Genetics, and Pathology. Eaton Publishing, Natick, MA, USA, 2000.

Koolman J, Roehm KH: Color Atlas of Biochemistry, 2nd ed. Thieme, Stuttgart – New York, 2005.

MIM—McKusick, VA: Mendelian Inheritance in Man, 12th ed., Johns Hopkins University Press Baltimore, 1998, available online at www.ncbi.nlm.nih.gov/Omim.

Scriver CR, Beaudet AL, Sly WS, Valle D (eds): The Metabolic and Molecular Bases of Inherited Disease, 8th ed. McGraw-Hill, New York, 2001.

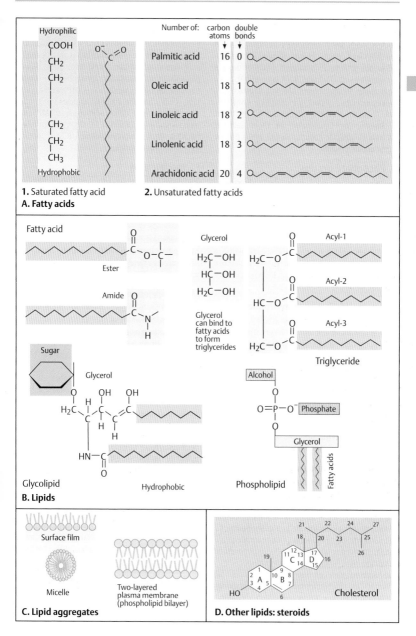

A. Fatty acids

Hydrophilic

COOH

Hydrophobic

1. Saturated fatty acid

	Number of: carbon atoms	double bonds	
Palmitic acid	16	0	
Oleic acid	18	1	
Linoleic acid	18	2	
Linolenic acid	18	3	
Arachidonic acid	20	4	

2. Unsaturated fatty acids

B. Lipids

Fatty acid

Ester

Amide

Glycerol

Glycerol can bind to fatty acids to form triglycerides

Acyl-1

Acyl-2

Acyl-3

Triglyceride

Sugar

Glycerol

Glycolipid

Hydrophobic

Alcohol

Phosphate

Glycerol

Fatty acids

Phospholipid

C. Lipid aggregates

Surface film

Micelle

Two-layered plasma membrane (phospholipid bilayer)

D. Other lipids: steroids

Cholesterol

Nucleotides and Nucleic Acids

Nucleic acids are macromolecules that as DNA and RNA are central to the storage and transmission of genetic information. Nucleotides are the subunits of DNA and RNA (see p. 44). They participate in numerous biological processes, convey energy, are part of essential coenzymes, and regulate numerous metabolic functions. Nucleotides are composed of three integral parts: a phosphate group, a sugar, and a purine or pyrimidine base.

A. Phosphate groups

Phosphate groups occur in nucleic acids and nucleotides as monophosphates (one P atom), diphosphates (2), or triphosphates (3).

B. Sugar residues

The carbohydrate residues in nucleotides are usually derived from either ribose as β-D-ribose (in ribonucleic acid, RNA) or β-D-deoxyribose (in deoxyribonucleic acid, DNA) (ribonucleoside or deoxyribonucleoside).

C. Nucleotide bases of pyrimidine

Cytosine (C), thymine (T), and uracil (U) are the three pyrimidine nucleotide bases. They differ from each other in their side chains ($-NH_2$ on C4 of cytosine, $-CH_3$ on C5 in thymine, and O at C4 in uracil). In addition, cyotsine has a double bond between N3 and C4.

D. Nucleotide bases of purine

Adenine (A) and guanine (G) are the two nucleotide bases of purine. They differ in their side chains and in having a double bond between N1 and C6 (present in adenine, absent in guanine).

E. Nucleosides and nucleotides

A *nucleoside* is a compound of a sugar residue (ribose or deoxyribose) and a nucleotide base. The bond is between the C atom in position 1 of the sugar and an N atom of the base (N-glycosidic bond). A *nucleotide* is a compound of a five-C-atom sugar residue (ribose or deoxyribose) attached to a nucleotide base (pyrimidine or purine base) and a phosphate group.

The nucleosides of the various bases are grouped as ribonucleosides or deoxyribonucleosides, e.g., adenosine or deoxyadenosine, guanosine or deoxyguanosine, uridine (occurs only as a ribonucleoside), cytidine or deoxycy-tidine. Thymidine occurs only as a deoxynucleoside.

Nucleotides are the subunits of nucleic acids. The nucleotides of the individual bases are referred to as follows: adenylate (AMP, adenosine monophosphate), guanylate (guanosine monophosphate, GMP), uridylate (UMP), and cytidylate (CMP) for the ribonucleotides (5′ monophosphates) and deoxyadenylate (dAMP), deoxyguanylate (dGMP), deoxythymidylate (dTMP), and deoxycytidylate (dCMP) for the deoxyribonucleotides.

F. Nucleic acid

A nucleic acid consists of a series of nucleotides. A phosphodiester bridge between the 3′ C atom of one nucleotide and the 5′ C atom of the next joins two nucleotides. The linear sequence is usually given in the 5′ to 3′ direction with the abbreviations of the respective nucleotide bases. For instance, ATCG would signify the sequence adenine (A), thymine (T), cytosine (C), and guanine (G) in the 5′ to 3′ direction.

Medical relevance

Examples of human hereditary disorders in purine and pyrimidine metabolism are:
Hyperuricemia and gout: A group of disorders resulting from genetically determined excessive synthesis of purine precursors (MIM 240000).
Lesch–Nyhan syndrome: A variable, usually severe infantile X-chromosomal disease with marked neurological manifestations resulting from hypoxanthine – guanine phosphoribosyltransferase deficiency (MIM 308000).
Adenosine deaminase deficiency: A heterogeneous group of disorders resulting in severe infantile immunodeficiency. Different autosomal recessive and X-chromosomal types exist (MIM 102700).

References

Gilbert-Barness E, Barness L: Metabolic Diseases. Foundations of Clinical Management, Genetics, and Pathology. Eaton Publishing, Natick, MA, USA, 2000.

Koolman J, Roehm KH: Color Atlas of Biochemistry, 2nd ed. Thieme, Stuttgart – New York, 2005.

Scriver CR, Beaudet AL, Sly WS, Valle D (eds): The Metabolic and Molecular Bases of Inherited Disease, 8th ed. McGraw-Hill, New York, 2001.

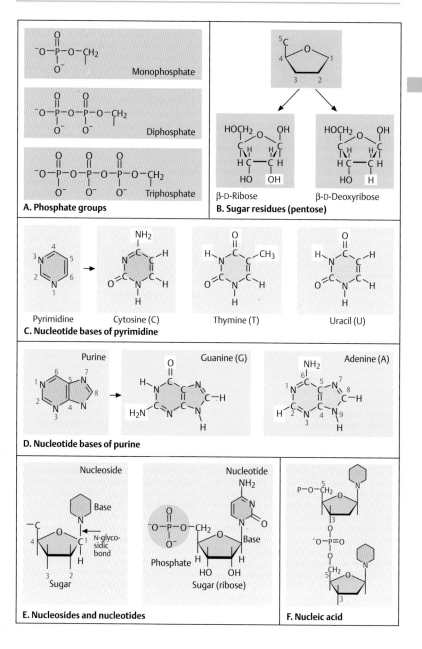

A. Phosphate groups

Monophosphate

Diphosphate

Triphosphate

B. Sugar residues (pentose)

β-D-Ribose

β-D-Deoxyribose

C. Nucleotide bases of pyrimidine

Pyrimidine

Cytosine (C)

Thymine (T)

Uracil (U)

D. Nucleotide bases of purine

Purine

Guanine (G)

Adenine (A)

E. Nucleosides and nucleotides

Nucleoside

Base

N-glyco-sidic bond

Sugar

Nucleotide

Phosphate

Sugar (ribose)

F. Nucleic acid

Amino Acids

Amino acids (2-aminocarboxylic acids) are the basic structural units of proteins. An amino acid consists of a central α carbon (C_α) bonded to four different chemical groups: one bond to an amino group ($-NH_2$), one to a carboxyl group ($-COOH$), one to a hydrogen atom (H), and one to a variable side chain (R). The side chain is the major determinant of the individual functional property of each amino acid in a protein. The α carbon is asymmetric except in glycine; therefore, amino acids exist in two mirror-image forms, D (dextro) and L (levo) isomers. Only the L forms occur in proteins, with rare exceptions.

Amino acids are ionized in neutral solutions, the amino group taking on a proton ($-NH_3^+$) and the carboxyl group dissociating ($-COO^-$). Amino acids are classified according to their side chains and chemical reactivity. Each amino acid has its own three-letter and one-letter abbreviations. Essential amino acids for humans are valine (Val), leucine (Leu), isoleucine (Ile), phenylalanine (Phe), tryptophan (Trp), methionine (Met), threonine (Thr), and lysine (Lys). These have to be supplied by food intake.

A. Neutral amino acids

Simple amino acids have an aliphatic side chain, e.g., glycine has a hydrogen atom ($-H$) and alanine a methyl group ($-CH_3$), or a larger, hydrophobic (water-repellent) side chain as in valine, leucine, and isoleucine. Proline has an aliphatic side chain that is bound to both the central carbon and to the amino group in a ring structure. Hydrophobic aromatic side chains occur in phenylalanine (a phenyl group bound via a methylene [$-CH_2-$] group) and tryptophan (an indol ring bound via a methylene group). Two hydrophobic amino acids contain sulfur (S) atoms: cysteine with a sulfhydryl group ($-SH$) and methionine with a thioether ($-S-CH_3$). The sulfhydryl group in cysteine is very reactive and forms stabilizing disulfide bonds ($-S-S-$). These play an important role in stabilizing the three-dimensional forms of proteins. Selenocysteine is a cysteine analogue occurring in a few proteins such as the enzyme glutathione peroxidase.

B. Hydrophilic amino acids

Serine, threonine, and tyrosine contain hydroxyl groups ($-OH$). Thus, they are hydrolyzed forms of glycine, alanine, and phenylalanine. The hydroxyl groups make them hydrophilic and more reactive than the nonhydrolyzed forms. Asparagine and glutamine both contain an amino and an amide group. At physiological pH their side chains are negatively charged.

C. Charged amino acids

These amino acids have either two ionized amino groups (basic) or two carboxyl groups (acidic). Basic amino acids (positively charged) are arginine, lysine, and histidine. Histidine has an imidazole ring and can be uncharged or positively charged, depending on its surroundings. It is frequently found in the reactive centers of proteins, where it takes part in alternating bonds (e.g., in the oxygen-binding region of hemoglobin). Aspartic acid and glutamic acid each have two carboxyl groups ($-COOH$) and thus are usually acidic. Seven of the 20 amino acids have slightly ionizable side chains, making them highly reactive (Asn, Glu, His, Cys, Tyr, Lys, Arg).

Medical relevance

The amino acids glycine, phenylalanine, tyrosine, histidine, proline, lysine, and the branched chain amino acids valine, leucine, and isoleucine are predominantly involved in various genetic diseases with toxic metabolic symptoms that result when their plasma concentration is too high or too low.

Phenylketonuria: Disorders of phenylalanine hydroxylation result in variable clinical signs and severity, caused by a spectrum of mutations in the responsible gene (MIM 261600, see p. 384).

Maple syrup urine disease: A variable disorder due to deficiency of branched chain α-keto acid dehydrogenase, which leads to accumulation of valine, leucine, and isoleucine (MIM 248600). The classic severe form results in severe neurological damage to the infant.

References

Gilbert-Barness E, Barness L: Metabolic Diseases. Foundations of Clinical Management, Genetics, and Pathology. Eaton Publishing, Natick, MA, USA, 2000.

Koolman J, Roehm KH: Color Atlas of Biochemistry, 2nd ed. Thieme, Stuttgart – New York, 2005.

Scriver CR, Beaudet AL, Sly WS, Valle D (eds): The Metabolic and Molecular Bases of Inherited Disease, 8th ed. McGraw-Hill, New York, 2001.

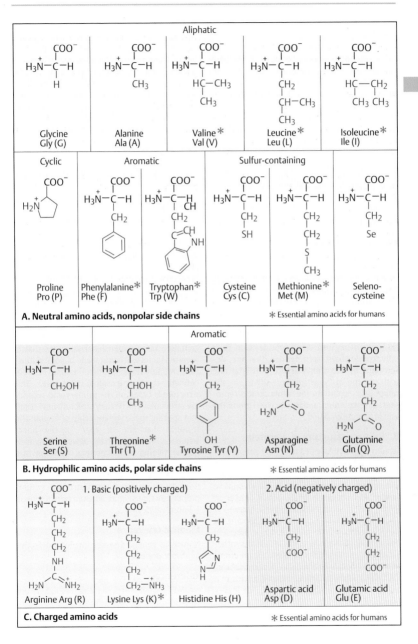

Aliphatic

Glycine
Gly (G)

Alanine
Ala (A)

Valine ∗
Val (V)

Leucine ∗
Leu (L)

Isoleucine ∗
Ile (I)

Cyclic — **Aromatic** — **Sulfur-containing**

Proline
Pro (P)

Phenylalanine∗
Phe (F)

Tryptophan∗
Trp (W)

Cysteine
Cys (C)

Methionine ∗
Met (M)

Seleno-
cysteine

A. Neutral amino acids, nonpolar side chains   ∗ Essential amino acids for humans

Aromatic

Serine
Ser (S)

Threonine ∗
Thr (T)

Tyrosine Tyr (Y)

Asparagine
Asn (N)

Glutamine
Gln (Q)

B. Hydrophilic amino acids, polar side chains   ∗ Essential amino acids for humans

1. Basic (positively charged) — **2. Acid (negatively charged)**

Arginine Arg (R)

Lysine Lys (K) ∗

Histidine His (H)

Aspartic acid
Asp (D)

Glutamic acid
Glu (E)

C. Charged amino acids   ∗ Essential amino acids for humans

Proteins

Proteins are linear macromolecules (polypeptides) consisting of amino acids joined by peptide bonds, arranged in a complex three-dimensional structure that is specific for each protein. Proteins are involved in all chemical processes in living organisms. As enzymes, they drive chemical reactions that in living cells would not occur spontaneously. They serve to transport small molecules, ions, or metals and have important functions in cell division during growth and in cell and tissue differentiation. Proteins control the coordination of movements by regulating muscle cells and the production and transmission of impulses within and between nerve cells; they control blood homeostasis (blood clotting) and immune defense. They have mechanical functions in skin, bone, blood vessels, and other areas.

A. Peptide bonds

Amino acids are easily joined together owing to their dipolar ionization (zwitterions). The carboxyl group of one amino acid bonds to the amino group of the next (a peptide bond, sometimes also referred to as an amide bond). When many amino acids are bound together by peptide bonds, they form a polypeptide chain. Each polypeptide chain has a defined direction, determined by the amino group ($-NH_2$) at one end and the carboxyl group ($-COOH$) at the other. By convention, the amino group represents the beginning, and the carboxyl group the end of a peptide chain.

B. Primary structure of a protein

Insulin is an example of a relatively simple protein consisting of two polypeptide chains, an A chain of 21 amino acids and a B chain of 30 amino acids. The determination of its complete amino acid sequence by Frederick Sanger in 1955 was a landmark accomplishment. It showed for the first time that a protein, in genetic terms a gene product, has a precisely defined amino acid sequence. Insulin is synthesized from two precursor molecules, preproinsulin and proinsulin. Preproinsulin consists of 110 amino acids, including 24 amino acids of a leader sequence at the amino end. The leader sequence directs the molecule to the correct site within the cell, where it is removed to yield proinsulin with 86 amino acids. From this, a connecting (C) peptide is removed (amino acids number 31–65). This yields the two chains, B (amino acids no. 1–30) and A (amino acids 1–21). The A and B chains are connected by two disulfide bridges, which respectively join the cysteines in positions 7 and 19 of the B chain with positions 6 and 20 of the A chain. The A chain contains a disulfide bridge between positions 6 and 11. The linear sequence of the amino acids is called the primary structure. It yields important information about the function and evolutionary origin of a protein. The positions of the disulfide bridges reflect the spatial arrangements of the amino acids, called the secondary structure.

C. Secondary structural units

The secondary structure of a protein refers to regions with a defined spatial arrangement. Two basic units of global proteins are α helix formation (α helix) and a flat sheet (β-pleated sheet). Insulin is made up of 57% α-helical areas, 6% β-pleated sheets, 10% β-turns, and 27% other areas without a defined secondary structure. (Figure adapted from Stryer, 1995.)

D. Tertiary structure

The tertiary structure of a protein is the complete three-dimensional structure that is required for its biochemical and biological function. All functional proteins assume a well-defined three-dimensional structure. This structure is based on the primary and secondary structures. The tertiary structure may result in a specific spatial relationship of amino acid residues that are far apart in the linear sequence. The quaternary structure involves further folding of the protein, resulting in a specific three-dimensional spatial arrangement of different subunits that affects their interactions. The correct quaternary structure ensures proper function. (Figure adapted from Koolman & Röhm, 2005.)

Medical relevance

Numerous genetic diseases involve a defective or absent protein.

References

Koolman J, Röhm K-H: Color Atlas of Biochemistry, 2nd ed. Thieme, Stuttgart–New York, 2005.
Stryer L: Biochemistry, 4th ed. WH Freeman & Co, New York, 1995.

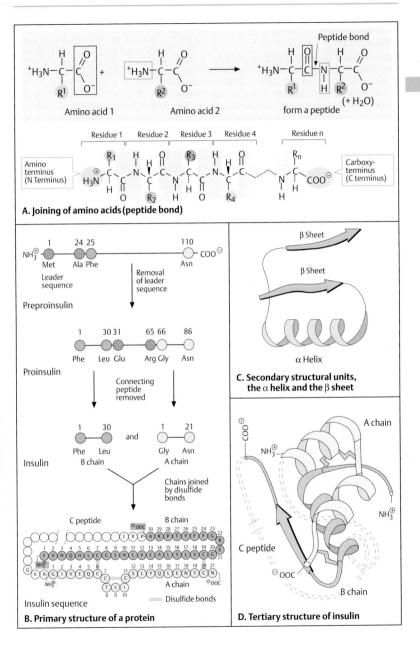

A. Joining of amino acids (peptide bond)

B. Primary structure of a protein

C. Secondary structural units, the α helix and the β sheet

D. Tertiary structure of insulin

DNA as a Carrier of Genetic Information

When Friedrich Miescher in 1869 discovered a new, acidic, phosphorus-containing substance made up of very large molecules, which he named "nuclein," its biological role was not apparent. The term "nucleic acid," later to be recognized as deoxyribonucleic acid (DNA), was introduced by Richard Altmann in 1889. By 1900, the purine and pyrimidine bases were known, and twenty years later, the two kinds of nucleic acid, RNA and DNA, were distinguished. An incidental but precise observation in1928 and further investigations in 1944 indicated that DNA could be the carrier of genetic information.

A. Griffith's observation

In 1928, the English microbiologist Fred Griffith made a remarkable observation. While investigating various strains of *Pneumococcus* bacteria (*Streptomyces pneumoniae*, a cause of inflammation of the lungs, pneumonia), he determined that mice injected with strain S (smooth) died (**1**). On the other hand, animals injected with strain R (rough) survived (**2**). When he inactivated the lethal S strain by heat, the animals also survived (**3**). Surprisingly, a mixture of the nonlethal R strain and the heat-inactivated S strain had the same lethal effect as the original S strain (**4**). When he found normal living pneumococci of the S strain in the animals' blood, he concluded that cells of the R strain must have changed into cells of the S strain. This is called bacterial transformation. For some time, this surprising result could not be explained and was met with skepticism. Its relevance for genetics was not apparent.

B. The transforming principle is DNA

Griffith's findings formed the basis for investigations by Avery, MacLeod, and McCarty (1944) at the Rockefeller Institute in New York. They determined that the chemical basis of the transforming principle was DNA. From cultures of an S strain (**1**) they produced an extract of lysed cells (cell-free extract, **2**). After all its proteins, lipids, and polysaccharides had been removed, the extract still retained the ability to transform pneumococci of the R strain to pneumococci of the S strain (transforming principle, **3**).

With further studies, Avery and co-workers determined that this was due to the DNA alone. Thus, the DNA must contain the corresponding genetic information, which explained Griffith's observation. Heat had left the DNA of the bacterial chromosomes intact. The section of the chromosome with the gene responsible for capsule formation (S gene) could be released from the destroyed S cells and taken up by some R cells in subsequent cultures. After the S gene was incorporated into its DNA, an R cell was transformed into an S cell (**4**). This observation is based on the ability of bacteria to take up foreign DNA, which alters (transforms) some of their genetic attributes correspondingly.

C. Genetic information is transmitted by DNA only

The final evidence that DNA, and no other molecule, transmits genetic information was provided by Hershey and Chase in 1952. They labeled the capsular protein of bacteriophages (see p. 98) with radioactive sulfur (^{35}S) and the DNA with radioactive phosphorus (^{32}P). When bacteria were infected with the labeled bacteriophage, only ^{32}P (DNA) entered the cells, and not the ^{35}S (capsular protein). The subsequent formation of new, complete phage particles in the cell proved that DNA was the exclusive carrier of the genetic information needed to form new phage particles, including their capsular protein. (Figures in A and B adapted from Stent & Caldendar, 1978.)

References

Avery OT, MacLeod CM, McCarty M: Studies on the chemical nature of the substance inducing transformation of pneumococcal types. J Exp Med 79: 137–158, 1944.

Griffith F: The significance of pneumoccocal types. J Hyg 27: 113–159, 1928.

Hershey AD, Chase M: Independent functions of viral protein and nucleic acid in growth of bacteriophage. J Gen Physiol 36: 39–56, 1952.

Judson MF: The Eighth Day of Creation. Makers of the Revolution in Biology. Expanded Edition. Cold Spring Harbor Laboratory Press, New York, 1996.

McCarty M: The Transforming Principle. Discovering that Genes are made of DNA. WW Norton & Co, New York–London, 1985.

Stent GS, Calendar R: Molecular Genetics. An Introductory Narrative 2nd ed. WH Freeman, San Francisco, 1978.

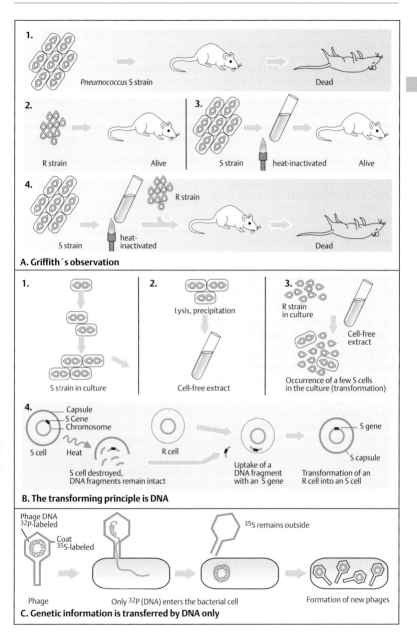

1.
Pneumococcus S strain   Dead

2.
R strain   Alive

3.
S strain   heat-inactivated   Alive

4.
S strain   heat-inactivated   R strain   Dead

A. Griffith´s observation

1.
S strain in culture

2.
Lysis, precipitation
Cell-free extract

3.
R strain in culture
Cell-free extract
Occurrence of a few S cells in the culture (transformation)

4.
Capsule
S Gene
Chromosome
S cell   Heat
S cell destroyed, DNA fragments remain intact
R cell
Uptake of a DNA fragment with an S gene
S gene
S capsule
Transformation of an R cell into an S cell

B. The transforming principle is DNA

Phage DNA
^{32}P-labeled
Coat
^{35}S-labeled
^{35}S remains outside
Phage   Only ^{32}P (DNA) enters the bacterial cell   Formation of new phages

C. Genetic information is transferred by DNA only

DNA and Its Components

Deoxyribonucleic acid (DNA) stores genetic information (see previous pages). Its components, two types of nucleotide base (purines and pyrimidines), deoxyribose, and a phosphate group, are arranged in a specific chemical relationship. They determine the three-dimensional structure of DNA, from which it derives its functional consequences (see next page).

A. Nucleotide bases

The nucleotide bases in DNA are heterocyclic molecules derived from either purine or pyrimidine. Five bases occur in the two types of nucleic acids, DNA and RNA. The purine bases are adenine (A) and guanine (G). The pyrimidine bases are thymine (T) and, in DNA, cytosine (C) or, in RNA, uracil (U). The nucleotide bases are part of a subunit of DNA, the nucleotide. This consists of one of the four nucleotide bases, a sugar (deoxyribose), and a phosphate group. The nitrogen atom in position 9 of a purine or in position 1 of a pyrimidine is bound to the carbon in position 1 of the sugar (N-glycosidic bond).

Ribonucleic acid (RNA) differs from DNA in two respects: it contains ribose instead of deoxyribose (unlike the latter, ribose has a hydroxyl group on the position 2 carbon atom) and uracil instead of thymine. Uracil does not have a methyl group at position C5.

B. Nucleotide chain

DNA is a linear polymer of deoxyribonucleotide units. The nucleotide chain is formed by joining a hydroxyl group on the sugar of one nucleotide to the phosphate group attached to the sugar of the next nucleotide. The sugars linked together by the phosphate groups form the invariant part of the DNA. The variable part is in the sequence of the nucleotide bases A, T, C, and G. A DNA nucleotide chain is polar. The polarity results from the way the sugars are attached to each other. The phosphate group at position C5 (the 5' carbon) of one sugar joins to the hydroxyl group at position C3 (the 3' carbon) of the next sugar by means of a phosphate diester bridge. Thus, one end of the chain has a 5' phosphate group free and the other end has a 3' hydroxy group free (5' end and 3' end, respectively). By convention, the sequence of nucleotide bases is written in the 5' to 3' direction.

C. Hydrogen bonds between bases

The chemical structure of the nucleotide bases determines a defined spatial relationship. A purine (adenine or guanine) always lies opposite to a pyrimidine (thymine or cytosine). Three hydrogen-bond bridges form between cytosine (C) and guanine (G). Two hydrogen bonds form between adenine (A) and thymine (T). Therefore, either guanine and cytosine or adenine and thymine are posed opposite each other, forming complementary base pairs G–C and A–T. Other spatial relationships are not possible. The distance between two bases is 2.90 or 3.00 Å.

D. DNA double strand

DNA consists of two opposing double strands in a double helix (see next page). As a result of the spatial relationships of the nucleotide bases, a cytosine will always lie opposite to a guanine and a thymine opposite to an adenine. The sequence of the nucleotide bases on one strand of DNA (in the 5' to 3' direction) is complementary to the nucleotide base sequence (or simply the base sequence) of the other strand in the 3' to 5' direction. The specificity of base pairing is the most important structural characteristic of DNA. Most DNA is a right-handed helix with a diameter of 20 Å (2×10^{-6} mm). Bases on the same strand are 0.36 nm (3.6 Å) apart. A helical turn repeats itself at intervals of 3.6 nm, about 10.5 base pairs per turn. This type is the *B form* of DNA (see next two pages). Genetic information is determined by the sequence of the base pairs (bp).

References

Alberts B et al: Molecular Biology of the Cell, 4th ed. Garland Publishing Co, New York, 2002.

Koolman J, Roehm KH: Color Atlas of Biochemistry, 2nd ed. Thieme, Stuttgart – New York, 2005.

Lodish H et al: Molecular Cell Biology, 5th ed. WH Freeman, New York, 2004.

Stryer L: Biochemistry, 4th ed. WH Freeman & Co, New York, 1995.

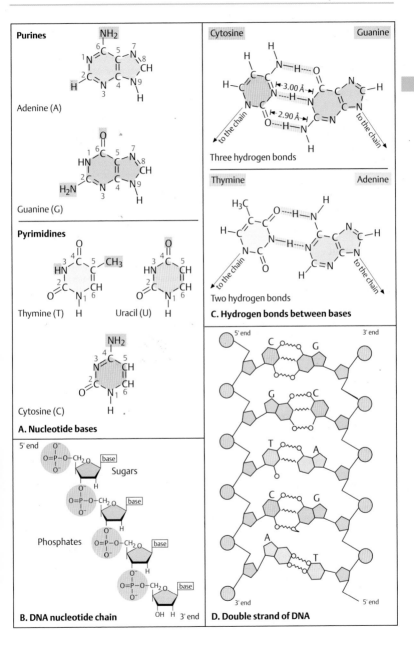

Purines

Adenine (A)

Guanine (G)

Pyrimidines

Thymine (T)   Uracil (U)

Cytosine (C)

A. Nucleotide bases

5' end

Sugars

Phosphates

OH   H   3' end

B. DNA nucleotide chain

Cytosine   Guanine

3.00 Å

2.90 Å

to the chain   to the chain

Three hydrogen bonds

Thymine   Adenine

to the chain   to the chain

Two hydrogen bonds

C. Hydrogen bonds between bases

5' end   3' end

C   G

G   C

T   A

C   G

A   T

3' end   5' end

D. Double strand of DNA

DNA Structure

The elucidation of the structure of deoxyribonucleic acid (DNA) as a double helix by Watson and Crick in 1953 is considered as a cornerstone of modern genetics. This was not immediately recognized. The novel features of this structure explain two fundamental genetic mechanisms: storage of genetic information in a linear, readable pattern, and replication of genetic information to ensure its faithful transmission from generation to generation.

A. DNA double helix

Two helical polynucleotide chains wind around each other along a common axis. The nucleotide bases lie within in pairs, (A–T or G–C, see previous page). The sugar-phosphate backbone lies on the outside.

B. Replication

Because of the complementary arrangement of the base pairs, each strand can serve as a template to generate a new strand. By opening the double helix, two new, identical molecules can be generated (DNA replication).

C. Denaturation and renaturation

Although each of the noncovalent hydrogen bonds between the nucleotide base pairs is weak, DNA is stable at physiological temperatures because the two strands are held together by the many hydrogen bonds along the up to several hundred million nucleotides. The two strands can separate by unwinding when exposed in solution to increasing temperature or weak chemical reagents (e.g., alkali, formamide, or urea). This reversible process is called *denaturation* or melting. The melting temperature (T_m) depends on many factors, including the proportion of G-C pairs, because the three hydrogen bonds between G and C are more stable than the two between AT pairs. The single-stranded DNA molecules resulting form random coils and do not maintain their helical structure, as shown. Lowering the temperature, increasing the ion concentration, or neutralizing the pH will reassociate the two strands into a double helix (*renaturation*), but only if they are complementary. If identical, they hybridize rapidly; if closely related, they hybridize slowly; if unrelated, they do not hybridize. Denaturation and renaturation are the basis for analyzing whether the sequence of the nucleotides of two single strands of DNA are related (DNA hybridization). This is an important principle in the analysis of genes. DNA can also be hybridized with RNA (ribonucleic acid, a structurally related molecule).

D. Transmission of genetic information

The sequence of the nucleotide base pairs (A–T or G–C) determines the contents of the genetic information, corresponding to a linear text of letters and words. This is decoded and read in two steps called *transcription* and *translation*. In transcription, the sequence of one of the two DNA strands is converted into a complementary sequence of bases in a similar molecule, RNA (ribonucleic acid), called messenger RNA (mRNA). One DNA strand, the one in the 3′ to 5′ direction (coding strand), serves as the template. In translation, this sequence is converted into a defined sequence of amino acids according to the genetic code. This dictates that a triplet sequence, three base pairs called a codon, encodes one of the twenty amino acids. Beginning with a defined start point (methionine), the nucleotide sequence of the mRNA molecule is translated into a corresponding sequence of amino acids, aligned in the same sequence as originally contained in the DNA sequence. DNA and RNA differ with respect to one nucleotide: RNA contains uracil (U) instead of the structurally related thymine (T, see previous page). The mechanisms and biochemical reactions of transcription and translation are complex (not shown here).

References

Crick F: What Mad Pursuit. A Personal View of Scientific Discovery. Basic Books Inc, New York, 1988.

Judson HF: The Eighth Day of Creation. Makers of the Revolution in Biology. Expanded Edition. Cold Spring Harbor Laboratory Press, New York, 1996.

Stent GS (ed): The Double Helix. Weidenfeld & Nicolson, London, 1981.

Watson JD: The Double Helix. A Personal Account of the Structure of DNA. Atheneum, New York, 1968.

Watson JD, Crick FHC: Molecular structure of nucleic acid. Nature 171: 737–738, 1953.

Watson JD, Crick FHC: Genetic implications of the structure of DNA. Nature 171: 964–967, 1953.

Wilkins MFH, Stokes AR, Wilson HR: Molecular structure of DNA. Nature 171: 738–740, 1953.

Wilkins M: The Third Man of the Double Helix. Oxford University Press, Oxford, 2003.

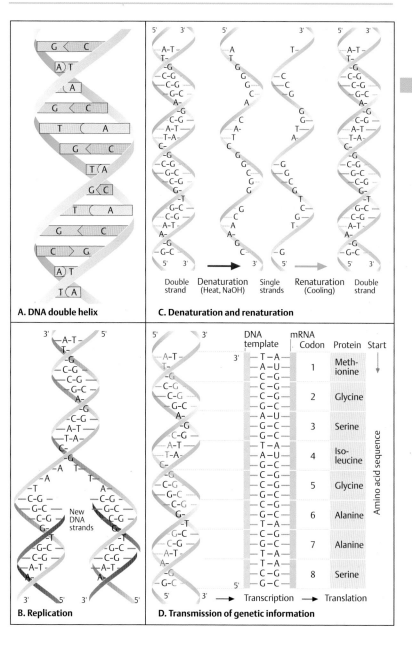

A. DNA double helix

C. Denaturation and renaturation

Double strand — Denaturation (Heat, NaOH) — Single strands — Renaturation (Cooling) — Double strand

B. Replication

New DNA strands

D. Transmission of genetic information

DNA template	mRNA Codon	Protein	Start
T – A	A – U	1	Meth-ionine
A – U			
C – G			
C – G	C – G	2	Glycine
C – G			
G – C			
A – U	A – U	3	Serine
G – C			
G – C			
T – A	A – U	4	Iso-leucine
A – U			
G – C			
C – G	C – G	5	Glycine
C – G			
G – C			
C – G	G – C	6	Alanine
G – C			
T – A			
C – G	G – C	7	Alanine
G – C			
T – A			
T – A	C – G	8	Serine
C – G			
G – C			

Amino acid sequence

Transcription → Translation

Alternative DNA Structures

In addition to the B form of DNA shown schematically on the previous page, two other structures of the double helix have been recognized, the *A form* and the *Z form*, as shown here.

A. Three forms of DNA

The original classic form, determined by Watson and Crick in 1953, is B-DNA, a right-handed helix. In very low humidity, the B structure changes to the A form. RNA/DNA and RNA/RNA helices assume this form in cells and *in vitro*. The A form is compact and has 11 bases per turn instead of 10.5 in the B form. A-DNA is rare. It exists only in the dehydrated state and differs from the B form by a 20-degree rotation of the perpendicular axis of the helix. A-DNA has a deep major groove and a flat minor groove.

The Z form is left-handed instead of right-handed. Its helix has a higher-energy form. This leads to a greater distance (0.77 nm) between the base pairs than in B-DNA and a zigzag form of the sugar-phosphate skeleton when viewed from the side (thus the designation Z-DNA). The single Z-DNA groove has a greater density of negatively charged molecules. A segment of B-DNA consisting of GC pairs can be converted into Z-DNA when the bases are rotated 180 degrees. Normally, Z-DNA is thermodynamically relatively unstable. However, transition to Z-DNA is facilitated when cytosine is methylated in position 5 (C5). The modification of DNA by methylation of cytosine is frequent in certain regions of the DNA of eukaryotes.

Z-DNA has biological roles. Sequences favoring the formation of Z-DNA occur frequently near the promoter region, where Z-DNA stimulates transcription (Ha et al., 2005). Four families of specific Z-DNA binding proteins of defined three-dimensional structure interact with Z-DNA. These are the editing enzyme ADAR1, an interferon-inducible protein DLM-1, an orthologue of the interferon-induced kinase PKR, and the N-terminal domain of the pox virulence factor protein E3L. The E3 L protein is required for pathogenicity in mice. Its sequence is similar to other members of the family of Z-DNA binding proteins (Kim et al., 2003; Ha et al., 2005).

In addition to the three forms shown here, a triple-stranded DNA structure forms when synthetic polymers of poly(A) and polydeoxy(U) are mixed in a test tube. (Figure adapted from Koolman & Röhm, 2005.)

B. Major and minor grooves in B-DNA

The essential structural characteristic of B-DNA is the formation of two grooves, one large (major groove) and one small (minor groove). The base pairing in DNA (adenine – thymine, AT, and guanine – cytosine, GC) leads to the formation of a large and a small groove because the glycosidic bonds to deoxyribose (dRib) are not diametrically opposed. In B-DNA, the purine and pyrimidine rings are 0.34 nm apart. B-DNA has 10.5 base pairs per turn. The distance from one complete turn to the next is 3.4 nm. In this way, localized curves arise in the double helix. The result is a somewhat larger and a somewhat smaller groove.

C. Physical dimensions of the double helix (B form)

The usual B form of the DNA double helix has defined physical dimensions. It has a diameter of 20 Å (2×10^{-6} mm). A helical turn repeats itself at intervals of 3.6 nm (36 Å), about 10.5 base pairs per turn. Bases on the same strand are 0.36 nm (3.6 Å) apart.

References

Ha SC et al: Crystal structure of a junction between B-DNA and Z-DNA reveals two extruded bases. Nature 437: 1183–1186, 2005.

Kim Y-G et al: A role for Z-DNA binding in vaccinia virus pathogenesis. Proc. Nat Acad Sci 100: 6974–6979, 2003.

Koolman J, Röhm KH: Color Atlas of Biochemistry, 2nd ed. Thieme Medical Publishers, Stuttgart-New York, 2005.

Lodish H et al.: Molecular Biology of the Cell, 5th ed. WH Freeman, New York, 2004.

Rich A, Zhang S: Z-DNA: The long road to biological function. Nature Rev Genet 4: 566–572, 2003.

Rich A, Nordheim A, Wang AH: The chemistry and biology of left-handed Z-DNA. Ann Rev Biochem 53: 791–846, 1984.

Stryer L: Biochemistry, 4th ed. WH Freeman & Co, New York, 1995.

Wang AH-J et al: Molecular structure of a left-handed double helical DNA fragment at atomic resolution. Nature 282: 680–686, 1979.

Watson JD et al: Molecular Biology of the Gene, 3rd ed. Benjamin/Cummings Publishing Co, Menlo Park, California, 1987.

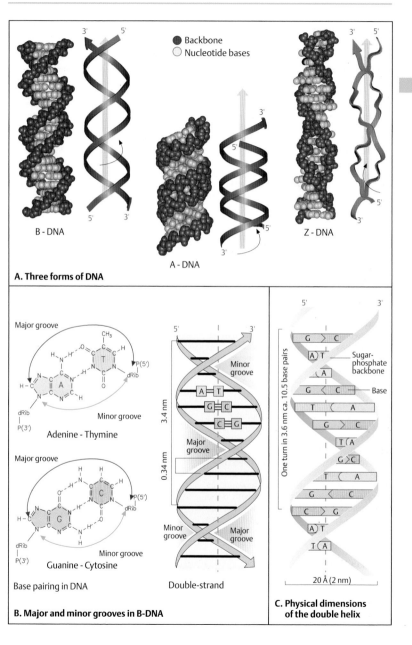

A. Three forms of DNA

B - DNA

A - DNA

Z - DNA

● Backbone
○ Nucleotide bases

Major groove

Adenine - Thymine

Minor groove

Major groove

Guanine - Cytosine

Minor groove

Base pairing in DNA

3.4 nm

0.34 nm

Minor groove

Major groove

Minor groove

Major groove

Double-strand

B. Major and minor grooves in B-DNA

One turn in 3.6 nm ca. 10.5 base pairs

Sugar-phosphate backbone

Base

20 Å (2 nm)

C. Physical dimensions of the double helix

DNA Replication

DNA replication refers to the process of copying each DNA strand into a new complementary strand. This occurs at every cell division and ensures that genetic information is transmitted to both daughter cells. DNA replication requires the highly coordinated action of many proteins in a complex called the replisome. Precision and speed are required. The two new DNA chains are assembled at a rate of about 1000 nucleotides per second in E. coli. During replication, each pre-existing strand of DNA serves as a template for the formation of a new strand. This is referred to as semiconservative replication. DNA replication also occurs during recombination and repair of damaged DNA (see DNA repair).

A. Prokaryotic replication begins at one site

In prokaryote cells, replication begins at a defined point in the ring-shaped bacterial chromosome, the origin of replication (1). From here, new DNA is formed at the same speed in both directions until the DNA has been completely duplicated and two chromosomes are formed. Replication can be visualized by autoradiography after the newly replicated DNA has incorporated tritium (^{3}H)-labeled thymidine (2).

B. Eukaryotic replication begins at several sites

In eukaryote cells DNA synthesis occurs during a defined phase of the cell cycle (S phase). This would take a very long time if there were only one starting point. However, replication of eukaryotic DNA begins at numerous sites (replicons) (1). At each site where replication is initiated, the parental strands are unwound by helicases. Replication proceeds in both directions from each replicon until neighboring replicons fuse (2) and the entire DNA is duplicated (3). The electron micrograph (4) shows replicons at three sites.

C. Replication fork

During DNA replication a characteristic structure forms at the site of the opened double helix where the new strands are synthesized, the replication fork (1). It moves along the double helix as the parental strands are unwound by heli-

cases. Before this, an enzyme called topoisomerase I binds to DNA at random sites and relieves torsional stress by breaking a phosphodiester bond in one strand (making a nick). Each of the preexisting strands serves as a template for the synthesis of a new strand. However, this differs, because new DNA is synthesized in the 5′ to 3′ direction only, not in the 3′ to 5′ direction. At the 3′ end nucleotides can be attached continuously to the leading strand (2). At the 5′ end (the lagging strand), this is not possible.

At the site of single-stranded DNA a specialized RNA polymerase, called a primase, serves as a short RNA primer complementary to the template strand. At the leading strand, synthesis of new DNA proceeds continuously from a single primer in the 5′ to 3′ direction. In contrast, at the lagging strand DNA is synthesized in small segments of 1000–2000 bases (Okazaki fragments) in the opposite direction. Each fragment requires its own primer as it is synthesized in the 5′ to 3′ direction (2). Subsequently, the primers are removed and replaced by DNA chain growth from the neighboring fragment, and the gaps are closed by DNA ligase. The enzyme responsible for DNA synthesis, DNA polymerase III, is complex and comprises several subunits. There are different enzymes for the leading and lagging strands in eukaryotes. During replication, mistakes are eliminated by a complex proofreading mechanism that removes any incorrectly incorporated bases and replaces them with the correct ones. (Figure in 1 adapted from Alberts, 2003.)

References

Alberts B: DNA replication and recombination. Nature 421: 431–435, 2003.

Albert B et al: Molecular Biology of the Cell, 4th ed. Garland Science, New York, 2002.

Cairns J: The bacterial chromosome and its manner of replication as seen by autoradiography. J Mol Biol 6: 208–213, 1963.

Lodish H et al: Molecular Cell Biology, 5th ed. Scientific American Books, FH Freeman & Co, New York, 2004.

Marx J: How DNA replication originates. Science 270: 1585–1587, 1995.

Meselson M, Stahl FW: The replication of DNA in Escherichia coli. Proc Natl Acad Sci 44: 671–682, 1958.

Watson JD et al: Molecular Biology of the Gene, 3rd ed. Benjamin/Cummings Publishing Co, Menlo Park, California, 1987.

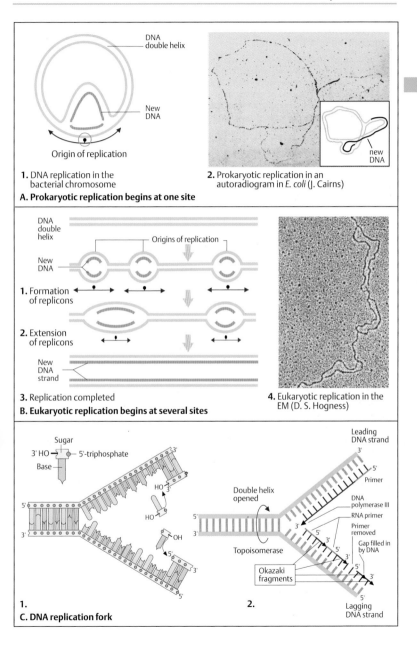

1. DNA replication in the bacterial chromosome

2. Prokaryotic replication in an autoradiogram in *E. coli* (J. Cairns)

A. Prokaryotic replication begins at one site

1. Formation of replicons

2. Extension of replicons

3. Replication completed

4. Eukaryotic replication in the EM (D. S. Hogness)

B. Eukaryotic replication begins at several sites

1.

2.

C. DNA replication fork

The Flow of Genetic Information: Transcription and Translation

The information contained in the nucleotide sequence of a gene is converted into useful biological function in two major steps: *transcription* and *translation*. This flow of genetic information is unidirectional. First, the information of the coding sequences of a gene is transcribed into an intermediary RNA molecule, which is synthesized in sequences that are precisely complementary to those of the coding strand of DNA (transcription). Second, the sequence information in the messenger RNA molecule (mRNA) is translated into a corresponding sequence of amino acids (translation). In eukaryotes RNA is not used directly for translation. It first is processed into a messenger RNA molecule (mRNA) before it can be used. In prokaryotes mRNA is directly transcribed from DNA.

A. Transcription

Here the nucleotide sequence of one strand of DNA is transcribed into a complementary molecule of RNA (messenger RNA, mRNA). The DNA helix is opened by a complex set of proteins. The DNA strand in the 3' to 5' direction (coding strand) serves as the template for the transcription of DNA into RNA by RNA polymerase. RNA is synthesized in the 5' to 3' direction. This strand is the RNA sense strand. RNA transcribed under experimental conditions from the opposing DNA strand is called antisense RNA. It inhibits regular transcription.

B. Translation

Translation refers to the process by which the nucleotide sequence of the mRNA is used to construct a chain of amino acids (a polypeptide chain, see p. 40) in the sequence encoded in the DNA. Translation occurs in a reading frame, which is defined at the start of translation (start codon, AUG). Translation involves two further types of RNA molecule aside from mRNA, transfer RNA (tRNA) and ribosomal RNA (rRNA). tRNA deciphers the codons. Each amino acid has its own set of tRNAs, which bind the amino acid and carry it to the end of the growing polypeptide chain. Each tRNA has a region, called anticodon, which is complementary to a codon of the mRNA. The figure shows codons 1, 2, 3, and 4 of the mRNA, which have been translated

into the amino acid sequence methionine (Met), glycine (Gly), serine (Ser), and isoleucine (Ile). Glycine and alanine are added next in this example.

C. Stages of translation

Translation is accomplished in three stages. First, at initiation (**1**) an initiation complex comprising mRNA, a ribosome, and tRNA is formed. This requires a number of initiation factors (IF1, IF2, IF3, etc., not shown). Then elongation follows (**2**): a further amino acid, determined by the next codon, is attached. A three-phase elongation cycle develops, with codon recognition, peptide binding of the next amino acid residue, and movement (translocation) of the ribosome three nucleotides further in the 3' direction of the mRNA. Translation ends with termination (**3**), when one of three mRNA stop codons (UAA, UGA, or UAG) is reached.

Translation (protein synthesis) in eukaryotes occurs outside of the cell nucleus in ribosomes in the cytoplasm (see p. 208). The biochemical processes of the stages shown here have been greatly simplified.

D. Structure of transfer RNA (tRNA)

Transfer RNA has a characteristic, cloverleaf-like structure, illustrated here by yeast phenylalanine tRNA (**1**). It has three single-stranded loop regions and four double-stranded "stem" regions. The three-dimensional structure (**2**) is complex, but various functional areas can be differentiated, such as the recognition site (anticodon) for the mRNA codon and the binding site for the respective amino acid (acceptor stem) on the 3' end (acceptor end). (Figure adapted from Alberts et al., 2002, and Lodish et al., 2004.)

References

Alberts B et al: Molecular Biology of the Cell, 4th ed. Garland Sciene, New York, 2002.

Brenner S, Jacob F, Meselson M: An unstable intermediate carrying information from genes to ribosomes for protein synthesis. Nature 190: 576–581, 1961.

Ibba M, Söll D: Quality control mechanisms during translation. Science 286: 1893–1897, 1999.

Lodish H et al: Molecular Cell Biology, 5th ed. WH Freeman, New York, 2004.

Watson JD et al: Molecular Biology of the Gene, 3rd ed. Benjamin/Cummings Publishing Co, Menlo Park, California, 1987.

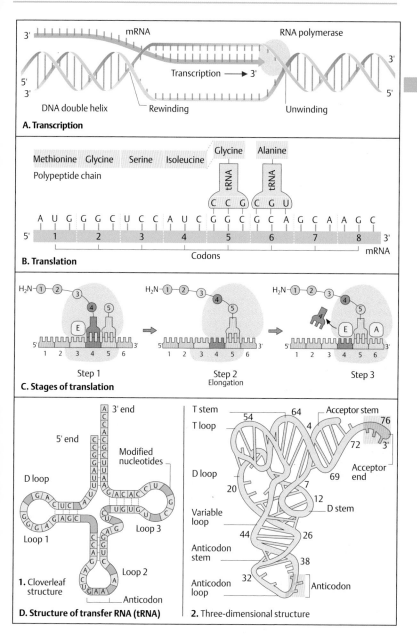

A. Transcription

3′ — mRNA — RNA polymerase — 3′

Transcription → 3′

5′
3′

DNA double helix — Rewinding — Unwinding

5′

B. Translation

Methionine Glycine Serine Isoleucine | Glycine | Alanine

Polypeptide chain

tRNA tRNA

C C G | C G U

5′ A U G G G C U C C A U C G G C G C A G C A A G C 3′

1 2 3 4 5 6 7 8

Codons — mRNA

C. Stages of translation

H_2N–①–② ③ ④ ⑤ E
5′ 1 2 3 4 5 6 3′
Step 1

H_2N–①–② ③ ④ ⑤
5′ 1 2 3 4 5 6 3′
Step 2
Elongation

H_2N–①–② ③ ④ ⑤ E A
5′ 1 2 3 4 5 6 3′
Step 3

D. Structure of transfer RNA (tRNA)

A 3′ end
A
C
C
A
5′ end — C G
C G
G C
G C
U A
A U Modified
U A nucleotides
U A
G A C A C C U U
D loop
G A C U C G U G U G U C G
G G A G C C U
G G A U
Loop 1 C C A G Loop 3
C A G
A G
A G Loop 2
C A A A
U G A A
Anticodon

1. Cloverleaf structure

T stem 54 64 Acceptor stem
T loop 4 76
72
3′
D loop
20 7 Acceptor end
69
12
Variable D stem
loop
44 26
Anticodon
stem 38
32 Anticodon
Anticodon
loop

2. Three-dimensional structure

Genes and Mutation

The information transmitted during replication and transcription is arranged in units called *genes*. This term was introduced in 1909 by the Danish biologist Wilhelm Johannsen (along with the terms genotype and phenotype). Until it was realized that a gene consists of DNA, it was defined in somewhat abstract terms as a factor (Mendel's term) that confers certain heritable properties to a plant or an animal. Mutation, a term introduced by H. de Vries in 1901, refers to a process by which a structural alteration changes the biological function of a gene. The discovery that mutations also occur in bacteria and other microorganisms paved the way to understanding how genes and mutations are related (see p. 54).

A. Transcription in prokaryotes and eukaryotes

Transcription differs in unicellular organisms without a nucleus, such as bacteria (prokaryotes, **1**), and multicellular organisms (eukaryotes, **2**), which have a cell nucleus. In prokaryotes, the mRNA serves directly as a template for translation. The sequences of DNA and mRNA correspond in a strict 1 : 1 relationship, i.e., they are colinear. In eukaryotic cells a primary transcript of RNA is first formed. The mature mRNA is formed by removing noncoding sections from the primary transcript before it leaves the nucleus to act as a template for the synthesis of a polypeptide (RNA processing, p. 58).

B. DNA and mutation

The systematic analysis of mutations in microorganisms provided the first evidence that coding DNA and its corresponding polypeptides are colinear. Yanofsky et al. showed in 1964 that the position of the mutation in the *E. coli* gene encoding the protein tryptophan synthetase A corresponds to the position of the resulting change in the sequence of amino acids (**1**). Panel B shows mutations at four positions. At position 22, phenylalanine (Phe) is replaced by leucine (Leu); at position 49, glutamic acid (Glu) by glutamine (Gln); at position 177, Leu by arginine (Arg). Each mutation has a defined position. Whether it leads to incorporation of a different amino acid depends on how the corresponding codon has been altered. Different mutations at one position (one codon) in different DNA molecules are possible (**2**): two different mutations were observed at position 211: glycine (Gly) to arginine (Arg) and Gly to glutamic acid (Glu). Normally (in the wild type), codon 211 is GGA and codes for glycine (**3**). A mutation of GGA to AGA leads to a codon for arginine; a mutation to GAA leads to a codon for glutamic acid (**4**).

C. Types of point mutation

Three different types of mutation, i.e., changes from the usual or so-called wild types, involving single nucleotides (point mutations) can be distinguished: (i) substitution (exchange of one nucleotide base for another, altering a codon), (ii) deletion (loss of one or more bases), and (iii) insertion (addition of one or more bases). With substitution, the consequences depend on how a codon has been altered. Two types of substitution are distinguished: *transition* (exchange of one purine for another purine or of one pyrimidine for another) and *transversion* (exchange of a purine for a pyrimidine, or vice versa). A substitution may alter a codon so that a wrong amino acid is present at this site but has no effect on the reading frame (*missense mutation*). The resulting polypeptide change contains a wrong amino acid at the site of the mutation. A deletion or insertion causes a shift of the reading frame (frameshift mutation). Thus, the sequence that follows no longer corresponds to the normal sequence of codons. No functional gene product is produced (*nonsense mutation*).

D. Different mutations at the same site

Different mutations may occur at the same site. In the example in B at position 211 glycine is replaced by either arginine or glutamic acid.

References

Alberts B et al: Molecular Biology of the Cell, 4th ed. Garland Publishing, New York, 2002.

Alberts B et al: Essential Cell Biology. An Introduction to the Molecular Biology of the Cell. Garland Publishing, New York, 1998.

Lodish H et al: Molecular Cell Biology, 5th ed. FH Freeman & Co, New York, 2004.

Watson JD et al: Molecular Biology of the Gene, 3rd ed. Benjamin/Cummings Publishing Co, Menlo Park, California, 1987.

Yanofsky C et al: On the colinearity of gene structure and protein structure. Proc Nat Acad Sci 51: 261–272, 1964.

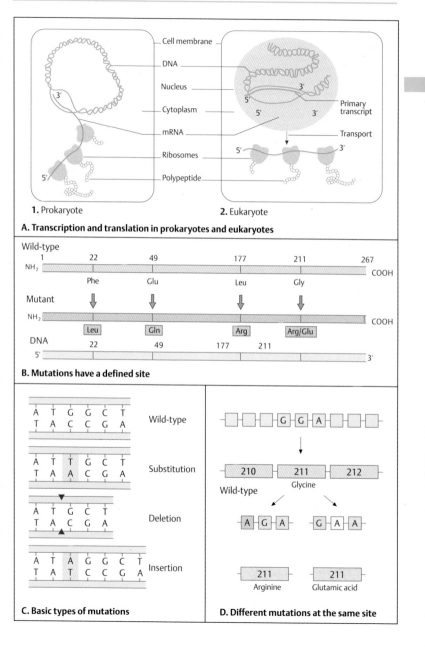

1. Prokaryote   2. Eukaryote

A. Transcription and translation in prokaryotes and eukaryotes

B. Mutations have a defined site

C. Basic types of mutations

D. Different mutations at the same site

Genetic Code

The genetic code is the set of biological rules by which DNA nucleotide base pair sequences are translated into corresponding sequences of amino acids. The genetic code is a triplet code. Each code word (codon) for an amino acid consists of a sequence of three nucleotide base pairs. The genetic code also specifies the beginning (start codon) and the end (stop codon) of the coding region. The genetic code is universal; all organisms, bacteria and viruses, animals and plants, use the same code, with few exceptions.

A. Genetic code in mRNA

The first, second, and third nucleotide bases in the 5′ to 3′ direction determine the codon for one of the twenty amino acids. Each codon corresponds to one amino acid. The genetic code usually is written in the code language of mRNA, using uracil (U) instead of thymine (T) in DNA. Some amino acids are encoded by more than one codon. This is referred to as the redundancy of the code. For example, two codons define the amino acid phenylalanine: UUU and UUC. Six codons define the amino acid serine: UCU, UCC, UCA, UCG, AGU, and AGC. The third position (at the 3′ end of the triplet) of a codon is the most variable. The only amino acids that are encoded by a single codon are methionine (AUG) and tryptophan (UGG). The start codon is AUG (methionine). Stop codons are UAA, UAG, and UGA.

The genetic code was elucidated in 1966 by analyzing how triplets transmit information from the genes to proteins. mRNA added to bacteria could be directly converted into a corresponding protein.

Some deviations from the universal genetic code are listed in the table below.

B. Abbreviated code

In order to save space, long sequences of amino acids are usually written using abbreviations, each amino acid designated by just one letter, as shown.

C. Overlapping reading frames

A reading frame is a nucleotide sequence from the start codon to the stop codon. The interval between a start and a stop codon containing genetic information is called an open reading frame (ORF). An ORF does not contain a further stop codon. Normal reading frames do not overlap. Of the possible three reading frames A, B, or C, only one is correct (the open reading frame A). The reading frames B and C are interrupted by a stop codon after three and five codons, respectively, and cannot serve to code sequences.

References

Alberts B et al: Molecular Biology of the Cell, 4th ed. Garland Publishing, New York, 2002.

Crick FHC et al: General nature of the genetic code for proteins. Nature 192: 1227–1232, 1961.

Lodish H et al: Molecular Cell Biology, 5th ed. FH Freeman & Co, New York, 2004.

Rosenthal N: DNA and the genetic code. New Eng J Med 331: 39–41, 1995.

Singer M, Berg P: Genes and Genomes: A changing perspective. Blackwell Scientific Publications, Oxford–London, 1991.

Deviations from the universal genetic code

Codon	Universal	Deviation	Occurrence
UGA	Stop	Tryptophan	*Mycoplasma*, mitochondria of some species
CUG	Leucine	Threonine	Mitochondria in yeast
UAA, UAG	Stop	Glycine	*Acetabularia, Tetrahymena, Paramecium*
UGA	Stop	Cysteine	*Euplotes*

(Data from Lodish et al., 2004, p. 121)

Nucleotide base					
First	Second				Third
	Uracil (U)	Cytosine (C)	Adenine (A)	Guanine (G)	
Uracil (U)	F Phenylalanine (Phe) F Phenylalanine (Phe) L Leucine (Leu) L Leucine (Leu)	S Serine (Ser) S Serine (Ser) S Serine (Ser) S Serine (Ser)	Y Tyrosine (Tyr) Y Tyrosine (Tyr) Stop Codon Stop Codon	C Cysteine (Cys) C Cysteine (Cys) Stop Codon W Tryptophan (Trp)	U C A G
Cytosine (C)	L Leucine (Leu) L Leucine (Leu) L Leucine (Leu) L Leucine (Leu)	P Proline (Pro) P Proline (Pro) P Proline (Pro) P Proline (Pro)	H Histidine (His) H Histidine (His) Q Glutamine (Gln) Q Glutamine (Gln)	R Arginine (Arg) R Arginine (Arg) R Arginine (Arg) R Arginine (Arg)	U C A G
Adenine (A)	I Isoleucine (Ile) I Isoleucine (Ile) I Isoleucine (Ile) Start (Methionine)	T Threonine (Thr) T Threonine (Thr) T Threonine (Thr) T Threonine (Thr)	N Asparagine (Asn) N Asparagine (Asn) K Lysine (Lys) K Lysine (Lys)	S Serine (Ser) S Serine (Ser) R Arginine (Arg) R Arginine (Arg)	U C A G
Guanine (G)	V Valine (Val) V Valine (Val) V Valine (Val) V Valine (Val)	A Alanine (Ala) A Alanine (Ala) A Alanine (Ala) A Alanine (Ala)	D Aspartic acid (Asp) D Aspartic acid (Asp) E Glutamic acid (Glu) E Glutamic acid (Glu)	G Glycine (Gly) G Glycine (Gly) G Glycine (Gly) G Glycine (Gly)	U C A G

A. Genetic code for all amino acids in mRNA

Start	AUG	F (Phe)	UUU UUC	L (Leu)	CUU CUC CUG CUA UUG UUA	R (Arg)	CGU CGC CGG CAA AGG AGA	V	GUU GUC GUG GUA
Stop	UAA UAG UGA	G (Gly)	GGU GGC GGG GGA	M (Met)	AUG	S (Ser)	UCU UCC UCG UCA AGU AGC	W (Trp)	UGG
A (Ala)	GCU GCC GCG GCA			N (Asn)	AAU AAC			Y (Tyr)	UAU UAC
C (Cys)	UGU UGC	H (His)	CAU CAC	P (Pro)	CCU CCC CCG CCA			B (Asx)	Asn or Asp
D (Asp)	GAU GAC	I (Ile)	AUU AUC AUA			T (Thr)	ACU ACC ACG ACA		
E (Glu)	GAG GAA	K (Lys)	AAG AAA	Q (Gln)	CAG CAA			Z (Glx)	Gln or Glu

B. Abbreviated code

A — GCA — AAU — AAG — GUA — GAC — CAU —
   — Ala — Asn — Lys — Val — Asp — His —   ORF not interrupted

B GC — AAA — UAA — GGU — AGA — CCA — UA
                Stop   ORF interrupted by stop codon

C GG — CAA — AUA — AGG — UAG — ACC — AU
                      Stop   ORF interrupted by stop codon

C. Overlapping reading frames

Processing of RNA

In eukaryotic genes the coding sequences are interrupted by noncoding sequences of variable length. The coding sequences are called *exons*; the noncoding, *introns*, two terms introduced in 1978 by W. Gilbert. The introns are removed before translation can begin. This process, the removal of introns and joining together (splicing) of exons, is called RNA processing. At first glance it seems to be an unnecessary burden to carry DNA without obvious functions within a gene. However, it has been recognized that this has great evolutionary advantages. During evolution different parts of genes were rearranged and transferred to new chromosomal sites. In this way a new gene could be constructed from parts of previously existing genes. Furthermore, some introns contain sequences necessary for the regulation of gene activity.

A. Exons and introns

In 1977, it was unexpectedly found that the DNA of a eukaryotic gene is longer than its corresponding mRNA. When mRNA is hybridized to its complementary single-stranded DNA, loops of single-stranded DNA remain, as shown in the electron micrograph (**1**). mRNA hybridizes only with certain sections of the single-stranded DNA because it is shorter than its corresponding DNA coding strand (**2**). Here, seven loops (A to G) and eight hybridizing sections are shown (1 to 7, and the leading section L). Of the total 7700 DNA base pairs of this gene (**3**), only 1825 hybridize with mRNA. Each hybridizing segment is an exon, here a total of seven. The single-stranded segments that do not hybridize correspond to the introns. The size and arrangement of exons and introns are characteristic for every eukaryotic gene (exon/intron structure). (Electron micrograph from Watson et al., 1987)

B. Intervening DNA sequences (introns)

In prokaryotes, DNA is colinear with mRNA and contains no introns (**1**). In eukaryotes, mature mRNA is complementary to only certain sections of DNA because the latter contains introns (**2**). (Figure adapted from Stryer, 1995)

C. Basic eukaryotic gene structure and transcript

Exons and introns are numbered in the 5′ to 3′ direction of the coding strand. Both exons and introns are transcribed into a precursor RNA (primary transcript). The first and the last exons usually contain sequences that are not translated. These are called the 5′ untranslated region (5′ UTR) of exon 1 and the 3′ UTR, at the 3′ end of the last exon. After transcription, RNA processing begins by removing the noncoding segments (introns) from the primary transcript. Then the exons are connected by a process called *splicing*. Splicing must be very precise to avoid an undesirable change of the correct reading frame. Introns almost always start with the nucleotides GT in the 5′ to 3′ strand (GU in RNA) and end with AG. The sequences at the 5′ end of the intron beginning with GT are called the splice donor site. The splice acceptor site is at the 3′ end of the intron. Mature mRNA is modified at its 5′ end by adding a stabilizing structure called a "cap" and at its 3′ end by adding many adenines (polyadenylation) (see p. 58).

D. Splicing pathway in GU–AG introns

RNA splicing is a complex process mediated by a large RNA-containing protein called a spliceosome. This consists of five types of small nuclear RNA molecules (snRNA) and more than 50 proteins (small nuclear riboprotein particles). The basic mechanism of splicing schematically involves autocatalytic cleavage at the 5′ end of the intron, resulting in lariat formation. This is an intermediate circular structure formed by connecting the 5′ terminus (UG) to a base (A) within the intron. This site is called the branch site. In the next stage, cleavage at the 3′ site releases the intron in lariat form. At the same time the right exon is ligated (spliced) to the left exon. The lariat is debranched to yield a linear intron, and this is rapidly degraded. The branch site identifies the 3′ end for precise cleavage at the splice acceptor site. It lies 18–40 nucleotides upstream (in the 5′ direction) of the 3′ splice site. (Figure adapted from Strachan and Read, 2004.)

References

Lewin B: Genes VIII. Pearson Prentice Hall, London, 2004.

Strachan T, Read AP: Human Molecular Genetics, 3rd ed. Garland Science, London-New York, 2004.

Stryer L: Biochemistry, 4th ed. WH Freeman & Co, New York, 1995.

Watson JD et al: Molecular Biology of the Gene, 3rd ed. Benjamin/Cummings Publishing Co, Menlo Park, California, 1987.

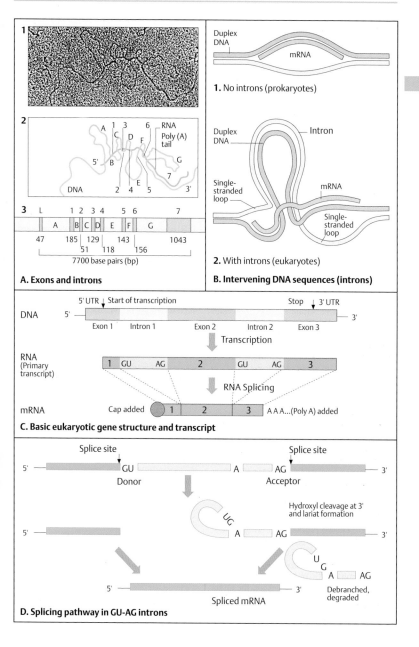

A. Exons and introns

A 1 3 6 RNA
C D F Poly (A) tail
5' B G
DNA 7
2 4 5

L 1 2 3 4 5 6 7
A B C D E F G
47 185 129 143 1043
51 118 156
7700 base pairs (bp)

B. Intervening DNA sequences (introns)

Duplex DNA
mRNA

1. No introns (prokaryotes)

Duplex DNA Intron
Single-stranded loop
mRNA
Single-stranded loop

2. With introns (eukaryotes)

C. Basic eukaryotic gene structure and transcript

5' UTR Start of transcription   Stop 3' UTR
DNA 5' _____ 3'
Exon 1 Intron 1 Exon 2 Intron 2 Exon 3

Transcription

RNA (Primary transcript)
1 GU AG 2 GU AG 3

RNA Splicing

mRNA
Cap added 1 2 3 A A A...(Poly A) added

D. Splicing pathway in GU-AG introns

Splice site Splice site
5' _____ GU _____ A ___ AG _____ 3'
Donor Acceptor

Hydroxyl cleavage at 3' and lariat formation

5' _____ UG A ___ AG _____ 3'

U G
A ___ AG
Debranched, degraded

5' _____ 3'
Spliced mRNA

DNA Amplification by Polymerase Chain Reaction (PCR)

The introduction of cell-free methods for multiplying DNA fragments of defined origin from a complex mixture has greatly facilitated the molecular analysis of genes. Such a method of DNA amplification, the polymerase chain reaction (PCR), was introduced in 1985. It is a cell-free, rapid, and sensitive method for multiplying DNA fragments that can be employed using an automated machine (called thermo cycler).

A. Polymerase chain reaction (PCR)

Standard PCR is an in vitro procedure for amplifying defined target DNA sequences from small amounts of DNA of different origins. This selective amplification requires some prior information about DNA sequences flanking the target DNA. Based on this information, two oligonucleotide primers of about 15–25 base pairs in length are designed. The primers are complementary to sequences outside the 3′ ends of the target site on the two DNA strands and bind specifically to these.

During PCR double-stranded DNA molecules are alternately denatured, each single strand used as template for synthesis of a new strand, and renatured (annealed) with a complementary strand under controlled conditions. PCR is a chain reaction of about 25–35 cycles. Each cycle, involving three precisely time-controlled and temperature-controlled reactions in automated thermal cyclers, takes about 1–5 min. The three steps in each cycle are (1) denaturation of double-stranded DNA at about 93–95°C for human DNA, (2) primer annealing at about 50–70°C depending on the expected melting temperature of the duplex DNA, and (3) DNA synthesis using heat-stable DNA polymerase (from microorganisms living in hot springs, such as *Thermophilus aquaticus*, Taq polymerase), typically at about 70–75°C. At each subsequent cycle the template (shown in blue) and the DNA newly synthesized during the preceding cycle (shown in red) act as templates for another round of synthesis. The first cycle results in newly synthesized DNA of varied lengths (shown with an arrow) at the 3′ ends because synthesis continues beyond the target sequences. However, DNA strands of fixed length at both ends rapidly outnumber those of variable length because synthesis cannot proceed past the terminus of the primer at the opposite template DNA. At the end, at least 10^5 copies of the specific target sequence are present. This can be visualized as a distinct band of a specific size after gel electrophoresis. In addition to the standard reaction, a wide variety of PCR-based methods have been developed for different purposes.

B. Reverse PCR (RT-PCR)

This approach utilizes mRNA as starting material. After the first primer is attached, complementary new DNA is synthesized by reverse transcription (cDNA, see p. 70). This is used as a template for a new DNA strand. Subsequently multiple copies of cDNA are produced by PCR.

C. Allele-specific PCR

This is designed to amplify a DNA sequence from one allele only and to exclude the other allele. For example, if allele 1 contains an AT base pair at a particular site (**1**) and allele 2 a CG pair (**2**), the two alleles can be distinguished by allele-specific PCR (**3** and **5**). If a mutation has changed the T to C in allele 1, the allele-specific oligonucleotide primer does not bind perfectly, making amplification impossible (**4**). Similarly, if a C has been replaced by an A in allele 2, allele-specific amplification is not possible (**6**). Reverse transcriptase PCR (RT-PCR) can be used when the known exon sequences are widely separated within a gene. With rapid amplification of cDNA ends (RACE-PCR), the 5′ and 3′ end sequences can be isolated from cDNA. Other variations of PCR are Alu-PCR, anchored PCR, real-time PCR, and others (see Strachan & Read, 2004, p. 124).

References

Brown TA: Genomes, 2nd ed. Bios Scientific Publ, Oxford, 2002.

Erlich HA, Gelfand D, Sninsky JJ: Recent advances in the polymerase chain reaction. Science 252: 1643–1651, 1991.

Erlich HA, Arnheim N: Genetic analysis with the polymerase chain reaction. Ann Rev Genet 26: 479–506, 1992.

Lodish H: Molecular Cell Biology, 5th ed. WH Freeman, New York, 2004.

Strachan T, Read AP: Human Molecular Genetics, 3rd ed. Garland Science, London-New York, 2004.

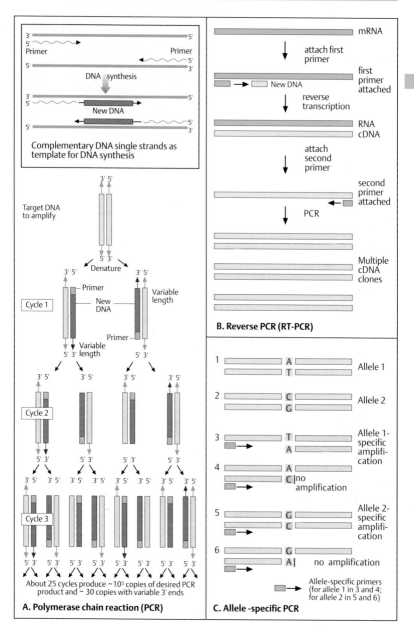

Complementary DNA single strands as template for DNA synthesis

Target DNA to amplify

Denature

Cycle 1

Primer

New DNA

Variable length

Primer

Variable length

Cycle 2

Cycle 3

About 25 cycles produce ~10⁵ copies of desired PCR product and ~ 30 copies with variable 3' ends

A. Polymerase chain reaction (PCR)

mRNA

attach first primer

first primer attached

New DNA

reverse transcription

RNA
cDNA

attach second primer

second primer attached

PCR

Multiple cDNA clones

B. Reverse PCR (RT-PCR)

1   A   Allele 1
     T

2   C   Allele 2
     G

3   T   Allele 1-
     A   specific
              amplifi-
              cation

4   A   no
     C   amplification

5   G   Allele 2-
     C   specific
              amplifi-
              cation

6   G   no amplification
     A

Allele-specific primers
(for allele 1 in 3 and 4;
for allele 2 in 5 and 6)

C. Allele -specific PCR

DNA Sequencing

The primary goal of genetic analysis of a gene or a whole genome is to determine the sequence of its nucleotide bases. Thus, the development in the 1970s of relatively simple methods for sequencing DNA has had a great impact on genetics. Two basic methods for DNA sequencing have been developed: a chemical cleavage method (A. M. Maxam and W. Gilbert, 1977) and an enzymatic method (F. Sanger, 1981). A brief outline of the underlying principles follows, although automated techniques are available today (see next page).

A. Sequencing by chemical degradation

This method utilizes base-specific cleavage of DNA by certain chemicals. Four different chemicals are used in four reactions, one for each base. Each reaction produces a set of DNA fragments of different sizes. The sizes of the fragments in a reaction mixture are determined by positions in the DNA of the nucleotide that has been cleaved. A double-stranded or single-stranded fragment of DNA to be sequenced (1) is processed to obtain a single strand labeled with a radioactive isotope at the 5′ end (2). This DNA strand is exposed to one of the four chemicals in one of the four reactions. Here the reaction at guanine sites (G) by dimethyl sulfate (DMS) is shown (3). Dimethyl sulfate attaches a methyl group to the purine ring of G nucleotides. The amount of DMS used is limited so that on average just one G nucleotide per strand is methylated, not the others (shown here in four different positions of G). When a second chemical, piperidine, is added, the nucleotide purine ring is removed and the DNA molecule is cleaved at the phosphodiester bond just upstream of the site without the base. The overall procedure results in a set of labeled fragments of defined sizes according to the positions of G in the DNA sample being sequenced (4). The four reaction mixtures, one for each of the bases, are run in separate lanes of a polyacrylamide gel electrophoresis, named a sequencing gel (5). Each of the four lanes represents one of the four bases G, A, T, or C. The smallest fragment will migrate the farthest downward, the next a little less far, etc. One can then read the sequence in the direction opposite migration to obtain the sequence in the 5′ to 3′ direction (here TAGTCGCAGTACCGTA, 7).

B. Sequencing by chain termination

This method, now much more widely used than the chemical cleavage method, rests on the principle that DNA synthesis is terminated when instead of a normal deoxynucleotide (dATP, dTTP, dGTP, dCTP), a dideoxynucleotide (ddATP, ddTTP, ddGTP, ddCTP) is used. A dideoxynucleotide (ddNTP) is an analogue of the normal dNTP. It differs by lack of a hydroxyl group at the 3′ carbon position. When a dideoxynucleotide is incorporated during DNA synthesis, no bond between its 3′ position and the next nucleotide is possible because the ddNTP lacks the 3′ hydroxyl group. Thus, synthesis of the new chain is terminated at this site. The DNA fragment to be sequenced has to be single-stranded (1). DNA synthesis is initiated using a primer and one of the four dNTPs labeled with ^{32}P in the phosphate groups (2). Here an example of chain termination using ddATP is shown (3). Wherever a thymine (T) occurs in the sequence, the dideoxyadenine triphosphate will cause termination of the new DNA chain being synthesized. This will produce a set of different DNA fragments whose sizes are determined by the positions of the thymine residues occurring in the fragment to be sequenced. The four parallel reactions (4) will yield a set of fragments with defined sizes according to the positions of the nucleotides where the new DNA synthesis has been terminated. The fragments are separated according to size by gel electrophoresis as in the chemical method (5). The sequence gel is read in the direction from small fragments to large fragments to derive the nucleotide sequence in the 5′ to 3′ direction. An example of an actual sequencing gel is shown between panel A and B. Similar reactions are done for the other three nucleotides.

References

Alberts B et al: Molecular Biology of the Cell, 4th ed. Garland Science, New York, 2002.

Brown TA: Genomes, 2nd ed. Bios Scientific Publ, Oxford, 2002.

Lodish H et al: Molecular Cell Biology, 5th ed. WH Freeman, New York, 2004.

Rosenthal N: Fine structure of a gene—DNA sequencing. New Eng J Med 332: 589–591, 1995.

Strachan T, Read AP: Human Molecular Genetics, 3rd ed. Garland Science, London-New York, 2004.

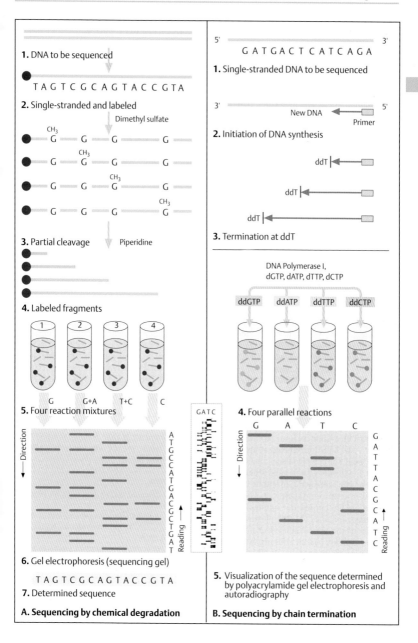

1. DNA to be sequenced

T A G T C G C A G T A C C G T A

2. Single-stranded and labeled

Dimethyl sulfate

CH₃

3. Partial cleavage   Piperidine

4. Labeled fragments

G   G+A   T+C   C

5. Four reaction mixtures

A T G C C A T G A C G C T G A T

6. Gel electrophoresis (sequencing gel)

T A G T C G C A G T A C C G T A

7. Determined sequence

A. Sequencing by chemical degradation

5'   3'
G A T G A C T C A T C A G A

1. Single-stranded DNA to be sequenced

3'   5'
New DNA
Primer

2. Initiation of DNA synthesis

ddT
ddT
ddT

3. Termination at ddT

DNA Polymerase I,
dGTP, dATP, dTTP, dCTP

ddGTP   ddATP   ddTTP   ddCTP

4. Four parallel reactions

G   A   T   C

G A T T A C G C A T C

5. Visualization of the sequence determined
by polyacrylamide gel electrophoresis and
autoradiography

B. Sequencing by chain termination

Automated DNA Sequencing

Large-scale DNA sequencing utilizes automated procedures developed during the 1980s. They are based on fluorescence labeling of DNA and suitable detection systems. The direct fluorescent labels used in automated sequencing are fluorophores. These are molecules that emit a distinct fluorescent color when exposed to UV light of a specific wavelength. Examples of fluorophores used in sequencing are fluorescein, which fluoresces pale green when exposed to a wavelength of 494 nm; rhodamine, which fluoresces red at 555 nm; and aminomethylcumarin acetic acid, which fluoresces blue at 399 nm. In addition, a combination of different fluorophores can be used to produce a fourth color. Thus, each of the four bases can be distinctly labeled.

Several variant techniques for sequencing have been devised. Other rapid methods not requiring electrophoresis have been developed recently (Margulies et al, 2005; Shendure et al., 2005).

A. Automated DNA sequencing

Automated DNA sequencing involves four fluorophores, one for each of the four nucleotide bases. The resulting fluorescent signal is recorded at a fixed point when DNA passes through a capillary containing an electrophoretic gel. The base-specific fluorescent labels are attached to appropriate dideoxynucleotide triphosphates (ddNTP). Each ddNTP is labeled with a different color, e.g., ddATP green, ddCTP blue, ddGTP black, and ddTTP red (1). (The actual colors for each nucleotide may be different.) All chains terminated at an adenine (A) will yield a green signal; all chains terminated at a cytosine (C) will yield a blue signal, and so on. The sequencing reactions based on this kind of chain termination at labeled nucleotides (2) are carried out automatically in sequencing capillaries (3). The electrophoretic migration of the ddNTP-labeled chains in the gel in the capillary pass in front of a laser beam focused on a fixed position (4). The laser induces a fluorescent signal that is dependent on the specific label representing one of the four nucleotides. The sequence is electronically read and recorded and is visualized as alternating peaks in one of the four colors, representing the alternating nucleotides in their sequence posi-

tions (5). (Figure adapted from Brown, 2002, and Strachan & Read, 2004; figure in 5 courtesy of D. Lohmann, Essen.)

B. Thermal cycle sequencing

This method has the advantage that double-stranded rather than single-stranded DNA can be used as the starting material. And since small amounts of template DNA are sufficient, the DNA to be sequenced does not have to be cloned beforehand. The DNA to be sequenced is contained in vector DNA (1). The primer, a short oligonucleotide with a sequence complementary to the site of attachment on the single-stranded DNA, is used as a starting point. For sequencing short stretches of DNA, a universal primer is sufficient. This is an oligonucleotide that will bind to vector DNA adjacent to the DNA to be sequenced. However, if the latter is longer than about 750 bp, only part of it will be sequenced. Therefore, additional internal primers are required. These anneal to different sites and amplify the DNA in a series of contiguous, overlapping chain termination experiments (2). Here, each primer determines which region of the template DNA is being sequenced.

In thermal cycle sequencing (3), only one primer is used to carry out PCR reactions, each with one dideoxynucleotide (ddA, ddT, ddG, or ddC) in the reaction mixture. This generates a series of different chain-terminated strands, each dependent on the position of the particular nucleotide base where the chain is being terminated (4). After many cycles and with electrophoresis, the sequence can be read as shown in the previous plate. One advantage of thermal cycle sequencing is that double-stranded DNA can be used as starting material. (Figure adapted from Brown, 2002.)

References

Brown TA: Genomes, 2nd ed. Bios Scientific Publ, Oxford, 2002.

Margulies M et al: Genome sequencing in microfabricated high-density picolitre reactors. Nature 337: 376–380, 2005.

Shendure J et al: Accurate multiplex polony sequencing of an evolved bacterial genome. Science 309: 1728–1732, 2005.

Strachan T, Read AP: Human Molecular Genetics, 3rd ed. Garland Science, London-New York, 2004.

Wilson RK et al: Development of an automated procedure for fluorescent DNA sequencing. Genomics 6: 626–636, 1990.

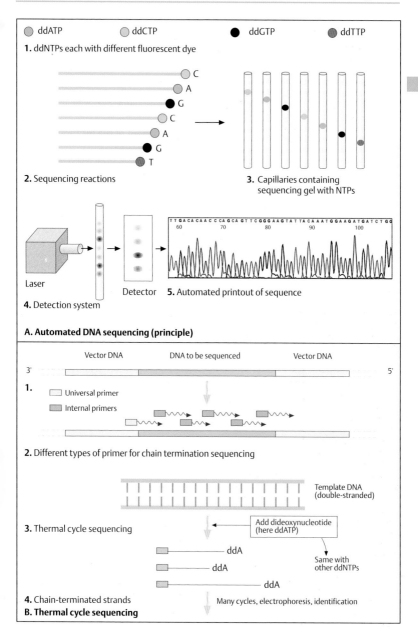

1. ddNTPs each with different fluorescent dye

ddATP   ddCTP   ddGTP   ddTTP

2. Sequencing reactions

3. Capillaries containing sequencing gel with NTPs

Laser

Detector

4. Detection system

5. Automated printout of sequence

A. Automated DNA sequencing (principle)

Vector DNA   DNA to be sequenced   Vector DNA

1.
☐ Universal primer
▨ Internal primers

2. Different types of primer for chain termination sequencing

Template DNA (double-stranded)

3. Thermal cycle sequencing

Add dideoxynucleotide (here ddATP)

Same with other ddNTPs

ddA
ddA
ddA

4. Chain-terminated strands

Many cycles, electrophoresis, identification

B. Thermal cycle sequencing

Restriction Mapping

The molecules of DNA composing the genomes of living organisms are too long to be analyzed directly. However, they can be cleaved into relatively small fragments in a reproducible manner. For this purpose, about 400 different restriction endonucleases have been derived from various bacteria. Restriction endonucleases (restriction enzymes) cleave DNA at defined sites. Such enzymes protect bacteria from invading foreign DNA, which is cut into small pieces. A given enzyme typically recognizes a specific sequence of 4–8 (usually 6) nucleotides, called a restriction site, where it cleaves the DNA. The sizes of the DNA fragments produced depend on the distribution of the restriction sites.

A. DNA cleavage by restriction nucleases

The recognition site of a common restriction enzyme, *Eco*RI, derived from the bacterium *Escherichia coli* restriction enzyme I, is 5′-GAATTC-3′ (**1**). The enzyme cleaves double-stranded DNA asymmetrically to produce fragments with single-stranded ends (**2**). On one fragment, the single-stranded 3′-5′ end has four nucleotides (3′-TTAA) overhanging, and on the other fragment the 5′-3′ overhang is AATT-3′. This common asymmetric cleavage pattern is called palindromic because it reads the same in opposite directions. Some restriction enzymes have a symmetric recognition site (**3**) and produce blunted ends (**4**), such as *Hae*III (5′-CGCG-3′). The ends of fragments with single-stranded overhangs can be easily connected in different ways, within molecules by cyclization or between molecules to form linear concatemers. This ligation requires the enzyme ligase.

B. Examples of restriction enzymes

Restriction enzymes can be classified according to the type of ends they produce: (**1**) 5′ overhangs (e.g., *Eco*RI, see above), (**2**) 3′ overhangs (e.g., *Pst*I), (**3**) blunted ends (e.g., *Alu*I, *Hae*III [see above], *Hpa*I), or (**4**) nonpalindromic ends (*Mln*I). Some have a bipartite recognition sequence with different numbers of nucleotides at the single-stranded ends (e.g., *Bst*I). In *Hin*dII it suffices that the two middle nucleotides are a pyrimidine and a purine (GTPy-PuAC), and it does not matter whether the former is thymine (T) or cytosine (C) or whether the latter is adenine (A) or guanine (G). Such recognition sites occur frequently and produce many relatively small fragments. Rare-cutters recognize long sites of 10 and more nucleotides. Consequently they produce large fragments, which is useful for many purposes. Some enzymes have cutting sites with limited specificity.

C. Determining of the location of restriction sites

Since the fragment sizes reflect the relative positions of the cutting sites, they can be used to characterize a DNA segment (restriction map). For example, if a 10-kb DNA segment cleaved by two enzymes, A and B, results in three fragments, of 2 kb, 3 kb, and 5 kb, then the relative location of the cleavage sites can be determined by using enzymes A and B alone in further experiments. If enzyme A yields two fragments of 3 kb and 7 kb, and enzyme B two fragments of 2 kb and 8 kb, then the two recognition sites of enzymes A and B must be located 5 kb apart. The recognition site for A must be 3 kb from the left end, and that for B 2 kb from the right end (red arrows). This establishes a restriction map to characterize this fragment.

D. Restriction map

A given DNA segment can be characterized by the distribution pattern of restriction sites. In the example shown, a DNA segment is characterized by the distribution of the recognition sites for enzymes E (*Eco*RI) and H (*Hin*dIII). The individual sites are separated by intervals defined by the size of the fragments after digestion with the enzyme. A restriction map is a linear sequence of restriction sites at defined intervals along the DNA. Restriction mapping is of considerable importance in medical genetics and evolutionary research.

References

Alberts B, Johnson A, Lewis J, Raff M, Roberts K, Walter P: Molecular Biology of the Cell, 4th ed. Garland Publishing Co, New York, 2002.

Brown TA: Genomes, 2nd ed. Bios Scientific Publishers, Oxford, 2002.

Strachan T, Read AP: Human Molecular Genetics, 3rd ed. Garland Publishers, New York, 2004.

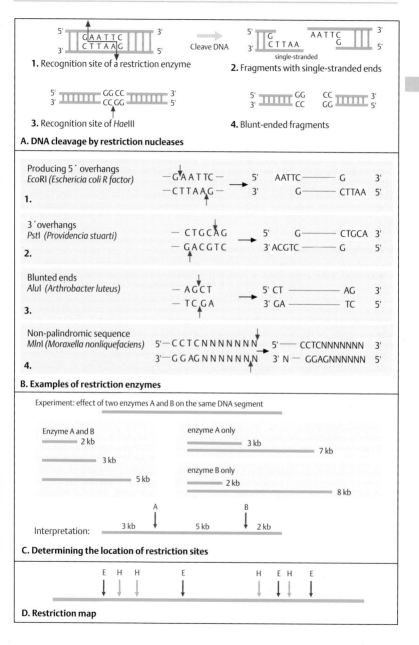

1. Recognition site of a restriction enzyme

Cleave DNA
single-stranded

2. Fragments with single-stranded ends

3. Recognition site of *Hae*III

4. Blunt-ended fragments

A. DNA cleavage by restriction nucleases

Producing 5′ overhangs
*Eco*RI *(Eschericia coli R factor)*

1.

3′ overhangs
*Pst*I *(Providencia stuarti)*

2.

Blunted ends
*Alu*I *(Arthrobacter luteus)*

3.

Non-palindromic sequence
*Mln*I *(Moraxella nonliquefaciens)*

4.

B. Examples of restriction enzymes

Experiment: effect of two enzymes A and B on the same DNA segment

Enzyme A and B
2 kb

3 kb

5 kb

enzyme A only
3 kb

7 kb

enzyme B only
2 kb

8 kb

Interpretation:
3 kb   5 kb   2 kb
A   B

C. Determining the location of restriction sites

E H H   E   H E H   E

D. Restriction map

DNA Cloning

Genetic analysis at the DNA level requires large quantities of material derived from the gene under study. DNA cloning selectively amplifies specific DNA sequences.

To identify the correct DNA fragments, the specific hybridization of complementary single-stranded DNA (molecular hybridization) is utilized. A short segment of single-stranded DNA, a probe, originating from the sequence to be studied, will hybridize to its complementary sequences after these have been denatured (made single-stranded). After the hybridized sequence has been separated from other DNA, it can be cloned. The selected DNA sequences can be amplified in two basic ways: in cells (cell-based cloning) or by cell-free cloning (see polymerase chain reaction, PCR, p. 60).

A. Principle of cell-based DNA cloning

Cell-based DNA cloning requires a series of consecutive steps. First, a collection of different DNA fragments (here labeled 1, 2, and 3) are obtained from the desired DNA (target DNA) by cleaving it with a restriction enzyme (see previous page) (**1**). Fragments with short single-stranded ends resulting from restriction enzyme cleavage are ligated to DNA fragments containing the origin of replication (OR) of a replicon, enabling them to replicate, and a selectable marker, e.g., a DNA sequence containing an antibiotic resistance gene (**2**). The recombinant DNA molecules are transferred into host cells (bacterial or yeast cells), where they can replicate independently of the host cell genome (**3**). Host cells that have incorporated the fragment to be cloned are said to be transformed (here fragment 1, a brown circle containing the number 1). Usually the host cell takes up only one (although occasionally more than one) foreign DNA molecule. The host cells transformed by recombinant (foreign) DNA are grown in culture, where they multiply (propagation, **4**). Selective growth of transformed cells containing the desired DNA fragment produces multiple copies of the fragment (**5**). The resulting recombinant DNA clones form a homogeneous population (**6**). They are used to build a large collection of cloned DNA fragments, called a clone library (**7**) (see next two pages). In cell-based cloning, the replicon-containing DNA molecules are referred to as vector molecules. (Figure adapted from Strachan and Read, 2004.)

B. A vector for cloning

Many different DNA vector systems exist for cloning DNA fragments of different sizes. Plasmid vectors are used to clone small fragments (for plasmids, see p. 94). The experiment is designed so that the plasmid incorporating the fragment to be cloned confers antibiotic resistance to its bacterial host, which will be grown in culture medium containing the antibiotic.

One of the first vectors to be developed was the plasmid vector pBR322 (Bolivar et al., 1977). It is small, 4363 bp, and has an origin of replication and two genes encoding antibiotic resistance, against ampicillin and tetracycline (**1**). Since it was constructed by ligating restriction fragments from three *E. coli* plasmids that occur naturally, it contains recognition sites for seven restriction enzymes, as shown. Resistance to two antibiotics is used to provide selectable markers. They can serve to distinguish cells containing recombinant from those containing nonrecombinant pBR322 plasmids. Bacteria that have been exposed to pBR322 are plated on a medium containing both antibiotics. Only bacteria containing plasmids that have taken up new DNA at the *Bam*HI recognition site lose the tetracycline resistance, because it is inserted at the site for this resistance (insertional activation) (**2**). Cells containing the nonrecombinant pBR322 plasmid remain resistant to tetracycline. If the enzyme *Pst*I is used to incorporate a fragment, ampicillin resistance is lost (the bacterium becomes ampicillin-sensitive), but tetracycline resistance is retained. Thus, with the help of replica plating, recombinant plasmids containing the DNA fragment to be cloned can be distinguished from nonrecombinant plasmids by altered antibiotic resistance. Cloning in plasmids (bacteria) has become less important since yeast artificial chromosomes (YACs) became available for cloning relatively large DNA fragments (see p. 68).

References

Bolivar F et al: Construction and characterization of new cloning vectors. II. A multi-purpose cloning system. Gene 2: 95–113, 1977.

Brown TA: Genomes, 2nd ed. Bios Scientific Publ, Oxford, 2002.

Strachan T, Read AP: Human Molecular Genetics, 3rd ed. Garland Sciene, London-New York, 2004.

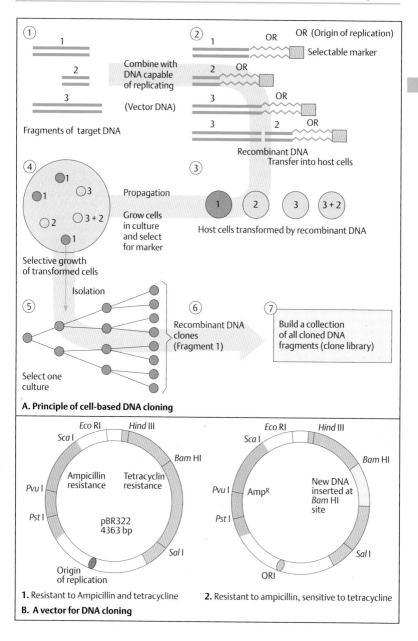

① Fragments of target DNA

② Combine with DNA capable of replicating (Vector DNA)

OR (Origin of replication)
Selectable marker

Recombinant DNA
Transfer into host cells

③ Propagation
Grow cells in culture and select for marker

Host cells transformed by recombinant DNA

④ Selective growth of transformed cells

⑤ Isolation
Select one culture

⑥ Recombinant DNA clones (Fragment 1)

⑦ Build a collection of all cloned DNA fragments (clone library)

A. Principle of cell-based DNA cloning

Eco RI   Hind III
Sca I
Bam HI
Pvu I   Ampicillin resistance   Tetracyclin resistance
Pst I
pBR322 4363 bp
Sal I
Origin of replication

1. Resistant to Ampicillin and tetracycline

Eco RI   Hind III
Sca I
Bam HI
Pvu I   Amp^R New DNA inserted at Bam HI site
Pst I
Sal I
ORI

2. Resistant to ampicillin, sensitive to tetracycline

B. A vector for DNA cloning

cDNA Cloning

A DNA copy of RNA is called complementary DNA (cDNA). It is single-stranded and derived from a coding DNA segment of an active (expressed) gene. It can be synthesized by reverse transcription and cloned. A collection of cDNA clones is called a cDNA library (see next page). Its advantage is that it corresponds to coding parts of the gene. Its disadvantage is that it does not yield information about the exon/intron structure of a gene. From the cDNA sequence, essential inferences can be made about a gene and its gene product. It can be used as a probe (cDNA probe) to recognize structural rearrangements of a gene. Thus, the preparation and cloning of cDNA are important.

A. Preparation of cDNA

This is based on the use of reverse transcriptase. Reverse transcriptase is an enzyme system present in retroviruses (see section on viruses). It synthesizes DNA from RNA as a template. cDNA is prepared from mRNA. Therefore, a tissue is chosen in which the respective gene is transcribed and mRNA is produced in sufficient quantities. First, mRNA is isolated. Then a primer is attached to it so that the enzyme reverse transcriptase can form complementary DNA (cDNA) from the mRNA. Since mRNA contains poly(A) at its $3'$ end, a primer of poly(T) can be attached. From here, the enzyme reverse transcriptase can start forming cDNA in the $5'$ to $3'$ direction. The RNA is then removed by ribonuclease. The cDNA serves as a template for the formation of a new strand of DNA, making the cDNA double-stranded. This requires the enzyme DNA polymerase. The result is a double strand of DNA, one strand of which is complementary to the original mRNA. To this DNA, single sequences (linkers) are attached that are complementary to the single-stranded ends produced by the restriction enzyme to be used. The same enzyme is used to cleave the DNA of the cloning vector, e.g., a plasmid into which the cDNA is incorporated for cloning (see previous page). (Figure adapted from Watson et al., 1987.)

B. Cloning vectors

The cell-based cloning of DNA fragments of different sizes is facilitated by a wide variety of vector systems. Plasmid vectors are used to clone small cDNA fragments in bacteria. Their main disadvantage is that only 5–10 kb of foreign DNA can be cloned. A plasmid cloning vector that has taken up a DNA fragment to be cloned (recombinant vector), e.g., pUC8 with 2.7 kb of DNA, has to be distinguished from one that has not. First, ampicillin resistance (*Amp*⁺) serves to distinguish bacteria that have taken up plasmids from those that have not. Several unique restriction sites in the plasmid DNA segment, where a DNA fragment might be inserted, serve as markers along with a marker gene, such as the *lacZ* gene encoding β-galactosidase. β-Galactosidase cleaves an artificial sugar (5-bromo-4-chloro-3-indolyl-β-D-galactopyranoside) similar to lactose, the natural substrate for this enzyme, into two sugar components, one of which is blue. Colonies with active β-galactosidase appear blue, those with inactive β-galactosidase are white. The uptake of a DNA fragment by the plasmid vector disrupts the gene for β-galactosidase. Thus, all white colonies represent bacteria that contain the recombinant plasmid with a cDNA fragment. (Figure adapted from Brown, 2002.)

C. cDNA cloning

The white colonies are grown in a medium containing ampicillin. Only those bacteria that have incorporated the recombinant plasmid are ampicillin-resistant, and only the white colonies contain a cDNA fragment. By further propagation of these bacteria, the cDNA fragments can be cloned until there is enough material to be studied. Subsequently, a clone library can be constructed. (Figure adapted from Lodish et al., 2004.)

References

Brown TA: Genomes, 2nd ed. Bios Scientific Publ, Oxford, 2002.

Lodish H: Molecular Cell Biology, 5th ed. WH Freeman, New York, 2004.

Watson JD et al: Molecular Biology of the Gene, 3rd ed. Benjamin/Cummings Publishing Co, Menlo Park, California, 1987.

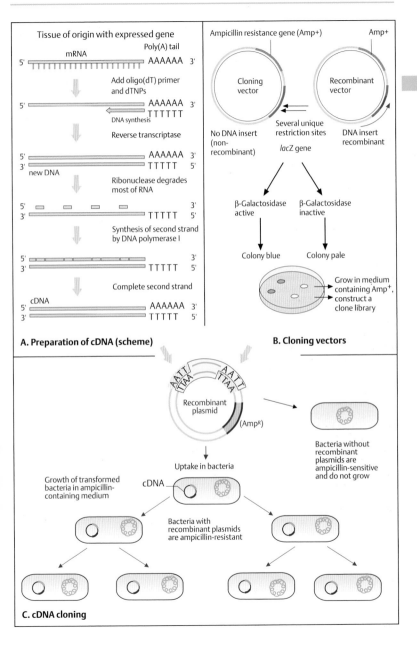

A. Preparation of cDNA (scheme)

Tissue of origin with expressed gene

mRNA
Poly(A) tail

5' AAAAAA 3'

Add oligo(dT) primer and dTNPs

5' AAAAAA 3'
TTTTT
DNA synthesis

Reverse transcriptase

5' AAAAAA 3'
3' TTTTT 5'
new DNA

Ribonuclease degrades most of RNA

5' 3'
3' TTTTT 5'

Synthesis of second strand by DNA polymerase I

5' 3'
3' TTTTT 5'

Complete second strand

cDNA
5' AAAAAA 3'
3' TTTTT 5'

B. Cloning vectors

Ampicillin resistance gene (Amp+)   Amp+

Cloning vector   Recombinant vector

No DNA insert (non-recombinant)

Several unique restriction sites

DNA insert recombinant

lacZ gene

β-Galactosidase active   β-Galactosidase inactive

Colony blue   Colony pale

Grow in medium containing Amp+, construct a clone library

C. cDNA cloning

A A T T / T T A A
A A T T / T T A A
Recombinant plasmid
(Amp^R)

Bacteria without recombinant plasmids are ampicillin-sensitive and do not grow

Uptake in bacteria

cDNA

Growth of transformed bacteria in ampicillin-containing medium

Bacteria with recombinant plasmids are ampicillin-resistant

DNA Libraries

A DNA library is a random collection of individual DNA fragments that in their entirety represent part or all of an organism's genome. Two types of DNA libraries exist: (i) genomic DNA libraries and (ii) cDNA libraries. The first is a collection of DNA clones containing all unique nucleotide sequences of the genome. A sufficient number of clones must be present so that every segment of the genome is represented at least once. A cDNA library is a collection of cDNA that is representative of all mRNA molecules produced by the organism. A DNA library may be the starting point for cloning a gene of unknown chromosomal location.

A. Genomic DNA library

Clones of genomic DNA are copies of DNA fragments from all of the chromosomes (**1**). They contain coding and noncoding sequences. Restriction enzymes cleave the genomic DNA into many fragments. Four fragments containing two genes, A and B, are schematically shown here (**2**). Fragments of the genes and their surrounding DNA are incorporated into vectors, e.g., into phage DNA, and are cloned in bacteria or yeast cells. For example, partial digestion with a 4-bp cutter such as MboI, which has a short recognition sequence (GATC) and occurs about every 280 bp in the human genome, will yield overlapping clones representing the genome. In eukaryotes, a genomic library will contain hundreds of thousands to more than a million of individual DNA clones. A screening procedure is required to find a particular gene (see C).

B. cDNA library

Since a cDNA library consists only of coding DNA sequences, it is smaller than a genomic library. The starting material is usually total RNA from a specific tissue or a particular developmental stage of embryogenesis. A limitation is that mRNA can be obtained only from cells in which the respective gene is transcribed, i.e., in which mRNA is produced (**1**).

Unlike a genomic library, which is complete and contains coding and noncoding DNA, a cDNA library contains only coding DNA. This specificity offers considerable advantages over genomic DNA. However, it requires that mRNA be available and does not yield information about the structure of the gene. In eukaryotes, the primary transcript RNA undergoes splicing to form mRNA (**2**, see p. 58). Complementary DNA (cDNA) is derived from mRNA by reverse transcription (**3**, see previous page). The subsequent steps, incorporation into a vector and replication in bacteria, correspond to those of the procedure for producing a genomic library. The sequence of amino acids in a protein can be determined from cloned and sequenced cDNA. Furthermore, large amounts of a protein can be produced when the cloned gene is expressed in bacteria or yeast cells (proteome library).

C. Screening of a DNA library

To identify clones carrying a gene, parts of a gene, or another DNA region of interest, a screening procedure is required. Two principal approaches can be used: (i) detection using oligonucleotide probes that bind to the sequences of interest, and (ii) detection based on protein produced by the gene in question. Screening with oligonucleotide probes rests on the principle of hybridization. Single-stranded DNA or RNA molecules will specifically hybridize to their complementary single-stranded sequences (see p. 70).

In the membrane-based assay, colonies of cultured bacteria (**1**), some of which will have taken up a recombinant vector, are transferred to a filter paper or a membrane (**2**), lysed, and their DNA is denatured (made single-stranded). Complementary single-stranded DNAs or RNAs are used as probes. Each probe is radioactively labeled. It will hybridize only with complementary DNA or RNA (**4**). After hybridization, a signal appears on the membrane (**5**). DNA complementary to the labeled probe is located here; its exact position in the culture corresponds to that of the signal on the membrane (**5**). A sample is taken from the corresponding area of the culture (**6**). Bacteria from such colonies have taken up the vectors; they are grown on an agar-coated Petri dish to produce multiple copies of (clone) the desired DNA fragments.

References

Lodish H et al: Molecular Cell Biology, 5th ed. WH Freeman, New York, 2004.

Rosenthal N: Stalking the gene - DNA libraries. New Eng J Med 331: 599–600, 1994.

Strachan T, Read AP: Human Molecular Genetics, 3rd ed. Garland Science, London-New York, 2004.

Watson JD et al: Recombinant DNA, 2nd ed. Scientific American Books, New York, 1992.

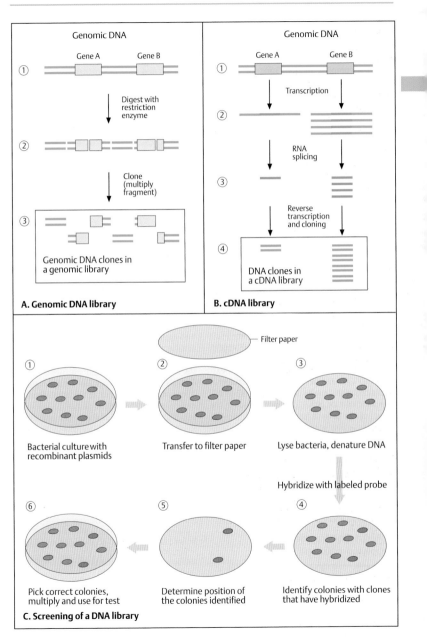

A. Genomic DNA library

Genomic DNA

Gene A   Gene B

①

Digest with
restriction
enzyme

②

Clone
(multiply
fragment)

③

Genomic DNA clones in
a genomic library

B. cDNA library

Genomic DNA

Gene A   Gene B

①

Transcription

②

RNA
splicing

③

Reverse
transcription
and cloning

④

DNA clones in
a cDNA library

C. Screening of a DNA library

Filter paper

① Bacterial culture with
recombinant plasmids

② Transfer to filter paper

③ Lyse bacteria, denature DNA

Hybridize with labeled probe

④ Identify colonies with clones
that have hybridized

⑤ Determine position of
the colonies identified

⑥ Pick correct colonies,
multiply and use for test

Southern Blot Hybridization

Southern blot refers to a sensitive method of detecting one or more specific DNA fragments within a complex, random mixture of other DNA fragments. The procedure is named after E.M. Southern, who developed this method in 1975. A corresponding method for assaying mRNA is called Northern blot hybridization (a word play on Southern, not named after a Dr. Northern). Immunoblotting (Western blot) detects proteins by an antibody-based procedure. A restriction enzyme cleaves DNA only at its specific recognition sequence. Owing to the uneven distribution of recognition sites, the DNA fragments differ in size. A starting mixture of DNA fragments is sorted according to size.

A. Southern blot hybridization

Total DNA is extracted from white blood cells or other cells (**1**). The DNA is isolated and digested with a restriction enzyme (**2**). The fragments are sorted by size in a gel (usually agarose) in an electric field, by electrophoresis (**3**). The smallest fragments migrate fastest from the cathode to the anode; the largest fragments migrate slowest. Next, the blot is carried out: the fragments contained in the gel are transferred to a nitrocellulose or nylon membrane (**4**). The DNA is denatured (made single-stranded) with alkali and fixed to the membrane by moderate heating ($\sim 80^{\circ}$C) or UV cross-linkage. The sample is incubated with a probe of single-stranded DNA (genomic DNA or cDNA) complementary to the region of the gene to be studied (**5**). The probe hybridizes solely with the complementary fragment being sought, and not with others. Since the probe is radiolabeled, the fragment being sought induces a signal on an X-ray film placed on the membrane. Here it becomes visible as a black band on the film after development (autoradiogram) (**6**). This band, thus identified, is taken from the original gel at the corresponding position and processed for further analysis.

B. Restriction fragment length polymorphism (RFLP)

Restriction fragment length polymorphism (RFLP) utilizes the fact that the DNA nucleotide sequences of different individuals are not identical. Rather, about once in every 10^3 nucleotides the sequences differ by one nucleotide, usually in a noncoding region of DNA. An RFLP results when an individual difference of this type creates or eliminates a restriction enzyme recognition site. If an additional site is created, two smaller-than-usual fragments appear in the Southern blot hybridization. If a site is eliminated, one larger-than-usual fragment appears instead of two smaller ones.

An example is shown for a 5-kb (5000 base pairs) stretch of DNA. At the left it contains a restriction recognition site in the middle that is not present at the right. The first, with the polymorphic site, is arbitrarily called allele 1; the one at the right without this additional site is called allele 2.

Southern blot analysis distinguishes the two alleles. At the left, the restriction enzyme cleaves the 5-kb stretch into two fragments. A probe bridging the site will hybridize to both fragments, one 3 kb and the other 2 kb. At the left, a single 5-kb fragment will result. Thus, three possibilities, called the genotypes, can be distinguished: a person with two alleles 1, called homozygous 1–1; a person with one allele 1 and one allele 2, called heterozygous; and a person with two alleles 2, called homozygous 2–2 (for explanation of the terms homozygous and heterozygous, see section on Formal Genetics and the Glossary).

This approach can be used for indirect detection of a disease-causing mutation. It is important to understand that the RFLP itself is unrelated to the mutation. It simply distinguishes DNA fragments of different sizes from the same region. These can be used as markers within a family to determine who is likely to carry a disease-causing mutation and who is not. In addition to RFLPs, other types of DNA polymorphism can be detected by Southern blot hybridization, although polymerase chain reaction-based analysis of microsatellites is now used more frequently (see p. 60).

References

Botstein D et al: Construction of a genetic linkage map in man using restriction fragment length polymorphism. Am J Hum Genet 32: 314–331, 1980.

Brown TA: Genomes, 2nd ed. Bios Scientific Publ, Oxford, 2002.

Housman D: Human DNA polymorphism. New Engl J Med 332: 318–320, 1995.

Kan YW, Dozy AM: Antenatal diagnosis of sickle-cell anaemia by DNA analysis of amniotic-fluid cells. Lancet II: 910–912, 1978.

Strachan T, Read AP: Human Molecular Genetics, 3rd ed. Garland Science, London-New York, 2004.

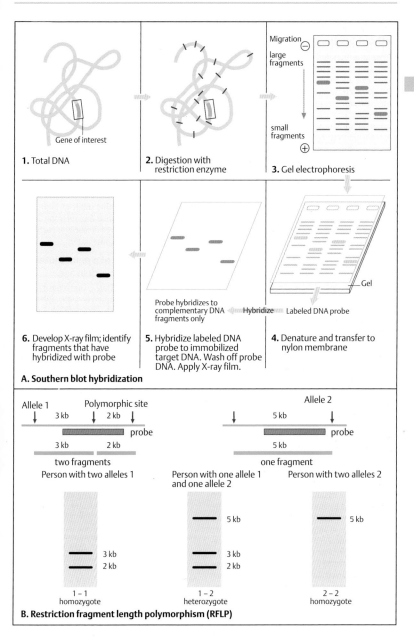

1. Total DNA

2. Digestion with restriction enzyme

3. Gel electrophoresis

Migration

large fragments

small fragments

6. Develop X-ray film; identify fragments that have hybridized with probe

5. Hybridize labeled DNA probe to immobilized target DNA. Wash off probe DNA. Apply X-ray film.

Probe hybridizes to complementary DNA fragments only

Hybridize

Labeled DNA probe

4. Denature and transfer to nylon membrane

Gel

Gene of interest

A. Southern blot hybridization

Allele 1

Polymorphic site

Allele 2

3 kb   2 kb

5 kb

probe

probe

3 kb   2 kb

5 kb

two fragments

one fragment

Person with two alleles 1

Person with one allele 1 and one allele 2

Person with two alleles 2

5 kb

5 kb

3 kb

3 kb

2 kb

2 kb

1 – 1
homozygote

1 – 2
heterozygote

2 – 2
homozygote

B. Restriction fragment length polymorphism (RFLP)

Detection of Mutations without Sequencing

Although automated sequencing methods exist, this is still cumbersome approach in view of the large size of many genes. Therefore, many methods have been devised to detect a mutation without sequencing. Some of these methods are based on differences in the hybridization of mutated and normal segments of DNA. Incomplete hybridization is determined by using short segments of single-stranded DNA (oligonucleotides) with a sequence complementary to the investigated region.

A. Detection of a point mutation by oligonucleotides

This method is designed to detect a difference in hybridization caused by a mutation. If, for example, the normal (**1**) and the mutant (**2**) DNA differ by one nucleotide, here an adenine (A) replacing a guanine (G), a different hybridization pattern results when different oligonucleotides are used as probes. Oligonucleotides are short segments of DNA of about 20 nucleotides. They hybridize to a complementary sequence only when they match each other perfectly. One oligonucleotide (called 1) hybridizes to the normal DNA because all its nucleotides match, including the G (**3**).

However, this oligonucleotide does not match the mutant DNA at A and hybridizes incompletely with it (**4**). In contrast, oligonucleotide 2 hybridizes completely with the mutant (**5**), but not with the normal (**6**) DNA. Such probes are called allele-specific oligonucleotides (ASO). ASO 1 hybridizes to normal DNA but not to mutant (an allele) and vice versa for ASO 2.

This difference in hybridization patterns is visualized in a dot blot analysis (**7**). Hybridization is indicated by a signal, a black dot on a roentgen film produced by the radiolabeled probes. If the mutation is present in both strands of DNA (in a homozygote), a single dot will appear after oligonucleotide 2 (ASO 2) is used for hybridization. If it is present in one strand, but not the other (in a heterozygote), two dots will appear when hybridized with both probes. Normal DNA in both strands will produce one dot when hybridized with ASO 1. Thus, all three possibilities (genotypes) can be distinguished.

B. Denaturing gradient gel electrophoresis (DGGE)

This method exploits differences in the stability of DNA segments with and without mutation. While double-stranded DNA of a control person is completely complementary (homoduplex), a mutation leads to a mismatch at the site of mutation (heteroduplex). This DNA is less stable than completely complementary DNA strands (it has a lower melting point). If normal DNA (control) and DNA with the mutation are placed in a gel with an increasing concentration gradient of formamide (denaturing gradient gel), the mutant and normal DNA can subsequently be differentiated in a Southern blot. The normal DNA remains stable at higher concentrations of formamide and migrates farther than mutant DNA, which dissociates earlier and therefore does not migrate as far.

C. Demonstration of a point mutation by ribonuclease A cleavage

The basis for this method is that a normal DNA strand hybridizes completely with mRNA from that region. If mRNA has completely hybridized to DNA, it is protected from the effects of the enzyme ribonuclease A, which cleaves single-stranded RNA. The difference of even one nucleotide at the site of mutation (here a G instead of an A) causes incomplete hybridization. This difference is detected by a ribonuclease protection assay. In the schematic Southern blot shown here, the normal DNA is represented by a 600-bp fragment, while the mutant DNA, cleaved at the site of the mutation by ribonuclease A, produced two fragments, one of 400 bp and one of 200 bp.

References

Beaudet AL et al: Genetics, biochemistry, and molecular bases of variant human phenotypes, pp 3–45. In: Scriver CR et al (eds): The Metabolic and Molecular Bases of Inherited Disease, 8th ed. McGraw-Hill, New York, 2001.

Caskey CT: Disease diagnosis by recombinant DNA methods. Science 236: 1223–1229, 1987.

Dean M: Resolving DNA mutations. Nature Genet 9: 103–104, 1995.

Mashal RD, Koontz J, Sklar J: Detection of mutations by cleavage of DNA heteroduplexes with bacteriophage resolvases. Nature Genet 9: 177–183, 1995.

Strachan T, Read AP: Human Molecular Genetics, 3rd ed. Garland Science, London-New York, 2004.

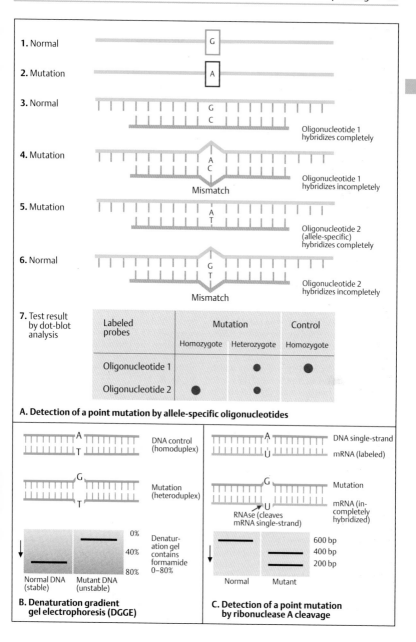

1. Normal

2. Mutation

3. Normal

Oligonucleotide 1
hybridizes completely

4. Mutation

Mismatch

Oligonucleotide 1
hybridizes incompletely

5. Mutation

Oligonucleotide 2
(allele-specific)
hybridizes completely

6. Normal

Mismatch

Oligonucleotide 2
hybridizes incompletely

7. Test result
by dot-blot
analysis

Labeled probes	Mutation		Control
	Homozygote	Heterozygote	Homozygote
Oligonucleotide 1		●	●
Oligonucleotide 2	●	●	

A. Detection of a point mutation by allele-specific oligonucleotides

DNA control
(homoduplex)

Mutation
(heteroduplex)

0%
40%
80%

Denaturation gel
contains
formamide
0–80%

Normal DNA
(stable)

Mutant DNA
(unstable)

B. Denaturation gradient gel electrophoresis (DGGE)

DNA single-strand

mRNA (labeled)

Mutation

RNAse (cleaves
mRNA single-strand)

mRNA (incompletely
hybridized)

600 bp
400 bp
200 bp

Normal   Mutant

C. Detection of a point mutation by ribonuclease A cleavage

DNA Polymorphism

Most genes occur in many different copies, called alleles. In particular, noncoding DNA varies among individuals. The existence of multiple alleles at one particular site is called a polymorphism. A DNA polymorphism is independent of gene function. Differences in a single nucleotide at one particular site are called a single nucleotide polymorphism (SNP). SNPs occur about once every 1000 bases in the human genome. Other types of DNA polymorphisms involve small blocks of repetitive DNA bases, called microsatellites and minisatellites (see below). These are widely used in genetic analysis to distinguish alleles and to establish a genetic map.

A. Single nucleotide polymorphism (SNP)

At a particular site, three possibilities for the maternal and paternal chromosomes can be distinguished: In (1), an adenine (A) is present at the site on both the maternal and the paternal chromosomes; in (2), an A is on the paternal and a guanine (G) on the maternal chromosome (or vice versa); and in (3), a G is on both. A series of such sites with different nucleotides, SNPs, serves to characterize a stretch of DNA. SNPs can be detected by methods based on a polymerase chain reaction (PCR, see p. 60), which does not require gel electrophoresis. The whole human genome contains more than 1.5 million SNPs, each with a defined chromosomal location (Hinds et al., 2005).

B. SNP, microsatellite, minisatellite

These are three types of common DNA polymorphisms. Aside from SNPs, short tandem repeats (STRs) exist. Microsatellites are variable blocks of short repeating nucleotide sequences, here CA repeats (5'-CACACA-3', with 3 repeats; 5'-CACACACACA-3', with five repeats etc.). They consist of units of 1, 2, 3, or 4 base pairs repeated from 2 to about 10 times. Each allele is defined by the number of repeats, e.g., 3 and 4, as shown. Minisatellites (also called variable number of tandem repeats, VNTRs) consist of repeat units of 20–500 base pairs. The size differences due to the number of repeats are determined by PCR. These allelic variants differing in the number of tandemly repeated short nucleotide sequences usually occur in noncoding DNA. They are referred to as repetitive DNA.

C. Genetic variability along a stretch of 100 000 bp

Along a typical stretch of DNA, most variability is represented by SNPs. Minisatellites are quite unevenly distributed and vary in density.
(Figures A–C adapted from Cichon et al, 2002.)

D. CEPH families

The inheritance patterns of DNA polymorphisms are best recognized in a collection of three-generation families with at least eight children in the third generation. DNA from such families has been collected by the *Centre pour l'Étude du Polymorphisme Humain* (CEPH) in Paris, now called the Centre Jean Dausset, after the founder. Immortalized cell lines are stored from each family. A CEPH family consists of four grandparents, the two parents, and eight children. The schematic figure shows the RFLP patterns of a family with four grandparents, two parents, and eight offspring. The four alleles present at a given locus analyzed by Southern blot are designated A, B, C, and D. Starting with the grandparents, the inheritance of each allele through the parents to the grandchildren can be traced. Of the four grandparents, three are heterozygous (AB, CD, BC) and one is homozygous (CC). Since the parents are heterozygous for different alleles (AD the father and BC the mother), all eight children are heterozygous: BD, AB, AC, or CD.

References

Brown TA: Genomes, 2nd ed. Bios Scientific Publ, Oxford, 2002.

Cichon S, Freudenberg J, Propping P, Nöthen MM: Variabilität im menschlichen Genom. Dtsch Ärztebl 99: A3091–3101, 2002.

Collins FS, Guyer MS, Chakravarti A: Variations on a theme: cataloguing human DNA sequence variation. Science 282: 682–689, 1998.

Dausset J et al: Centre d'étude du polymorphism human (CEPH): collaborative genetic mapping of the human genome. Genomics 6: 575–577, 1990.

Feuk L, Carson, AR, Scherer W: Structural variation in the human genome. Nature Rev Genet 7: 85–97, 2006 (with online links to databases).

Hinds DA et al: Whole-genome patterns of common DNA variation in three human populations. Science 307: 1072–1079, 2005.

Lewin B: Genes VIII. Pearson International, 2004.

Strachan T, Read AP: Human Molecular Genetics, 3rd ed. Garland Science, London-New York, 2004.

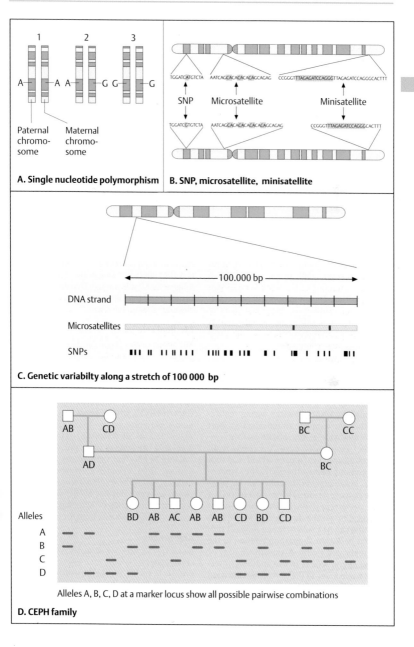

A. Single nucleotide polymorphism

B. SNP, microsatellite, minisatellite

C. Genetic variabilty along a stretch of 100 000 bp

Alleles A, B, C, D at a marker locus show all possible pairwise combinations

D. CEPH family

Mutations

When it was recognized that changes (mutations) in genes occur spontaneously (T. H. Morgan, 1910) and can be induced by X-rays (H. J. Muller, 1927), the mutation theory of heredity became a cornerstone of early genetics. The study of mutations is important for several reasons. Mutations cause diseases, including all forms of cancer. And without mutations, well-organized forms of life would not have evolved. The following two plates summarize the chemical nature of mutations.

A. Errors in replication

Errors in replication occur at a rate of about 1 in 10^5. Proofreading mechanisms reduce this rate to about 1 in 10^7 to 10^9. When an error in replication occurs before the next cell division, e.g., a cytosine (C) might be incorporated instead of an adenine (A) at the fifth base pair as shown here. If the error remains undetected, the next (second) division will result in a mutant molecule containing a CG instead of an AT pair at this position. This mutation will be perpetuated in all daughter cells. Depending on its location within or outside of the coding region of a gene, functional consequences due to a change in a codon could result.

B. Replication slippage

A different class of mutations does not involve an alteration of individual nucleotides, but results from incorrect alignment between allelic and nonallelic DNA sequences during replication. When the template strand contains short tandem repeats, e.g., CA repeats as in microsatellites (see previous page), the newly replicated strand and the template strand may shift their positions relative to each other (microsatellite instability). With replication or polymerase slippage, leading to incorrect pairing of repeats, some repeats are copied twice or not at all, depending on the direction of the shift. A distinction can be made between forward slippage and backward slippage in relation to the newly replicated strand. Backward slippage of the new strand results in the addition (insertion) of nucleotides to the new strand. Forward slippage of the new DNA strand results in the loss (deletion) of nucleotides from the new DNA. (Figures in A and B redrawn from Brown, 2002.)

C. Functional consequences of mutations

Different classes of mutation are known. Their analysis is referred to as molecular pathology. A principal goal is to understand the relationship between the genotype (the genetic situation at a given gene locus) and the phenotype (the observable effect, e.g., the manifestation pattern of a disease). The main class is loss of function of a normal allele. Here the gene product is reduced or has no function. When both alleles are necessary for normal function, but one is inactivated by a mutation, the result is called haploinsufficiency. The opposite is an undesirable functional effect of a new gene product resulting from a mutation; this is called a dominant negative effect. Overexpression of a normal gene product with undesirable effects is caused by a gain-of-function mutation. An epigenetic change is caused by an alteration other than of the DNA sequence, quite commonly a change in the DNA methylation pattern (see section on Epigenetic Modifications). Dynamic mutations result from the abnormal expansion of nucleotide repeats (see p. 88).

Medical relevance

Mutations causes about 3000 known individually defined diseases (see MIM, Mendelian Inheritance of Man, online version available at www.ncbi.nlm.nih.gov/omim).

Microsatellite instability is a characteristic feature of hereditary nonpolyposis cancer of the colon (HNPCC). HNPCC genes are localized on human chromosomes at 2p15–22 and 3p21.3. About 15% of all colorectal, gastric, and endometrial carcinomas show microsatellite instability. Replication slippage has to be distinguished from unequal crossing-over during meiosis. This is the result of recombination between adjacent, but not allelic, sequences on nonsister chromatids of homologous chromosomes. (Figures redrawn from Brown, 2002.)

References

Brown TA: Genomes, 2nd ed. Bios Scientific Publ, Oxford, 2002.

Lewin B: Genes VIII. Pearson International, 2004.

Strachan TA, Read AP: Human Molecular Genetics, 3rd ed. Garland Science, London-New York, 2004.

Vogel F, Rathenberg R: Spontaneous mutation in man. Adv Hum Genet 5: 223–318, 1975.

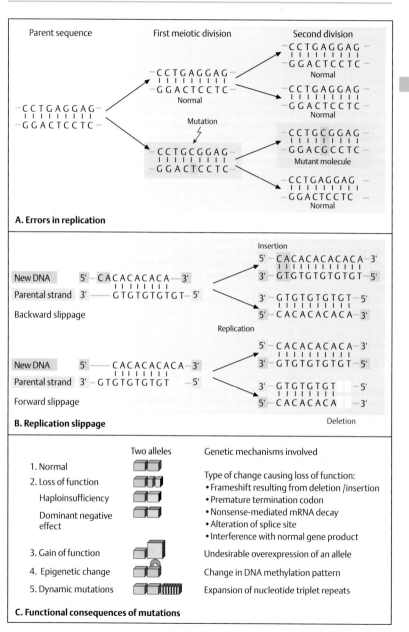

A. Errors in replication

Parent sequence   First meiotic division   Second division

—C C T G A G G A G—
 I I I I I I I I I
—G G A C T C C T C—
Normal

—C C T G A G G A G—
 I I I I I I I I I
—G G A C T C C T C—
Normal

—C C T G A G G A G—
 I I I I I I I I I
—G G A C T C C T C—
Normal

—C C T G A G G A G—
 I I I I I I I I I
—G G A C T C C T C—
Normal

Mutation

—C C T G C G G A G—
 I I I I I I I I I
—G G A C T C C T C—

—C C T G C G G A G—
 I I I I I I I I I
—G G A C G C C T C—
Mutant molecule

—C C T G A G G A G—
 I I I I I I I I I
—G G A C T C C T C—
Normal

B. Replication slippage

Insertion

New DNA   5'— C A C A C A C A C A —3'
Parental strand 3'——— G T G T G T G T G T —5'

Backward slippage

5'— C A C A C A C A C A —3'
 I I I I I I I I I I I
3'— G T G T G T G T G T —5'

3'— G T G T G T G T G T —5'
 I I I I I I I I I
5'— C A C A C A C A —3'

Replication

New DNA   5'——— C A C A C A C A C A —3'
Parental strand 3'— G T G T G T G T G T —5'

Forward slippage

5'— C A C A C A C A —3'
 I I I I I I I I I
3'— G T G T G T G T G T —5'

3'— G T G T G T G T —5'
 I I I I I I I
5'— C A C A C A —3'

Deletion

C. Functional consequences of mutations

Two alleles   Genetic mechanisms involved

1. Normal

2. Loss of function   Type of change causing loss of function:
                       • Frameshift resulting from deletion /insertion
   Haploinsufficiency • Premature termination codon
                       • Nonsense-mediated mRNA decay
   Dominant negative   • Alteration of splice site
   effect   • Interference with normal gene product

3. Gain of function   Undesirable overexpression of an allele

4. Epigenetic change   Change in DNA methylation pattern

5. Dynamic mutations   Expansion of nucleotide triplet repeats

Mutations Due to Different Base Modifications

A mutation can result from a chemical or a physical event that leads to modification of a nucleotide base. If this affects the base-pairing pattern, it will interfere with replication or transcription. A chemical substance able to induce such changes is called a mutagen. Mutagens cause mutations in different ways. Spontaneous oxidation, hydrolysis, uncontrolled methylation, alkylation, and ultraviolet irradiation result in alterations that modify nucleotide bases. DNA-reactive chemicals change nucleotide bases into different chemical structures or remove a base.

A. Deamination and methylation

Cytosine, adenine, and guanine each contain an amino group. When this is removed (deamination), a modified base with a different base-pairing pattern results. Nitrous acid typically removes the amino group. This also occurs spontaneously at a rate of 100 bases per genome per day (Alberts et al., 2002). Deamination of cytosine removes the amino group at position 4 (1). The resulting molecule is uracil (2). This pairs with adenine rather than guanine. Normally this change is efficiently repaired by uracil-DNA glycosylase. Deamination at the RNA level occurs in RNA editing (see Expression of genes). Methylation of the carbon atom at position 5 of cytosine results in 5-methylcytosine, containing a methyl group at position 5 (3). Deamination of 5-methylcytosine will result in a change to thymine, containing an oxygen at position 4 instead of an amino group (4). This mutation will not be corrected because thymine is a natural base. Adenine (5) can be deaminated at position 6 to form hypoxanthine, which contains an oxygen in this position instead of an amino group (6) and which pairs with cytosine instead of thymine. The resulting change after DNA replication is a cytosine instead of a thymine in the mutant strand.

B. Depurination

About 5000 purine bases (adenine and guanine) are lost per day from DNA in each cell (depurination) owing to thermal fluctuations. Depurination of DNA involves hydrolytic cleavage of the N-glycosyl linkage of deoxyribose to the guanine nitrogen in position 9. This leaves a depurinated sugar. The loss of a base pair will lead to a deletion after the next replication if not repaired in time (see DNA repair).

C. Alkylation of guanine

Alkylation is the introduction of a methyl or an ethyl group into a molecule. The alkylation of guanine involves reaction with the ketone group at position 6 to form 6-methylguanine. This cannot form a hydrogen bond and thus is unable to pair with cytosine. Instead, it will pair with thymine. Thus, after the next replication the opposite cytosine (C) is replaced by a thymine (T) in the mutant daughter molecule. As a result, this molecule contains an abornal GT pair instead of GC. Important alkylating agents are ethylnitrosourea (ENU), ethylmethane sulfonate (EMS), dimethylnitrosamine, and N-methyl-N-nitro-N-nitrosoguanidine.

D. Base analogs

Base analogs are purines or pyrimidines that are similar enough to the regular DNA nucleotide bases to be incorporated into the new strand during replication. 5-Bromodeoxyuridine (5-BrdU) is an analog of thymine. It contains a bromine atom instead of the methyl group in position 5. Thus, it can be incorporated into the new DNA strand during replication. However, the presence of the bromine atom causes ambiguous and often incorrect base pairing.

E. UV-light-induced thymine dimers

Ultraviolet irradiation at 260 nm wavelength induces covalent bonds between adjacent thymine residues at carbon positions 5 and 6. If located within a gene, this will interfere with replication and transcription unless repaired. Another important type of UV-induced change is a photoproduct consisting of a covalent bond between the carbons in positions 4 and 6 of two adjacent nucleotides, the 4–6 photoproduct (not shown). (Figures redrawn from Lewin, 2004.)

References

Brown TA: Genomes, 2nd ed. Bios Scientific Publ, Oxford, 2002.

Lewin B: Genes VIII. Pearson International, 2004.

Strachan T, Read AP: Human Molecular Genetics, 3rd ed. Garland Science, London-New York, 2004.

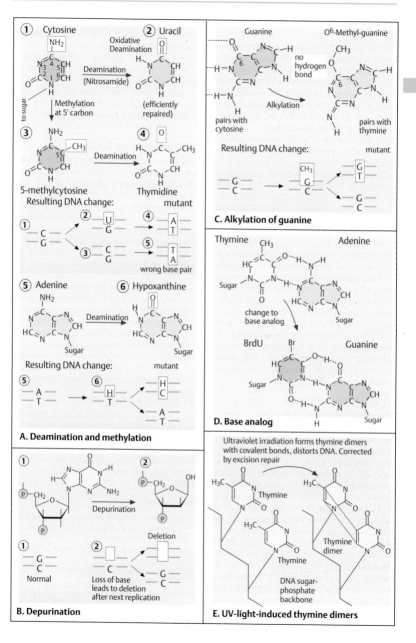

1 Cytosine

2 Uracil

Oxidative Deamination

Deamination (Nitrosamide)

(efficiently repaired)

Methylation at 5' carbon

to sugar

3 5-methylcytosine

4 Thymidine

Deamination

Resulting DNA change:
mutant

1 —C—
  —G—

2 —U—
  —G—

4 —A—
  —T—

3 —C—
  —G—

5 —T—
  —A—

wrong base pair

5 Adenine

6 Hypoxanthine

Deamination

Sugar

Sugar

Resulting DNA change:
mutant

5 —A—
  —T—

6 —H—
  —T—

—H—
—C—

—A—
—T—

A. Deamination and methylation

Guanine O^6-Methyl-guanine

no hydrogen bond

Alkylation

pairs with cytosine

pairs with thymine

Resulting DNA change:
mutant

—G—
—C—

—G—
—C—

—G—
—T—

—G—
—C—

C. Alkylation of guanine

Thymine   Adenine

Sugar

change to base analog

Sugar

BrdU   Br   Guanine

Sugar

Sugar

D. Base analog

B. Depurination

1

Depurination

2

1 —G—
  —C—

Normal

2

Deletion

Loss of base leads to deletion after next replication

—G—
—C—

E. UV-light-induced thymine dimers

Ultraviolet irradiation forms thymine dimers with covalent bonds, distorts DNA. Corrected by excision repair

H_3C Thymine

H_3C

H_3C Thymine

Thymine dimer

Thymine

DNA sugar-phosphate backbone

Recombination

Genetic recombination is an exchange between two homologous DNA molecules. Recombination provides the means to restructure genetic information. It confers an evolutionary advantage by helping to eliminate unfavorable mutations, to maintain and spread favorable mutations, and to endow each individual with a unique set of genetic information.

Recombination must occur between precisely corresponding sequences (homologous recombination) to ensure that not one base pair is lost or added. The newly combined (recombined) stretches of DNA have to retain their original structure in order to function properly. Two types of recombination can be distinguished: (1) generalized or homologous recombination, which in eukaryotes occurs at meiosis (see p. 118) and (2) site-specific recombination. A third process, transposition, utilizes recombination to insert one DNA sequence into another without regard to sequence homology (see next page). The examples here show homologous recombination, a complex biochemical reaction between two duplexes of DNA. The necessary enzymes, which can involve any pair of homologous sequences, are not described. Two general models are distinguished, recombination initiated from single-strand DNA breaks and recombination initiated from double-strand breaks.

A. Single-strand breaks

This model assumes that the process starts with breaks at corresponding positions at each of one of the strands of homologous DNA (same sequences of different parental origin, shown in blue and red) (1). A nick is made in each molecule by a single-strand-breaking enzyme (endonuclease) at corresponding sites (2). This allows the free ends of one nicked strand to join with the free ends of the other nicked strand from the other molecule, and allows a single strand exchange between the two duplex molecules at the recombination joint (3). The recombination joint moves along the duplex, a process called branch migration (4). This ensures a sufficient distance for a second nick at another site in each of the two strands (5). After the two other strands have joined and gaps have been sealed (6), a reciprocal recombinant molecule is generated (7).

Recombination involving DNA duplexes requires topological changes, i.e., either the molecules have to be free to rotate or the restraint has to be relieved in some other way. This structure is called a Holliday structure (not shown), first described in 1964. This model has an unresolved difficulty: How is it ensured that the single-strand nicks shown in step 2 occur at precisely the same position in the two double helix DNA molecules?

B. Double-strand breaks

The current model for recombination is based on initial double-strand breaks (DSB) in one of the two homologous DNA molecules (1). Both strands are cleaved by an endonuclease. The break is enlarged to a gap by an exonuclease. It removes the new 5′ ends of the strands at the break and leaves 3′ single-stranded ends (2). One free 3′ end recombines with a homologous strand of the other molecule, generating a D loop (displacement) (3). This consists of a displaced strand from the "donor" duplex. The D loop is extended by repair synthesis from the 3′ end (4). The displaced strand anneals to the single-stranded complementary homologous sequences of the recipient strand and closes the gap (5). DNA repair synthesis from the other 3′ end closes the remaining gap (6). In contrast to the single-strand exchange model, the reciprocal double-strand breaks result in heteroduplex DNA in the entire region that has undergone recombination (7).

Double-strand breaks occur in meiosis (see p. 118) and DNA repair (see p. 90). A disadvantage of this model is the temporary loss of information in the gaps after the initial cleavage. However, the ability to retrieve this information by resynthesis from the other duplex avoids permanent loss. (Figures adapted from Lewin, 2004.)

References

Alberts B et al: Essential Cell Biology. An Introduction to the Molecular Biology of the Cell. Garland Publishing, New York, 1998.

Brown TA: Genomes, 2nd ed. Bios Scientific Publ, Oxford, 2002.

Holliday R: A mechanism for gene conversion in fungi. Genet Res 5: 282–304, 1964.

Kanaar KS et al: Genetic recombination: from competition to collaboration. Nature 391: 335–337, 1998.

Lewin, B.: Genes VIII. Pearson International, 2004.

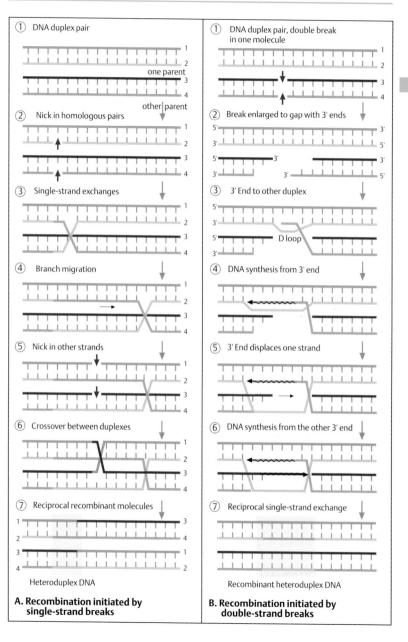

① DNA duplex pair

1
2
one parent
3
other parent
4

② Nick in homologous pairs

1
2
3
4

③ Single-strand exchanges

1
2
3
4

④ Branch migration

1
2
3
4

⑤ Nick in other strands

1
2
3
4

⑥ Crossover between duplexes

1
2
3
4

⑦ Reciprocal recombinant molecules

1 → 3
2 → 4
3 → 1
4 → 2

Heteroduplex DNA

A. Recombination initiated by single-strand breaks

① DNA duplex pair, double break in one molecule

1
2
3
4

② Break enlarged to gap with 3' ends

5' — 3'
3' — 5'
5' — 3' — 3'
3' — 3' — 5'

③ 3' End to other duplex

5'
3'
5' — D loop
3'

④ DNA synthesis from 3' end

⑤ 3' End displaces one strand

⑥ DNA synthesis from the other 3' end

⑦ Reciprocal single-strand exchange

Recombinant heteroduplex DNA

B. Recombination initiated by double-strand breaks

Transposition

Transposition is a widespread spontaneous process in living organisms by which a DNA sequence inserts itself at a new location in the genome. This type of DNA sequence is called a transposon or transposable element. It does not have any sequence relationship with the target site. Transposons are a major source of genetic variation. They play an important role in the evolution of genomes. Tranposition utilizes recombination, but does not result in an exchange. Instead, a transposon moves directly from one site of the genome to another without an intermediary such as plasmid or phage DNA (see section on prokaryotes). This results in rearrangements that create new sequences and change the functions of target sequences. In some cases, they cause disease, when inserted into a functioning gene.

Three examples of the different classes of transposons are presented below: insertion sequences (IS), replicative and nonreplicative transposons (Tn), and transpositions of retroelements via RNA intermediates.

A. Insertion sequences (IS) and transposons (Tn)

The host DNA contains a target site of about 4–10 base pairs (bp) (**1**). The selection of the target site of the host DNA is either random or selective for particular sites. The insertion sequence (IS) consists of about 700–1500 base pairs (bp), depending on the particular class. It contains a transposase gene encoding the enzyme responsible for transposition of mobile sequences. It is flanked by inverted repeats of about 9 bp at both ends. This is a characteristic feature of IS transposition. The IS inserts itself at the target site by means of the transposase activity (**2**). Transposons (Tn) may contain other genes, such as for antibiotic resistance, and have direct (**3**) or inverted (**4**) repeats at either end. Direct repeats are identical or closely related sequences oriented in the same direction. Inverted repeats are oriented in opposite directions.

B. Replicative and nonreplicative transposition

In replicative transposition (**1**) the donor transposon remains in place and creates a new copy of itself, which inserts into a recipient site elsewhere. This mechanism leads to an increase in the number of copies of the transposon in the genome. It involves two enzymatic activities: a transposase, acting on the ends of the original transposon, and resolvase, acting on the duplicated copies. In nonreplicative transposition (**2**) the transposing element itself moves as a physical entity directly to another site. The donor site is either repaired (in eukaryotes) or may be destroyed (in bacteria) if more than one copy of the chromosome is present.

C. Transposition of retroelements

Retrotransposition requires synthesis of an RNA copy of the inserted retroelement. Retroviruses, including the human immunodeficiency virus and RNA tumor viruses, are important retroelements (see p. 106). The first step in retrotransposition is the synthesis of an RNA copy of the inserted retroelement, followed by reverse transcription up to the polyadenylation sequence in the 3′ long terminal repeat (3′ LTR). Three important classes of mammalian transposons that undergo or have undergone retrotransposition through an RNA intermediary are shown. Endogenous retroviruses (**1**) are sequences that resemble retroviruses but cannot infect new cells and are restricted to one genome. Nonviral retrotransposons (**2**) lack LTRs and usually other parts of retroviruses. Both types contain reverse transcriptase and are therefore capable of independent transposition. Processed pseudogenes (**3**) or retropseudogenes lack reverse transcriptase and cannot transpose independently. They contain two groups: low copy number of processed pseudogenes transcribed by RNA polymerase II and high copy number of mammalian SINE sequences, such as human *Alu* and the mouse B1 repeat families. One in 600 mutations are estimated to arise from retrotransposon-mediated insertion (Kazazian, 1999). (Figures adapted from Lewin, 2004, and Brown, 2002.)

References

Brown TA: Genomes, 2nd ed. Bios Scientific Publ, Oxford, 2002.

Kazazian jr HH: An estimated frequency of endogenous insertional mutations in humans. Nature Genet 22: 130, 1999.

Lewin B: Genes VIII. Pearson International, 2004.

Lodish H et al: Molecular Cell Biology, 5th ed. WH Freeman, New York, 2004.

Strachan T, Read AP: Human Molecular Genetics, 3rd ed. Garland Science, London-New York, 2004.

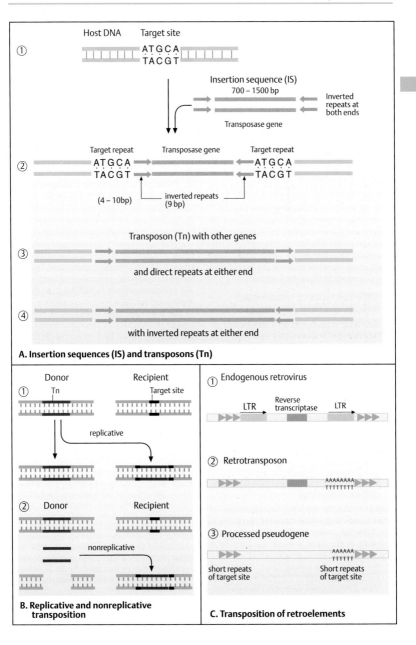

A. Insertion sequences (IS) and transposons (Tn)

B. Replicative and nonreplicative transposition

C. Transposition of retroelements

Trinucleotide Repeat Expansion

A new class of mutations was discovered in 1991: *unstable mutations*. The human genome contains tandem repeats of trinucleotides (triplets). These can expand abnormally within or near certain genes, interfere with gene expression, and cause disease depending on the gene involved (*triplet diseases*). Normally triplets occur in groups of 5–35 repeats. Although usually transmitted stably, they can become unstable and expand to pathological lengths. Two types exist: (i) very large noncoding expansions outside the coding region of a gene, and (ii) modest expansions within a gene. Most of these diseases affect the central nervous system. Once the normal, variable length has expanded, the number of repeats tends to increase even more when passed through the germline. This causes an earlier onset of the disease than in the preceding generations, an observation called *anticipation*. In some diseases the repeat consists of more than three nucleotides.

A. Different types of trinucleotide repeats and their expansions

Trinucleotide repeats can be distinguished according to their location with respect to a gene. Very long expansions occur outside of genes (**1**). The increase in the number of these repeats can be drastic, up to 1000 or more repeats. The first stages of expansion do not usually lead to clinical signs of a disease, but they do predispose to increased expansion of the repeat in the offspring of a carrier (premutation). Within coding regions, expansions are more moderate (**2**). However, their effect is dramatic, as in several severe neurological diseases, because they result in expanded glutamine tracts.

B. Unstable trinucleotide repeats in different diseases

Disorders due to pathological expansion of trinucleotide repeats are classified according to the type of trinucleotide repeat, i.e., the sequence of the three nucleotides, their location with respect to the gene involved, and their clinical features. All involve the central or the peripheral nervous system. Type I trinucleotide diseases are characterized by CAG trinucleotide expansions within the coding regions of different genes. The triplet CAG codes for glutamine. About 20 CAG repeats occur normally in these

genes, so that about 20 glutamines occur in the gene product. In the disease state the number of glutamines is greatly increased in the protein. Hence, they are collectively referred to as polyglutamine disorders.

Type II trinucleotide diseases are characterized by expansions of CTG, GAA, GCC, or CGG trinucleotides within a noncoding region of the gene involved, either at the 5′ end (GCC in fragile X syndrome type A, FRAXA), at the 3′ end (CGG in FRAXE; CTG in myotonic dystrophy), or in an intron (GAA in Friedreich ataxia). A brief review of these disorders is given on p. 404.

C. Principle of laboratory diagnosis

The laboratory diagnosis compares the sizes of the trinucleotide repeats in the two alleles of the gene examined. The schematic figure shows 11 lanes, each representing one individual: normal controls in lanes 1–3; confirmed patients in lanes 4–7 and 10. A family is represented in lanes 7–11: an affected father (lane 7), an affected son (lane 10), an unaffected mother (lane 11), and two unaffected children, a son (lane 8) and a daughter (lane 9). Size markers are shown at the left. Each lane represents a polyacrylamide gel and the (CAG) repeat of the Huntington locus amplified by polymerase chain reaction shown as a band of defined size. The two alleles are shown for each individual. In the affected individuals the band representing the abnormal allele is located above the threshold in the expanded region (in practice the bands may be blurred because the exact repeat size varies in DNA from different cells).

References

Kremer EJ et al: Mapping of DNA instability at the fragile X to a trinucleotide repeat sequence p(CCG)n. Science 252: 1711–1714, 1991.

Oberlé I et al: Instability of a 550-base pair DNA segment and abnormal methylation in fragile X syndrome. Science 252: 1097–1102, 1991.

Rosenberg RN: DNA-triplet repeats and neurologic disease. New Eng J Med 335: 1222–1224, 1996.

Strachan T, Read AP: Human Molecular Genetics, 3rd ed. Garland Science, London-New York, 2004.

Zoghbi HY: Spinocerebellar ataxia and other disorders of trinucleotide repeats, pp. 913–920. In: Jameson JC (ed) Principles of Molecular Medicine, Humana Press, Totowa, NJ, 1998.

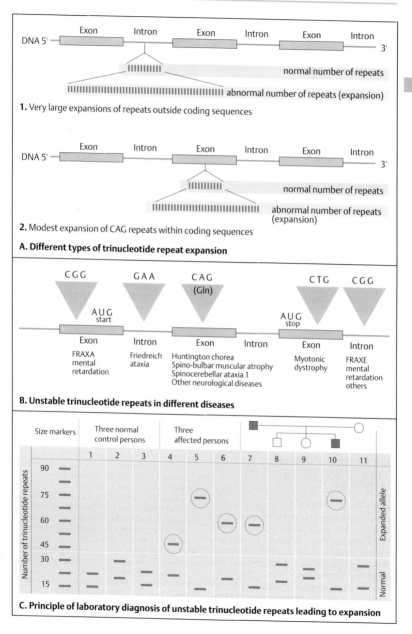

1. Very large expansions of repeats outside coding sequences

2. Modest expansion of CAG repeats within coding sequences

A. Different types of trinucleotide repeat expansion

B. Unstable trinucleotide repeats in different diseases

C. Principle of laboratory diagnosis of unstable trinucleotide repeats leading to expansion

DNA Repair

The following types of DNA repair can be distinguished: (i) excision repair, removing one strand with a damaged DNA site; (ii) mismatch repair, correcting errors of replication by excising a stretch of single-stranded DNA containing one or more wrong bases; (iii) recombination-repair systems, using recombination to replace a damaged double-stranded site; and (iv) transcription-coupled repair in active genes.

A. Excision repair

Excision repair systems recognize the damaged strand of DNA because it is distorted, for example by ultraviolet light. Three proteins, the UvrA, UvrB, and UvrC endonucleases in prokaryotes and XPA, XPB, and XPC in human cells, detect the damaged site and form a repair protein complex. An exonuclease cleaves the damaged strand at two sites, before and after the damaged site. About 12 or 13 nucleotides in prokaryotes and 27 to 29 nucleotides in eukaryotes on the damaged strand are removed. DNA repair synthesis restores the missing stretch, and a DNA ligase closes the gap.

B. Mismatch repair

Mismatch repair corrects errors of replication (see p. 80). *E. coli* has three mismatch repair systems: long patch, short patch, and very short patch. The long patch system can replace 1 kb of DNA and more. The most important mismatch repair proteins are MutH, MutL, and MutS in bacteria and their homologues hMSH1, hMLH1, and hMSH2 in man, among others. MutS/hMSH2 bind to mismatched bases pairs. MutI/hMLH1 and MutH/hMSH1 cleave DNA and remove the strand with erroneous bases. DNA synthesis by DNA polymerase III replaces the damaged strand.

C. Replication repair

DNA damage interferes with replication and transcription. In replication it especially affects the leading strand. Without repair, long stretches would remain unreplicated beyond the damaged site (in the 3′ direction of the new strand). The lagging strand is less affected because Okazaki fragments (about 100 nucleotides in length) of newly synthesized DNA can form beyond the damaged site. However, it would lead to an asymmetric replication fork and single-stranded regions of the leading strand. Transcription would stall at a damaged site because RNA polymerase cannot use it as a template. XPV (XP variant) relieves the block in replication and repairs the affected strand. A mutation in the gene encoding XPV results in an error-prone bypass with unreplicated sections.

D. Double-strand repair

Double-strand damage is a common consequence of γ-radiation. An important human pathway for repair requires three central proteins, encoded by the genes *ATM*, *BRCA1*, and *BRCA2*, and others. Their names are derived from important diseases that result from mutations in these genes: ataxia-telangiectasia (see p. 340) and hereditary predisposition to breast cancer (BRCA1 and BRCA2, see p. 332). ATM, a member of a protein kinase family, is activated in response to DNA damage (**1**). Its active form phosphorylates BRCA1 at specific sites (**2**). Phosphorylated BRCA1 induces homologous recombination in cooperation with BRCA2 and RAD51, the mammalian homologue of *E. coli* RecA repair protein (**3**). This is required for efficient DNA double-break repair. Phosphorylated BRCA1 is also involved in transcription and transcription-coupled DNA repair (**4**). (Figure adapted from Ventikaraman, 1999.)

Medical relevance

Mutations in repair genes lead to many important hereditary diseases and certain types of cancer (see Bootsma, 2002; Wood, 2001).

References

Bootsma D et al: Nucleotide excision repair syndromes. Xeroderma pigmentosum, Cockayne syndrome, and trichothiodystrophy, pp. 211–237. In: Vogelstein B, Kinzler KW (eds) The Genetic Bases of Human Cancer, 2nd ed. McGraw-Hill, New York, 2002.

D'Andrea ADD, Grompe M: The Fanconi anaemia/BRCA pathway. Nature Rev Cancer 3: 23–34, 2003.

Masutani C et al: The XPV (xeroderma pigmentosum variant) gene encodes human DNA polymerase. Nature 399: 700–704, 1999.

O'Driscoll M, Jeggo PA: The role of double-strand break repair – insights from human genetics. Nature Rev Genet 7: 45–54, 2006.

Ventikaraman AR: Breast cancer genes and DNA repair. Science 286: 1100–1101, 1999.

Wood RD et al: Human repair genes. Science 291: 1284–1289, 2001.

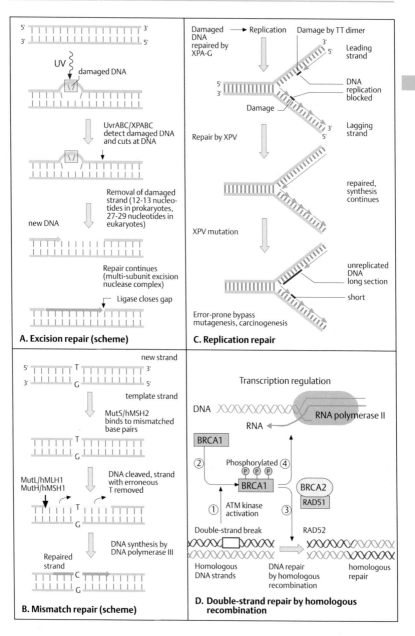

A. Excision repair (scheme)

5' [ ] 3'
3' [ ] 5'

UV — damaged DNA

UvrABC/XPABC detect damaged DNA and cuts at DNA

Removal of damaged strand (12–13 nucleotides in prokaryotes, 27–29 nucleotides in eukaryotes)

new DNA

Repair continues (multi-subunit excision nuclease complex)

Ligase closes gap

B. Mismatch repair (scheme)

new strand
5' [T] 3'
3' [G] 5'
template strand

MutS/hMSH2 binds to mismatched base pairs

MutL/hMLH1 MutH/hMSH1

DNA cleaved, strand with erroneous T removed

DNA synthesis by DNA polymerase III

Repaired strand

C. Replication repair

Damaged DNA repaired by XPA-G → Replication   Damage by TT dimer

Leading strand

DNA replication blocked

Damage

Lagging strand

Repair by XPV

repaired, synthesis continues

XPV mutation

unreplicated DNA long section

short

Error-prone bypass mutagenesis, carcinogenesis

D. Double-strand repair by homologous recombination

Transcription regulation

DNA

RNA polymerase II

RNA

BRCA1

② Phosphorylated ④

Ⓟ Ⓟ Ⓟ

BRCA1   BRCA2
RAD51

① ATM kinase activation   ③

Double-strand break   RAD52

Homologous DNA strands   DNA repair by homologous recombination   homologous repair

Xeroderma Pigmentosum

Xeroderma pigmentosum (XP, MIM 278700–80) is a heterogeneous group of genetically determined cancer-prone skin diseases with unusual sensitivity to ultraviolet light. It manifests as dryness and pigmentation of sun-exposed regions of the skin (xeroderma pigmentosum: "dry, pigmented skin" as first decribed by F. von Hebra and M. Kaposi in 1874). Many tumors develop in exposed areas of the skin. The basic defect is DNA excision repair deficiency, as shown by Cleaver in 1968. The cause is mutation in a gene encoding one of the several excision repair proteins: XPA, -B, -C, -D, -E, -F, -G, or XPV. These genes are highly conserved in bacteria, yeast, and mammals.

A. Clinical phenotype

The skin changes are limited to UV-exposed areas (**1**, **2**). Unexposed areas show no changes. Thus it is important to protect patients from UV light. An especially important feature is the tendency for multiple skin tumors to develop in the exposed areas (**3**). These may occur as early as childhood or early adolescence. The types of tumor are the same as those occurring in healthy individuals after prolonged UV exposure.

B. Cellular phenotype

The UV sensitivity is evident in cultured fibroblast cells. When exposed to UV light, XP cells show a distinct dose-dependent decrease in survival rate compared with normal cells (**1**). Such cells have reduced or nearly absent UV-light induced DNA synthesis. When cultured in the presence of ^{3}H-thymidine and exposed to UV light, an autoradiograph (a film that, placed on the cells, develops black dots at sites where radiolabed thymine has been incorporated during DNA synthesis) shows few dots over the nucleus of repair-deficient XP cells (here, one cell with XP type A and two cells with XP type D, **2**). On the other hand, a cell containing a nucleus from an XPA and an XPD cell (a heterokaryon) shows a normal pattern. The reason is that the XPA and the XPD cells correct each other. (Photograph courtesy of Bootsma & Hoeijmakers, 1999.)

C. Genetic complementation in cell hybrids

A normal cell fused with a xeroderma pigmentosum (XP) cell corrects the mutant cell (**1**). Genetic analysis by cell fusion (cell hybrids) reveals the different complementation groups of xeroderma pigmentosum. When mutant cells with different defects are fused to form cell hybrids, they correct each other (**2**). If the mutant cells have the same defect (**3**), they do not correct each other (**4**). Each of the seven XP complementation groups, characterized by the ability to correct each other, corresponds to one of the genes mutated in the different types of XP (XPA-G, and XPV).

In addition to XP, two other excision repair diseases are known, Cockayne syndrome (CS, MIM 216400, 21641, 133540) and trichothiodystrophy (TTD, MIM 234050, 278730, 601675) in a photosensitive and a nonphotosensitive form. CS comprises two different complementation groups (CS-A, CS-B) and TTD three, which partially overlap with XP groups (TTD-A, XP-B, and XP-G).Most of the genes involved in XP show homology with repair genes of other organisms, including yeast and bacteria.

References

Berneburg M et al: UV damage causes uncontrolled DNA breakage in cells from patients with combined features of XP-D and Cockayne syndrome. Embo J 19: 1157–1166, 2000.

Bootsma DA, Hoeijmakers JHJ: The genetic basis of xeroderma pigmentosum. Ann Génét 34: 143–150, 1991.

Cleaver JE: Defective repair replication in xeroderma pigmentosum. Nature 218: 652–656, 1968.

Cleaver JE et al: A summary of mutations in the UV-sensitive disorders: xeroderma pigmentosum, Cockayne syndrome, and trichothiodystrophy. Hum Mutat 14: 9–22, 1999.

Cleaver JE: Common pathways for ultraviolet skin carcinogenesis in the repair and replication defective groups of xeroderma pigmentosum. J Dermatol Sci 23: 1–11, 2000.

de Boer J, Hoeijmakers JH: Nucleotide excision repair and human syndromes. Carcinogenesis 21: 453–460, 2000.

Taylor EM et al: Xeroderma pigmentosum and trichothiodystrophy are associated with different mutations in the XPD (ERCC2). Proc Natl Acad Sci 94: 8658–8663, 1997.

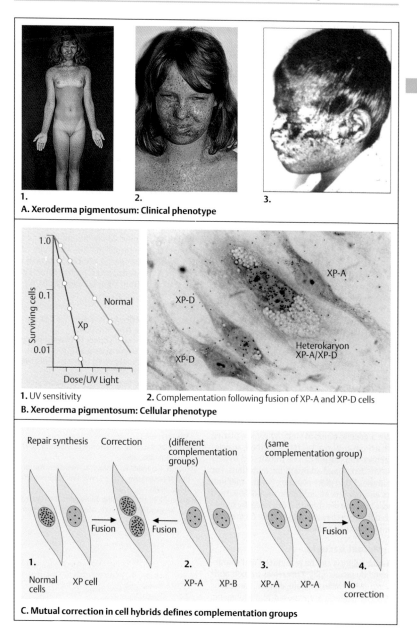

1.

2.

3.

A. Xeroderma pigmentosum: Clinical phenotype

1. UV sensitivity

2. Complementation following fusion of XP-A and XP-D cells

B. Xeroderma pigmentosum: Cellular phenotype

Repair synthesis   Correction   (different complementation groups)

(same complementation group)

Fusion   Fusion   Fusion

1.   **2.**   **3.**   **4.**

Normal cells   XP cell   XP-A   XP-B   XP-A   XP-A   No correction

C. Mutual correction in cell hybrids defines complementation groups

Bacteria in the Study of Genetics

Bacteria are unicellular prokaryotic organisms surrounded by a lipid membrane. Most contain a single circular DNA molecule with about 500–10000 genes. The cytoplasma contains numerous small auxiliary DNA molecules, *plasmids*. Systematic studies of bacteria began in 1943 when S. E. Luria and M. Delbrück demonstrated mutations in bacteria. Bacteria as prokaryotic organisms have several advantages for genetic analyses over eukaryotic organisms: they are haploid and have an extremely short generation time. Mutations in bacteria can be easily identified by changes in their ability to grow in the presence or absence of certain substances in culture. Without great difficulty, it is possible to detect one mutant in 10^7 colonies. In addition to vertical transmission of genes through successive generations, bacteria exchange DNA readily by *lateral gene transfer*.

A. Basic genetic features of bacteria

Most bacteria are about 1–5 μm long and 0.5–1 μm in diameter. They have a simple organization with a plasma membrane (bacterial wall), one circular double-stranded DNA molecule (some with more), and no nucleus. Bacteria contain additional, small DNA molecules. These either are independent structures, *plasmids* or are inserted into the bacterial genome.

B. Replica plating

In 1952, Joshua and Esther Lederberg developed an important method for analyzing bacteria from a genetic point of view, by replica plating of bacterial cultures. With this method, an original bacterial culture can be transferred to new culture dishes and analyzed for mutations. For example, if one plate contains normal medium and another one an antibiotic in addition, then those bacteria resistant to it can be easily recognized because they are able to grow on medium containing the antibiotic.

C. Mutant bacteria

Mutant bacteria can be recognized and characterized by auxotrophy. This refers to their ability to grow under certain culture conditions only. The bacteria are exposed to a mutagenic substance and first grown in a plate containing normal growth medium. The culture is then transferred to different plates by replica plating. The new plates contain either normal medium as control, minimal medium, or minimal medium to which one substance has been added. In the example shown, two mutants do not grow on minimal medium (gray circles). However, if the amino acid threonine (Thr) is added to one minimal medium plate, an additional colony grows (red dot). This indicates that bacteria of this colony are auxotrophic for threonine (need Thr for growth), because owing to a mutation they cannot synthesize threonine. On the right, a dish is shown to which the amino acid arginine (Arg) has been added to the minimal medium. Here the other mutant unable to grow in minimal medium can grow (red dot in the middle). Thus, by this simple procedure, two specific mutants are identified, one unable to synthesize threonine, the other unable to synthesize arginine. Many mutant bacteria are defined by auxotrophism. The wild-type cells that do not have special additional growth requirements are called prototrophs. (Figures in B and C adapted from Stent & Calendar, 1978.)

References

Alberts B et al: Molecular Biology of the Cell, 4th ed. Garland Science, New York, 2002.

Hacker J, Hentschel U, Dobrindt U: Prokaryotic chromosomes and disease. Science 301: 790–793, 2003.

Lederberg J: Infectious history. Pathways of discovery. Science 288: 287–293, 2000.

Lederberg J, Lederberg EM: Replica plating and indirect selection of bacterial mutants. J Bacteriol 63: 399–406, 1952.

Luria SE, Delbrück M: Mutations in bacteria from virus sensitivity to virus resistance. Genetics 28: 491–511, 1943.

Sherratt DJ: Bacterial chromosome dynamics. Science 301: 780–785, 2003.

Stent GS, Calendar R: Molecular Genetics. An Introductory Narrative, 2nd ed. WH Freeman, San Francisco, 1978.

Watson JD et al: Molecular Biology of the Gene, 4th ed. Benjamin/Cummings, Menlo Park, California, 1987.

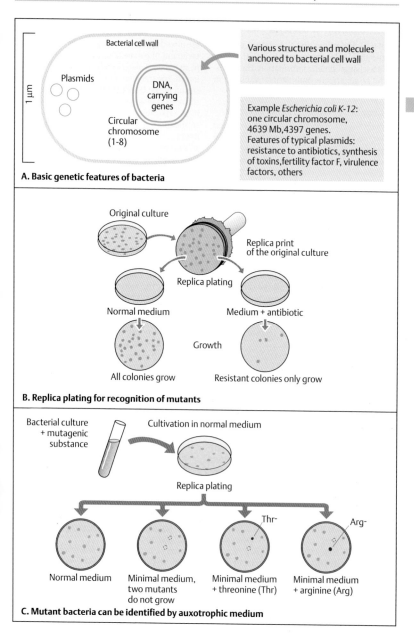

Bacterial cell wall

Various structures and molecules anchored to bacterial cell wall

Plasmids

DNA, carrying genes

Circular chromosome (1-8)

1 μm

Example *Escherichia coli K-12*: one circular chromosome, 4639 Mb, 4397 genes. Features of typical plasmids: resistance to antibiotics, synthesis of toxins, fertility factor F, virulence factors, others

A. Basic genetic features of bacteria

Original culture

Replica print of the original culture

Replica plating

Normal medium

Medium + antibiotic

Growth

All colonies grow

Resistant colonies only grow

B. Replica plating for recognition of mutants

Bacterial culture + mutagenic substance

Cultivation in normal medium

Replica plating

Normal medium

Minimal medium, two mutants do not grow

Thr‑

Minimal medium + threonine (Thr)

Arg‑

Minimal medium + arginine (Arg)

C. Mutant bacteria can be identified by auxotrophic medium

Recombination in Bacteria

In 1946, J. Lederberg and E. L. Tatum demonstrated that different mutant bacterial strains could exchange genetic information. The mechanism is genetic recombination resulting in bacteria with new genetic properties.

A. Genetic recombination in bacteria

In their classic experiment, Lederberg and Tatum used two different auxotrophic bacterial strains. One (A) was auxotrophic for methionine (Met⁻) and biotin (Bio⁻). This strain requires methionine and biotin in the medium, but not threonine and leucine (being prototrophic Thr⁺, Leu⁺). Bacterial strain B required threonine and leucine in the medium (being auxotrophic Thr⁻, Leu⁻), but not methionine and biotin (being prototrophic Met⁺, Bio⁺).

When strain A and strain B were placed together into one culture, some colonies grew in minimal medium lacking the four amino acids met, bio, thr, and leu. Although this occurred rarely, in only about 1 in 10^7 plated cells, these few colonies consisted of bacteria with altered genetic properties. It could be concluded that genetic recombination had taken place between strain A and strain B. (Figure adapted from Stent & Calendar, 1978.)

B. Bacterial gene transfer

Conjugation is the temporary close contact between two single-celled organisms leading to a genetic exchange. In bacteria the exchange is unidirectional, one cell being the donor of genetic material (DNA), the other the recipient. This form of bacterial "sexuality," described by Cavalli et al. and by Hayes in 1953, differs from that of eukaryotes. It is a lateral gene transfer, unidirectional, and incomplete. Only variable parts of genetic information are transferred. Bacterial cells can be distinguished by the presence or absence of a supernumerary chromosome called the F (fertility) factor. When present (F⁺, cell, "male"), the bacterium is capable of transferring genes to an F⁻ cell ("female"), without an F factor. The F factor is a circular double-stranded DNA molecule of about 94 × 10^3 base pairs, present in one copy per cell. This size corresponds to about 1/40th of the total genetic information. The F factor contains genes involved in (i) the transfer of the F factor (19 *tra* genes), (ii) formation of a sex pilus (see below),

and (iii) replication or fertility inhibition. Only the F factor is transferred. When F⁺ and F⁻ cells are mixed together, conjugal pairs form with attachment of a male (F⁺) sex pilus to the surface of an F⁻ cell (F pilus). Only the F factor is transferred. The transfer of the F factor begins after a strand of the DNA double helix is opened. One strand is transferred to the acceptor cell. There it is replicated, so that it becomes double-stranded. The DNA strand remaining in the donor cell is likewise restored to a double strand by replication. When the process is completed, the acceptor cell is also an F⁺ cell.

C. Conjugation in bacteria

Conjugation was subsequently visualized by light microscopy (Brinton et al., 1964). (Photograph: Science 257 : 1037, 1992, with permission.)

D. Integration of the F factor

The F factor can be integrated into the bacterial chromosome by means of specific crossing-over. After the factor is integrated, the original bacterial chromosome with the sections a, b, and c contains additional genes, the F factor genes (e, d). This type of chromosome is called an Hfr chromosome (Hfr, high frequency of recombination) owing to its high rate of recombination with genes of other cells as a result of conjugation. (Figures in B and D adapted from Watson et al., 1987.)

References

Brinton CCP, Gemski P, Carnahan J: A new type of bacterial pilus genetically controlled by fertility factor of *E. coli* K12. Proc Nat Acad Sci 52: 776–783, 1964.

Cavalli LL, Lederberg J, Lederberg EM: An infective factor controlling sex compatibility in Bacterium coli. J Gen Microbiol. 8 : 89–103, 1953.

Hayes W: Observations on a transmissible agent determining sexual differentiation in bacterium-coli. J Gen Microbiol 8: 72–88, 1953.

Kohiyama M, Hiraga S, Matic I, Radman M: Bacterial sex: playing voyeurs 50 years later. Science 301: 802–803, 2003.

Lederberg J, Tatum EL: Novel genotypes in mixed cultures of biochemical mutants in bacteria. Cold Spring Harbor Symp Quant Biol 11: 113–114, 1946.

Lederberg J, Tatum EL: Gene recombination in *Escherichia coli*. Nature 158: 558, 1946.

Stent GS, Calendar R: Molecular Genetics. An Introductory Narrative, 2nd ed. WH Freeman, San Francisco, 1978.

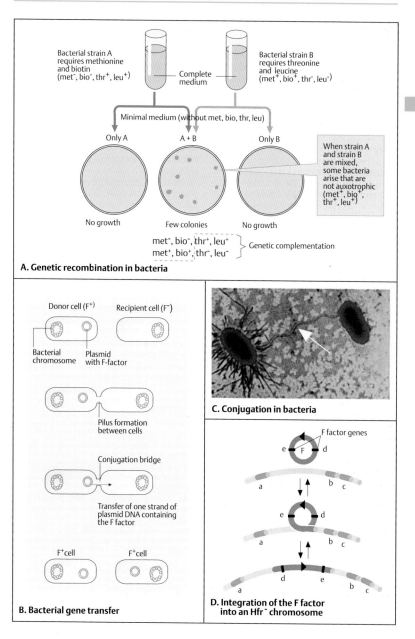

A. Genetic recombination in bacteria

Bacterial strain A requires methionine and biotin (met⁻, bio⁻, thr⁺, leu⁺)

Bacterial strain B requires threonine and leucine (met⁺, bio⁺, thr⁻, leu⁻)

Complete medium

Minimal medium (without met, bio, thr, leu)

Only A

A + B

Only B

When strain A and strain B are mixed, some bacteria arise that are not auxotrophic (met⁺, bio⁺, thr⁺, leu⁺)

No growth

Few colonies

No growth

met⁻, bio⁻, thr⁺, leu⁺
met⁺, bio⁺, thr⁻, leu⁻ } Genetic complementation

B. Bacterial gene transfer

Donor cell (F⁺)   Recipient cell (F⁻)

Bacterial chromosome   Plasmid with F-factor

Pilus formation between cells

Conjugation bridge

Transfer of one strand of plasmid DNA containing the F factor

F⁺cell   F⁺cell

C. Conjugation in bacteria

D. Integration of the F factor into an Hfr⁻ chromosome

F factor genes

e   F   d

a   b c

e   d

a   b c

d   e

a   b c

Bacteriophages

Bacteriophages (or phages) are bacterial viruses able to insert their DNA into bacteria and multiply in them. Their discovery by F. d'Herelle in 1917 as viruses that can attack *Salmonella typhimurium* caused a sensation in medical biology, as it spurred hopes for possible treatment, although these were disappointed. Modern phage research began in 1938 when Max Delbrück became the focus of a new school of phage workers (Stent and Calendar, 1978, p. 295). Phages differ greatly in genome size and complexity. The first electron micrograph was taken by Ruska in 1940. The first electron micrograph of a phage (T2) by S. E. Luria and T.F. Anderson in 1942 showed a polyhedral body and a tail. Unlike viruses that infect plant or animal cells, phages can be analyzed relatively easily in their host cells.

A. Attachment of a bacteriophage

A phage consists of DNA, a protein coat, and terminal filaments for attachment. A phage enters a bacterial cell by attaching to a receptor on the outer cell membrane surface of the bacterium. Numerous different phages that interact with particular bacteria are known, e.g., phages T1, T2, P1, F1, lambda, T4, T7, phiX174, and others for *Escherichia coli* and *Salmonella*.

B. Lytic and lysogenic cycles

A phage whose DNA has been inserted into a bacterial cell has one of two options. The DNA may be integrated to the host DNA to begin a *lysogenic cycle*. Here phage DNA is replicated along with the bacterial DNA without obvious consequences. Phage DNA that has been integrated into the bacterial chromosome is designated a *prophage*. Bacteria containing prophages are designated *lysogenic* bacteria. The corresponding phages are termed *lysogenic phages*.

The more common alternative is the *lytic cycle*. Phages do not reproduce by cell division like bacteria, but by intracellular formation and assembly of different phage components. Phage DNA that has entered the bacterial cell serves as a template to synthesize new phage DNA, which is transcribed into mRNA encoding phage coat proteins. These are assembled into new phage particles. Cellular enzymes provide translation. The change from a lysogenic to a lytic cycle is rare. It requires induction by external influences and complex genetic mechanisms.

C. Integration of bacteriophage λ DNA

Lambda (λ) phage, discovered by E.M. Lederberg in 1950, is a double-stranded DNA virus that infects the bacterium *E. coli*. Study of this phage yielded the first indication that a bacterium can absorb pure DNA, a new mechanism by which bacterial genes can move horizontally from one cell to another. This process is called *transformation*. Like other phages, phage λ can follow a lytic or a lysogenic pathway. In the lysogenic pathway it inserts itself by crossing-over (see p. 118). A virus-specific enzyme, lambda integrase, is synthesized after the phage enters the bacterial cell. This enzyme mediates the covalent joining at specific bacterial and phage attachment sites with sequence similarity. Here the virus is integrated into the bacterial chromosome and replicated along with the bacterial chromosome. The recognition of the two related but different DNA sequences of the phage and the bacterial chromosome is the key event of the integrase reaction leading to site-specific recombination. The phage may be released by a reverse procedure, and a lytic cycle be induced. The integrase is a specialized topoisomerase type I. Topoisomerases are enzymes that mediate breakage-reunion reactions. Topoisomerase I makes a nick in one strand of DNA, passes through the intact strand, and reseals the gap. (Figures adapted from Watson et al., 1987.)

References

Alberts B et al. Molecular Biology of the Cell, 4th ed. Garland Science, New York, 2002.

Brown TA: Genomes, 2nd ed. Bios Scientific, Oxford, 2002.

Cairns J, Stent GS, Watson JD (eds) Phage and the Origins of Molecular Biology. Cold Spring Harbor Laboratory Press, New York, 1966.

Kwon HJ et al: Flexibility in DNA recombination: structure of the lambda integrase catalytic core. Science 276: 126–131, 1997.

Watson JD et al: Molecular Biology of the Gene, 3rd ed. Benjamin/Cummings, Menlo Park, California, 1987.

Online information:
The Bacteriophage Ecology Group at http://www.phage.org/.

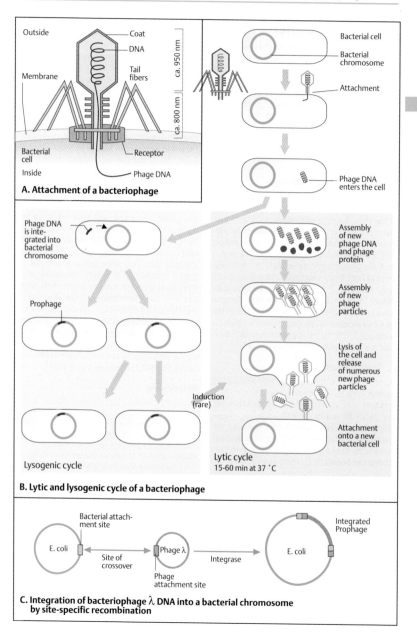

Outside
Coat
DNA
Membrane
Tail fibers
ca. 950 nm
ca. 800 nm
Bacterial cell
Inside
Receptor
Phage DNA

A. Attachment of a bacteriophage

Bacterial cell
Bacterial chromosome
Attachment
Phage DNA enters the cell
Assembly of new phage DNA and phage protein
Assembly of new phage particles
Lysis of the cell and release of numerous new phage particles
Attachment onto a new bacterial cell

Phage DNA is integrated into bacterial chromosome
Prophage
Induction (rare)
Lysogenic cycle
Lytic cycle
15-60 min at 37 °C

B. Lytic and lysogenic cycle of a bacteriophage

Bacterial attachment site
E. coli
Site of crossover
Phage λ
Phage attachment site
Integrase
Integrated Prophage
E. coli

C. Integration of bacteriophage λ DNA into a bacterial chromosome by site-specific recombination

DNA Transfer between Cells

In prokaryotes, a lateral or horizontal transfer of DNA is common. The three principal mechanisms for gene flow between prokaryotes are *conjugation* between bacteria (see p. 96); *transduction,* involving the transfer of a small segment of DNA from one bacterium to another by a bacteriophage; and *transformation,* the uptake of a DNA fragment from a donor cell by a recipient cell. The DNA transfer induced experimentally between eukaryotic cells in vitro is called *transfection.*

A. Transduction by viruses

In 1952, N. Zinder and J. Lederberg described a new type of recombination between two strains of bacteria. Bacteria previously unable to produce lactose (*lac⁻*) acquired the ability to produce lactose after being infected with phages that had replicated in bacteria containing a gene for producing lactose (*lac⁺*). A small segment of DNA from a bacterial chromosome had been transferred by a phage to another bacterium. This is called transduction. General transduction (insertion of phage DNA into the bacterial genome at any unspecified location) is distinguished from special transduction (insertion at a particular location).

B. Transformation by plasmids

Plasmids are small, autonomously replicating, circular DNA molecules separate from the chromosome in a bacterial cell. Since they often contain genes for antibiotic resistance (e.g., ampicillin), their incorporation into a sensitive cell renders the cell resistant to the antibiotic. Only these bacteria can grow in culture medium containing the antibiotic (selective medium).

C. Multiplication of a DNA segment in transformed bacteria

A DNA fragment to be studied can be inserted into a vector and multiplied in rapidly dividing cells. A plasmid containing one or more genes conferring antibiotic resistance can be used for selecting bacteria that have taken up this plasmid. In an antibiotic-containing medium (selective medium) only those bacteria can grow that have incorporated a recombinant plasmid containing the DNA to be investigated.

D. Transfection by DNA

In the 1970s it was discovered that pure DNA added to a bacterial culture in the presence of a high concentration of calcium ions (Ca^{++}) can transform the bacteria. This discovery allowed any piece of DNA to be introduced into any bacterial genome and the genetic consequences to be observed. DNA can also be introduced into mammalian cells by a process called *transfection.* On the left, a DNA transfer experiment is shown in a culture of mouse fibroblasts; on the right, in a culture of human tumor cells (Weinberg, 1985, 1987). The mouse fibroblast culture (see p. 128) is exposed to the chemical carcinogen methylcholanthrene (left). DNA from these cells is precipitated with calcium phosphate, extracted, and then taken up by a normal culture (transfection). About 2 weeks later, cells appear that have lost contact inhibition (transformed cells). When these cells are injected into mice that lack a functional immune system (nude mice), tumors develop. DNA from cultured human tumor cells (right) can also transform normal cells after several transfer cycles. Detailed studies of cancer-causing genes (oncogenes) in eukaryotic cells were first carried out using transfection. Many eukaryotic cell lines have been generated from tumor cells. (Figures in A–C adapted from Watson et al. 1987, D from Weinberg 1987.)

References

Brown TA: Genomes, 2nd ed. Bios Scientific, Oxford, 2002.

Lwoff A: Lysogeny. Bacterial Rev 17: 269–337, 1953.

Smith HO, Danner DB, Deich RA: Genetic transformation. Ann Rev Biochem 50: 41–68, 1981.

Watson JD et al: Molecular Biology of the Gene, 4th ed. Benjamin/Cummings, 1987.

Weinberg RA: Oncogenes of spontaneous and chemically induced tumors. Adv Cancer Res 36: 149–163, 1982.

Weinberg RA: The action of oncogenes in the cytoplasm and nucleus. Science 230: 770–776, 1985.

Zinder N, Lederberg J: Genetic exchange in *Salmonella.* J Bacteriol 64: 679–699, 1952.

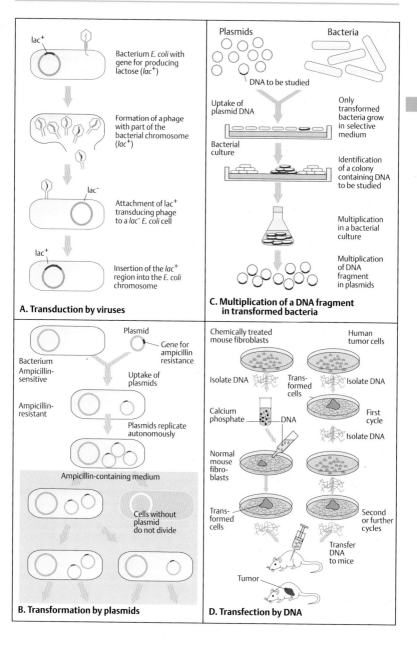

A. Transduction by viruses

Bacterium *E. coli* with gene for producing lactose (*lac*⁺)

Formation of a phage with part of the bacterial chromosome (*lac*⁺)

Attachment of lac⁺ transducing phage to a lac⁻ *E. coli* cell

Insertion of the *lac*⁺ region into the *E. coli* chromosome

B. Transformation by plasmids

Plasmid

Gene for ampicillin resistance

Bacterium Ampicillin-sensitive

Ampicillin-resistant

Uptake of plasmids

Plasmids replicate autonomously

Ampicillin-containing medium

Cells without plasmid do not divide

C. Multiplication of a DNA fragment in transformed bacteria

Plasmids   Bacteria

DNA to be studied

Uptake of plasmid DNA

Only transformed bacteria grow in selective medium

Bacterial culture

Identification of a colony containing DNA to be studied

Multiplication in a bacterial culture

Multiplication of DNA fragment in plasmids

D. Transfection by DNA

Chemically treated mouse fibroblasts   Human tumor cells

Isolate DNA   Transformed cells   Isolate DNA

Calcium phosphate — DNA   First cycle

Isolate DNA

Normal mouse fibroblasts

Transformed cells   Second or further cycles

Transfer DNA to mice

Tumor

Classification of Viruses

A virus is an ultramicroscopic intracellular parasite incapable of autonomous replication. It can only replicate inside a cell. Viruses range in size from about 20 to 250 nm. A virus consists of a small molecule of nucleic acid, either DNA or RNA (single-stranded or double-stranded), surrounded by proteins. Some viruses have an outer-membrane envelope. The small genome of a virus contains from a few to several hundred genes for viral-specific information but lacks genes to encode the many enzymes required for metabolism. The extracellular form of a virus particle includes a protein coat (*capsid*), which encloses the genome of DNA or RNA. The capsids contain multiple units of one or a few different protein molecules encoded by the virus genome. Capsids usually have an almost spherical, icosahedral (20 plane surfaces), or occasionally a helical structure.

Viruses are important pathogens in plants and animals, including man. The complete infectious viral particle is called a *virion*. The term *virus* is a Latin word meaning "slimy liquid" or "poison" derived from usage of this term in ancient Rome to mean "poison" of animal origin. Examples of representative viruses important in human diseases are listed in the appendix.

A. A virus with an envelope

The herpesvirus shown is a large, complex virus with a double-stranded DNA genome of about 250 kb contained in a core of nucleoproteins in a capsid. The structural unit of nucleic acid, nucleoprotein, and capsid is called a *nucleocapsid*. A matrix or tegument protein fills the space between the nucleocapsid and the envelope. The capsid contains multiple units of one or a few different protein molecules encoded by the virus genome. The envelope contains virus-encoded glycoproteins. These serve to make contact with an uninfected cell. (Figure adapted from Wang & Kieff, 2005.)

B. Structure of a human retrovirus

A retrovirus has a genome of RNA with a unique replication cycle. The term retrovirus reflects the way information is encoded and decoded: an RNA-dependent DNA polymerase, called a reverse transcriptase, directs the synthesis of a DNA form of the viral genome after infection of a cell. Retroviruses form a large family of retroviridae.

Retroviruses are similar in structure, genome organization, and replication but have a wide range of biological consequences after infection. Their sizes range between 70–130 nm in diameter. The core contains two identical copies of single-stranded RNA of 8–10 kb and an associated reverse transcriptase and tRNA. The RNA has a cap site at the 5′ end and a polyadenylation site at the 3′ end, both characteristic features of mRNA. The envelope of the human immunodeficiency virus (HIV-1) contains an outer membrane surface glycoprotein of 120 kDa (gp120), a transmembrane protein gp41, a capsid p24 core protein, and several other proteins. (Figure adapted from Longo & Fauci, 2005.)

C. Major virus families

Viruses can be classified on the basis of their type of genome (DNA viruses and RNA viruses), presence or absence of a lipid envelope, shape of the capsid (viral protein coat), and their organ or tissue specificity. Viruses with genomes of single-stranded RNA are classified according to whether their genome is a positive (plus RNA) or negative (minus RNA) RNA strand. Only an RNA plus strand can serve as a template for translation (5′ to 3′ orientation).

Poxviruses are among the largest, with a genome of 270 kb and 450 nm external length. Among the smallest are parvoviruses, with a genome of 5 kb and 20 nm external length. The major virus families with species infecting humans are shown by genome type and approximately to scale. (Figure adapted from Wang & Kieff, 2005.)

References

Brock TD, Madigan MT: Biology of Microorganisms 6th ed. Prentice Hall, Englewood Cliffs, New Jersey, 1991.

Lederberg J: Infectious history. Pathways to discovery. Science 288 : 287–293, 2000.

Longo DL, Fauci AS: The human retroviruses, pp 1071–1075. In: Kasper DL et al (eds) Harrison's Principles of Internal Medicine, 16th ed. McGraw-Hill, New York, 2005.

Wang F, Kieff E: Medical virology, pp 1019–1027. In: Kasper DL et al (eds) Harrison's Principles of Internal Medicine, 16th ed. McGraw-Hill, New York, 2005.

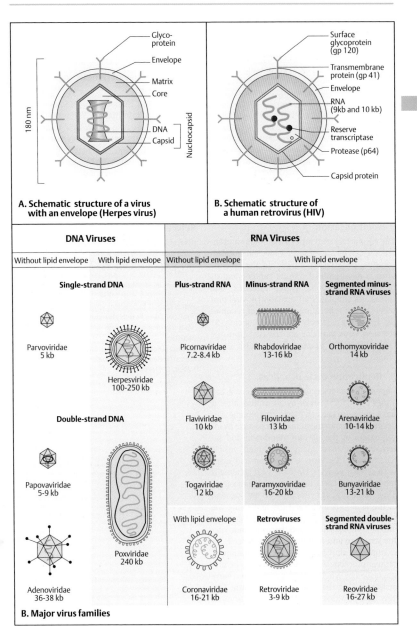

A. Schematic structure of a virus with an envelope (Herpes virus)

- Glyco-protein
- Envelope
- Matrix
- Core
- DNA
- Capsid
- Nucleocapsid
- 180 nm

B. Schematic structure of a human retrovirus (HIV)

- Surface glycoprotein (gp 120)
- Transmembrane protein (gp 41)
- Envelope
- RNA (9kb and 10 kb)
- Reserve transcriptase
- Protease (p64)
- Capsid protein

DNA Viruses		RNA Viruses		
Without lipid envelope	With lipid envelope	Without lipid envelope	With lipid envelope	
Single-strand DNA		**Plus-strand RNA**	**Minus-strand RNA**	**Segmented minus-strand RNA viruses**
Parvoviridae 5 kb	Herpesviridae 100–250 kb	Picornaviridae 7.2–8.4 kb	Rhabdoviridae 13–16 kb	Orthomyxoviridae 14 kb
Double-strand DNA		Flaviviridae 10 kb	Filoviridae 13 kb	Arenaviridae 10–14 kb
Papovaviridae 5–9 kb	Poxviridae 240 kb	Togaviridae 12 kb	Paramyxoviridae 16–20 kb	Bunyaviridae 13–21 kb
Adenoviridae 36–38 kb		With lipid envelope	**Retroviruses**	**Segmented double-strand RNA viruses**
		Coronaviridae 16–21 kb	Retroviridae 3–9 kb	Reoviridae 16–27 kb

B. Major virus families

Replication of Viruses

Viruses depend on their host cells for propagation. They encode only information for virus-specific proteins. They replicate in different ways, but the basic pattern of replication is similar: First, the infectious virus particle is disassembled in the host cell; then the viral genome is replicated using host cell enzymes; next, viral proteins are synthesized using host cell translation machinery; and finally the viral genome and proteins are reassembled into progeny virus particles. These are released from the host cell. A single virion (virus particle) infecting a cell can produce thousands of progeny virions in a short time.

A. Principal stages of virus multiplication

The virus attaches to the cell membrane and enters the cell by one of several mechanisms (adsorption, penetration, **1**). Within the cell, the virus is disassembled into its genome and its proteins. The genome is replicated (**2**) and serves as a template for translation of viral proteins (**3**) and replication of new viral genomes (**4**). Following assembly of the components (**5**), the new virus particles are released from the cell (**6**). (Figure adapted from Lodish et al, 2004.)

B. Replication cycle of a virus

Depending on its type, a virus enters a cell either by endocytosis or by adsorption after it has fused with the cell plasma membrane. Endocytosis refers to the uptake of a particle by invagination of the plasma membrane and internalization in a membrane-bound vesicle (coated pit). The virus is uncoated and fuses with cellular vesicles. Next the viral genome is replicated and viral proteins are synthesized. Three basic types of viral proteins are encoded in the viral genome: those for replicating the viral genome, those for packaging the genome into new viral particles, and those for modifying the structure and function of the host cell, for example by shutting down the translation of cellular proteins. Viral proteins can also be distinguished by whether they are expressed early or later during the viral replication cycle. The new viral particles leave the cell by exocytosis or by budding. Exocytosis is the process by which the virion is released from the cell, packaged in membrane-bound vesicles. Budding refers to the release with a lipid membrane. (Figure adapted from Wang & Kieff, 2005; and Doerr & Ehrlich, 2002.)

C. Efficient use of genetic information by viruses

Viruses utilize their small genomes to maximize genetic information. Some viruses use both strands of DNA or RNA for transcription (**1**). The reading frames may overlap (**2**, e.g., in phage ΦX174, p. 246). Alternative RNA splicing occurs in some viruses (**3**), e.g., adenoviruses. Some viruses are integrated into the host cell genome and participate in nonlytic viral growth cycles, e.g., retroviruses. (Figure adapted from Doerr & Ehrlich, 2002.)

References

Alberts B et al: Molecular Biology of the Cell, 4th ed. Garland Science, New York, 2002.

Doerr HW, Ehrlich WH, eds: Medizinische Virologie. Thieme Verlag, Stuttgart, 2002.

Lodish H et al: Molecular Cell Biology, 5th ed. WH Freeman, New York, 2004.

Wang F, Kieff E: Medical virology, pp 1019–1027. In: Kasper DL et al (eds) Harrison's Principles of Internal Medicine, 16th ed. McGraw-Hill, New York, 2005.

Watson JD et al: Molecular Biology of the Gene, 4th ed. The Benjamin/Cummings Publishing Co, Menlo Park, California, 1987.

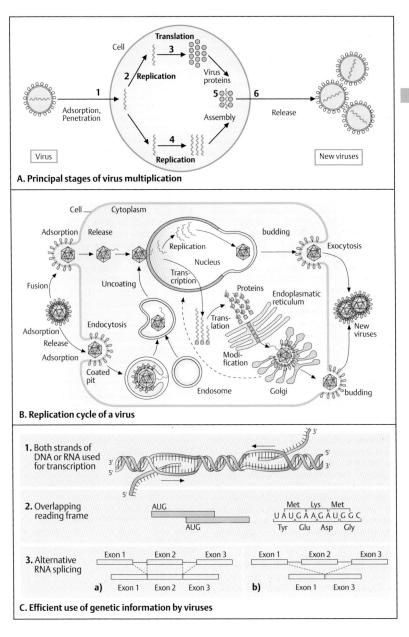

A. Principal stages of virus multiplication

Virus

1 — Adsorption, Penetration

Cell

2 — Replication

Translation

3 → Virus proteins

5 — Assembly

6 — Release

4 — Replication

New viruses

B. Replication cycle of a virus

Cell — Cytoplasm

Adsorption   Release   budding   Exocytosis

Fusion

Uncoating   Replication   Nucleus

Trans-cription

Adsorption   Endocytosis   Proteins   Endoplasmatic reticulum

Release   Trans-lation

Adsorption   New viruses

Coated pit   Modi-fication

Endosome   Golgi   budding

C. Efficient use of genetic information by viruses

1. Both strands of DNA or RNA used for transcription

3'   5'
5'   3'
5'

2. Overlapping reading frame

AUG

AUG

Met   Lys   Met
U A U G A A G A U G G C
Tyr   Glu   Asp   Gly

3. Alternative RNA splicing

a)
Exon 1   Exon 2   Exon 3
Exon 1   Exon 2   Exon 3

b)
Exon 1   Exon 2   Exon 3
Exon 1   Exon 3

Retroviruses

Retroviruses are RNA viruses with a unique replication cycle in which double-stranded DNA is transcribed from the viral RNA genome by reverse transcription (RNA ⇒ DNA). Retroviruses belong to the family of Retroviridae, which includes three subfamilies, the Oncovirinae to which the human T-cell lymphotropic virus (HTLV) belongs, the Lentivirinae with HIV-1 and HIV-2, and the Spumavirinae (foamy viruses, which are not associated with a known disease). Several retroviruses cause tumors in mice (mouse leukemia) or chickens (Rous sarcoma).

A. Genomic structure of retroviruses

Retroviruses have a characteristic genome structure that differs slightly among the various types. The murine leukemia virus (MuLV) has the typical three structural genes: *gag*, *pol*, and *env* (**1**). The *gag* region encodes three proteins, matrix (MA), capsid (CA), and nucleic acid-binding protein (NC). All retroviral genomes include noncoding sequences at their 5′ and 3′ ends with redundant, identical sequences called long terminal repeats (LTRs). The LTRs contain sequences for initiation of the expression of viral RNAs. A primer-binding site is located outside the LTRs (not shown). The human immunodeficiency virus 1 (HIV-1) resembles the MuLV, but has six accessory genes: *tat*, *rev*, *vif*, *nef*, *vpr*, and *vpu*. HIV-2 has *vpx* instead of *vpu*.

The *pol* (polymerase) region encodes reverse transcriptase (RT) and integrase (IN). The *env* (envelope) region encodes the surface proteins gp120 and pg41. Tat is a protein that augments virus expression from the LTR; Rev (regulator of expression of virion proteins) regulates RNA splicing; Nef protein (negative factor) downregulates cellular CD4 (see section on immune system); and Vif protein (virion infectivity factor) assembles the HIV nucleoprotein core. Vpr and Vpu are viral proteins that support transport of the provirus into the cell nucleus and induce cellular G_2 cell cycle growth arrest and other changes. (Figure adapted from Lono & Fauci, 2005.)

B. Replication cycle of the HIV-1 virus

This replication cycle can be divided into 10 schematic steps. After attaching to a cell surface receptor, the virion is taken into the cell. Immediately after entry into the cell, the viral RNA genome is transcribed into double-stranded DNA by reverse transcriptase. The RNAs transcribed from this DNA copy by cellular RNA polymerase II serve either as mRNA for the synthesis of viral proteins or as new viral genomes. Newly formed virions leave the cell by a specific process called exocytosis. (Figure adapted from Longo & Fauci, 2005.)

C. DNA synthesis of a retrovirus

The RNA genome of a typical retrovirus contains short repetitive (R) and unique (U) sequences at both ends (RU5 at the 5′ end and U3R at the 3′ end). The 5′ end of the genome contains a nucleotide sequence that is complementary to the 3′ end of a host cell tRNA. This nucleotide sequence binds to tRNA, which serves as a primer for the synthesis by reverse transcriptase of viral DNA from the virus genome RNA.

The first step of replication of a retrovirus genome is initiated by a primer of host cell tRNA, which attaches to the 5′ end of the viral RNA plus strand genome (**1**). After synthesis of the first DNA strand and removal of the tRNA primer, synthesis of the second DNA strand (plus strand) begins at the RU5 region (**2**). Here, the previously formed minus DNA strand serves as the primer (**3**). As DNA synthesis is continued, the remaining RNA is degraded (**4**), and the DNA plus strand is synthesized to completion (**5**, **6**). The double-stranded DNA copy of the virus contains long terminal repeats (LTRs) at both ends. These enable the viral DNA intermediary step to be integrated into the host cell DNA, and they contain the necessary regulatory sequences for transcribing the provirus DNA. (Figure adapted from Watson et al., 1987.)

References

Alberts B et al: Molecular Biology of the Cell, 4th ed. Garland Science, New York, 2002.

Longo DL, Fauci AS: The human retroviruses, pp 1071–1075. In: Kasper DL et al (eds) Harrison's Principles of Internal Medicine, 16th ed. McGraw-Hill, New York, 2005.

Watson JD et al: Molecular Biology of the Gene, 3rd ed. Benjamin/Cummings Publishing Co, Menlo Park, California, 1987.

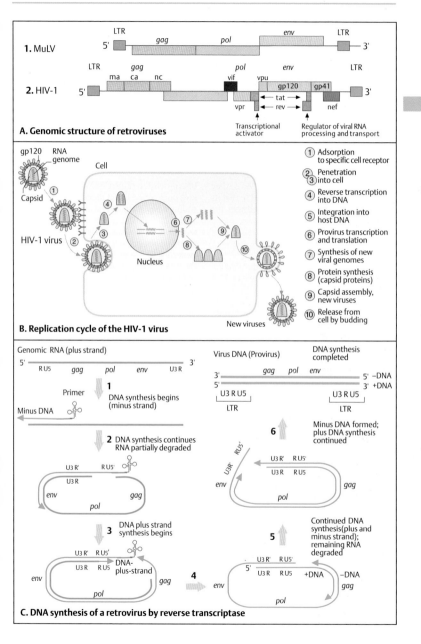

A. Genomic structure of retroviruses

1. MuLV

LTR   *gag*   *pol*   *env*   LTR
5'   3'

2. HIV-1

LTR   *gag*   *pol*   *env*   LTR
ma   ca   nc   vif   vpu   gp120   gp41
5'   3'
vpr   tat   nef
rev

Transcriptional activator

Regulator of viral RNA processing and transport

B. Replication cycle of the HIV-1 virus

gp120   RNA genome   Cell

Capsid

HIV-1 virus

Nucleus

New viruses

(1) Adsorption to specific cell receptor
(2) Penetration
(3) into cell
(4) Reverse transcription into DNA
(5) Integration into host DNA
(6) Provirus transcription and translation
(7) Synthesis of new viral genomes
(8) Protein synthesis (capsid proteins)
(9) Capsid assembly, new viruses
(10) Release from cell by budding

C. DNA synthesis of a retrovirus by reverse transcriptase

Genomic RNA (plus strand)

5'   R U5   *gag*   *pol*   *env*   U3 R   3'

Primer

1 DNA synthesis begins (minus strand)

Minus DNA

Virus DNA (Provirus)   DNA synthesis completed

3'   *gag*   *pol*   *env*   5' −DNA
5'   3' +DNA
U3 R U5   U3 R U5
LTR   LTR

2 DNA synthesis continues RNA partially degraded

U3 R'   R U5'
U3 R
env   gag
pol

6 Minus DNA formed; plus DNA synthesis continued

U3 R'   R U5'
U3 R   R U5
env   gag
pol

3 DNA plus strand synthesis begins

U3 R'   R U5'
U3 R   R U5   DNA-plus-strand
env   gag
pol

4

5 Continued DNA synthesis (plus and minus strand); remaining RNA degraded

U3 R'   R U5'
5'   U3 R   R U5   +DNA
env   −DNA
pol   gag

Retrovirus Integration and Transcription

Integration of the DNA copy of a retrovirus into the host cell DNA occurs at a random location. This may alter a cellular gene (insertional mutation). The viral genes of provirus DNA are transcribed by cellular RNA polymerase II. The resulting mRNA serves either for translation or for the production of new RNA genomes, which are packaged into new virions. Some retrovirus genomes may contain an additional viral oncogene (v-*onc*). Viral oncogenes are parts of cellular genes (c-*onc*) previously taken up by the virus. If they enter a cell with the virus, they may change (transform) the host cell so that its cell cycle is altered and the cell becomes the origin of a tumor.

A. Retrovirus integration into cellular DNA

In the nucleus of the host cell, the double-stranded DNA is synthesized from the viral RNA genome by reverse transcription (**1**). The initial DNA is circularized by joining the LTRs (long terminal repeats) RU5 and RU3 by virtue of their sequence identity (**2**). Recognition sequences in the LTRs and in the cellular DNA (**3**) allow the circular viral DNA to be opened at a specific site (**4**) and the viral DNA to be integrated into the host DNA (**5**). Viral genes can then be transcribed from the integrated provirus (**6**).

The genomes of vertebrates (including man) contain numerous DNA sequences that consist of endogenous proviruses. The genomes of higher organisms also contain LTR-like sequences that are very similar to those of an endogenous retrovirus. These sequences can change their location in the genome (mobile genetic elements or transposons). Since many of them have the fundamental structure of a retrovirus (LTR genes), they are designated retrotransposons (see p. 86).

B. Control of retrovirus transcription

The LTRs are important not only for integration of the virus into cellular DNA, but also because they contain all regulatory signals necessary for efficient transcription of a viral gene. Typical transcription signals are the so-called "CCAAT" and "TATA" sequences of promoters, which are respectively located about 80 and 25 base pairs above the 5' end of the sequence to be transcribed. Further upstream (in the 5' direction) are nucleotide sequences that can increase the expression of viral genes (enhancers). Similar regulatory sequences are located at the 5' end of eukaryotic genes (see section on gene transcription). Newly synthesized viral RNA is structurally modified at the 5' end (formation of a cap). Furthermore, numerous adenine residues are added at the 3' end (polyadenylated, poly(A), see p. 17).

C. Viral protein synthesis by posttranscriptional modification of RNA

DNA provirus is transcribed into RNA by cellular RNA polymerase II and modified into mRNA by adding a cap at the 5' end and polyadenylation at the 3' end. The mRNA virus-encoded proteins, e.g., gag and pol as shown, are first translated into one polypeptide. This is subsequently cleaved into the individual proteins. Some of the RNA is spliced to form new mRNAs that encode other proteins, e.g., the coat protein env. (Figures adapted from Watson et al., 1987.)

References

Alberts B et al: Molecular Biology of the Cell, 4th ed. Garland Science, New York, 2002.

Longo DL, Fauci AS: The human retroviruses, pp 1071–1075. In: Kasper DL et al (eds) Harrison's Principles of Internal Medicine, 16th ed. McGraw-Hill, New York, 2005.

Watson JD et al: Molecular Biology of the Gene, 3rd ed. Benjamin/Cummings Publishing Co, Menlo Park, California, 1987.

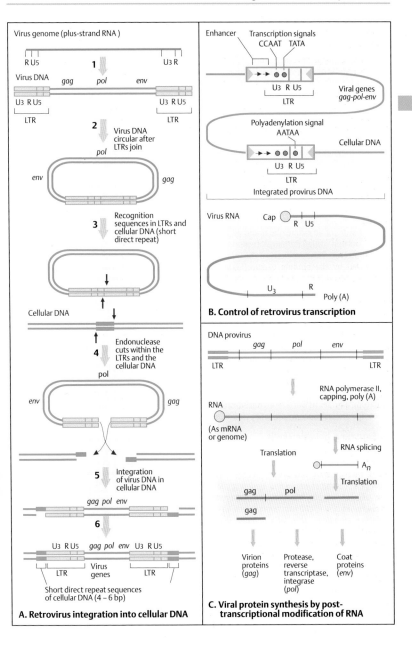

A. Retrovirus integration into cellular DNA

Virus genome (plus-strand RNA)

R U5   U3 R

1

Virus DNA

gag   pol   env

U3 R U5   U3 R U5

LTR   LTR

2 Virus DNA circular after LTRs join

pol

env   gag

3 Recognition sequences in LTRs and cellular DNA (short direct repeat)

Cellular DNA

4 Endonuclease cuts within the LTRs and the cellular DNA

pol

env   gag

5 Integration of virus DNA in cellular DNA

gag pol env

6

U3 R U5   gag pol env   U3 R U5

LTR   Virus   LTR
           genes

Short direct repeat sequences of cellular DNA (4–6 bp)

B. Control of retrovirus transcription

Enhancer   Transcription signals
           CCAAT TATA

U3 R U5   Viral genes
                          gag-pol-env
LTR

Polyadenylation signal
AATAA

U3 R U5   Cellular DNA
LTR

Integrated provirus DNA

Virus RNA   Cap

R U5

U3   R
                          Poly (A)

C. Viral protein synthesis by post-transcriptional modification of RNA

DNA provirus

gag   pol   env

LTR   LTR

RNA polymerase II, capping, poly (A)

RNA

(As mRNA or genome)

Translation   RNA splicing

A_n

Translation

gag   pol

gag

Virion   Protease,   Coat
proteins   reverse   proteins
(gag)   transcriptase, (env)
             integrase
             (pol)

Cell Communication

Multicellular organisms depend on cell-to-cell communication for processes such as growth (e.g., embryonic development), cell differentiation into the different types, regulation of the cell cycle, and other important functions. Cells communicate via a vast variety of proteins that mediate cell-specific responses. These proteins include extracellular signal molecules, cell surface receptors, intracellular receptors, and intracellular signal molecules that transmit signals.

A. Principle of signal transduction

The transduction of a signal elicits a cell-specific effect involving several types of interacting proteins. A cell membrane-bound receptor, consisting of an extracellular and an intracellular portion (called domains), responds to a signal molecule. The specificity of the response is achieved by the binding of specific signal molecule (called a ligand) to the extracellular domain of the receptor. This in turn activates a series of further signaling proteins downstream. Generally, one activated protein activates the next by a specific biochemical reaction, called a signaling cascade. Only two are shown here (here designated signaling proteins 1 and 2), but quite often many more are involved. The final steps of a signaling cascade reach the target protein and elicit the desired cellular response. Extracellular signal molecules typically act at very low concentrations (about 10^{-8} molar concentration, M).

B. Signaling between cells

Cells transmit signals by different means. A signal molecule may remain attached to the surface of the signaling cell when binding to the target cell (**1**, *contact-dependent signaling*). This type of signaling is common in embryonic development and in the immune system. In (**2**) *paracrine signaling*, the signal molecule is secreted and released into the extracellular space. Here it acts over a short distance on cells in the neighborhood. Many growth and differentiation factors act in this manner. Some important long-distance signaling is mediated via hormones in the blood circulation (**3**, *endocrine signaling*). In this type of signaling, specialized cells, called *endocrine cells*, secrete a substance, called a *hormone*, into the bloodstream. From there it can reach the target cells at a distance in another part of the body. *Synaptic signaling* refers to nerve cells or the junction of nerve and muscle cells (**4**). In this case, a specialized cell, a nerve cell (neuron), sends electrical impulses along a cellular extension, the *axon*, which can be quite long. At the end of the axon, a chemical signal, called a neurotransmitter, is secreted at the junction (the synapse) between the signaling cell (the neuron) and the postsynaptic target cell. In some cases the same types of signaling molecules are used in paracrine, endocrine, and synaptic signaling, but in different contexts of selectivity. (Figure adapted from Lodish et al, 2004.)

The term hormone is derived from a Greek word meaning to spur on. It was first used in 1904 by William Bayliss and Ernest Starling to describe the action of a secreted molecule (Stryer, 1995, p. 342). Five major classes of hormones can be defined: (i) amino acid derivatives (e.g., catecholamine, dopamine, thyroxine), (ii) small neuropeptides (e.g., thyrotropin-releasing hormone, somatostatin, vasopressin), (iii) proteins (e.g. insulin, luteinizing hormone), (iv) steroid hormones, derived from cholesterol (e.g., cortisol, sex hormones), and (v) vitamin derivatives (e.g., retinoids [vitamin A], peptide growth hormones).

Medical relevance

Mutations in the genes encoding proteins involved in signal transduction cause a vast array of human genetic disorders (Jameson, 2005).

References

Alberts B et al: Molecular Biology of the Cell, 4th ed. Garland Science, New York, 2002.

Jameson, JL: Principles of endocrinology, pp 2067–2075. In: Kasper DL et al (eds) Harrison's Principles of Internal Medicine, 16th ed. McGraw-Hill, New York, 2005.

Lewin, B: Genes VIII. Pearson Educational International, Prentice Hall, Upper Saddle River, NJ, 2004.

Lodish H et al: Molecular Cell Biology, 5th ed. WH Freeman, New York, 2004.

Mapping Cellular Signaling. Special Issue. Science 296: 1557–1752, 2002.

Stryer, L: Biochemistry, 4th ed. W.H. Freeman, New York, 1995.

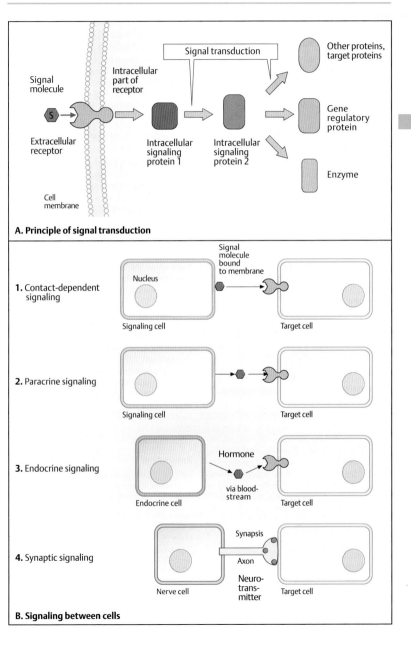

A. Principle of signal transduction

Signal molecule

Extracellular receptor

Intracellular part of receptor

Cell membrane

Signal transduction

Intracellular signaling protein 1

Intracellular signaling protein 2

Other proteins, target proteins

Gene regulatory protein

Enzyme

B. Signaling between cells

1. Contact-dependent signaling

Nucleus

Signal molecule bound to membrane

Signaling cell

Target cell

2. Paracrine signaling

Signaling cell

Target cell

3. Endocrine signaling

Hormone

via blood-stream

Endocrine cell

Target cell

4. Synaptic signaling

Synapsis

Axon

Neuro-trans-mitter

Nerve cell

Target cell

Yeast: Eukaryotic Cells with a Diploid and a Haploid Phase

Yeast is a single-celled eukaryotic fungus with a genome of individual linear chromosomes enclosed in a nucleus and with cytoplasmic organelles such as endoplasmic reticulum, Golgi apparatus, mitochondria, peroxisomes, and a vacuole analogous to a lysosome. About 40 different types of yeast are known. Baker's yeast, *Saccharomyces cerevisiae* (budding yeast) consists of oval cells of about 5 μm diameter. A cell can divide by budding every 90 minutes under good nutritional conditions. Fission yeast, *Schizosaccharomyces pombe*, has rod-shaped cells that divide by elongation at the ends.

The haploid genome of *S. cerevisiae* contains about 6200 genes in 1.4×10^7 DNA base pairs distributed in 16 chromosomes (Goffeau, 1996). They are involved in the following functions: cell structure 250 (4%), DNA metabolism 175 (3%), transcription and translation 750 (13%), energy production and storage 175 (3%), biochemical metabolism 650 (11%), and transport 250 (4%). The *S. cerevisiae* genome is very compact compared with other eukaryotic genomes, with about one gene every 2 kb. Nearly half of the human proteins known to be defective in hereditary disease have amino acid similarities to a yeast protein.

A. Yeast life cycle through a haploid and a diploid phase

The life cycle of yeast passes through a haploid or a diploid phase. Haploid cells of opposite types can fuse (mate) to form a diploid cell. Haploid cells are of one of two possible mating types, called **a** and α. The mating is mediated by a small secreted polypeptide called a pheromone or mating factor. A cell-surface receptor recognizes the pheromone secreted by cells of the opposite type, i.e., **a** cell receptors bind only α factor and α cell receptors bind only **a** factor. Mating and subsequent mitotic divisions occur in favorable conditions for growth. In starvation conditions, a diploid yeast cell undergoes meiosis and forms four haploid spores (sporulation), two of type **a** and two of type α. (Photographs: *S. pombe* at www.steve.gb.com/science/model_organisms.html; *S. cerevisiae* from Maher, BA: Rising to the occasion. The Scientist 17, Supplement, page S9, June 2, 2003.)

B. Switch of mating type

The switch of mating type (mating-type conversion) is initiated by a double-strand break in the DNA at the MAT locus (recipient) and may involve the boundary to either of the flanking donor loci (HMR or HML). This is mediated by an HO endonuclease through site-specific DNA cleavage.

C. Cassette model for mating type switch

Mating type switch is regulated at three gene loci near the centromere (cen) of chromosome III of *S. cerevisiae*. The central locus is MAT (mating-type locus), which is flanked by loci HMLα (left) and HMRa (right). Only the MAT locus is active and transcribed into mRNA. Transcription factors regulate other genes responsible for the **a** or the α phenotype. The HMLα and HMRa loci are repressed (silenced). DNA sequences from either the HMLα or the HMLa locus are transferred into the MAT locus once during each cell generation by a specific recombination event called gene conversion. The presence of HMRα sequences at the MAT locus determines the **a** cell phenotype. When HMLα sequences are transferred (switch to an α cassette), the phenotype is switched to α. Any gene placed by recombinant DNA techniques near the yeast mating-type silencer is repressed. Apparently the HML and HMR loci are permanently repressed, because they are inaccessible to proteins (transcription factors and RNA polymerase) owing to the condensed chromatin structure near the centromere.

(Figures adapted from Lodish et al, 2004.)

References

Botstein D, Chervitz SA, Cherry JM: Yeast as a model organism. Science 277: 1259–1260, 1997.

Brown TA: Genomes, 2nd ed. Bios Scientific Publishers, Oxford, 2002.

Goffeau A et al: Life with 6000 genes. Science 274: 562–567, 1996.

Haber JE: A locus control region regulates yeast recombination. Trends Genet 14: 317–321, 1998.

Lewin B: Genes VIII. Pearson International, 2004.

Lodish H et al: Molecular Cell Biology, 5th ed. WH Freeman, New York, 2004.

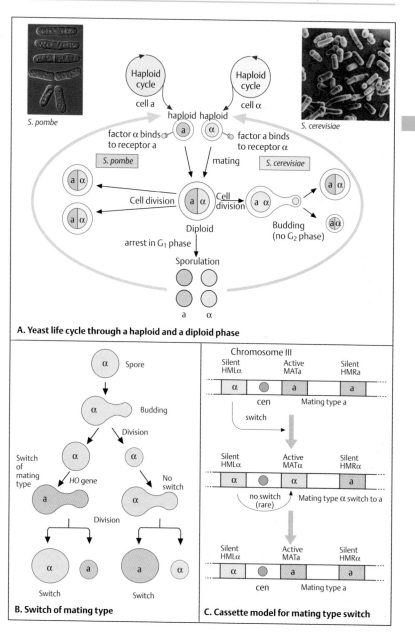

A. Yeast life cycle through a haploid and a diploid phase

B. Switch of mating type

C. Cassette model for mating type switch

Mating Type Determination in Yeast Cells and Yeast Two-Hybrid System

Yeast cells *(S. cerevisiae)* are unicellular eukaryotes with three different cell types: haploid **a** and α cells and diploid **a**/α cells. Owing to their relative simplicity compared with multicellular animals and plants, yeast serves as a model for understanding the underlying control mechanisms specifying cell types. The generation of many different cell types in different tissues of multicellular organisms probably evolved from mechanisms that determine cell fate in unicellular organisms such as yeast.

A. Regulation of cell-type specificity in yeast

Each of the three *S. cerevisiae* cell types expresses cell-specific genes. The different resulting combinations of DNA-binding proteins determine the specific cell types. Haploid cells of the a type produce a regulatory protein a1. Haploid cells of the α type produce two regulatory proteins, α1 and α2 with different effects. Protein α1 has no effect on cells of the a type, but has an effect on haploid cells (see below). Protein α1 as a transcription factor activates haploid α-specific genes, whereas protein α2 suppresses a-cell-specific genes. Both together activate haploid-specific genes corresponding to their own type. Following fusion of a cell of the a type with an α cell, the combination of the three proteins α1, α2, and a1 in a diploid cell (a/α) follows a completely different pattern: α2 inactivates genes of the a type, but in combination with α1 it now also inactivates all haploid-specific genes.

The principle is that each of the three cell types is determined by a cell-specific set of transcription factors acting as activators or suppressors, depending on the regulatory sequences to which they bind. These regulatory proteins are encoded at the MAT locus together with a general transcription factor called Mcm1. Mcm1 is expressed in all three cell types. Cells of type a only express a-specific genes. In diploid (a/α) cells, haploid-specific genes are suppressed. Mcm1 is a dimeric general transcription factor that binds to a-specific upstream regulatory sequences (URSs). This stimulates transcription of the a-specific genes, but it does not bind too efficiently to the α-specific URSs when α1 protein is absent.

B. Yeast two-hybrid system

Defining the function of a newly isolated gene may be approached by determining whether its protein specifically reacts with another protein of known function. Yeast cells can be used to assay protein–protein interactions. The two-hybrid method rests on observing whether two different proteins, each hybridized to a different protein domain required for transcription factor activity, are able to interact and thereby reassemble the transcription factor. When this occurs, a reporter gene is activated. Neither of the two hybrid proteins alone is able to activate transcription. Hybrid 1 consists of protein X, the protein of interest (the "bait"), attached to a transcription factor DNA-binding domain (BD). This fusion protein alone cannot activate the reporter gene because it lacks a transcription factor activation domain (AD). Hybrid 2, consisting of a transcription factor AD and an interacting protein, protein Y (the "prey"), lacks the BD. Therefore, hybrid 2 alone also cannot activate transcription of the reporter gene. Different ("prey") proteins expressed from cDNAs in vectors have been tested. Fusion genes encoding either hybrid 1 or hybrid 2 are produced using standard recombinant DNA methods. Cells are cotransfected with the genes. Only cells producing the correct hybrids, i.e., those in which the X and Y proteins interact and thereby reconnect AD and BD to form an active transcription factor, can initiate transcription of the reporter gene. This can be observed as a color change or by growth in selective medium. (Figure adapted from Oliver, 2000, and Frank Kaiser, Lübeck, personal communication.)

References

Li T et al: Crystal structure of the MATa1/MATα2 homeodomain heterodimer bound to DNA. Science 270: 262–269, 1995.

Lodish et al: Molecular Cell Biology, 4th ed. WH Freeman & Co, New York, 2000.

Oliver S: Guilt-by-association goes global. News & Views. Nature 403: 601–603, 2000.

Strachan T, Read AP: Human Molecular Genetics, 2nd ed. Bios Scientific Publishers, Oxford, 1999.

Uetz P et al: A comprehensive analysis of protein-protein interaction in *Saccharomyces cerevisiae*. Nature 403: 623–627, 2000 (and at http://www.curatools.curagen.com).

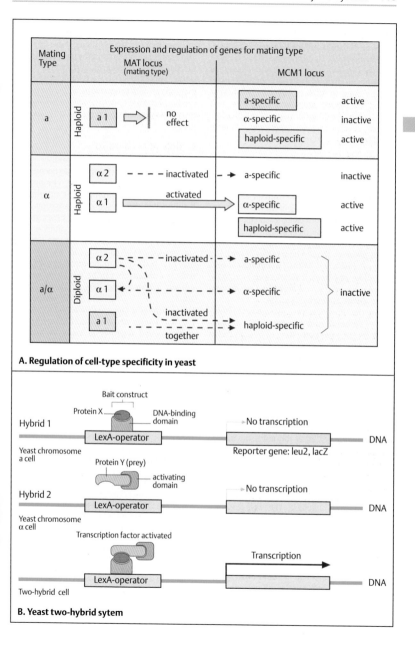

A. Regulation of cell-type specificity in yeast

B. Yeast two-hybrid sytem

Cell Division: Mitosis

Mitosis is the process of cell division. This term, introduced by W. Flemming in 1882, is derived from the Greek word mitos, a thread. Threadlike structures in dividing cells were first observed by Flemming in 1879. In 1884, E. Strasburger coined the terms *prophase*, *metaphase*, and *anaphase* for the different stages of cell division. Mitosis results in two genetically identical daughter cells.

A. Mitosis

A cell divides during a series of consecutive phases. The period between two divisions defines the cell cycle. The time from the end of one mitosis until the beginning of the next is the *interphase*. In eukaryotic cells, each cell division begins with a phase of DNA synthesis, which lasts about 8 hours (*S phase*). This is followed by the G_2 phase of about 4 hours (gap 2) until the onset of *mitosis* (M). Mitosis in eukaryotic cells lasts about one hour. It follows the interphase (G_1) of extremely varied duration. Cells that no longer divide are in the G_0 phase.

During the transition from interphase to mitosis, the chromosomes become visible as elongated threads, the first phase of mitosis called prophase. In early *prophase*, each chromosome is attached to a specific site on the nuclear membrane and appears as a double structure (sister chromatids), the result of the foregoing DNA synthesis. The chromosomes contract during late prophase to become thicker and shorter (chromosomal condensation). In late prophase, the nuclear membrane disappears and *metaphase* begins. At this point, the mitotic spindle becomes visible as thin threads. It begins at two polelike structures, the centrioles. The chromosomes become arranged on the equatorial plate, but homologous chromosomes do not pair. In late metaphase during the transition into *anaphase*, the chromosomes divide at the centromere region. The two chromatids of each chromosome migrate to opposite poles, and telophase begins with the formation of the nuclear membranes. Finally the cytoplasm also divides (cytokinesis). In early interphase the individual chromosomal structures become invisible in the cell nucleus.

B. Metaphase chromosomes

Waldeyer (1888) coined the term chromosome for the stainable threadlike structures visible during mitosis. A metaphase chromosome consists of two chromatids (*sister chromatids*) and the centromere, which holds them together. The regions at both ends of the chromosome are the telomeres. The point of attachment to the mitotic spindle fibers is the kinetochore. During metaphase and prometaphase, chromosomes can be visualized under the light microscope as discrete elongated structures, 3–7 μm long (see p. 76).

C. Role of condensins

The first sign that a cell is about to enter mitosis is that the replicated chromosomes become visible under the light microscope. The progressive compaction of chromosomes entering mitosis is called chromosome condensation. A mitotic chromosome is about 50 times shorter than during interphase. Chromosomes are condensed by proteins called condensins. Condensins consist of five subunits (not shown). They use ATP hydrolysis to promote chromosome coiling. Condensins can be visualized along a mitotic chromosome (Fig. 23.35 in Lewin, 2004).

When chromosomes are duplicated in S phase, the two copies of each chromosome remain tightly bound together as sister chromatids. They are held together by multiunit proteins called cohesins. Cohesins are structurally related to condensins. Mutations in fission yeast condensins interfere with mitosis. (Figure adapted from Uhlmann, 2002.)

Medical relevance

A mutation in any of the five genes encoding condensin subunits results in a severe growth and malformation syndrome, the Roberts syndrome (MIM 268300; Vega et al., 2005).

References

Karscenti E, Vernos I: The mitotic spindel: A self-made machine. Science 294: 543–547, 2001.

Nurse P: The incredible life and times of biological cells. Science 289: 1711–1716, 2000.

Rieder CL, Khodjakov A: Mitosis through the microscope: Advances in seeing inside live dividing cells. Science 300: 91–96, 2003.

Vega H et al: Roberts syndrome is caused by mutations in *ESCO2*, a human homolog of yeast *ECO1* that is essential for establishment of sister chromatid cohesion. Nature Genet 37: 468–470, 2005.

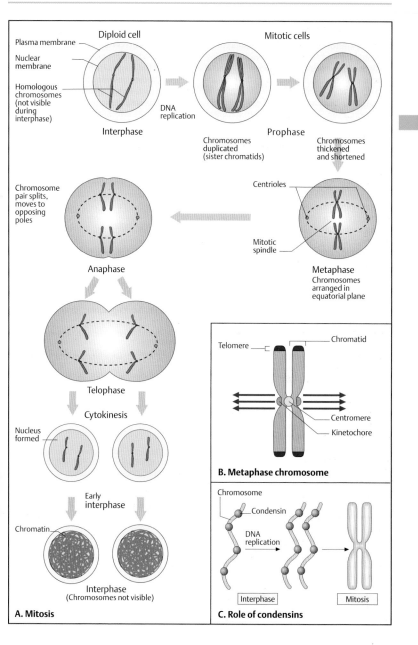

A. Mitosis

Plasma membrane

Nuclear membrane

Homologous chromosomes (not visible during interphase)

Diploid cell

Interphase

DNA replication

Mitotic cells

Prophase

Chromosomes duplicated (sister chromatids)

Chromosomes thickened and shortened

Centrioles

Mitotic spindle

Metaphase
Chromosomes arranged in equatorial plane

Chromosome pair splits, moves to opposing poles

Anaphase

Telophase

Cytokinesis

Nucleus formed

Early interphase

Chromatin

Interphase
(Chromosomes not visible)

B. Metaphase chromosome

Telomere

Chromatid

Centromere

Kinetochore

C. Role of condensins

Chromosome

Condensin

DNA replication

Interphase

Mitosis

Meiosis in Germ Cells

Meiosis, a term introduced by Strasburger in 1884 and derived from a Greek work meaning diminution (or maturation division), is a special kind of cell division that occurs in the production of egg and sperm cells. Meiosis consists of two nuclear divisions but only one round of DNA replication. As a result, the four daughter cells are haploid, i.e., they contain only one chromosome of each pair.

Meiosis differs from mitosis fundamentally in genetic and cytological respects. First, homologous chromosomes pair at prophase of the first division. Second, exchanges between homologous chromosomes (*crossing-over*) occur regularly. This results in chromosomes having segments of both maternal and paternal origin, the production of new combinations of genetic information being called *genetic recombination*. Third, the chromosome complement is reduced to half during the first cell division, meiosis I. Thus, the daughter cells resulting from this division are haploid. Each chromosome of a pair is distributed to its daughter cell independently of the other pairs (independent assortment).

Meiosis is a complex cellular and biochemical process. The cytologically observable course of events and the genetic consequences do not correspond exactly in time. A genetic process occurring in one phase usually becomes visible cytologically at a later phase.

A. Meiosis I

A gamete-producing cell goes through two cell divisions at meiosis, meiosis I and meiosis II. The relevant genetic events, genetic recombination by means of crossing-over and reduction to the haploid chromosome complement, occur in meiosis I. Meiosis begins with DNA replication. Initially the chromosomes in late interphase are visible only as threadlike structures. At the beginning of prophase I, the chromosomes are doubled. The pairing allows an exchange between homologous chromosomes (crossing-over), made possible by juxtapositioning homologous chromatids. At certain sites a chiasma forms. As a result of crossing-over, chromosome material of maternal and paternal origin is exchanged between two chromatids of homologous chromosomes. After the homologous chromosomes migrate to opposite poles, the cell enters anaphase I.

B. Meiosis II

Meiosis II consists of longitudinal division of the duplicated chromosomes (chromatids) and a further cell division. Each daughter cell is haploid, as it contains one chromosome of a pair only. On each chromosome, recombinant and nonrecombinant sections can be identified. The genetic events relevant to these changes have occurred in the prophase of meiosis I (see next page).

The independent distribution of chromosomes (independent assortment) during meiosis explains the segregation (separation or splitting) of observable traits according to the rules of Mendelian inheritance (1 : 1 segregation, see p. 140).

Medical relevance

Errors in the correct distribution of the chromosomes, called nondisjunction, result in gametes with an extra chromosome or a chromosome missing. After fertilization, the zygote will have either three homologous chromosomes (*trisomy*) or only one (*monosomy*). Both result in embryonic developmental disturbances (see p. 412).

References

Carpenter ATC: Chiasma function. Cell 77: 959–962, 1994.

Kitajima TS et al: Distinct cohesin complexes organize meiotic chromosome domains. Science 300: 1152–1155, 2003.

McKim KS, Hawley RS: Chromosomal control of meiotic cell division. Science 270: 1595–1601, 1995.

Moens PB (ed): Meiosis. Academic Press, New York, 1987.

Page SL, Hawley RS: Chromosome choreography: The meiotic ballet. Science 301: 785–789, 2003.

Petronczki M, Simons MF, Nasmyth K: Un ménage à quatre: The molecular biology of chromosome segragation in meiosis. Cell 112: 423–440, 2003.

Whitehouse LHK: Towards an Understanding of the Mechanism of Heredity, 3rd ed. Edward Arnold, London, 1973.

Zickler D, Kleckner N: Meiotic chromosomes: Integrating structure and function. Ann Rev Genet 33: 603–754, 1999.

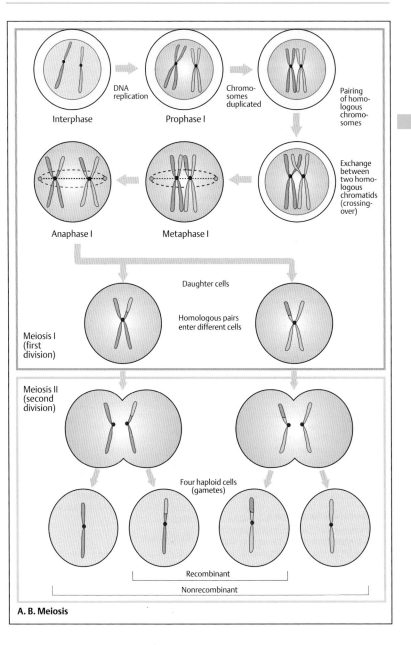

Interphase

DNA replication

Prophase I

Chromo-
somes
duplicated

Pairing
of homo-
logous
chromo-
somes

Exchange
between
two homo-
logous
chromatids
(crossing-
over)

Anaphase I

Metaphase I

Daughter cells

Homologous pairs
enter different cells

Meiosis I
(first
division)

Meiosis II
(second
division)

Four haploid cells
(gametes)

Recombinant

Nonrecombinant

A. B. Meiosis

Meiosis Prophase I

The decisive cytological and genetic events take place in the prophase of meiosis I. In prophase I, exchanges between homologous chromosomes occur regularly by crossing-over. Crossing-over, a term introduced by Morgan and Cattell in 1912, is an elaborate cytological process by which parts of chromosomes of maternal and paternal origin exchange stretches of DNA. This results in new combinations of chromosome segments (genetic recombination).

A. Prophase of meiosis I

The prophase of meiosis I passes through consecutive stages that can be differentiated schematically, although they proceed continuously. The first is the *leptotene* stage. Here the chromosomes first become visible as fine threadlike structures (in A only one chromosome pair is shown). Next is *zygotene*: each chromosome is visible as a paired structure, the result of DNA replication prior to the beginning of prophase. Consequently, each chromosome has been doubled and consists of two identical chromatids (sister chromatids). These are held together at the centromere. Each chromatid contains a DNA double helix. Two homologous chromosomes that have paired are referred to as a bivalent. In the *pachytene* stage, the bivalents become thicker and shorter. In *diplotene* the two homologous chromosomes separate, but still remain attached to each other at a few points, each called a chiasma (see below). In the next phase, *diakinesis*, each of the chromosome pairs has separated further, although they still remain attached to each other at the ends. A chiasma corresponds to a region at which crossing-over has previously taken place. However, in late diakinesis the chiasmata shift distally, called chiasma terminalization. The mechanisms of meiosis II correspond to those of a mitosis.

B. Synaptonemal complex

The synaptonemal complex, independently observed in spermatocytes by D. Fawcett und M.J. Moses in 1956, is a complex structure formed during meiotic prophase I. It consists of two chromatids (1 and 2) of maternal origin (mat) and two chromatids (3 and 4) of paternal origin (pat). It initiates chiasma formation and is the prerequisite for crossing-over and subsequent recombination. (Figure adapted from Alberts et al, 2002.)

C. Chiasma formation

Chiasma is the term introduced by F.A. Janssens in 1909 for the cytological manifestation of crossing-over during meiotic prophase I. A chiasma forms between one chromatid of a chromosome of maternal origin (chromatids 1 and 2 in the figure) and one chromatid of a chromosome of paternal origin (chromatids 3 and 4). Either of the two chromatids of one chromosome can cross over with one of the chromatids of the homologus chromosome (e.g., 1 and 3, 2 and 4, etc.). Chiasma formation is also important for the separation (segregation) of chromosomes.

D. Genetic recombination

Through crossing-over, new combinations of chromosome segments arise (recombination). As a result, recombinant and nonrecombinant chromosome segments can be differentiated. In the diagram, the areas A–E (shown in red) of one chromosome and the corresponding areas a – e (shown in blue) of the homologous chromosome become respectively a-b-C–D–E and A–B–c-d–e in the recombinant chromosomes.

E. Pachytene and diakinesis under the microscope

During pachytene and diakinesis the individual chromosomes can be readily visualized by light and electron microscopy. Here diakinesis under the light microscope (a) and pachytene under the electron microscope are shown. In (a) an extra chromosome 21 (red arrow) present in a man with trisomy 21 does not pair. In the male meiosis, the X and the Y chromosomes form an XY body. Pairing of the X and Y is limited to the extreme end of the short arms (see p. 256). In (b) the thickened (duplicated) chromosomes and the XY bivalent are visible.

(Figures kindly provided by Dr. R. Johannisson, Lübeck, Germany; (a) from Johannisson et al, 1983.)

References

Alberts B: Molecular Biology of the Cell, 4th ed. Garland Science, 2002.

Johannisson R et al: Down's syndrome in the male. Reproductive pathology and meiotic studies. Hum Genet 63: 132–138, 1983.

Miller OJ, Therman E: Human Chromosomes, 4th ed. Springer, New York-Berlin, 2001.

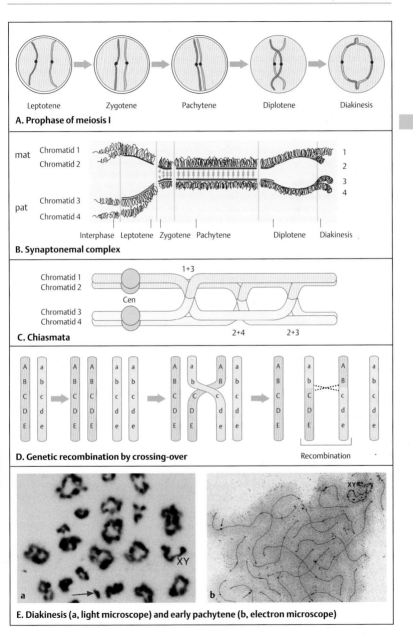

Leptotene   Zygotene   Pachytene   Diplotene   Diakinesis

A. Prophase of meiosis I

mat   Chromatid 1   1
      Chromatid 2   2
                                                                                      3
pat   Chromatid 3   4
      Chromatid 4

Interphase Leptotene Zygotene Pachytene   Diplotene   Diakinesis

B. Synaptonemal complex

Chromatid 1   1+3
Chromatid 2
                    Cen
Chromatid 3
Chromatid 4   2+4   2+3

C. Chiasmata

D. Genetic recombination by crossing-over   Recombination

E. Diakinesis (a, light microscope) and early pachytene (b, electron microscope)

Formation of Gametes

Gametes (germ cells) are produced in the gonads. In females the process is called oogenesis (formation of oocytes) and in males, spermatogenesis (formation of spermatozoa). Primordial germ cells migrate during early embryonic development from the genital ridge to the gonads. Here they increase in number by mitotic divisions. The actual formation of germ cells (gametogenesis) begins with meiosis. Gametogenesis in males and females differs in duration and results.

A. Spermatogenesis

Spermatogonia are diploid cells that go through mitotic divisions in the gonads of male animals. The primary spermatocytes result from the first meiotic division beginning at the onset of puberty. At the completion of meiosis I, one primary spermatocyte gives rise to two secondary spermatocytes. Each has a haploid set of duplicated chromosomes. In meiosis II, each secondary spermatocyte divides to form two spermatids. Thus, one primary spermatocyte forms four spermatids, each with a haploid chromosome complement. The spermatids differentiate into mature spermatozoa in about 6 weeks. Male spermatogenesis is a continuous process. In human males, the time required for a spermatogonium to develop into a sperm cell is about 90 days.

B. Oogenesis

Oogenesis is the formation of eggs (oocytes) in females. It differs from spermatogenesis in timing and result. During early embryogenesis the germ cells migrate from the genital ridge to the ovary, where they form oogonia by repeated mitoses. A primary oocyte results from the first meiotic division of an oogonium. In human females, meiosis I begins about 4 weeks before birth. Then meiosis I is arrested in a stage of prophase designated *dictyotene*. The primary oocyte persists in this stage until ovulation. Only then is meiosis I continued.

In primary oocytes the cytoplasm divides asymmetrically in both meiosis I and meiosis II. The result each time is two cells of unequal size. One cell is larger and will eventually form the egg; the other, smaller becomes the polar body I. When the secondary oocyte divides, the daughter cells again differ; one secondary oocyte and another polar body (polar body II)

are the result. The polar bodies degenerate and do not develop. On rare occasions when this does not occur, a polar body may become fertilized. This can give rise to an incompletely developed twin. In the secondary oocyte, each chromosome still consists of two sister chromatids. These do not separate until the next cell division (meiosis II), when they enter into two different cells. In most vertebrates, maturation of the secondary oocyte is arrested in meiosis II. At ovulation the secondary oocyte is released from the ovary, and if fertilization occurs, meiosis is then completed.

The maximal number of germ cells in the ovary of the human fetus at about the 5th month is 6.8×10^6. By the time of birth, this has been reduced to 2×10^6, and by puberty to about 200 000. Of these, about 400 eventually go through ovulation.

Medical relevance

Most new mutations occur during gametogenesis. The difference in time in the formation of gametes during oogenesis and spermatogenesis is reflected in the difference in germline cell divisions. The number of cell divisions in spermatogenesis and oogenesis differs considerably. On average about 380 chromosome replications have taken place in the progenitor cells of spermatozoa by age 30 years, and about 610 chromosome replications by age 40. Altogether, 25 times more cell divisions occur during spermatogenesis than during oogenesis (Crow, 2000). This probably accounts for the higher mutation rate in males, especially with increased paternal age. In the female, on average 22 mitotic cell divisions occur before meiosis, resulting in a total of 23 chromosome replications.

Faulty distribution of the chromosomes (nondisjunction) during meiosis I or meiosis II is the cause of aberrations of the chromosome number (see p. 412).

References

Alberts B et al: Molecular Biology of the Cell, 4th ed. Garland Science, New York, 2002.

Crow JF: The origins, patterns and implications of human spontaneous mutation. Nature Rev Genet 1: 40–47, 2000.

Hurst LD, Ellegren H: Sex biases in the mutation rate. Trends Genet 14: 446–452, 1998.

Miller OJ, Therman E: Human Chromosomes, 4th ed. Springer, New York-Berlin, 2001.

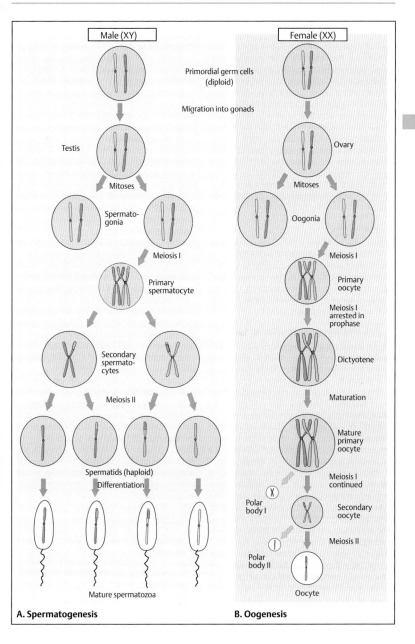

A. Spermatogenesis   **B. Oogenesis**

Cell Cycle Control

The cell cycle as first defined by A. Howard and S. R. Pelc in 1953 has two main phases, interphase and mitosis. During DNA replication and cell division multiple types of errors can occur. A faulty cell division would have grave consequences for the daughter cells and the whole organism. An elaborate control system has evolved that can detect errors, eliminate them, or abandon defective cells. Cell cycle control mechanisms include complex sets of interacting proteins. These guide the cell through its cycle by regulating the sequential cyclical events. These are coordinated with extracellular signals and result in cell division at the right time.

A. Yeast cells

Genetic studies in yeast cells (see p. 112) have provided important insights into cell cycle control. Baker's yeast (*Saccharomyces cerevisiae*) and fission yeast (*Schizosaccharomyces pombe*) have cell cycle control mechanisms that are similar to those of higher eukaryotes. (Images obtained from Google, July 2005.)

B. Cell division cycle models in yeast

Mitotic division in budding yeast (baker's yeast) results in one large and one small daughter cell. Since a microtubule mitotic spindle forms very early during the S phase, there is practically no G_2 phase (**1**). In contrast, fission yeast *(S. pombe)* forms a mitotic spindle at the end of the G_2 phase, then proceeds to mitosis to form two daughter cells of equal size (**2**). Unlike vertebrate cells, the nuclear envelope remains intact during mitosis. An important regulator of yeast cell division is cdc2 protein (cell division cycle-2). Absence of cdc2 activity (cdc2 mutants) in *S. pombe* results in cycle delay and prevents entry into mitosis (**3**). Thus, too large a cell with only one nucleus results. Increased activity of cdc2 (dominant mutant *cdc^D*) results in premature mitosis and cells that are too small (wee phenotype, from the Scottish word for small). Normally a yeast cell has three options: (i) halt the cell cycle if the cell is too small or nutrients are scarce, (ii) mate (see p. 114), or (iii) enter mitosis. (Figures adapted from Lodish et al., 2004.)

B. Cell cycle control systems

The eukaryotic cell cycle is driven by cell cycle "engines," a set of interacting proteins, the cyclin-dependent kinases (Cdks). An important member of this family of proteins is cdc2 (also called Cdk1). Other proteins act as rate-limiting steps in cell cycle progression and are able to induce cell cycle arrest at defined stages (checkpoints). The cell is induced to progress through G_1 by growth factors (mitogens) acting through receptors that transmit signals to proceed towards the S phase. D-type cyclins (D1, D2, D3) are produced, which associate with and activate Cdks 4 and 6. Other proteins can induce G_1 arrest. The detection of DNA damage and subsequent cell cycle arrest due to activated p53 is an important mechanism for preventing the cell from entering the S phase.

In early G_1 phase, cdc2 is inactive. It is activated in late G_1 by associating with G_1 cyclins, such as cyclin E. Once the cell has passed the G_1 restriction point, cyclin E is degraded and the cell enters the S phase. This is initiated, among many other activities, by cyclin A binding to Cdk2 and phosphorylation of the RB protein (retinoblastoma protein, see p. 334). The cell passes through the mitosis checkpoint only if no damage is present. Cdc2 (Cdk1) is activated by association with mitotic cyclins A and B to form the mitosis-promoting factor (MPF).

During mitosis, cyclins A and B are degraded, and an anaphase-promoting complex forms (details not shown). When mitosis is completed, cdc2 is inactivated by the S-phase inhibitor Sic1 in yeast. At the same time the retinoblastoma (RB) protein is dephosphorylated. Cells can progress to the next cell cycle stage only when feedback controls have ensured the integrity of the genome. (This figure is an overview only; it omits many important protein transactions.)

References

Hartwell L, Weinert T: Checkpoints: Controls that ensure the order of cell cycle events. Science 246: 629–634, 1989.

Howard A, Pelc S: Synthesis of deoxyribonucleic acid in normal and irradiated cells and its relation to chromosome breakage. Heredity 6 (Suppl.): 261–273, 1953.

Nurse P: A long twentieth century of the cell cycle and beyond. Cell 100: 71–78, 2000.

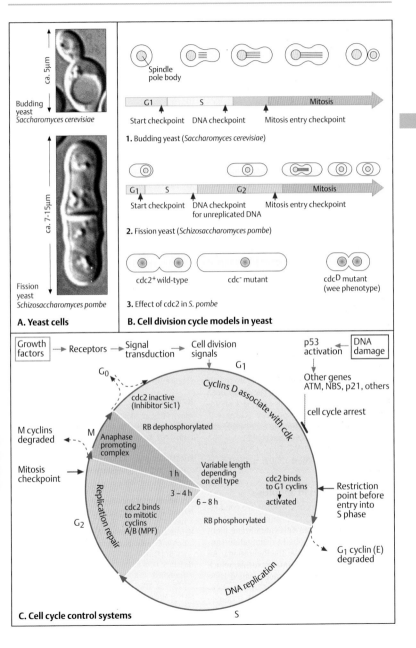

A. Yeast cells

Budding yeast
Saccharomyces cerevisiae

ca. 5µm

Fission yeast
Schizosaccharomyces pombe

ca. 7–15µm

B. Cell division cycle models in yeast

Spindle pole body

G1 | S | Mitosis

Start checkpoint | DNA checkpoint | Mitosis entry checkpoint

1. Budding yeast (*Saccharomyces cerevisiae*)

G1 | S | G2 | Mitosis

Start checkpoint | DNA checkpoint for unreplicated DNA | Mitosis entry checkpoint

2. Fission yeast (*Schizosaccharomyces pombe*)

cdc2+ wild-type | cdc− mutant | cdcD mutant (wee phenotype)

3. Effect of cdc2 in *S. pombe*

C. Cell cycle control systems

Growth factors → Receptors → Signal transduction → Cell division signals

G1

p53 activation ← DNA damage

Other genes ATM, NBS, p21, others

cell cycle arrest

G0

Cyclins D associate with cdk

cdc2 inactive (Inhibitor Sic1)

RB dephosphorylated

M cyclins degraded

M

Anaphase promoting complex

Mitosis checkpoint

Variable length depending on cell type

1 h

3 – 4 h

6 – 8 h

cdc2 binds to G1 cyclins activated

Restriction point before entry into S phase

Replication repair

G2

cdc2 binds to mitotic cyclins A/B (MPF)

RB phosphorylated

G1 cyclin (E) degraded

DNA replication

S

Programmed Cell Death

At certain stages in the development of a multi-cellular organism, some cells must die. This well-regulated process is called called *apoptosis* (programmed cell death), as suggested by Kerr in 1972. The importance of this biological phenomenon was first realized in studies of a tiny worm, the soil nematode *C. elegans* (see p. 304). If apoptosis does not occur, developmental failure of the organism or cancer may result. Apoptosis is regulated in the apoptosis pathway by many proteins, which either trigger or prevent apoptosis. Apoptosis can be triggered by a variety of stimuli that act either from outside the cell (extrinsic pathway) or from within the cell (intrinsic pathway). External stimuli may be irradiation, withdrawal of essential growth factors, or glucocorticoids. An intrinsic stimulus may be spontaneous damage to the DNA of the cell.

A. Importance of apoptosis

Apoptosis occurs mainly during development. For example, the digits in the developing mammalian embryo are sculptured by apoptosis (**1**). The paws (hands in humans) start out as spade-like structures. The formation of digits requires that cells between them die (here shown as bright green dots on the left). More staggering is the amount of apoptosis in the developing vertebrate nervous system. Normally up to half of the nerve cells die soon after they have been formed. In the embryos of mice that lack an important gene regulating apoptosis (caspase 9, see below), neurons proliferate excessively and the brain protrudes above the face (**2**).
(Illustration in 1 modified from Alberts et al., 2002; in 2 from Gilbert, 2003, according to Kaida et al., 1998.)

B. Cellular events in apoptosis

The first visible signs of apoptosis are condensation of chromatin and shrinking of the cell. The cell membrane shrivels (membrane blebbing), and the cell begins to disintegrate (nuclear segmentation, DNA fragmentation). An apoptotic body of cell remnants forms, which eventually dissolves by a process called lysis. (Figure modified from Dr. A. J. Cann; Microbiology, Leicester University, displayed at Google Images, 22 March, 2005.)

C. Regulation of apoptosis

Specialized cysteine-containing aspartate proteinases, called caspases, play a central role. They activate or inactivate each other in a defined sequence. Binding of a ligand, Fas, of a cytotoxic T cell (see section on immune system) to the Fas receptor (also called CD95) activates an intracellular adaptor protein, FADD (Fas-associated death domain). This binds to and activates procaspase 8 into active caspase 8. Caspase 8 causes release of cytochrome *c* in mitochondria (see p. 130) and activates several different effector caspases. Downstream of caspase 8, two pathways exist. In type I cells (in thymocytes and fibroblasts), caspase 8 directly activates caspase 3. In type II cells, such as hepatocytes, caspase 8 cleaves Bid, a member of the Bcl-2 family.

The mouse and human genomes contain 13 caspase genes (1–12 and 14; see table in the appendix). Human caspases 3 and 6–10 are involved in apoptosis, the others in inflammation (Nagata, 2005). Caspase 8 also serves as a selective signal transducer for nuclear factor KB (NF-kB) during the early genetic response to an antigen (Su et al., 2005). Other regulators of apoptosis are members of the Bcl-2 family (the name Bcl is derived from B-cell lymphoma, a human malignant tumor originating from B-lymphocytes and caused by mutations in this gene). (Figure from Koolman & Röhm, 2005.)

References

Alberts B et al: Molecular Biology of the Cell, 4th ed. Garland Science, New York, 2002.

Danial NN, Korsmeyer SJ: Cell death: Critical control points. Cell 116: 205–219, 2004.

Friedlander RM: Apoptosis and caspases in neurodegenerative diseases. New Eng J Med 348: 1365–1375, 2003.

Gilbert SF: Developmental Biology, 7th ed. Sinauer, Sunderland, Massachusetts, 2003.

Hengartner MD: The biochemistry of apoptosis. Nature 407: 770–776, 2000.

Kerr JF et al: Apoptosis: a basic biological phenomenon with wide ranging implications in tissue kinetics. Br J Cancer 26: 239–257, 1972.

Nagata S: DNA degradation in development and programmed cell death. Ann Rev Immunol. 23: 821–852, 2005.

Su H et al: Requirement for caspase-8 in NF-kB activation by antigen receptor. Science 307: 1465–1468, 2005.

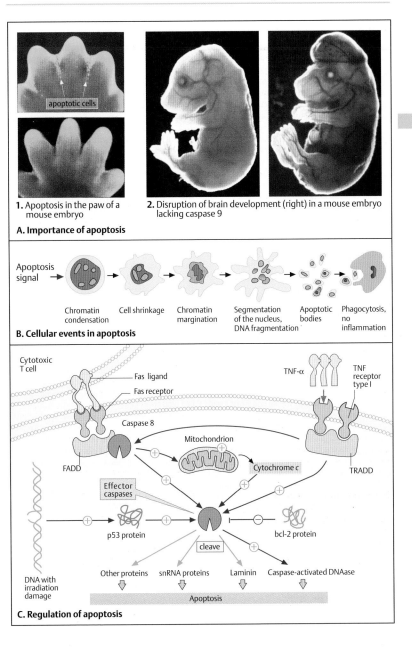

1. Apoptosis in the paw of a mouse embryo

2. Disruption of brain development (right) in a mouse embryo lacking caspase 9

apoptotic cells

A. Importance of apoptosis

Apoptosis signal →

Chromatin condensation | Cell shrinkage | Chromatin margination | Segmentation of the nucleus, DNA fragmentation | Apoptotic bodies | Phagocytosis, no inflammation

B. Cellular events in apoptosis

Cytotoxic T cell

Fas ligand

Fas receptor

Caspase 8

Mitochondrion

TNF-α

TNF receptor type I

FADD

Cytochrome c

TRADD

Effector caspases

p53 protein

bcl-2 protein

cleave

DNA with irradiation damage

Other proteins | snRNA proteins | Laminin | Caspase-activated DNAase

Apoptosis

C. Regulation of apoptosis

Cell Culture

Cells of animals and plants can live and multiply in a tissue-culture dish as a cell culture at 37°C in a medium containing vitamins, sugar, serum (containing numerous growth factors and hormones), the nine essential amino acids for vertebrates (His, Ile, Leu, Lys, Met, Phe, Thr, Tyr, Val), and usually also glutamine and cysteine. A great variety of growth media is available for culturing mammalian cells. Modern applications of cell cultures began 1940 when W. Earle established a permanent mouse cell strain, and T.T. Puck and co-workers grew clones of human cells *in vitro*. Cell cultures have been widely used since 1965 in *somatic cell genetics*.

The predominant cell type that grows from a piece of mammalian tissue in culture is the fibroblast. Cultured skin fibroblasts have a finite life span (Hayflick, 1997). Human cells have a capacity for about 30 doublings until they reach a state called senescence. Cells derived from adult tissues have a shorter life span than those derived from fetal tissues.

Cultured cells are highly sensitive to increased temperature and do not survive above about 39°C, whereas in special conditions they can be stored alive in vials kept in liquid nitrogen at -196°C. They can be thawed after many years or even decades and cultured again.

A. Skin fibroblast culture

To initiate a culture, a small piece of skin (2 × 4 mm) is obtained in sterile conditions and cut into smaller pieces, which are placed in a culture dish. The pieces have to attach to the bottom of the dish. After about 8–14 days, cells begin to grow out from each piece and begin to multiply. They grow and multiply only when adhering to the bottom of the culture vessel (adhesion culture due to anchorage dependency of the cells). When the bottom of the culture vessel is covered with a dense layer of cells, they stop dividing owing to contact inhibition (this is lost in tumor cells). When transferred into new culture vessels (subcultures), the cells will resume growing until they again become confluent. With a series of subcultures, several million cells can be obtained for a given study.

B. Hybrid cells for study

Cells in culture can be induced to fuse by exposure to polyethylene glycol or Sendai virus. If parental cells from different species are fused, interspecific (from different animal species) hybrid cells can be derived. The hybrid cells can be distinguished from the parental cells by using parental cells deficient in thymidine kinase (TK⁻) or hypoxanthine phosphoribosyltransferase (HPRT⁻). When cell cultures of parental type A (TK⁻, **1**) and type B (HPRT⁻, **2**) are cultured together (co-cultivation, **3**), some cells fuse (**4**). In a selective medium containing hypoxanthine, aminopterin, and thymidine (HAT, Littlefield, 1964) only fused cells with a nucleus from each of the parental cells (**1** and **2**) can grow. Cells that have not fused cannot grow in HAT medium (**5**). The reason for this is that TK⁻ cell cannot synthesize thymidine monophosphate; HPRT⁻ cells cannot synthesize purine nucleoside monophosphates. The fused cells complement each other. The two nuclei of fused cells (heterokaryon) will also fuse (**6**). This forms a hybrid cell (**7**). Hybrid cells containing two different sets of chromosomes, one from each of the parental cells, are cultured. During cell divisions in culture, each cell loses chromosomes at random. At the end, cells with different sets of chromosomes result.

C. Radiation hybrids

A radiation hybrid is a rodent cell containing small fragments of chromosomes from another organism (McCarthy, 1996). When human cells are irradiated with lethal roentgen doses of 3–8 Gy, the chromosomes break into small pieces (**1**) and the cells cannot divide in culture. However, if these cells are fused with nonirradiated rodent cells (**2**), some human chromosome fragments will be integrated into the rodent chromosomes (**3**). Cells containing human DNA can be identified by human chromosome-specific probes.

References

Alberts B et al: Molecular Biology of the Cell, 4th ed. Garland Science, New York, 2002.

Brown TA: Genomes, 2nd ed. Bios Scientific Publishers, Oxford, 2002.

Hayflick L: Mortality and immortality at the cellular level. Biochemistry 62: 1180–1190, 1997.

Lodish H et al: Molecular Cell Biology, 5th ed. WH Freeman & Co, New York, 2000.

McCarthy L: Whole genome radiation hybrid mapping. Trends Genet 12: 491–493, 1996.

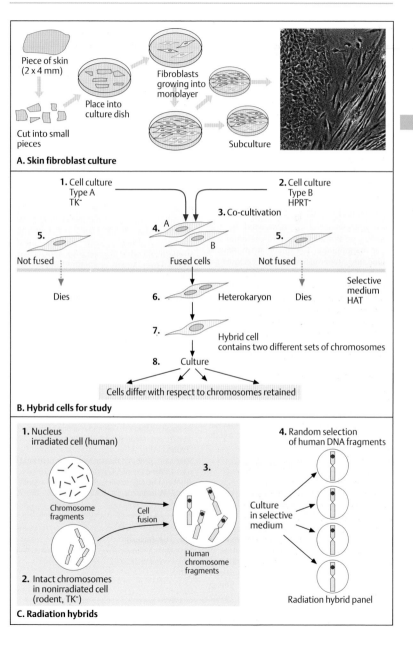

A. Skin fibroblast culture

Piece of skin (2 x 4 mm)

Cut into small pieces

Place into culture dish

Fibroblasts growing into monolayer

Subculture

B. Hybrid cells for study

1. Cell culture Type A TK⁻

2. Cell culture Type B HPRT⁻

3. Co-cultivation

4. A / B

5. Not fused

Fused cells

5. Not fused

Dies

Dies

Selective medium HAT

6. Heterokaryon

7. Hybrid cell contains two different sets of chromosomes

8. Culture

Cells differ with respect to chromosomes retained

C. Radiation hybrids

1. Nucleus irradiated cell (human)

Chromosome fragments

Cell fusion

3.

Human chromosome fragments

2. Intact chromosomes in nonirradiated cell (rodent, TK⁻)

4. Random selection of human DNA fragments

Culture in selective medium

Radiation hybrid panel

Mitochondria: Energy Conversion

A mitochondrion is a semiautonomous, self-reproducing organelle in the cytoplasm of eukaryotic cells. It has a diameter of $1–2\,\mu m$ and contains multiple copies of circular mitochondrial DNA (mtDNA) of 16569 base pairs in man. The number of mitochondria per cell and their shape differ in different cell types and can change. An average eukaryotic cell contains $10^3–10^4$ copies of mitchondria. Mitochondria in animal cells and chloroplasts in plant cells are the sites of essential energy-delivering processes, chloroplasts also being the sites of photosynthesis. Human mtDNA encodes 13 proteins of the respiratory chain.

A. Principal events in mitochondria

Each mitochondrion is surrounded by two highly specialized membranes, the outer and inner membranes. The inner membrane is folded into numerous cristae and encloses the matrix space.

The essential energy-generating process in mitochondria is oxidative phosphorylation (OXPHOS). Relatively simple energy carriers such as NADH and $FADH_2$ (nicotinamide adenine dinucleotide in the reduced form and flavin adenine dinucleotide in the reduced form) are produced from the degradation of carbohydrates, fats, and other foodstuffs by oxidation. The important energy carrier adenosine triphosphate (ATP) is formed by oxidative phosphorylation of adenosine diphosphate (ADP) through a series of biochemical reactions in the inner membrane of mitochondria (respiratory chain). Another important function is intracellular oxygen transfer.

B. Oxidative phosphorylation in mitochondria

Adenosine triphosphate (ATP) plays a central role in the conversion of energy in biological systems. It is formed from NADH (nicotinamide adenine dinucleotide) and adenosine diphosphate (ADP) by oxidative phosphorylation (OXPHOS). ATP is a nucleotide consisting of adenine, a ribose, and a triphosphate unit. It is energy-rich because the triphosphate unit contains two phospho-anhydride bonds. Energy (free energy) is released when ATP is hydrolyzed to form ADP. The energy contained in ATP and bound to phosphate is released, for example, during muscle contraction.

C. Electron transfer in the inner mitochondrial membrane

The genomes of mitochondria and chloroplasts contain genes for the formation of the different components of the respiratory chain and oxidative phosphorylation. Three enzyme complexes regulate electron transfer: the NADH-dehydrogenase complex, the $b–c_1$ complex, and the cytochrome oxidase complex (C). Intermediaries are quinone (Q) derivatives such as ubiquinone and cytochrome c. Electron transport leads to the formation of protons (H^+). These lead to the conversion of ADP and P_i (inorganic phosphate) into ATP (oxidative phosphorylation). ATP represents a phosphate-bound reservoir of energy, which serves as an energy supplier for all biological systems. This is the reason why genetic defects in mitochondria become manifest primarily as diseases with reduced muscle strength and other degenerative signs. (Figures adapted from Alberts et al., 1998, and Koolman & Röhm, 2005.)

References

Alberts B et al: Essential Cell Biology. An Introduction to the Molecular Biology of the Cell. Garland Publishing, New York, 1998.

Chinnery PE, Turnbull DM: The epidemiology and treatment of mitochondrial diseases. J Med Genet 106: 94–101, 2001.

Johns DR: Mitochondrial DNA and disease. New Eng J Med 333: 638–644, 1995.

Kogelnik AM et al: MITOMAP: a human mitochondrial genome database—1998 update. Nucl Acids Res 26: 112–115, 1998.

Koolman J, Röhm KH: Color Atlas of Biochemistry, 2nd ed. Thieme Medical Publishers, Stuttgart-New York, 2005.

Strachan T, Read AP: Human Molecular Genetics, 3rd ed. Garland Science, London-New York, 2004.

Turnball DM, Lighttowlers RN: An essential guide to mtDNA maintenance. Nature Genet 18: 199–200, 1998.

Wallace DC: Mitochondrial diseases in man. Science 283: 1482–1488, 1999.

Online information:
MITOMAP: A human mitochondrial genome database (www.mitomap.org/).

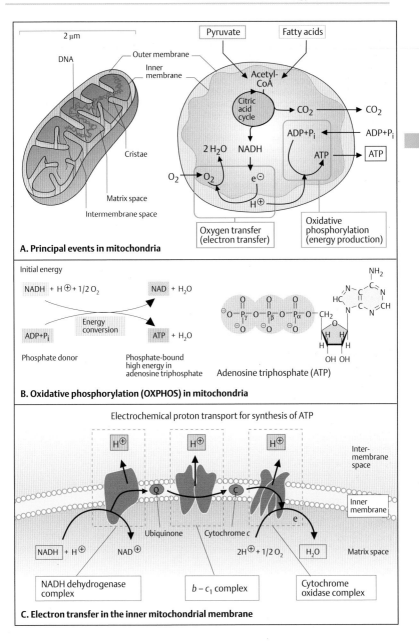

2 μm

DNA

Outer membrane

Inner membrane

Cristae

Matrix space

Intermembrane space

Pyruvate   Fatty acids

Acetyl-CoA

Citric acid cycle

CO_2 CO_2

ADP+P_i ADP+P_i

$2 H_2O$ NADH   ATP   ATP

O_2 O_2 e⊖

H⊕

Oxygen transfer (electron transfer)   Oxidative phosphorylation (energy production)

A. Principal events in mitochondria

Initial energy

NADH + H⊕ + 1/2 O_2 NAD + H_2O

Energy conversion

ADP+P_i ATP + H_2O

Phosphate donor   Phosphate-bound high energy in adenosine triphosphate

Adenosine triphosphate (ATP)

B. Oxidative phosphorylation (OXPHOS) in mitochondria

Electrochemical proton transport for synthesis of ATP

H⊕   H⊕   H⊕   Inter-membrane space

Q   C   Inner membrane

e

NADH + H⊕   NAD⊕   Ubiquinone   Cytochrome c $2H⊕ + 1/2 O_2$ H_2O Matrix space

NADH dehydrogenase complex $b - c_1$ complex   Cytochrome oxidase complex

C. Electron transfer in the inner mitochondrial membrane

Chloroplasts and Mitochondria

Chloroplasts are organelles of plant cells. In contrast to mitochondria, chloroplasts contain a third membrane, the *thylakoid* membrane. On this membrane, photosynthesis takes place. Chloroplasts and mitochondria of eukaryotic cells contain genomes of circular DNA. Chloroplast DNA ranges from about 120000 to 160000 base pairs, depending on the species. It encodes about 120 genes, of which half are involved in DNA-processing functions (transcription, translation, rRNAs, tRNAs, RNA polymerase subunits, and ribosomal proteins). The base sequences of several chloroplast DNAs have been determined. About 12000 base pairs (12 kb) of the genomes of chloroplasts and mitochondria are homologous. Chloroplasts are assumed to be descendants of endosymbiotic cyanobacteria.

A. Genes in the chloroplasts of a moss

Genes in the chloroplast genomes are interrupted and contain introns. Each chloroplast contains about 20–40 copies of chloroplast DNA (ctDNA), and there about are 20–40 chloroplasts per cell. The *M. polymorphia* chloroplast genome contains about 120 genes. Among these are genes for two copies each of four ribosomal RNAs (16 S rRNA, 23 S rRNA, 4.5 S rRNA, and 5 S rRNA). The genes for ribosomal RNA are located in two DNA segments with opposite orientation (inverted repeats), which are characteristic of chloroplast genomes. An 18–19-kb segment with short single-gene copies lies between the two inverted repeats. The genomes of chloroplasts contain genetic information for about 30 tRNAs and about 50 proteins. The proteins belong to photosystem I (two genes), photosystem II (seven genes), the cytochrome system (three genes), and the H^+-

ATPase system (six genes). The NADH dehydrogenase complex is coded for by six genes; ferredoxin by three genes, and ribulose by one gene. Many of the ribosomal proteins are homologous to those of *E. coli*. (Figure adapted from Alberts et al, 1994.)

B. Mitochondrial genes in yeast

The mitochondrial genome of yeast is large (120 kb). Its genes contain introns. It contains genes for the tRNAs, for the respiratory chain (cytochrome oxidase 1, 2, and 3; cytochrome *b*), for 15 S and 21 S rRNA, and for subunits 6, 8, and 9 of the ATPase system. The yeast mitochondrial genome is remarkable because its ribosomal RNA genes are separated. The gene for 21 S rRNA contains an intron. About 25% of the mitochondrial genome of yeast contains AT-rich DNA without a coding function.

The genetic code of the mitochondrial genome differs from the universal code in nuclear DNA with respect to usage of some codons. The nuclear stop codon UGA codes for tryptophan in mitochondria, while the nuclear codons for arginine (AGA and AGG) function as stop codons in mammalian mitochondria.

References

Alberts B et al: Molecular Biology of the Cell, 3rd ed. Garland Publishing, New York, 1994.

Borst P, Grivell LA: The mitochondrial genome of yeast. Cell 15: 705–723, 1978.

Foury F, Roganti T, Lecrenier N, Purnelle B: The complete sequence of the mitochondrial genome of *Saccharomyces cerevisiae*. FEBS Letters 440: 325–331, 1998.

Ohyama K et al: Chloroplast gene organization deduced from complete sequence of liverwort *M. polymorpha* chloroplast DNA. Nature 322: 572–574, 1986.

Differences between the universal genetic code and mitochondrial codes

Codon	Universal code	Mitochondrial codes			
		Mammals	Invertebrates	Yeasts	Plants
UGA	Stop	Trp	Trp	Trp	Stop
AUA	Ile	Met	Met	Met	Ile
CUA	Leu	Leu	Leu	Thr	Leu
AGA/AGG	Arg	Stop	Ser	Arg	Arg

(Data from Alberts et al, 2002)

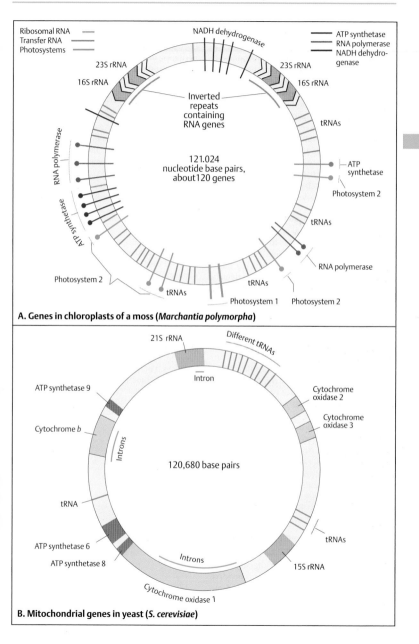

A. Genes in chloroplasts of a moss (*Marchantia polymorpha*)

B. Mitochondrial genes in yeast (*S. cerevisiae*)

The Mitochondrial Genome of Man

The mitochondrial genome in mammals is small and compact. It contains no introns, and in some regions the genes overlap, so that practically every base pair is part of a coding gene. The mitochondrial genomes of humans and mice have been sequenced and contain extensive homologies. Each consists of about 16.5 kb, i.e., they are considerably smaller than a yeast mitochondrial or a chloroplast genome. In germ cells, mitochondria are almost exclusively present in oocytes, whereas spermatozoa contain few. Thus, they are inherited from the mother, through an oocyte (maternal inheritance).

A. Mitochondrial genes in man

The human mitochondrial genome, sequenced in 1981 by Andersen et al., has 16 569 base pairs. Each mitochondrion contains 2–10 DNA molecules. A heavy (H) and a light (L) single strand can be differentiated by a density gradient. Human mtDNA contains 13 protein-coding regions for four metabolic processes: (i) for NADH dehydrogenase; (ii) for the cytochrome c oxidase complex (subunits 1, 2, and 3); (iii) for cytochrome b; and (iv) for subunits 6 and 8 of the ATPase complex. Unlike that of yeast, mammalian mitochondrial DNA contains seven subunits for NADH dehydrogenase (ND1, ND2, ND3, ND4 L, ND4, ND5, and ND6). Of the mitochondrial coding capacity, 60% is taken up by the seven subunits of NADH reductase (ND).

Most genes are found on the H strand. The L strand codes for a protein (ND subunit 6) and 8 tRNAs. From the H strand, two RNAs are transcribed, a short one for the rRNAs and a long one for mRNA and 14 tRNAs. A single transcript is made from the L strand. A 7 S RNA is transcribed in a counterclockwise manner close to the origin of replication (ORI), located between 11 and 12 o'clock on the circular structure.

B. Cooperation between mitochondrial and nuclear genome

Many mitochondrial proteins are aggregates of gene products of nuclear and mitochondrial genes. These gene products are transported into the mitochondria after nuclear transcription and cytoplasmic translation. In the mitochondria, they form functional proteins from subunits of mitochondrial and nuclear gene products. This explains why a number of mitochondrial genetic disorders show Mendelian inheritance, while purely mitochondrially determined disorders show exclusively maternal inheritance.

C. Evolutionary relationship of mitochondrial genomes

Mitochondria probably evolved from independent organisms that were integrated into cells. Similarities in structure and function between DNA in mitochondria, nuclear DNA, and DNA in chloroplasts suggest evolutionary relationships, in particular from chloroplasts to mitochondria, and from both to nuclear DNA of eukaryotic organisms.

References

Anderson S et al: Sequence and organization of the human mitochondrial genome. Nature 290: 457–474, 1981.

Chinnery PF: Searching for nuclear-mitochondrial genes. Trends Genet 19: 60–62, 2003.

Lang BF et al: Mitochondrial genome evolution and the origin of eukaryotes. Ann Rev Genet 33: 351–397, 1999.

Singer M, Berg P: Genes and Genomes. Blackwell Scientific Publishers, Oxford, 1991.

Suomalainen A et al: An autosomal locus predisposing to deletions of mitochondrial DNA. Nature Genet 9: 146–151, 1995.

Wallace DC: Mitochondrial diseases: genotype versus phenotype. Trends Genet 9: 128–133, 1993.

Wallace DC: Mitochondrial DNA sequence variation in human evolution and disease. Proc Nat Acad Sci 91: 8739–8746, 1994.

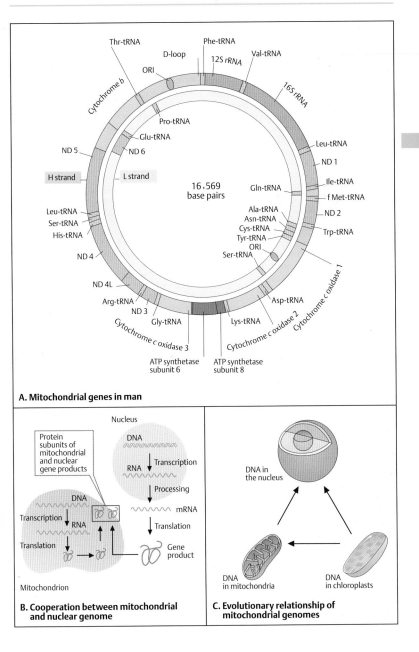

A. Mitochondrial genes in man

B. Cooperation between mitochondrial and nuclear genome

C. Evolutionary relationship of mitochondrial genomes

Mitochondrial Diseases

A large, complex, and heterogeneous group of diseases is caused by mutations or deletions in human mtDNA. The clinical spectrum and age of onset of mitochondrial diseases vary widely. Organs with high-energy requirements are particularly vulnerable to mitochondrial disorders: the brain, heart, skeletal muscle, eye, ear, liver, pancreas, and kidney. Normally, acquired mitochondrial mutations accumulate with age. Mitochondrial mutations are transmitted by maternal inheritance.

The mutation rate of mitochondrial DNA is ten times higher than that of nuclear DNA. Mitochondrial mutations are generated during oxidative phosphorylation through pathways involving reactive oxygen molecules. Mutations accumulate because effective DNA repair and protective histones are lacking. At birth most mtDNA molecules are identical (*homoplasmy*); later they differ as a result of mutations accumulated in different mitochondria (*heteroplasmy*).

A. Mutations and deletions in mitochondrial DNA in man

Both deletions and point mutations are causes of mitochondrial genetic disorders. Some are characteristic and recur in different, unrelated patients. Panel A and the table in the appendix show examples of important mutations and deletions and mitochondrial diseases. (Figure adapted from Wallace, 1999; MITOMAP; and Marie T. Lott and D. C. Wallace, personal communication.)

B. Maternal inheritance of a mitochondrial disease

Hereditary mitochondrial diseases are transmitted only through the maternal line, since spermatozoa contain hardly any mitochondria. Thus, the disease will not be transmitted from an affected man to his children.

C. Heteroplasmy for mitochondrial mutations

Many mutations or deletions in mitochondria are acquired during an individual's lifetime. Their proportion may be different in different tissues and influenced by age. This difference is referred to as heteroplasmy. This contributes to the considerable variability of mitochondrial diseases. A germline mutation may be present in all cells (homoplasmy). The proportion of defective mitochondria varies after repeated cell divisions.

References

Brandon MC, Lott MT, Nguyen KC, et al: MITOMAP: a human mitochondrial genome database—2004 update. Nucleic Acids Res 33 (Database issue): D611–613, 2005 (available online at http://www.mitomap.org).

Chinnery PF, Howell N, Andrews RM, Turnbull DM: Clinical mitochondrial genetics. J Med Genet 36: 425–436, 1999.

Estivill X, Govea N, Barcelo E, et al: Familial progressive sensorineural deafness is mainly due to the mtDNA A1555 G mutation and is enhanced by treatment with aminoglycosides. Am J Hum Genet 62: 27–35, 1998.

Gilbert-Barness E, Barness L: Metabolic Diseases. Foundations of Clinical Management, Genetics, and Pathology. Eaton Publishing, Natick, Massachusetts, 2000.

Harper PS: Practical Genetic Counselling, 6th ed. Edward Arnold, London, 2004.

Wallace DC: Mitochondrial diseases in man and mice. Science 283: 1482–1488, 1999.

Wallace DC et al: Mitochondria and neuro-ophthalmologic dieases. In: The Metabolic and Molecular Bases of Inherited Disease. 8th ed. CR Scriver et al, eds. McGraw-Hill, New York, 2001.

Wallace DC, Lott MT: Mitochondrial genes in degenerative diseases, cancer, and aging. In: Emery and Rimoin's Principles and Practice of Medical Genetics. 4th ed. DL Rimoin et al, eds. Churchill-Livingstone, Edinburgh, 2002.

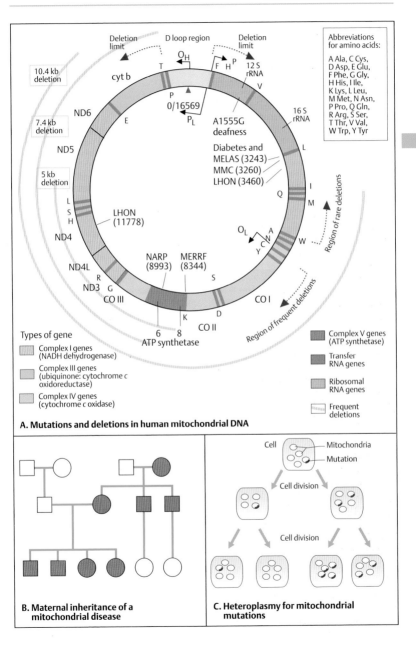

A. Mutations and deletions in human mitochondrial DNA

Types of gene
- Complex I genes (NADH dehydrogenase)
- Complex III genes (ubiquinone: cytochrome c oxidoreductase)
- Complex IV genes (cytochrome c oxidase)
- Complex V genes (ATP synthetase)
- Transfer RNA genes
- Ribosomal RNA genes
- Frequent deletions

Deletion limit
D loop region
O_H
Deletion limit
12 S rRNA

Abbreviations for amino acids:
A Ala, C Cys,
D Asp, E Glu,
F Phe, G Gly,
H His, I Ile,
K Lys, L Leu,
M Met, N Asn,
P Pro, Q Gln,
R Arg, S Ser,
T Thr, V Val,
W Trp, Y Tyr

10.4 kb deletion
cyt b
T
F H P
V
P
$0/16569$
P_L
A1555G deafness
16 S rRNA

ND6
E
Diabetes and
MELAS (3243)
MMC (3260)
LHON (3460)
L

7.4 kb deletion
ND5

5 kb deletion
L S H
ND4
LHON (11778)
Q
I
M
W

Region of rare deletions

ND4L
R
ND3 G
CO III
NARP (8993)
MERRF (8344)
O_L
A
N
C
Y
S
CO I

Region of frequent deletions

6 8
K
CO II
ATP synthetase
D

B. Maternal inheritance of a mitochondrial disease

C. Heteroplasmy for mitochondrial mutations

Cell
Mitochondria
Mutation
Cell division
Cell division

The Mendelian Traits

A scientific foundation for the rules underlying inheritance was established in 1865 when the Augustinian monk Gregor Mendel presented remarkable observations at the *Natur-geschichtliche Vereinigung von Brünn* (The Natural History Society of Brünn/Brno), published in 1866. In this work, entitled *Versuche über Pflanzenhybriden* (Experiments with Hybrid Plants) Mendel observed that certain traits in garden peas *(Pisum sativum)* are inherited independently of one another, and according to regular patterns. It was not until 1900 that H. de Vries, C. Correns, and E. Tschermak independently recognized the importance of Mendel's discovery for biology.

A. The pea plant (*Pisum sativum*)

The garden pea normally reproduces by self-fertilization. Pollen from the anther falls onto the stigma of the same blossom. However, one can easily cross-fertilize (cross-pollinate) pea plants. The plant (left) consists of a stem, leaves, blossoms, and seedpods. In the blossom (right), the female and male reproductive organs are visible. The female pistil comprises stigma, style, and ovule. The male organ is the stamen, comprising the anther and filament. For cross-fertilization, Mendel opened a blossom and removed the anther to avoid self-fertilization. Then he transferred pollen from another plant to the receptive stigma directly.

B. The observed traits (phenotypes)

Mendel observed seven characteristic traits: (**1**) height of the plants, (**2**) location of the blossoms on the stem of the plant, (**3**) the color of the pods, (**4**) the form of the pods, (**5**) the form of the seeds, (**6**) the color of the seeds, and (**7**) the color of the seed coat. Mendel observed these traits in defined proportions in the next plant generation in his experiments.

Deviations from the Mendelian pattern of inheritance

Mendelian traits can deviate from the proportions described on the next page. *Epistasis* is a nonreciprocal interaction between nonallelic genes. As a result, the effect of one gene masks the expression of alleles of another gene. W. Bateson described this phenomenon in 1902 in the recessive gene *apterous* (*ap*) in *Drosophila*. Homozygotes are wingless, but in addition other genes affecting wing morphology, such as *curled wing*, are masked (the *ap* gene is epistatic to *curled wing*). An example of *recessive epistasis* is the yellow coat color of Labrador dogs. Two alleles, *B* (black) and *b* (brown) are influenced by the allele *e* of another gene, resulting in a yellow coat color (see Griffiths et al, 2000). The Bombay blood group is another example (Bhende et al, 1952; Race & Sanger, 1975). *Meiotic drive* refers to the preferential transmission of one allele over others. As a result one trait occurs in offspring much more frequently than others. A striking example is the t complex of the mouse (about 99% instead of 50% of offspring of heterozygous t/+ male mice are also heterozygous) and Segregation Distorter (SD) in *Drosophila*.

In 1993, a mouse population in Siberia was described in which 85% and 65% of offspring were heterozygous for a chromosomal inversion. Homozygosity for the inversion leads to reduced fitness and is a selective disadvantage.

Possibly, deviations from Mendelian laws are more frequent than previously assumed. *Genomic imprinting* (see p. 232) is a further cause of deviation from the Mendelian pattern of inheritance.

References

Bhende YM et al: A "new" blood group character related to the ABO system. Lancet I: 903–904, 1952.

Brink RA, Styles ED: Heritage from Mendel. Univ of Wisconsin Press, Madison, 1967.

Corcos AF, Monaghan FV: Gregor Mendel's Experiments on Plant Hybrids. Rutgers Univ Press, New Brunswick, 1993.

Griffith AJF et al: An Introduction to Genetic Analysis, 7th ed. WH Freeman, New York, 2000.

Mendel G: Versuche über Pflanzenhybriden. Verh naturf Ver Brünn 4: 3–47, 1866.

Pomiankowski A, Hurst DL: Siberian mice upset Mendel. Nature 363 : 396–397, 1993.

Race RR, Sanger R: Blood Groups in Man, 6th ed. Blackwell, Oxford, 1975.

Vogel F, Motulsky AG: Human Genetics. Problems and Approaches, 3rd ed. Springer, Berlin-Heidelberg, 1997.

Weiling F: Johann Gregor Mendel: Der Mensch und Forscher. II Teil. Der Ablauf der Pisum Versuche nach der Darstellung. Med Genetik 2: 208–222, 1993.

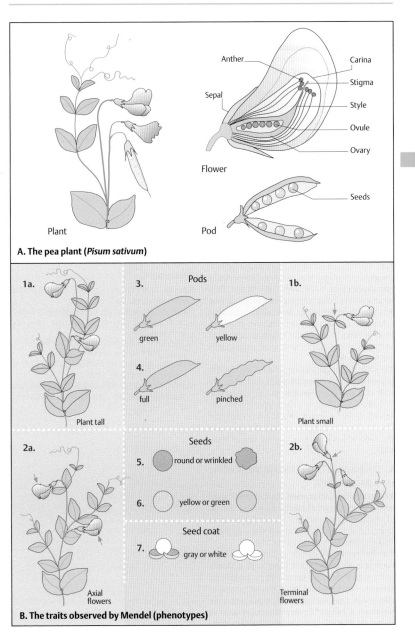

A. The pea plant (*Pisum sativum*)

Anther
Carina
Stigma
Sepal
Style
Ovule
Ovary

Flower

Plant

Pod

Seeds

B. The traits observed by Mendel (phenotypes)

1a. Plant tall

1b. Plant small

2a. Axial flowers

2b. Terminal flowers

3. Pods — green, yellow

4. full, pinched

Seeds

5. round or wrinkled

6. yellow or green

Seed coat

7. gray or white

Segregation of Mendelian Traits

Mendel observed that the traits of the pea plant (*Pisum sativum*) described on the previous page were transmitted to the next generation according to a defined pattern. He provided a meaningful biological interpretation, later called Mendel's laws of inheritance.

A. Segregation of dominant and recessive traits

In two different experiments, Mendel observed the shape (smooth or wrinkled) and the color (yellow or green) of the seeds. When he crossed the plants of the parental generation P, i.e., smooth and wrinkled or yellow and green, he observed that in the first filial (daughter) generation, F_1, all seeds were smooth and yellow.

In the next generation, F_2, which arose by self-fertilization, the traits observed in the P generation (smooth and wrinkled, or green and yellow, respectively) reappeared. Among 7324 seeds of one experiment, 5474 were smooth and 1850 were wrinkled. This corresponded to a ratio of 3 : 1. In the experiment with different colors (green vs. yellow), Mendel observed that in a total of 8023 seeds of the F_2 generation, 6022 were yellow and 2001 green, again corresponding to a ratio of 3 : 1.

The trait the F_1 generation showed exclusively (round or yellow), Mendel called *dominant*; the trait that did not appear in the F_1 generation (wrinkled or green) he called *recessive*. His observation that a dominant and a recessive pair of traits occur (segregate) in the F_2 generation in the ratio 3 : 1 is known as the first law of Mendel.

B. Backcross of an F_1 hybrid plant with a parent plant

When Mendel backcrossed the F_1 hybrid plant with a parent plant showing the recessive trait (**1**), both traits occurred in the next generation in a ratio of 1 : 1 (106 round and 102 wrinkled). This is called the second law of Mendel.

The interpretation of this experiment (**2**), the backcross of an F_1 hybrid plant with a parent plant, is that different germ cells (gametes) are formed. The F_1 hybrid plant (round) contains two traits, one for round (R, dominant over wrinkled, r) and one for wrinkled (r, recessive to round, R). This plant is a hybrid (*heterozygous*) and therefore can form two types of gametes (R and r).

In contrast, the other plant is *homozygous* for wrinkled (r). It can form only one type of gamete (r, wrinkled). Half of the offspring of the heterozygous plant receive the dominant trait (R, round), the other half the recessive trait (r, wrinkled). The resulting distribution of the observed traits is a ratio of 1 : 1, or 50 % each.

The observed trait is called the *phenotype* (the observed appearance of a particular characteristic). The composition of the two factors (genes) R and r, (Rr) or (rr), is called the *genotype*. The alternative forms of a trait (here, round and wrinkled) are called *alleles*. They are the result of different genetic information at one given gene locus.

If the alleles are different, the genotype is heterozygous; if they are the same, it is homozygous (this statement is always in reference to a single, given gene locus).

References

Brink RA, Styles ED: Heritage from Mendel. Univ of Wisconsin Press, Madison, 1967.

Corcos AF, Monaghan FV: Gregor Mendel's Experiments on Plant Hybrids. Rutgers Univ Press, New Brunswick, 1993.

Griffith AJF et al: An Introduction to Genetic Analysis, 7th ed. WH Freeman, New York, 2000.

Mendel G: Versuche über Pflanzenhybriden. Verh naturf Ver Brünn 4: 3–47, 1866.

Vogel F, Motulsky AG: Human Genetics. Problems and Approaches, 3rd ed. Springer, Berlin-Heidelberg, 1997.

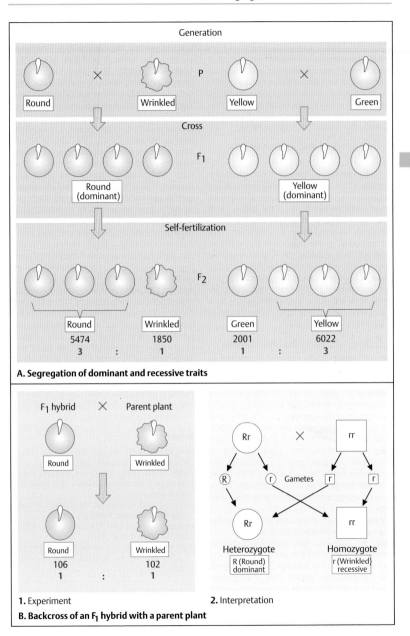

Generation

Round × Wrinkled   P   Yellow × Green

Cross

Round (dominant)   F₁   Yellow (dominant)

Self-fertilization

F₂

Round	Wrinkled		Green	Yellow
5474	1850		2001	6022
3	: 1		1 :	3

A. Segregation of dominant and recessive traits

F₁ hybrid × Parent plant

Round   Wrinkled

Round   Wrinkled

106	102
1	: 1

Rr × rr

R   r   Gametes   r   r

Rr   rr

Heterozygote
R (Round) dominant

Homozygote
r (Wrinkled) recessive

1. Experiment   **2.** Interpretation

B. Backcross of an F₁ hybrid with a parent plant

Independent Distribution of Two Different Traits

In a further experiment, Mendel observed that two different traits are inherited independently of each other. Each pair of traits shows the same 3 : 1 distribution of the dominant over the recessive trait in the F_2 generation as he had previously observed. The segregation of two pairs of traits again followed certain patterns.

A. Independent distribution of two traits

In one experiment, Mendel investigated the crossing of the trait pairs round/wrinkled and yellow/green. When he crossed plants with round and yellow seeds with plants with wrinkled and green seeds, only round and yellow seeds occurred in the F_1 generation. This corresponded to the original experiments as shown. Of 556 plants in the F_2 generation, the two pairs of traits occurred in the following distribution: 315 seeds yellow and round, 108 yellow and wrinkled, 101 green and round, 32 green and wrinkled. This corresponds to a segregation ratio of 9 : 3 : 3 : 1. This is referred to as the third Mendelian law.

B. Interpretation of the observations

Mendel's observations can be summarized as follows: If we assign the capital letter **G** to the dominant gene *yellow,* a lowercase **g** to the recessive gene *green,* the capital letter **R** to the dominant gene *round,* and the lowercase **r** to the recessive gene *wrinkled,* the following nine genotypes of these two traits can occur: **GGRR, GGRr, GgRR, GgRr** (all *yellow* and *round*); **GGrr, Ggrr** (*yellow* and *wrinkled*); **ggRR, ggRr** (*green* and *round*); and **ggrr** (*green* and *wrinkled*). The distribution of the traits shown in A is the result of the formation of gametes of different types, i.e., depending on which of the genes they contain.

The ratio of the dominant trait yellow (**G**) to the recessive trait green (**g**) is 12 : 4, or 3 : 1.

Also, the ratio of dominant round (**R**) to wrinkled (**r**) seeds is 12 : 4, i.e., 3 : 1.

The results can be visualized in a diagram, called the Punnett square. This is a checkerboard way of determining the types of zygotes produced when two gametes with a defined genotype fuse. It was first published in a book entitled *Mendelism* by R.C. Punnett in 1911.

This square shows the nine different genotypes that can be formed in the zygote after fertilization. Altogether there are 9/16 yellow round seeds (**GRGR, GRGr, GrGR, GRgR, gRGR, GRgr, GrgR, gRGr, grGR**), 3/16 green round (**gRgR, gRgr, grgR**), 3/16 yellow wrinkled (**GrGr, Grgr, grGr**), and 1/16 green wrinkled seeds (**grgr**). Each of the two traits (dominant yellow versus recessive green or dominant round versus recessive wrinkled) occurs in a 3 : 1 ratio (dominant vs. recessive).

Why were Mendel's observations fundamentally new and completely different from all other 19th-century attempts to understand heredity? First, Mendel simplified the experimental approach by selecting traits that could be easily observed; second, he assessed the pattern of transmission from one generation to the next quantitatively; third, he provided a biologically meaningful interpretation by noting that each pair of traits was inherited independently of the other pairs of traits in predictable pattern. This was a fundamentally new insight into the process of heredity. Since it distinctly diverged from the prevailing concepts about heredity at the time, its significance was not immediately recognized. Today we know that genetically determined traits are independently inherited (segregation) only when they are located on different chromosomes or are far enough apart on the same chromosome to be separated each time by recombination. This was case for the genes investigated by Mendel, which have been cloned and have had their molecular structures characterized.

References

Brink RA, Styles ED: Heritage from Mendel. Univ of Wisconsin Press, Madison, 1967.

Corcos AF, Monaghan FV: Gregor Mendel's Experiments on Plant Hybrids. Rutgers Univ Press, New Brunswick, 1993.

Griffith AJF et al: An Introduction to Genetic Analysis, 7th ed. WH Freeman, New York, 2000.

Mendel G: Versuche über Pflanzenhybriden. Verh naturf Ver Brünn 4: 3–47, 1866.

Vogel F, Motulsky AG: Human Genetics. Problems and Approaches, 3rd ed. Springer, Berlin-Heidelberg, 1997.

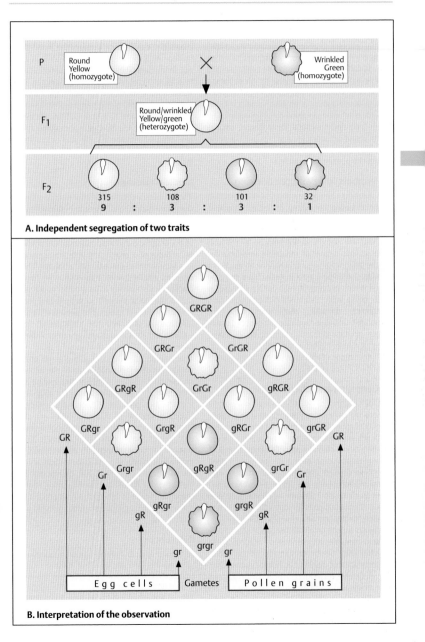

A. Independent segregation of two traits

B. Interpretation of the observation

Phenotype and Genotype

Formal genetic analysis in humans examines the genetic relationship of individuals based on their kinship. These relationships are presented in a pedigree (pedigree analysis). An observed trait is called the phenotype. This could be a disease, a blood group, a protein variant, or any other attribute determined by observation. The phenotype depends to a great degree on the method and accuracy of observation. The term genotype refers to the genetic information on which the phenotype is based.

A. Symbols in a pedigree drawing

The symbols shown here represent a common way of drawing a pedigree. Males are shown as squares, females as circles. Individuals of unknown sex (e.g., because of inadequate information) are shown as diamonds. In medical genetics, the degree of reliability in determining the phenotype, e.g., presence or absence of a disease, should be stated. In each case it must be stated which phenotype (e.g., which disease) is being dealt with. Established diagnoses (data complete), possible diagnoses (data incomplete), and questionable diagnoses (statements or data dubious) should be differentiated. A number of other symbols are used, e.g., for heterozygous females with X-chromosomal inheritance (see p. 148).

B. Genotype and phenotype

The definitions of genotype and phenotype refer to the genetic information at a given *gene locus*. The gene locus is the site on a chromosome at which the *gene*, the genetic information for the given trait, is located. Different forms of genetic information at a gene locus are called *alleles*. In diploid organisms—all animals and many plants—there are three possible genotypes with respect to two alleles at any one locus: (1) *homozygous* for one allele, (2) *heterozygous* for the two different alleles, and (3) *homozygous* for the other allele.

Alleles can be differentiated according to whether they can be recognized in the heterozygous state or only in the homozygous state. If they can be recognized in the heterozygous state, they are called *dominant*. If they can be recognized in the homozygous state only, they are *recessive*. The concepts dominant and recessive are an attribute of the accuracy in observation and do not apply at the molecular level. If the two alleles can both be recognized in the heterozygous state, they are designated codominant (e.g., the alleles A and B of the blood group system ABO; O is recessive to A and B). If there are more than two alleles at a gene locus, there will be correspondingly more genotypes. With three alleles there are six genotypes. For example, in the ABO blood group system there are three phenotypes, A, B, and O. They are due to six genotypes: AA, AO (both phenotype A), BB, BO (both phenotype B), AB, and OO (actually, there are more than three alleles in the ABO system).

Medical relevance

The Mendelian pattern of inheritance provides the foundation for genetic counseling of patients with monogenic diseases. *Genetic counseling* is a communication process relating to the diagnosis and assessment of the potential occurrence of a genetically determined disease in a family and in more distant relatives. The individual affected with a disease, who first attracted attention to a particular pedigree, is called the index patient (or proposita if female and propositus if male). The person who seeks information is called the consultand. Index patient and consultand are very often different persons.

The goal of genetic counseling is to provide comprehensive information about the expected course of the disease, medical care, and possible treatments or an explanation, for why treatment is not possible. Genetic counseling includes a review of possible decisions about family planning as a consequence of a genetic risk. Professional confidentiality must be observed. The counselor makes no decisions. The increasing availability of information about a disease based on a DNA test (predictive DNA testing) prior to disease manifestation requires the utmost care in establishing whether it is in the interest of a given individual to have a test carried out.

References

Griffith AJF et al: An Introduction to Genetic Analysis, 7th ed. WH Freeman, New York, 2000.

Harper PS: Practical Genetic Counselling, 6th ed. Edward Arnold, London, 2004.

Vogel F, Motulsky AG: Human Genetics. Problems and Approaches, 3rd ed. Springer, Berlin-Heidelberg, 1997.

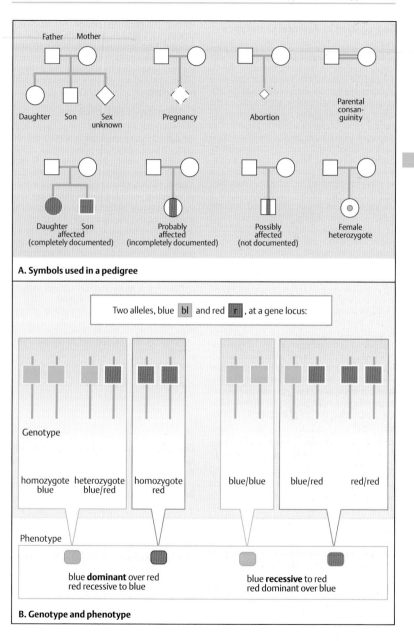

A. Symbols used in a pedigree

Father Mother

Daughter Son Sex unknown

Pregnancy

Abortion

Parental consanguinity

Daughter Son affected (completely documented)

Probably affected (incompletely documented)

Possibly affected (not documented)

Female heterozygote

Two alleles, blue **bl** and red **r**, at a gene locus:

Genotype

homozygote blue   heterozygote blue/red   homozygote red

blue/blue   blue/red   red/red

Phenotype

blue **dominant** over red
red recessive to blue

blue **recessive** to red
red dominant over blue

B. Genotype and phenotype

Segregation of Parental Genotypes

The segregation (distribution) of the genotypes of the parents (parental genotypes) in the offspring depends on the combination of the alleles in the parents. The Mendelian laws state the expected combination of alleles in the offspring of a parental couple. Depending on the effect of the genotype on the phenotype in the heterozygous state, an allele is classified as dominant or recessive. Hence, there are three basic modes of inheritance: (1) autosomal dominant, (2) autosomal recessive, and (3) X-chromosomal. For genes on the X chromosome, it is usually not important to distinguish dominant and recessive (see below). Since genes on the Y chromosome are always transmitted from the father to all sons and the Y chromosome bears very few disease-causing genes, Y-chromosomal inheritance can be disregarded when considering monogenic inheritance.

A. Possible mating types of genotypes

For a gene locus with two alleles, there are six possible combinations of parental genotypes (**1–6**). Here two alleles, blue (bl) and red (r), are shown, blue being dominant over red. In three of the parental combinations (1, 3, 4) neither of the parents is homozygous for the recessive allele (red). In three parental combinations (2, 5, 6), one or both parents manifest the recessive allele because they are homozygous. The distribution patterns of genotypes and phenotypes in the offspring of the parents are shown in B. In these examples, the sex of the parents is interchangeable.

B. Distribution pattern in the offspring of parents with two alleles, A and a

With three of the parental mating types for the two alleles **A** (dominant over **a**) and **a** (recessive to **A**), there are three combinations that lead to segregation (separation during meiosis) of allelic genes. These correspond to the parental combinations 1, 2, and 3 shown in A. In mating types 1 and 2, one of the parents is a heterozygote (**Aa**) and the other parent is a homozygote (**aa**). The distribution of observed genotypes expected in the offspring is 1:1; in other words, 50% (0.50) are **Aa** heterozygotes and 50% (0.50), **aa** homozygotes.

If both parents are heterozygous **Aa** (mating type 3 in A), the proportions of expected genotypes of the offspring (**AA, Aa, aa**) occur in a ratio of 1:2:1. In each case, 25% (0.25) of the offspring will be homozygous **AA**, 50% (0.50) heterozygous **Aa**, and 25% (0.25) homozygous **aa**. If the two parents are homozygotes for different alleles (**AA** and **aa**), all their offspring will be heterozygotes.

C. Phenotypes and genotypes

One dominant allele (in the first pedigree, **A**, in the father) can be expected in 50% of the offspring. If both parents are heterozygous, 25% of the offspring will be homozygous **aa**. If both parents are homozygous, one for the dominant allele **A**, the other for the recessive allele **a**, then all offspring are obligate heterozygotes (i.e., must necessarily be heterozygotes). It should be emphasized that the figures are percentages for expected distributions of the genotypes. The actual distribution may deviate from the expected one, especially with small numbers of children.

References

Griffith AJF et al: An Introduction to Genetic Analysis, 7th ed. WH Freeman, New York, 2000.

Harper PS: Practical Genetic Counselling, 6th ed. Edward Arnold, London, 2004.

Vogel F, Motulsky AG: Human Genetics. Problems and Approaches, 3rd ed. Springer, Berlin-Heidelberg, 1997.

Expected distribution of genotypes in the offspring of different parental genotype combinations

Parents	Offspring	Expected genotype proportion
AA x AA	AA	1 (100%)
AA x Aa	AA, Aa	1:1 (each 50%)
Aa x Aa	AA, Aa, aa	1:2:1 (25%, 50%, 25%)
AA x aa	Aa	1 (100%)
Aa x aa	Aa, aa	1:1 (50% each)
aa x aa	aa	1 (100%)

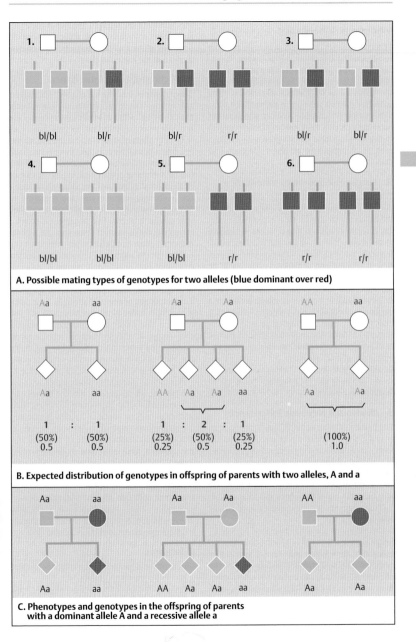

A. Possible mating types of genotypes for two alleles (blue dominant over red)

B. Expected distribution of genotypes in offspring of parents with two alleles, A and a

C. Phenotypes and genotypes in the offspring of parents with a dominant allele A and a recessive allele a

Monogenic Inheritance

When the phenotypic effect of a single gene can be recognized, it is referred to as monogenic inheritance. This is documented in a pedigree. For humans it is customary to designate consecutive generations by Roman numerals, starting with I for the first generation about which some information is available. Within a generation, each individual is assigned an Arabic numeral. Individuals can also be assigned non-overlapping combinations of numbers for computer calculations.

A. Autosomal dominant inheritance

The pattern of inheritance in a pedigree with an autosomal dominant trait is characterized by the following attributes: (i) affected individuals are directly related in one or more successive generations; (ii) both males and females are affected in a 1 : 1 ratio; (iii) the expected proportion of affected and unaffected offspring of an affected individual is 1 : 1 (0.50, or 50%) each. An important consideration in autosomal disorders is whether a new mutation is present in a patient without affected parents. Pedigrees 2 and 3 show a new mutation in generation II. In some autosomal dominant disorders, a carrier of the mutation does not manifest the disease. This is called *nonpenetrance*, but it is the exception rather than the rule. The degree of manifestation can vary within a family. This is called *variable expressivity*.

B. Autosomal recessive inheritance

The pattern of inheritance in a pedigree with an autosomal recessive trait is characterized as follows: (i) the expected segregation of genotypes of children of heterozygous patients is 1 : 2 : 1 (25% homozygous normal, 50% heterozygous, 25% homozygous affected); (ii) both sexes are affected with equal frequency at a ratio of 1 : 1; (iii) heterozygous parents have a risk of 25% of affected offspring.

In the examples shown, in pedigree 1 the unaffected parents (II-3 and II-4) must be heterozygous; the same holds true for I-1 and I-2 of pedigree 2. In pedigree 3, the ancestors of the affected child (IV-2) can be traced back to common ancestors (I-1 and I-2) of the parents, who are first cousins. Parental consanguinity (blood relationship) of the parents (III-1 and III-2) is indicated by a double line in the pedigree. From this observation it may be deduced that the ho-

mozygosity of the affected individual (IV-2) results from transmission of the mutant allele from one of the ancestors through both the paternal and the maternal lines.

C. X-chromosomal inheritance

The pattern of inheritance of genes located on the X chromosome can be observed by (i) only males being affected (rare exceptions are possible when a mother carries two mutant alleles); (ii) affected males inherit the mutant allele from the mother only; (iii) there is no male-to-male transmission. A female heterozygous for an X-chromosomal mutation has a risk of 50% for an affected son. She also will transmit the X chromosome carrying the mutation to 50% of her daughters, but as heterozygotes they will not be affected. The pedigree in panel 1 shows the distribution of three parental X chromosomes (one from the father, two from the mother) in the offspring (**1**). Panel 2 shows the corresponding distribution of the X chromosomes and the Y chromosome from the parents to the offspring (**2**).

The proportion of new mutations in X-chromosomal inheritance is relatively high (**3, 4**). The reason is that males affected with a severe X-chromosomal disease cannot reproduce to transmit the mutation to their offspring. Since one third of the X chromosomes in the population are present in males and since a severe X-chromosomal disease is incompatible with reproduction, these (male-carried) Xs will be eliminated from the population. However, their mutations will be replaced by new mutations. Therefore, one third of males with severe X-chromosomal disorders without a family history have a new mutation (Haldane's rule). Typical X-chromosomal inheritance (**5**) is easy to recognize.

Females with an affected son and an affected brother or with two affected sons must be heterozygous. They are *obligate heterozygotes*. Those who may or may not be heterozygous are *facultative heterozygotes* (e.g., III-5 and IV-2).

References

Griffiths AJF et al: An Introduction to Genetic Analysis, 7th ed. WH Freeman & Co, New York, 2000.

Harper PS: Practical Genetic Counselling, 6th ed. Edward Arnold, London, 2004.

Vogel F, Motulsky AG: Human Genetics. Problems and Approaches, 3rd ed. Springer Verlag, Heidelberg–New York, 1997.

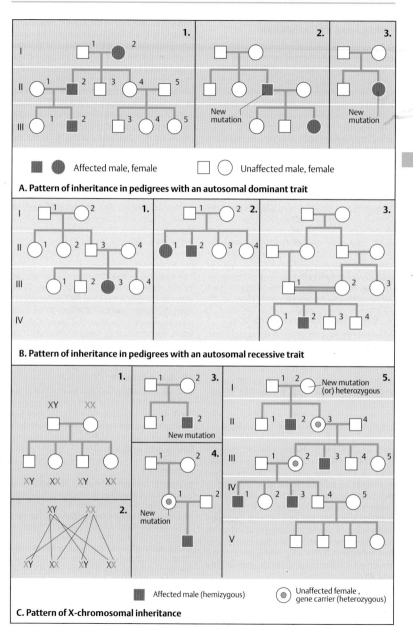

Affected male, female   **Unaffected male, female**

A. Pattern of inheritance in pedigrees with an autosomal dominant trait

B. Pattern of inheritance in pedigrees with an autosomal recessive trait

Affected male (hemizygous)   **Unaffected female, gene carrier (heterozygous)**

C. Pattern of X-chromosomal inheritance

Linkage and Recombination

Genetic linkage refers to the observation that two or more genes located on the same chromosome are transmitted together. This is in contrast to localization on different chromosomes or on the same chromosome far apart; in this case two genes will be distributed in a 1 : 1 ratio, independently of each other. Linkage was first reported in 1902 by Correns, and established by Bateson and coworkers in genes for colored flowers and long pollen shape of sweet peas. The ratio of being transmitted together versus being separated allows an estimate of their distance from each other.

Linkage relates to gene loci, not to specific alleles. Alleles at closely linked gene loci that are inherited together are called a *haplotype*. Alleles at different loci that are inherited together more frequently or less frequently than expected by their individual frequencies are said to show *linkage disequilibrium* (p. 164). *Synteny* (H. J. Renwick, 1971) refers to gene loci being located on the same chromosome without regard to linkage or the distance between them.

A. Recombination by crossing-over

Whether neighboring genes on the same parental chromosome remain together or become separated depends on the cytological events during meiosis. If there is no crossing-over between the two gene loci **A** and **B**, having the respective alleles **A, a** and **B, b**, then they remain together on the same chromosome. The gamete chromosomes formed during meiosis in this case are not recombinant and correspond to the parental chromosomes. However, if crossing-over occurs between the two gene loci, then the gametes formed are recombinant with reference to these two gene loci. The cytological events (**1**) are reflected in the genetic result (**2**). For two neighboring gene loci **A** and **B** on the same chromosome, the genetic result is one of two possibilities: not recombinant (gametes correspond to parental genotype) or recombinant (new combination). The two possibilities can be differentiated only when the parental genotype is informative for both gene loci (**Aa** and **Bb**).

B. Genetic linkage

The inheritance pattern of a mutation causing a disease may not be recognizable, e.g., in auto-somal recessive or X-chromosomal inheritance. However, if the disease locus is linked to a locus with polymorphic alleles, this can be assessed by analysis of the marker locus. The segregation of two linked gene loci in a family is shown here. There are two possibilities: 1, without recombination and 2, with recombination. If one locus (A) is a marker locus and locus B a disease locus, one can distinguish whether recombination has taken place by observing the haplotypes represented by the alleles A, a and B, b. In this example the father and three children (red symbols in the pedigree) are affected (the children are shown as diamonds, their genders ignored). All three affected children have inherited the mutant allele **B** as well as the marker allele **A** from their father. The three unaffected individuals have inherited the normal allele **b** and the marker allele **a** from their father. The paternal allele **a** indicates absence of the mutation (i.e., **B** not present). Thus, recombination has not occurred (**1**).

The situation differs if recombination has occurred, as in two individuals shown in (**2**). An affected individual has inherited alleles **a** and **b** from the father, instead of **A** and **B**; an unaffected individual has inherited allele **A** and allele **b**. This situation can only be observed if the father (affected parent) is heterozygous at the marker locus (A, a).

Medical relevance

DNA analysis of polymorphic marker loci linked to a disease locus can be used within a family to determine whether unaffected family members have inherited the mutant allele known to be present in an affected family member (indirect DNA genetic analysis).

References

Bateson W, Saunders ER, Punnett RC: Experimental studies in the physiology of heredity. Reports Evolut Comm Royal Soc 3: 1–53, 1906.

Correns C: Scheinbare Ausnahmen von der Mendelschen Spaltungsregel für Bastarde. Ber deutsch bot Ges 20: 97–172, 1902.

Griffiths AJF et al: An Introduction to Genetic Analysis. 7th ed. W.H. Freeman & Co., New York, 2000.

Harper PS: Practical Genetic Counselling. 6th ed. Edward Arnold, London, 2004.

Vogel F, Motulsky AG: Human Genetics. Problems and Approaches. 3rd ed. Springer Verlag, Heidelberg-New York, 1997.

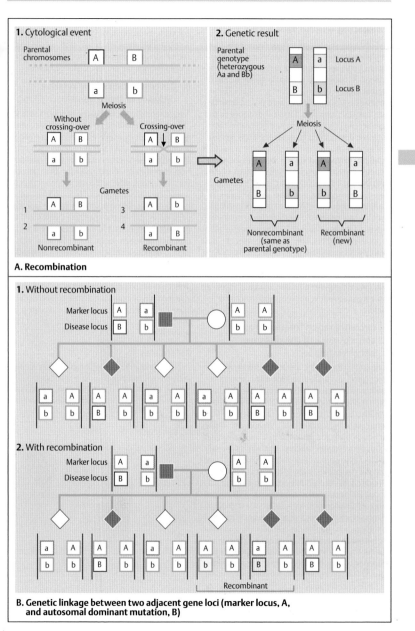

1. Cytological event

Parental chromosomes

Meiosis

Without crossing-over

Crossing-over

Gametes

1

2

Nonrecombinant

3

4

Recombinant

2. Genetic result

Parental genotype (heterozygous Aa and Bb)

Locus A

Locus B

Meiosis

Gametes

Nonrecombinant (same as parental genotype)

Recombinant (new)

A. Recombination

1. Without recombination

Marker locus

Disease locus

2. With recombination

Marker locus

Disease locus

Recombinant

B. Genetic linkage between two adjacent gene loci (marker locus, A, and autosomal dominant mutation, B)

Estimating Genetic Distance

The distance separating two or more linked loci can be estimated by the frequency of recombination between them. This distance can be expressed in two ways: genetic distance or physical distance.

Genetic distance is the frequency of recombination between two loci. Physical distance is the number of nucleotide base pairs between them. Both are determined in different ways, genetic distance by observing recombination in a set of families, physical distance by DNA analysis. Genetic distance is relative, dependent on the distribution of crossing-over with recombination along each chromosome. In mammals, recombination is more frequent in female meiosis than in male meiosis, so that the genetic distance in females is about 1.5 times greater than in males. Furthermore, crossing-over is not evenly distributed along all chromosomes. As a result, blocks of nonrecombinant stretches of DNA exist. For this reason, the genetic distance may not reflect the physical distance. Physical distance is absolute, but more difficult to determine.

A. Recombination frequency as a consequence of the distance between two loci

Estimating the genetic distance between two or more gene loci relates to how often genetic recombination occurs between the loci and how often it does not occur. Consider two loci A and B with respective alleles A, a and B, b. One parent has haplotypes AB, AB and the other parent, haplotypes AB and ab. In their offspring it can be observed whether the alleles located on the same chromosomes in the parents have remained together in the offspring or have been separated by recombination. In two of the offspring (1 and 2), the haplotypes differ from those of the parents, with (a "new") haplotype Ab occurring on one chromosome of child 1 and (another new) haplotype, aB, occurring on one chromosome of child 2. They are said to be recombinant. Offspring 3 and 4 have the same haplotypes as their parents; they are nonrecombinant. If the recombinant genotypes are present in 3% of observations in a sufficient number of families, the recombination frequency is 0.03 (3%). This distance (genetic distance) is expressed in units called Morgans (named after the geneticist T.H. Morgan). One Morgan (M) equals a recombination rate of 1 (100%). A recombination frequency of 0.01 (1%) corresponds to one centiMorgan (cM). In the example shown, the genetic distance is 3 cM.

B. Determination of the relative distance of three gene loci

The genetic distances and the order of three loci (1) can be determined by a test cross of parental genotypes, for example with three gene loci, A, B, and C of unknown distance from each other. In traditional experimental genetics, hybridization experiments were used for this purpose. For a test cross, different parental genotypes are used. The observed recombination frequencies indicate the relative distances of the loci from each other (2). In the example presented, the distance from locus A to locus C is 0.08 (8%); the distance between locus B and locus C is 0.23 (23%); and the distance between A and B is 0.31 (31%). Thus, they are located in the order A–C–B (3).

This type of indirect determination of the relative location of gene loci in their correct order in classical experimental genetics has been replaced by direct methods of recombinant DNA technology.

References

Griffiths AJF et al: An Introduction to Genetic Analysis, 7th ed. WH Freeman & Co, New York, 2000.

Vogel F, Motulsky AG: Human Genetics. Problems and Approaches, 3rd ed. Springer Verlag, Heidelberg–New York, 1997.

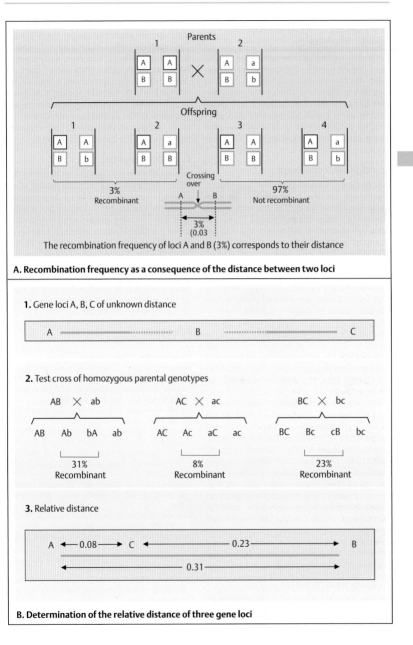

The recombination frequency of loci A and B (3%) corresponds to their distance

A. Recombination frequency as a consequence of the distance between two loci

1. Gene loci A, B, C of unknown distance

2. Test cross of homozygous parental genotypes

31%
Recombinant

8%
Recombinant

23%
Recombinant

3. Relative distance

B. Determination of the relative distance of three gene loci

Segregation Analysis with Linked Genetic Markers

Analysis of known linkage between a disease locus and one or more DNA marker loci can provide information about the genetic risk to members of a family in which a monogenic disease has occurred. This is called *segregation analysis* using linked polymorphic DNA markers (see p. 156). It is the basis for indirect gene diagnosis, which is used when the mutation is not known.

A. Autosomal dominant

Two pedigrees studied by marker analysis are shown, one without recombination between disease locus and marker (**1**), one with recombination (**2**). Below the pedigree in 1 (left) the lanes of a diagram of a Southern blot analysis representing each of the 8 individuals is shown. DNA samples from individuals affected with an autosomal dominant disease are in lanes 2, 4, and 5 (red symbols). The affected mother has two marker alleles, 1 and 2. This is indicated as a genotype 1–2. Her two affected children have the same marker genotype, 1–2. None of her unaffected children (lanes 3, 6, 7, and 8) has this genotype; they are 2–2. Since the father has the genotype 2–2 and must have transmitted marker allele 2 to all children, the other allele 2 in the unaffected children must have come from the mother. Thus, the diseases locus must be linked to marker 1.

In the pedigree on the right, an affected father (lane 1) and one affected child (lane 4) are shown (**2**). The genotypes of the parents are 1–2 (father) and 1–1 (mother), respectively. Allele 2, inherited from the father, is present in the affected child (1–2). Thus, allele 2 must represent the disease locus. In this family, however, an unaffected sib (lane 5) also has the genotype 1–2. One has to conclude that recombination has occurred between the disease locus and the marker locus. In this case the test result would be misleading. For this reason, very closely linked markers with very low recombination frequencies are used, preferably marker loci that flank the disease locus.

B. Autosomal recessive

The two affected individuals in the pedigree on the left are homozygous for allele 1 (1–1), inherited from each of their parents (**1**). Thus one allele 1 of the father and allele 1 of the mother represent the allele that carries the mutation. The unaffected sibs (individuals 3, 5, and 6) have received allele 2 from their mother and allele 1 from their father. Since allele 2 does not occur in the affected individuals, it cannot carry the mutation. In this case, it cannot be determined whether the unaffected children are heterozygous for the paternal mutant allele 1. In the pedigree on the right (**2**), recombination must have occurred in one child (lane 6).

C. X-chromosomal

Here an affected son (lane 4) carries the marker allele 1. His sister (lane 6) is the mother of six children (lanes 8–13). She is heterozygous 1–2 at the marker locus. Two of her affected sons also carry the marker allele 1. Her unaffected son (lane 11) carries allele 2. Thus, this allele is likely not to be linked with the disease locus. However, a third affected son (lane 13) carries allele 2. This can be explained by recombination of the disease locus and the marker locus in this son. In practice one tries to avoid such a misleading result by using several marker loci flanking the disease locus.

References

Harper PS: Practical Genetic Counselling, 6th ed. Edward Arnold, London, 2004.

Jameson JL, Kopp P: Principles of human genetics, pp 359–379. In: Kasper DL et al (eds) Harrison's Principles of Internal Medicine, 16th ed. (with online access). McGraw-Hill, New York, 2005.

Korf B: Molecular diagnosis. New Eng J Med 332: 1218–1220 and 1499–1502, 1995.

Miesfeldt S, Jameson JL: The practice of genetics in clinical medicine, pp 386–391. In: Kasper DL et al (eds) Harrison's Principles of Internal Medicine, 16th ed. (with online access). McGraw-Hill, New York, 2005.

Nussbaum RL et al: Thompson & Thompson Genetics in Medicine, 6th ed. WB Saunders, Philadelphia, 2001.

Richards CS, Ward PA: Molecular diagnostic testing, pp 83–88. In: Jameson JL (ed) Principles of Molecular Medicine. Humana Press, Totowa, New Jersey, 1998.

Strachan T, Read AP: Human Molecular Genetics, 3rd ed. Garland Science, London, 2004.

Turnpenny P, Ellard S: Emery's Elements of Medical Genetics, 12th ed. Elsevier–Churchill Livingstone, Edinburgh, 2005.

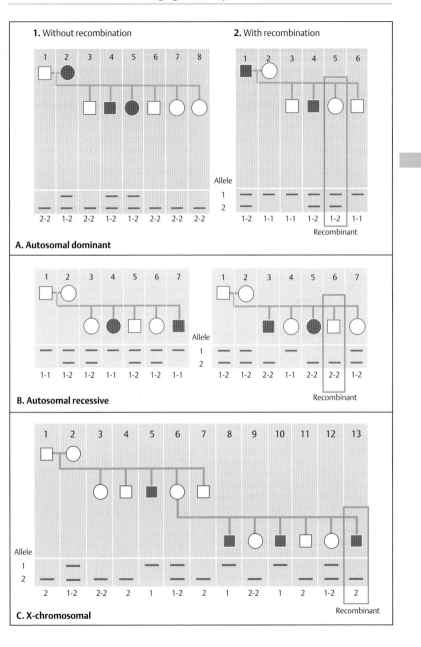

1. Without recombination

2. With recombination

Allele
1
2

A. Autosomal dominant

Allele
1
2

B. Autosomal recessive

Allele
1
2

C. X-chromosomal

Linkage Analysis

Linkage analysis is a set of different tests to determine the genetic distance between two or more gene loci. It is based on observations obtained in many families by determining how often the loci examined are inherited as a set and how often they are separated by recombination. This is expressed as a ratio of the probability for linkage to that of no linkage. It is obtained by a variety of statistical tests supported by computer programs.

Three situations can be distinguished: (i) linkage analysis of two loci, e.g., a locus of interest because a mutation of its gene causes a disease (disease locus) and a locus characterized by a detectable DNA polymorphism (marker locus); (ii) linkage analysis of several loci (multilocus analysis); and (iii) linkage analysis involving the entire genome by means of DNA markers (microsatellites) along each chromosome (genome scan, see p. 260). The three approaches require different procedures and are based on different assumptions. A simplified outline of the background is provided here.

A. LOD score

The recombination fraction, designated by the Greek letter *theta* (Θ), is the distance between two loci. This corresponds to the likelihood of being separated by recombination during meiosis. Loci that are not linked have a Θ of 0.5. If Θ is zero, the loci are identical.

Linkage of two gene loci is assumed when the probability of linkage divided by the probability of no linkage is equal to or greater than the ratio of $1000:1$ ($10^3:1$). The logarithm of this ratio, the odds, is called the LOD score (**l**ogarithm of the **od**ds). An LOD score of 3 corresponds to an odds ratio of $1000:1$. The closer two loci are next to each other, the higher the resulting LOD score. The table shows (in a simplified manner) the LOD score for close linkage, with a recombination fraction under 0.05 (a), high probability of linkage with a recombination fraction of 0.15 (b), weak linkage (c), and no linkage (d). If the LOD score is less than 0, linkage is excluded (not shown in the table). Nonparametric LOD scores are used to avoid an assumption about the mode of inheritance.

B. LOD scores in different recombination fractions

The diagram shows LOD scores based on the table in A. The blue curve with an LOD score above 3 at a recombination fraction of less than 0.05 indicates close linkage. The green curve b indicates high probability with a peak at a recombination fraction of 0.16 with an LOD score of 3.0. The red curve c indicates that linkage is unlikely. Linkage is excluded by the violet curve (d). (Figure adapted from Emery, 1986.)

C. Multilocus analysis

Linkage analysis today is usually carried out with multiple markers (multilocus analysis). With the chromosomal position of the marker taken into consideration, the localization score is determined as the logarithm of the probability quotient (likelihood ratio). The localization score of the locus being sought is noted with respect to each of the marker loci (A, B, C, D). Each of the four peaks expresses linkage. The highest peak marks the probable location of the gene being sought. If there is no peak, linkage is not present, and the locus being studied does not map to the region tested (exclusion mapping).

In contrast to linkage, which refers to the genetic distance between gene loci, the term association refers to the co-occurrence of alleles or phenotypes. If one particular allele occurs more frequently in people with a disease than expected from the individual frequencies, this is called an association.

References

Byerley WF: Genetic linkage revisited. Nature 340: 340–341, 1989.

Emery AEH: Methodology in Medical Genetics. 2nd ed. Churchill Livingstone, Edinburgh, 1986.

Lander ES, Kruglyak L: Genetic dissection of complex traits: guidelines for interpreting and reporting linkage results. Nature Genet 11: 241–247, 1995.

Morton NE: Sequential tests for detection of linkage. Am J Hum Genet 7: 277–318, 1955.

Ott J: Analysis of Human Genetic Linkage. Johns Hopkins University Press, Baltimore, 1991.

Strachan T, Read AP: Human Molecular Genetics. 3rd ed. Garland Science, London, 2004.

Terwilliger J, Ott J: Handbook for Human Genetic Linkage. Johns Hopkins University Press, Baltimore, 1994.

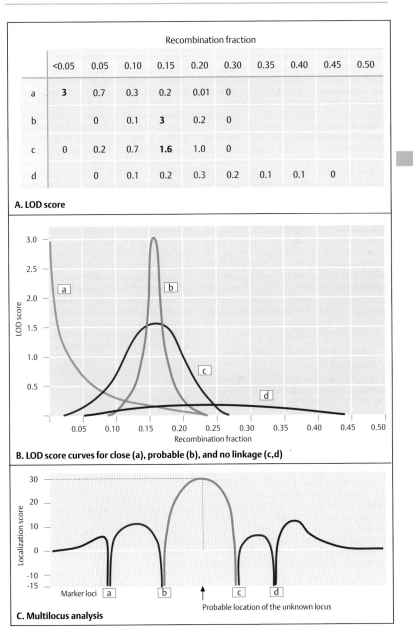

A. LOD score

B. LOD score curves for close (a), probable (b), and no linkage (c,d)

C. Multilocus analysis

Quantitative Differences in Genetic Traits

Most phenotypic variation among individuals and organisms is quantitative rather than qualitative in nature. Quantitative characteristics have a continuous distribution among different individuals. This differs from monogenic traits displaying a Mendelian mode of inheritance, which can be assigned to separate, distinct categories. The transmission of the underlying genes cannot be recognized. Examples of common quantitative genetic characteristics in humans are height, weight, eye and skin color, blood pressure, plasma concentrations of glucose, lipids, intellectual abilities, behavioral patterns, and others.

The term *quantitative genetics* was introduced by Francis Galton (1822–1911) in 1883. It was experimentally demonstrated in rye and wheat by Nilsson-Ehle in 1909. A primary goal of quantitative genetics is to distinguish the genetic and environmental contributions to a trait. Many quantitative genetic traits are difficult to define. The underlying genetic variation (genotypic variation) is interchangeably called polygenic (many genes), multigenic (several genes), or multifactorial (many factors).

A. Length of the corolla in *Nicotiana longiflora*

The plant *Nicotiana longiflora* is an example of how the genetic transmission of a quantitative trait can be analyzed experimentally. When parent plants with average corolla lengths of 40 cm and 90 cm are crossed, the next generation (F_1) shows a corolla length distribution that is longer than that of the short parent and shorter than that of the tall parent. In the next generation (F_2) the distribution spreads at both ends towards long and short. If plants from the short, middle, and long varieties are crossed, the mean distribution in the next generation (F_3) corresponds to the corolla length of the parental plants. They are short (shown on the left), long (shown on the right), or average (shown in the middle). This can be explained by a difference in the distribution of genes contributing to the variation in the trait. (Figure adapted from Ayala & Kiger, 1984.)

B. Influence of the number of gene loci on a quantitative trait

The number of gene loci involved in determining the phenotype of a quantitative trait may be relatively low. Four situations with 1, 2, 3, and 4 loci and two allele pairs A and a are shown for the parental generation P. Two phenotypes are shown (yellow bars). The hypothetical genotypes aa and AA, aabb and AABB, etc. appear uniform in the F1 generation. In the F2 generation, the number of distinct groups depends on the number of genes involved. With 4 loci, the difference in size between the groups is small. It begins to resemble a continuous distribution, as shown at the bottom of the diagram at the right. The smooth distribution curve corresponds to a bell-shaped normal distribution or Gaussian curve. The Y axis indicates the quantitative trait, e.g., height; the X axis shows the number of individuals. This is the total variance (V_T). Each individual's phenotype is the result of the contribution of each allele at the loci involved (their identity generally remains unknown). The variance of the phenotype (V_P) is the sum of the genetic variance (V_G) and the environmental variance (V_E). The ratio V_G/V_T is called heritability. However, in many cases these types of variance cannot be distinguished, especially in humans. A locus contributing to a quantitative characteristic is called a quantitative trait locus (QTL).

References

Ayala FJ, Kiger JA: Modern Genetics, 2nd ed. Benjamin/Cummings Publishing Co, Menlo Park, California, 1984.

Burns GW, Bottinger PJ: The Science of Genetics, 6th ed. Macmillan Publ Co, New York–London, 1989.

Falconer DS : Introduction to Quantitative Genetics, 2nd ed. Longman, London, 1981.

Griffith AJF et al: An Introduction to Genetic Analysis 5th ed. WH Freeman & Co, New York, 2000.

King R, Rotter J, Motulsky AG, eds: The Genetic Basis of Common Disorders, 2nd ed. Oxford Univ Press, Oxford, 2002.

Nilsson-Ehle H: Kreuzungsuntersuchungen an Hafer und Weizen. Lunds Universit Arsskr NF 5: 1–122, 1909.

Vogel F, Motulsky AG: Human Genetics and Approaches, 3rd ed. Springer Verlag, Heidelberg–New York, 1997.

Quantitative Trait Loci (QTL). Special Issue. Trends Genet 11: 463–524, 1995.

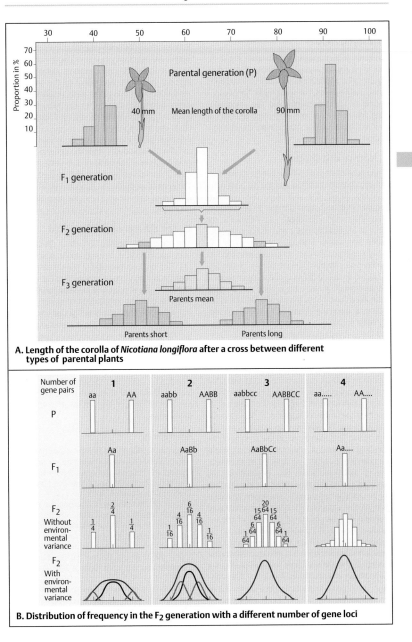

A. Length of the corolla of *Nicotiana longiflora* after a cross between different types of parental plants

B. Distribution of frequency in the F₂ generation with a different number of gene loci

Normal Distribution and Polygenic Threshold Model

Polygenic inheritance of quantitative traits occurs in all animal and plant species. Its analysis requires the application of statistical methods. These help assess the difference between a sample of measurements and the population from which the sample is derived and define how much confidence can be placed in the conclusions.

A. Normal distribution

When quantitative data from a large sample are plotted along the abscissa and the number of individuals along the ordinate, the resulting frequency distribution forms a bell-shaped Gaussian curve. The mean ($\bar{x}$) intersects the curve at its highest point and divides the area under the curve into two equal parts (**1**). Further perpendicular intersections can be placed one standard deviation (*s*) to the left ($-1s$) and to the right ($+1s$) of the mean to yield two additional areas, *c* to the left and *d* to the right. Areas *a* and *b* each comprise 34.13% of the total area under the curve (**2**). Further partitioning with perpendiculars to the abscissa at two and three standard deviations ($-2s$ and $-3s$ to the left and $+2s$ and $+3s$ to the right) results in further subsections (**3**).

The mean ($\bar{x}$) of a sample is determined by the sum of individual measurements (Σx) divided by the number of individuals (*n*) (formula 1). Many measurements yield the frequency f_x of observed individuals (formula 2). The population variance (σ^2) defines the variability of the population. It is expressed as the square of the sum (Σ) of individual measurements (x) minus the population mean (μ), divided by the number of individuals in the population (*N*) (formula 3). The variance of the population cannot be determined directly. Therefore, the variance of the sample (s^2) has to be estimated (formula 4). The square of the sum of the individual measurements (x) minus the mean ($\bar{x}$) is divided by the number of measurements (*n*). A correction factor $n/n - 1$ is introduced because the number of independent measurements is $n - 1$, resulting in a simplified formula 5. The standard deviation (*s*, formula 6), the square root of the sample variance (s^2), rests on a large number of successive samples. (Figure adapted from Burns & Bottino, 1989.)

B. Polygenic threshold model

Some continuously variable characteristics represent susceptibility to a disease. It was postulated that the disease becomes manifest at a certain threshold. According to the threshold model, populations differ with respect to susceptibility to the disease (**1**). The liability to disease in the offspring of affected individuals (first-degree relatives, **2**) or in second-degree relatives (**3**) is closer to the threshold than in unrelated individuals from the general population. In general, the liability of first-degree relatives is increased by about half of the mean of the general population ($^1/_2\,\bar{x}$) and by about one quarter of the mean ($^1/_4\,\bar{x}$) for second-degree relatives.

C. Sex differences in threshold

The threshold beyond which a disease will manifest may differ between males and females (**1**). This may be assumed if the disease occurs more frequently in one gender than the other. In families with at least one affected individual, a difference in the proportion of other affected individuals will be observed (**2**). In contrast to expectation, however, the proportion of affected first-degree relatives is higher if the index patient is female than if he is a male (**3**). This seeming paradox, known as the Carter effect after the medical geneticist Cedric O. Carter (London), is assumed to result from a greater or lesser influence of disease-causing genetic factors in the gender less often affected in the population.

References

Burns GW, Bottino PJ: The Science of Genetics, 6th ed. Macmillan Publishing Co, New York, 1989.

Comings DE: Polygenic disorders. Nature Encyclopedia of the Human Genome 4: 589–595, Nature Publishing Group, London, 2003.

Falconer DS: Introduction to Quantitative Genetics, 2nd ed. Longman, London, 1981.

Fraser CF: Evolution of a palatable multifactorial threshold model. Am J Hum Genet 32: 796–813, 1980.

Glazier AM, Nadeau JH, Aitman TJ: Finding genes that underlie complex traits. Science 298: 2345–2349, 2002.

King R, Rotter J, Motulsky AG (eds): The Genetic Basis of Common Disorders, 2nd ed. Oxford Univ Press, Oxford, 2002.

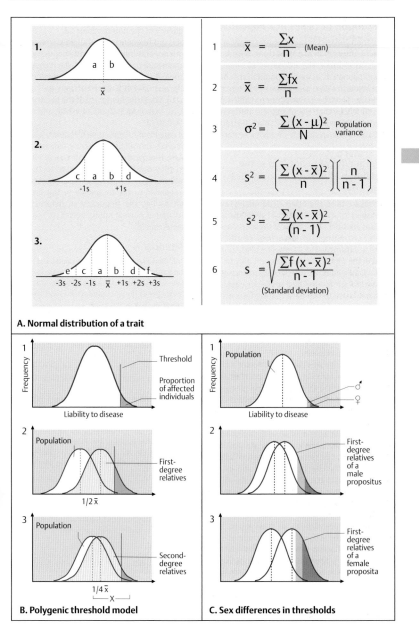

A. Normal distribution of a trait

1.
$$\bar{x} = \frac{\sum x}{n} \quad \text{(Mean)}$$

2.
$$\bar{x} = \frac{\sum fx}{n}$$

3.
$$\sigma^2 = \frac{\sum (x - \mu)^2}{N} \quad \text{Population variance}$$

4.
$$s^2 = \left[\frac{\sum (x - \bar{x})^2}{n} \right] \left[\frac{n}{n-1} \right]$$

5.
$$s^2 = \frac{\sum (x - \bar{x})^2}{(n-1)}$$

6.
$$s = \sqrt{\frac{\sum f (x - \bar{x})^2}{n-1}}$$
(Standard deviation)

B. Polygenic threshold model

C. Sex differences in thresholds

Distribution of Genes in a Population

Population genetics is the scientific study of the genetic composition of populations. A principal goal is to estimate the frequency of alleles at different gene loci in natural populations (*allele frequency*, also called *gene frequency*). From this, conclusions may be drawn about possible selective influences that might explain differences observed. A population can be characterized on the basis of the frequency of alleles at various gene loci.

A. Frequency of genotypes in the children of parents with various genotypes

With regard to an allele pair A (dominant) and a (recessive), six types of parental genotype matings are possible (**1-6**). Each of these has an expected distribution of genotypes in the offspring according to the Mendelian laws, as indicated in the figure. This pattern will only be observed if all genotypes can participate in mating and are not prohibited by a severe disease. The frequency with which each mating type occurs depends on the frequencies of the alleles in the population.

B. Allele frequency

The allele frequency (often called gene frequency) designates the proportion of a given allele at a given locus in a population. If an allele accounts for 20% of all alleles present (at a given locus) in the population, its frequency is 0.20. The allele frequency determines the frequencies of the individual genotypes in a population. For example, for a gene locus with two possible alleles **A** and **a,** three genotypes are possible: **AA, Aa** , or **aa**. The frequency of the two alleles together (p the frequency of **A** and q the frequency of **a**) is 1.0 (100%). If two alleles **A** and **a** are equally frequent (each 0.5), they have the frequencies of $p = 0.5$ for the allele **A** and $q = 0.5$ for the allele **a** (**1**). Thus, the equation $p + q = 1$ defines the population at this locus. The frequency distribution of the two alleles in a population follows a simple binomial relationship: $(p + q)^2 = 1$. Accordingly, the distribution of genotypes in the population corresponds to $p^2 + 2pq + q^2 = 1.0$. The expression p^2 corresponds to the frequency of the genotype **AA;** the expression $2pq$ corresponds to the frequency of the heterozygotes **Aa;** and q^2 corresponds to the frequency of the homozygotes **aa.**

When the frequency of an allele is known, the frequency of the genotype in the population can be determined. For instance, if the frequency p of allele **A** is 0.6 (60%), then the frequency q of allele **a** is 0.4 (40%, derived from q = 1 − p or 1 − 0.6). Thus, the frequency of the genotype **AA** is 0.36; that of **Aa** is 2 × 0.24 = 0.48; and that of **aa** is 0.16 (**2**).

And conversely, if genotype frequency has been observed, the allele frequency can be determined. If only the homozygotes **aa** are known (e.g., they can be identified owing to an autosomal recessive inherited disease), then q^2 corresponds to the frequency of the disorder. From $p = 1-q$, the frequency of heterozygotes ($2pq$) and of normal homozygotes (p^2) can also be determined.

References

Cavalli-Sforza LL, Bodmer WF: The Genetics of Human Populations. WH Freeman & Co, San Francisco, 1971.

Cavalli-Sforza LL, Menozzi P, Piazza A: The History and Geography of Human Genes. Princeton Univ Press, Princeton, New Jersey, 1994.

Eriksson AW et al (eds): Population Structure and Genetic Disorders. Academic Press, London, 1980.

Jorde L: Linkage disequilibrium and the search for complex diseases. Genome Res 10: 1435–1444, 2000.

Kimura M, Ohta T: Theoretical Aspects of Population Genetics. Princeton Univ Press, Princeton, New Jersey, 1971.

Kruglyak L: Prospects for whole-genome linkage disequilibrium mapping of common disease genes. Nature Genet 22: 139–144, 1999.

Terwilliger J, Ott J: Handbook for Human Genetic Linkage. Johns Hopkins Univ Press, Baltimore, 1994.

Vogel F, Motulsky AG: Human Genetics. Problems and Approaches, 3rd ed. Springer Verlag, Heidelberg–New York, 1997.

	Genotype	
	of parents	of offspring
1 AA □—○ AA / AA ◇	AA and AA	1.0 AA
2 AA □—○ Aa / AA ◇ Aa ◇	AA and Aa	0.50 AA 0.50 Aa
3 Aa □—○ Aa / AA ◇ Aa ◇ Aa ◇ aa ◇	Aa and Aa	0.25 AA 0.50 Aa 0.25 aa
4 Aa □—○ aa / Aa ◇ aa ◇	Aa and aa	0.50 Aa 0.50 aa
5 AA □—○ aa / Aa ◇	AA and aa	1.0 Aa
6 aa □—○ aa / aa ◇	aa and aa	1.0 aa

A. Expected frequency of genotypes in children of parents with different genotypes

1

Parents	0.5 A	0.5 a
0.5 A	AA 0.25	Aa 0.25
0.5 a	Aa 0.25	aa 0.25
	Offspring	

$p = 0.50$ (Frequency of A)
$q = 0.50$ (Frequency of a)

2

	A = 0.60	a = 0.40	
A 0.6	0.36 AA	0.24 Aa	} p
a 0.4	0.24 Aa	0.16 aa	} q

$$p^2 + 2pq + q^2 = 1$$
$$0.36 + 0.48 + 0.16 = 1.0$$
$$(AA) \quad (Aa) \quad (aa)$$

B. Allele frequency

Hardy–Weinberg Equilibrium Principle

The Hardy–Weinberg equilibrium principle states that in certain circumstances, the frequency of alleles will remain constant in a population from one generation to the other. This principle was formulated independently by the English mathematician G. F. Hardy and the German physician W. Weinberg in 1908. It assumes that any allele that causes a severe genetic disease incompatible with reproduction will be replaced by a new mutation, as long as the rate of mutations at this locus remains constant.

A. Constant allele frequency

An autosomal recessive allele (here referred to as allele **a**) that leads to a severe disorder in the homozygous state remains undetectable in the heterozygous state in a population. Only the homozygotes (**aa**) can be recognized because of their disease. The frequency of affected individuals (homozygotes **aa**) depends on the frequency of allele **a** (corresponding to q). The frequency of the three genotypes is determined by the binomial relationship $(p+q)^2 = 1$, where p represents the frequency of allele **A** and q, the frequency of allele **a** (see previous page). The homozygous alleles (**aa**) eliminated in one generation by illness are replaced by new mutations. This occurs in each generation, and results in an equilibrium between limination due to illness and frequency of the mutation.

B. Some factors influencing the allele frequency

The Hardy–Weinberg equilibrium principle is valid only in certain conditions: First, it applies only if there is no selection for one genotype. Selection for heterozygotes will increase the frequency of the allele with a selective advantage (see selective advantage of heterozygotes for hemoglobin disorders in regions where malaria is endemic, p. 174). Second, nonrandom matings (assortative mating) will change the allele frequency (proportion of p and q). Third, a change in the rate of mutations will increase the frequency of the allele resulting from mutations. Fourth, in a small population random fluctuation may change the frequency. This called *genetic drift*.

Other causes of a change in allele frequency may occur. If a population experiences a drastic reduction in size, followed by a subsequent increase in the number of individuals, an allele that was previously rare in this population may by chance subsequently become relatively common as the population expands again. This called *founder effect*.

Linkage disequilibrium (LD) is a nonrandom distribution of certain alleles of a haplotype (linked loci on the same chromosome), which deviates from the expected individual frequencies in the population. In the presence of LD, some of the haplotypes are more frequent and others less frequent than expected. LD may be due to one of several factors, such as selection. In other cases, LD may reflect the time elapsed (number of generations) since a mutation occurred in a population. (Detail from "Coney Island, 1938," by Weegee [Arthur Fellig J.].)

References

Cavalli-Sforza LL, Bodmer WF: The Genetics of Human Populations. WH Freeman & Co, San Francisco, 1971.

Cavalli-Sforza LL, Menozzi P, Piazza A: The History and Geography of Human Genes. Princeton Univ Press, Princeton, New Jersey, 1994.

Croucher PJP: Linkage disequlibrium. Nature Encyclopedia of the human Genome 3: 727–728. Nature Publishing Goup, London, 2003.

Eriksson AW et al (eds): Population Structure and Genetic Disorders. Academic Press, London, 1980.

Jorde L: Linkage disequilibrium and the search for complex diseases. Genome Res 10: 1435–1444, 2000.

Kimura M, Ohta T: Theoretical Aspects of Population Genetics. Princeton Univ Press, Princeton, New Jersey, 1971.

Kruglyak L: Prospects for whole-genome linkage disequilibrium mapping of common disease genes. Nature Genet 22: 139–144, 1999.

Vogel F, Motulsky AG: Human Genetics. Problems and Approaches, 3rd ed. Springer Verlag, Heidelberg–New York, 1997.

Zöllner S, Haeseler A von: Population history and linkage equilibrium. Nature Encyclopedia of the Human Genome 4: 628–637. Nature Publishing Goup, London, 2003.

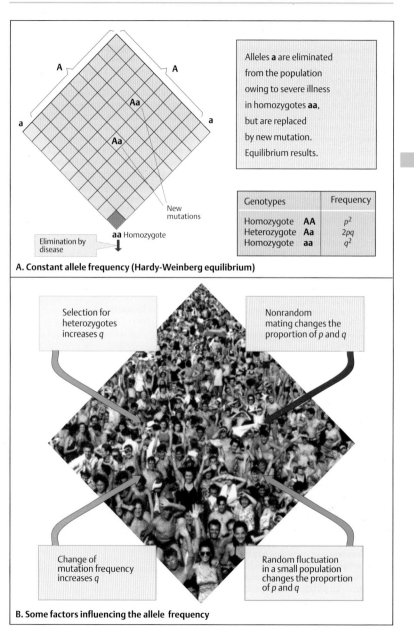

Alleles **a** are eliminated from the population owing to severe illness in homozygotes **aa**, but are replaced by new mutation. Equilibrium results.

Genotypes		Frequency
Homozygote	**AA**	p^2
Heterozygote	**Aa**	$2pq$
Homozygote	**aa**	q^2

New mutations

Elimination by disease

aa Homozygote

A. Constant allele frequency (Hardy-Weinberg equilibrium)

Selection for heterozygotes increases q

Nonrandom mating changes the proportion of p and q

Change of mutation frequency increases q

Random fluctuation in a small population changes the proportion of p and q

B. Some factors influencing the allele frequency

Consanguinity and Inbreeding

Parental consanguinity refers to parents who have at least one common ancestor (being of the "same blood") in the past four generations. Consanguinity increases the chance that an allele will become homozygous in a descendant. As both alleles will be identical, this is called identity by descent (IBD). For this reason consanguinity is often observed in vary rare autosomal recessive diseases. Although consanguineous marriage is widespread in some populations, where it may account for 25–40%, the usual rate is about 1–2%. Inbreeding refers to a population in which consanguinity is common. Incest refers to a relationship between first-degree relatives (brother–sister; parent–child).

The degree of consanguinity can be expressed by two measures: coefficient of kinship or relationship (Φ) and the inbreeding coefficient (F). Φ is the probability that an allele from A is identical by common descent with the allele from B at the same locus. The inbreeding coefficient expresses the probability that two alleles will be homozygous in offspring of consanguineous parents. F of an individual is the same as Φ of its parents. The coefficient of relationship (r) is the proportion of alleles in any two individuals that are identical by descent.

A. Simple types of consanguinity

A mating between brother and sister or father and daughter is called incest (**1**). For two alleles descending from both parents, here designated as A and B, the probability of transmission to each of two offspring, C and D, respectively, is 0.5 (1/2). The sibs (C and D) share half their genes, corresponding to a coefficient of relationship of $^1/_2$. The chance of homozygosity by descent at a given locus in their offspring is $^1/_4$ ($^1/_2$ each from C and D to E).
First cousins share $^1/_8$ of their genes (**2**), second cousins 1/32. The chance of homozygosity by descent in the offspring of first cousins is $^1/_{16}$. An uncle – niece union (**3**) has a coefficient of relationship of r = 1/4 because they share $^1/_4$ of their genes. The chance of homozygosity by descent in their offspring is $^1/_8$. The possibility that an unrelated individual E transmits a mutant allele at this locus can usually be disregarded.

B. Identity by descent (IBD)

IBD refers to regions in the genome that are identical because they are derived from a com-

mon ancestor. For example, for two alleles A and B in ancestor I, and C and D in ancestor II, the probability of transmission to the next generation is 0.5 (1/2) for each. The same applies to the next generation with individuals III and IV. Finally, the probability of transmission from III to V, and IV to V, is again 0.5 or 1/2 each. The number of steps (the probability of transmission) from I to III, and from I to IV, is $(1/2)^2$ each. However, for individual V to be homozygous requires that both alleles are transmitted from I to III and to IV. This corresponds to $(1/2)^4$. For individual V to be homozygous by IBD the probabilities $(1/2)^4$ have to be added and multiplied by the probability of $(1/2)$, which yields 1/16. The resulting probability of homozygosity at a given locus is the inbreeding coefficient, F. Alleles that are identical by descent from a common ancestor are called *autozygous*. IBD is used to identify regions in the genome shared by related individuals affected with a particular disease (homozygosity mapping). Two homozygous alleles that are not identical by descent are called *allozygous*.

Medical relevance

Consanguinity is a frequent reason for genetic counseling. First cousins have a probability of homozygosity by descent in their offspring (F) of 1/16. The risk of a harmful allele reaching their offspring in the homozygous state is 1/64. The total risk is 1/32 because the other common ancestor (both grandparents) also has to be taken into account. Although at first this risk of 3.125% seems high, it actually is not when compared with the risk in the general population. The overall risk for a newborn to have a disorder of any kind is estimated to be 1–2%.

References

Bittles AH, Neel JV: The costs of human inbreeding and their implications for variations at the DNA level. Nature Genet 8: 117–121, 1994.

Griffith AJF et al: An Introduction to Genetic Analysis, 7th ed. WH Freeman & Co, New York, 2000.

Harper PS: Practical Genetic Counselling, 6th ed. Edward Arnold, London, 2004.

Jaber L, Halpern GJ, Shohat M: The impact of consanguinity worldwide. Community Genet 1: 12–17, 1998.

Turnpenny P, Ellard S: Emery's Elements of Medical Genetics, 12th ed. Elsevier-Churchill Livingstone, Edinburgh, 1995.

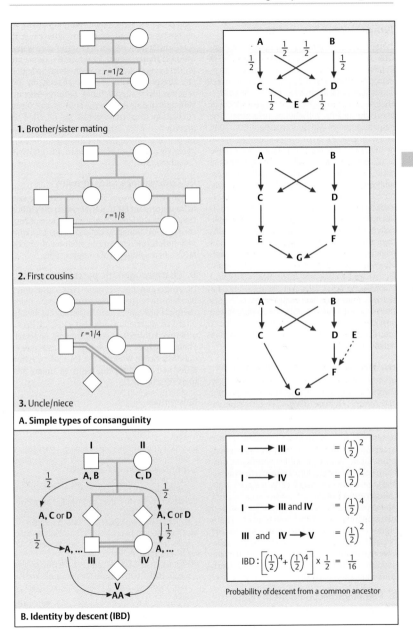

1. Brother/sister mating

$r = 1/2$

2. First cousins

$r = 1/8$

3. Uncle/niece

$r = 1/4$

A. Simple types of consanguinity

$I \longrightarrow III = \left(\frac{1}{2}\right)^2$

$I \longrightarrow IV = \left(\frac{1}{2}\right)^2$

$I \longrightarrow III$ and $IV = \left(\frac{1}{2}\right)^4$

III and $IV \longrightarrow V = \left(\frac{1}{2}\right)^2$

$IBD: \left[\left(\frac{1}{2}\right)^4 + \left(\frac{1}{2}\right)^4\right] \times \frac{1}{2} = \frac{1}{16}$

Probability of descent from a common ancestor

B. Identity by descent (IBD)

Twins

Twin, triplet, or quadruple pregnancies regularly occur in many species of animals. In humans, twinning is detected during pregnancy by ultrasonography in about 1 of 40 pregnancies, but in live births in 1 of 80. This difference results from the early intrauterine death of one twin and its subsequent resorption. The frequency varies widely from about 6 in 1000 births in Asia, 10–20 in 1000 in Europe, and 40 in 1000 in Africa. Twins can arise from a single fertilized egg (genetically identical monozygotic twins, MZ) or from two different eggs (dizygotic twins, DZ), a distinction first proposed by C. Dareste in 1874. The rate of monozygotic twins at birth is relatively constant. F. Galton initiated research on twins in 1876. Multiple systematic comparisons of MZ and DZ twins have attempted to disentangle the contribution of genetic and environmental factors in the etiology of diseases, disease susceptibility, intellectual and behavioral attributes, congenital malformations, etc. However, the resulting data remain controversial in many respects. Nonetheless, twin research may shed light on the possible genetic origin of complex traits, such as multifactorial diseases (see p. 158) and human behavior.

A. Types of human monozygotic twins

Monozygotic twins arise during very early stages of embryonic development by splitting of the inner cell mass (ICM) of the early embryo. Three stages in the timing of splitting can be distinguished: (1) after trophoblast formation, resulting in twins with individual amnions but a common chorion; (2) after amnion formation, resulting in twins in a single chorion and amnion; and (3) before formation of the trophoblast, resulting in twins each with its own chorion and amnion.

About 66% of MZ twins have one chorion and two amnions, suggesting a split after formation of the chorion at day 5, but before formation of the amnion at day 9. About 33% of MZ twins have two complete separate chorions and an individual amnion. (Figure adapted from Gilbert, 2003; and Goerke, 2002.)

B. Pathological conditions in twins

Conjoined twins result from late splitting after formation of the amnion as monochorionic, monoamniotic twins. Such twins are at risk of becoming conjoined (so-called Siamese twins). A relatively frequent form of incomplete separation is thoracopagus, in which the twins are joined to various extents at the thoracic region (1). Dizygotic twins may be affected by erroneous blood supply. If one twin receives insufficient blood due to a shunt in the blood circulation, this twin might be retarded in growth or die (2). Especially severe malformations may result from incompletely formed organs, e.g., absence of the heart in one twin (acardius) (3).

C. Concordance rates in twins

When twins show the same trait, they are said to be concordant for that trait; when they differ, they are discordant. Comparing the rate of concordance in monozygotic and dizygotic twins may reflect the relative contribution of genetic factors in the etiology of complex traits.

D. Pharmacogenetic pattern in twins

Dizygotic and monozygotic twins also differ biochemically. Due to genetic differences, many chemical substances used in therapy are metabolized or excreted at different rates, owing to different activities of corresponding enzymes. Phenylbutazone is excreted at the same rate in identical twins, whereas the rates of excretion differ between dizygotic twins or among siblings (Vesell, 1978).

References

Boomsma D, Busjahn A, Peltonen L: Classical twin studies and beyond. Nature Rev Genet 3: 872–882, 2002.

Bouchard TJ et al: Sources of human psychological differences: The Minnesota study of twins reared apart. Science 250: 223–228, 1990.

Gilbert SF: Developmental Biology, 7th ed. Sinauer, Sunderland, Massachusetts, 2003.

Goerke K: Taschenatlas der Geburtshilfe. Thieme Verlag, Stuttgart, 2002.

Hall JG: Twinning. Lancet 362: 735–743, 2003.

McGregor AJ et al: Twins. Novel uses to study complex traits and genetic diseases. Trends Genet 16: 131–134, 2000.

Phelan MC, Hall JG: Twins, pp 1377–1411. In: Stevenson RE, Hall JG (eds) Human Malformations and Related Anomalies, 2nd ed. Oxford Univ Press, Oxford, 2006

Vesell ES: Twin studies in pharmacogenetics. Hum Genet 1 (Suppl): 19–30, 1978.

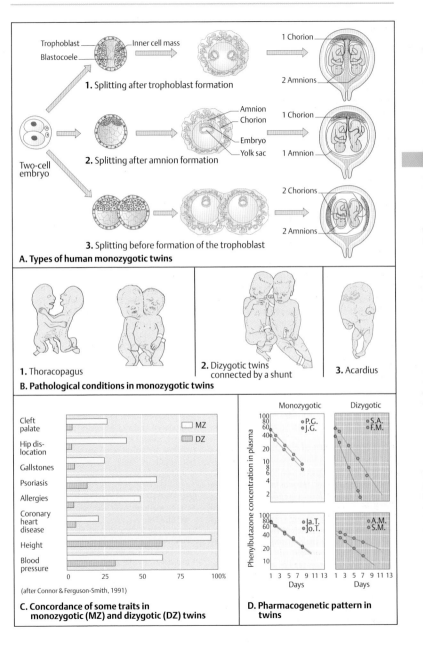

Trophoblast — Inner cell mass
Blastocoele

1. Splitting after trophoblast formation

1 Chorion

2 Amnions

Two-cell embryo

Amnion
Chorion
Embryo
Yolk sac

2. Splitting after amnion formation

1 Chorion

1 Amnion

3. Splitting before formation of the trophoblast

2 Chorions

2 Amnions

A. Types of human monozygotic twins

1. Thoracopagus

2. Dizygotic twins connected by a shunt

3. Acardius

B. Pathological conditions in monozygotic twins

Cleft palate
Hip dislocation
Gallstones
Psoriasis
Allergies
Coronary heart disease
Height
Blood pressure

☐ MZ
▨ DZ

0   25   50   75   100%

(after Connor & Ferguson-Smith, 1991)

C. Concordance of some traits in monozygotic (MZ) and dizygotic (DZ) twins

Monozygotic   Dizygotic

Phenylbutazone concentration in plasma

P.G.
J.G.

S.A.
F.M.

Ja.T.
Jo.T.

A.M.
S.M.

1 3 5 7 9 11 13   1 3 5 7 9 11 13
Days   Days

D. Pharmacogenetic pattern in twins

Polymorphism

Genetic polymorphism refers to genetic variation as observed in populations. Two types of variation are distinguished: *discontinuous* and *continuous* (see p. 158). Two or more common discontinuous variants in a natural population are called a polymorphism (many forms, from Greek). A gene locus is defined as polymorphic if a rare allele has a frequency of 0.01 (1%) or more, corresponding to a heterozygote frequency of 0.02 (2%). An allelic polymorphism often results in a different phenotype. A polymorphism is considered as neutral if the presence of a certain allele does not confer any selective advantage. On the other hand, it may represent a selective advantage. Natural selection refers to the differential rates of survival and reproduction. The relative probability of survival and rate of reproduction is called Darwinian (or reproductive) fitness. A fitness of 1.0 indicates a "normal" value, 0.0 indicates its absence. In this context, the term "fitness" must not be confused with its everyday sense. A polymorphism represents an advantage for a population when the resulting genetic variation contributes to individuals having a better reproductive fitness in the given environmental conditions.

Polymorphism can be observed at the level of the whole individual (phenotype), in variant forms of proteins and blood group substances (biochemical polymorphism), in morphological features of chromosomes (chromosomal polymorphism), or at the level of DNA in differences of nucleotides (DNA polymorphism, p. 78).

A. Polymorphism of the phenotype

An impressive example of phenotypic polymorphism is the color pattern on the wing sheaths of the Asian ladybug (*Harmonia axyridis*) (**1**). In the area of distribution, extending from Siberia to Japan, multiple variants can be distinguished (Ayala, 1978). The different color combinations are due to different alleles of the same gene. In the California king snake (*Lampropeltis getulus californiae*), color patterns differ to such an extent within the same species that they would seem to represent different species (**2**). (Figure in A1 adapted from F. J. Ayala, 1978, in A2 from J. H. Tashjian, San Diego Zoo.)

B. Polymorphism related to environmental conditions

Living organisms may differ in the frequency of some alleles in the population as a result of adaptation to environmental conditions. The differences may be gradual. For example, the average height of the yarrow plant on the slopes of the Sierra Nevada in California decreases as the altitude increases (**1**). By comparing the growth of plants whose seeds were obtained at different altitudes and sown in one garden, it can be shown that the average height of plants at the different altitudes is genetically determined. (Figure adapted from Campbell, 1990.)

One of the most impressive cases of natural selection for a polymorphic color pattern has been described by B. Kettlewell in the English peppered moth *Biston betularia* in the midlands of England in the late 19th century (**2**). It was observed in two varieties, an originally common light gray variety, *B. typica* and a rare dark gray variety, *B. carbonaria*, that live on lightly colored lichen covering tree barks and rocks. By about 1850, after industrialization turned the surfaces dark, the previously rare black variant (*B. carbonaria*) became common, and by 1900 was practically the only one present. It was assumed that the black variety was originally rare because predators could easily detect it on a light surface. With the beginning of industrialization, the light-colored lichen became rare. Now the dark variety, *B. carbonaria*, was assumed to have a selective advantage. After pollution was reduced in the 1950s, the light type reappeared and again became common. Unfortunately, doubt has been cast on this beautiful example of natural selection by genetic adaptation (Majerus, 1998; but see Coyne, 2002). It would be a pity.

References

Ayala FJ: Mechanisms of evolution. Scient Amer 329: 48–61, 1978.

Campbell NA: Biology, 2nd ed. Benjamin/Cummings Publishing Co, Menlo Park, California, 1990.

Coyne JA: A look at the controversy about industrial melansism in the peppered moth. Nature 418: 19–20, 2002.

Lewontin R: Adaptation. Scient Amer 239: 156–169, 1978.

Majerus, MEN: Melanism: Evolution in Action. Oxford Univ Press, Oxford, 1998. (reviewed in Nature 396: 35–36, 1998; doi:10.1038/23856).

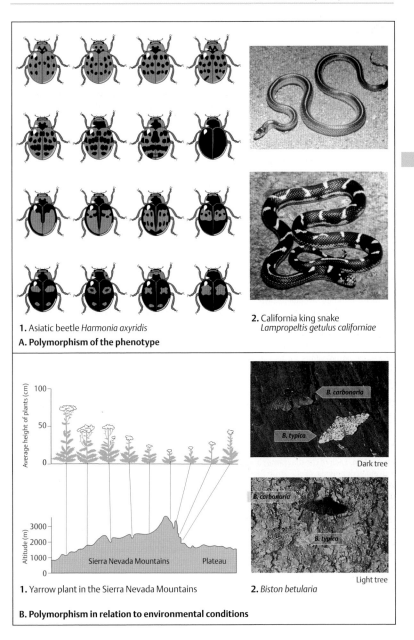

1. Asiatic beetle *Harmonia axyridis*

2. California king snake
Lampropeltis getulus californiae

A. Polymorphism of the phenotype

Average height of plants (cm)

Altitude (m)

Sierra Nevada Mountains   Plateau

B. carbonaria

B. typica

Dark tree

B. carbonaria

B. typica

Light tree

1. Yarrow plant in the Sierra Nevada Mountains

2. *Biston betularia*

B. Polymorphism in relation to environmental conditions

Biochemical Polymorphism

Biochemical polymorphism refers to a genetic polymorphism that causes differences in metabolic reactions detectable by laboratory methods. If a difference in the sequence of nucleotide bases of a gene changes a codon, a different amino acid may be incorporated at the corresponding site of its protein. This difference can be demonstrated by analyzing the electrophoretic pattern of the protein.

A. Recognition by gel electrophoresis

In 1966, R.C. Lewontin and J.L. Hubby first used gel electrophoresis to survey protein variants in a natural population of *Drosophila pseudoobscura*. Genetic polymorphism of a protein can be demonstrated when one variant form differs from the others by the presence of an amino acid with a different electric charge. In this case, the allelic forms of the gene product can be distinguished owing to their different speeds of migration in an electrical field (electrophoresis). If the difference results from two variant forms encoded by two alleles, the three genotypes can be detected. In the figure, six lanes are shown, lanes 1–3 for one, and lanes 4–6 for another protein. The top row shows the starting point of an electrophoresis. The lower bands represent homozygous forms (lanes 1 and 2, and lanes 4 and 5, respectively) and heterozygous forms in lanes 3 and 6.

B. Polymorphism in gene products

Here, polymorphism is shown to be frequent in three typical gene products, the enzymes phosphoglucomutase (**1**), malate dehydrogenase (**2**), and acid phosphatase (**3**), in different species of *Drosophila*. Each of the diagrams shows a starch gel electrophoresis of 12 fruit flies, each gel being specifically stained for the respective protein. The gene locus for phosphoglucomutase (**1**) is heterozygous in lanes 2, 4, and 10 by virtue of two bands migrating at different speeds.

Malate dehydrogenase is a dimeric protein. Therefore, heterozygous flies show three bands in lanes 4, 5, 6, and 8. Acid phosphatase (**3**) is encoded by four alleles and shows a complex pattern, with heterozygous flies represented by lanes 1, 3, 4, 6, and 9. The overall heterozygosity is about 8–15%, a new insight at the time, in 1966. (Figure adapted from Ayala & Kieger, 1984.)

C. Frequency of polymorphism

Genetic polymorphism is common. In a study of the average heterozygosity in *Drosophila willistoni*, 17.7% of 180 gene loci were found be heterozygous (**1**). The average heterozygosity is the proportion of heterozygous individuals in a population with reference to the number of analyzed loci. This is determined by the sum of the proportion of heterozygotes per total number of individuals per gene locus. Most common is genetic polymorphism at the DNA level (see p. 78). Here, a Southern blot following restriction enzyme digestion of DNA from a gene locus near the H chain of the immunoglobulin J region produces several polymorphic bands due to individual differences in the fragment sizes in 16 individuals (**2**). (Figure in 2 adapted from White et al, 1986.)

D. Genetic diversity and evolution

Genetic diversity is selected for by evolution. A genetically diverse population may have a higher probability of survival in a changing environment than a homogeneous one. An early study compared two populations of *Drosophila serrata* kept for 25 generations in separate closed bottles with limited availability of space and nourishment. A genetically diverse population appeared to thrive better than a homogeneous population. At the time, it was interpreted that the population with the greater genetic diversity was able to adjust better to the environmental conditions than the homogenous population. However, today this may not be considered the only interpretation, because random fluctuations cannot be excluded. (Figure adapted from Ayala & Kieger, 1984.)

References

Ayala FJ: Genetic polymorphisms: From electrophoresis to DNA sequences. Experentia 39: 813–823, 1983.

Ayala FJ, Kieger JA: Modern Genetics, 2nd ed. Benjamin/Cummings Publishing Co, Menlo Park, California, 1984.

Beaudet AL et al: Genetics, biochemistry, and molecular basis of variant human phenotypes, pp 3–45. In: Scriver CR et al (eds) The Metabolic and Molecular Bases of Inherited Disease, 8th ed. McGraw-Hill, New York, 2001.

White R et al: Construction of human genetic linkage images. I. Progress and perspectives. Cold Spring Harbor Symp Quant Biol 51: 29–38, 1986.

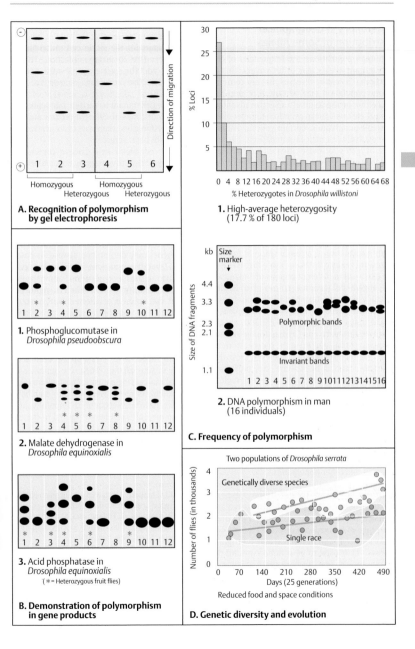

A. Recognition of polymorphism by gel electrophoresis

Homozygous — Heterozygous — Homozygous — Heterozygous

Direction of migration

B. Demonstration of polymorphism in gene products

1. Phosphoglucomutase in *Drosophila pseudoobscura*

2. Malate dehydrogenase in *Drosophila equinoxialis*

3. Acid phosphatase in *Drosophila equinoxialis*
(∗ = Heterozygous fruit flies)

1. High-average heterozygosity
(17.7 % of 180 loci)

% Loci

% Heterozygotes in *Drosophila willistoni*

2. DNA polymorphism in man
(16 individuals)

Size of DNA fragments

Size marker

Polymorphic bands

Invariant bands

C. Frequency of polymorphism

Two populations of *Drosophila serrata*

Genetically diverse species

Single race

Number of flies (in thousands)

Days (25 generations)

Reduced food and space conditions

D. Genetic diversity and evolution

Differences in Geographical Distribution of Some Alleles

In human populations, the frequencies of certain mutant alleles may differ widely. That is, autosomal recessive diseases that are common in one population may be rare in others. This may either be the result of a selective advantage of an allele or be due to a random founder effect in small populations. Certain genetic diseases are relatively frequent in populations, because heterozygotes for those diseases have a selective advantage. Two examples are presented here (for others, see Turnpenny and Ellard, 2005).

A. Different frequencies

Finland represents a small population in which several rare recessive diseases occur at a much higher frequency than elsewhere. The most likely explanation is a combination of founder effect and genetic isolation. Three diseases that are clustered in different regions of Finland are (1) congenital flat cornea (cornea plana 2, MIM 217300) in the western part of the country, (2) the Finnish type of congenital nephrosis, a severe renal disorder (MIM 256300) in the southwestern part, and (3) diastrophic skeletal dysplasia (MIM 222600) in the southeastern region. The recessive mutant alleles in these diseases are identical by descent. They must have arisen independently in the different regions because the grandparents live predominantly there. There is no known basis for a selective advantage of these mutant alleles in the heterozygotes to explain their frequencies. The different distributions merely reflect the places and relative points in time of the mutations. Similar examples are found in many other regions of the world and in other populations.

B. Malaria and hemoglobin disorders

The distribution of the parasitic disease malaria overlaps closely with the distribution of different types of hemoglobin disorder (see p. 344). Malaria is common in tropical and subtropical regions (1). It used to be common around the Mediterranean sea, but here the frequency of malaria has been reduced due to successful prevention. In the same regions with endemic malaria, several types of hemoglobin disorder are prevalent (2, 3). Typical examples are sickle cell anemia (MIM 141900) and different types of thalassemia (MIM 187550, see p. 350). In 1954 A. C. Allison proposed that individuals who are heterozygous for the sickle cell mutation are less susceptible to malaria infection. This is the first and the best example of a heterozygous selective advantage known in man (see p. 346).

The sickle cell mutant has arisen independently at least four times in different regions and has become established there due to a selective advantage for heterozygotes.

Another red blood cell disease confers on heterozygotes, an advantage against malaria infection: glucose-6-phosphate dehydrogenase deficiency (MIM 305900), an X-chromosomal anemia disorder leading to severe anemia in hemizygous males, whereas heterozygous females are normal and relatively protected against malaria.

The selective advantage of heterozygotes for these genetically determined diseases is based on less favorable conditions for the malaria parasite than in the blood of normal homozygotes (see p. 346). The protection of heterozygotes occurs at the expense of affected homozygotes, who suffer from one of the severe hemoglobin disorders. The benefits are at the population level. More than 400 million people are infected with malaria each year in Africa, Asia, and South America; about 1 to 3 million die each year, mainly in sub-Saharan Africa. The genomes of the malaria-causing parasite, *Plasmodium falciparum*, and its vector, the mosquito *Anopheles gambiae*, have been sequenced. Hopefully, preventive measures can be introduced in the endemic areas.

References

Cavalli-Sforza LL, Menozzi P, Piazza A: The History and Geography of Human Genes. Princeton Univ Press, Princeton, 1994.

Marshall E: Malaria: A renewed assault on an old and deadly foe. Science 290: 428–430, 2000.

Norio R: Diseases of Finland and Scandinavia. In: Rothschild HR (ed) Biocultural Aspects of Disease. Academic Press, New York, 1981.

Norio R: The Finnish disease heritage. III. The individual diseases. Hum Genet 112: 470–526, 2003.

Turnpenny P, Ellard S: Emery's Elements of Medical Genetics, 12th ed. Elsevier-Churchill Livingstone, Edinburgh, 2005.

Weatherall DJ, Clegg JB: Thalassemia - a global public health problem. Nature Med 2: 847–849, 1996.

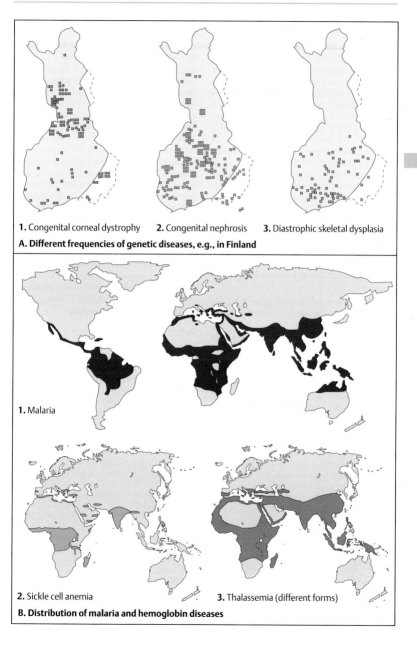

1. Congenital corneal dystrophy   **2.** Congenital nephrosis   **3.** Diastrophic skeletal dysplasia

A. Different frequencies of genetic diseases, e.g., in Finland

1. Malaria

2. Sickle cell anemia   **3.** Thalassemia (different forms)

B. Distribution of malaria and hemoglobin diseases

Chromosomes in Metaphase

Chromosomes are the units into which the genes are organized. In eukaryotic cells, each chromosome consists of a continuous thread of DNA double helix and associated proteins.

The term chromosome, derived from Greek for "colored thread," was introduced by W. Waldeyer in 1888. In prokaryotes, the single chromosome usually has a circular structure. In eukaryotes, the chromosomes are located in the nucleus at defined positions. As first noted by W. Flemming in 1879, chromosomes can be visualized under a light microscope during mitosis as separate, individual structures. Each organism has a defined number of chromosomes with distinct morphological appearances. In 1956, Tjio and Levan in Lund, and Ford and Hamerton in Oxford established that man has 46 chromosomes, not 48 as had been previously assumed.

A. The metaphase chromosomes of man

A metaphase is shown here at about 2800-fold magnification. Chromosomes differ from each other in length, position of the centromere, and size and arrangement of their transverse light and dark bands (banding pattern). Each chromosome and parts of it can be identified by its banding pattern. In a typical metaphase preparation, about 300–550 distinct bands can be distinguished. In prometaphase, the chromosomes are longer than in metaphase and show more bands. Thus, for certain purposes chromosomes are also studied in prometaphase.

B. Types of metaphase chromosome

Each chromosome is classified as *submetacentric*, *metacentric*, or *acrocentric* according to the location of its centromere. This appears as a constriction, the point of attachment of the spindle during mitosis. The centromere divides a submetacentric chromosome into a short arm (p arm, derived from the French *petit*) and a long arm (q, next letter after p). In metacentric chromosomes, the short and long arms are about the same length. Acrocentric chromosomes carry as their short arm a dense appendage called a satellite (not to be confused with satellite DNA) at the end of a stalk.

C. Karyotype

The karyotype is the chromosomal complement of a cell, an individual, or a species. It is characteristic for each species. The karyotype describes the light-microscopic appearance of chromosomes in metaphase according to their morphology. A karyogram displays all chromosomes in homologous pairs, one from the mother and one from the father, arranged according to their relative lengths and the positions of their centromeres. Chromosomes are arranged and numbered according to a convention. Man *(Homo sapiens)* has 22 pairs of chromosomes (autosomes, pairs 1–22) and in addition either two X chromosomes, in females, or an X and a Y chromosome, in males (karyotype 46,XX or 46,XY, respectively). In front of the comma, the karyotype formula gives the total number of chromosomes present, and after the comma, the composition of the sex chromosomes. The 22 pairs of autosomes in man are divided into seven groups (A–G).

The karyotypes of different species of animals show differences and similarities depending on their evolutionary relationships.

References

Caspersson T, Zech L, Johansson C: Differential binding of alkylating fluorochromes in human chromosomes. Exp Cell Res 60: 315–319, 1970.

Dutrillaux B: Chromosomal evolution in primates: tentative phylogeny for *Microcebus murinus* (prosimian) to man. Hum Genet 48: 251–314, 1979.

Dutrillaux B, Lejeune J: Sur une nouvelle technique d'analyse du caryotype humain. CR Acad Sci Paris D 272: 2638–2640, 1971.

Ford CE, Hamerton JL: The chromosomes of man. Nature 178: 1020–1023, 1956.

ISCN 2005. An International System for Human Cytogenetic Nomenclature. Shaffer LG, Tommerup N (eds). Karger, Basel, 2005.

Lewin B: Genes VIII. Pearson Educational International, 2004.

Miller OJ, Therman E: Human Chromosomes, 4th ed. Springer, New York, 2001.

Riddihough G: Chromosomes through space and time. Science 301: 779, 2003.

The Dynamic Chromosome. Science, special issue, 301: 717–876, 2003.

Tjio JH, Levan A: The chromosome number of man. Hereditas 42: 1–6, 1956.

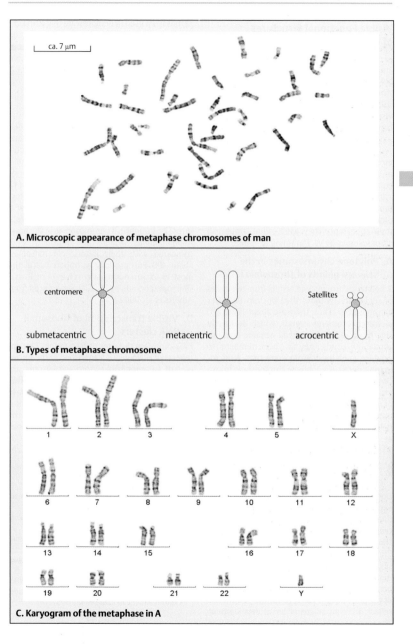

A. Microscopic appearance of metaphase chromosomes of man

centromere

submetacentric

metacentric

Satellites

acrocentric

B. Types of metaphase chromosome

ca. 7 µm

1 2 3 4 5 X

6 7 8 9 10 11 12

13 14 15 16 17 18

19 20 21 22 Y

C. Karyogram of the metaphase in A

Visible Functional Structures of Chromosomes

In certain cells and tissues of some insects and amphibians, chromosomal structures can be observed that relate to their function. Polytene ("many threads") chromosomes, formed as a result of repeated DNA synthesis without cell division, have a distinct pattern of chromosome banding readily visible under the light microscope. These chromosomes, first observed in cells of insect salivary glands (*Drosophila melanogaster* and *Chironomus*) by E.G. Balbiani in 1881, show regions of temporary localized enlargement called Balbiani rings.

In the oocytes of some animals, fine loops protrude from the chromosomes during the diplotene phase of meiosis (see p. 118). Because of their appearance they were called lampbrush chromosomes by W. Flemming in 1882.

A. Polytene chromosomes in the salivary glands of *Drosophila* larvae

A polytene chromosome results from ten cycles of replication without division into daughter chromosomes. Thus, there are about 1024 (2^{10}) identical chromatid strands, which lie strictly side by side. The *Drosophila* genome contains about 5000 bands. They have been numbered to produce a polytene chromosome map. A micrographic detail of a polytene chromosome from a *Drosophila* salivary gland shows the characteristic banding pattern (shown at the bottom of part A). The dark bands are the result of chromatin condensation in the large interphase polytene chromosomes. (Figure from Alberts et al., 2002, modified from T.S. Painter, J. Hered. 25: 465–476, 1934.)

B. Functional stages in polytene chromosomes

During the larval development of *Drosophila*, expansions (puffs) appear and recede at defined positions in temporal stages in the polytene chromosomes (1). Each is active for only a short developmental period of about 22 hours. Chromosome puffs are decondensed, expanded segments that represent active chromosomal regions, i.e., regions that contain genes that are being transcribed. The locations and durations of the puffs occur in a reproducible pattern and reflect different stages of larval development. The incorporation of radioactively labeled RNA

(2) has been used to demonstrate that RNA synthesis occurs in these regions as a sign of gene activity (transcription). (Figure from Alberts et al., 2002; adapted from M. Ashburner et al., Cold Spring Harbor Symp. Quant. Biol. 38: 655–662, 1974.)

C. Lampbrush chromosomes in oocytes

Lampbrush chromosomes are greatly extended chromosome bivalents during the diplotene phase of meiosis in oocytes of certain amphibians. This stage can last several months. A meiotic bivalent of two pairs of sister chromatids can be seen under the light miscrocope (1). They are held together at points of chiasma formation. The interpretation is that loops of a paired chromosome form mirror-image structures (2). Lampbrush chromosomes in the newt *Notophthalmus viridens* are unusually large compared with their mitotic chromosomes, about 400–800 μm long as opposed to at the most 15–20 μm during later stages of meiosis. (Microphotograph by J.G. Gall, reproduced from Alberts et al., 2002.)

D. Visible transcription of ribosomal RNA clusters

Ribosomes are the sites of translation. Tandem repeats of rRNA genes transcribed in the nucleolus of an amphibian, *Triturus viridescens*, are shown here. Along each gene many rRNA molecules are synthesized by RNA polymerase I. The growing RNA molecules extend from a backbone of DNA, the shorter ones being at the start site of transcription and the longer ones having been completed. (Photograph by O.L. Miller and B.A. Hamkalo, reproduced from Griffiths et al., 2000.)

References

Alberts B et al: Molecular Biology of the Cell, 4th ed. Garland Publishing, New York, 2002.

Callan HG: Lampbrush chromosomes. Proc R Soc Lond B Biol Sci 214: 417–448, 1982.

Gall JG: On the submicroscopic structure of chromosomes. Brookhaven Symp Biol 8: 17–32, 1956.

Griffiths AJF et al: An Introduction to Genetic Analysis, 7th ed. WH Freeman, New York, 2000.

Lewin B: Genes VIII. Prentice Hall, New Jersey, 2004.

Miller Jr OL: The visualization of genes in action. Scientific American 228: 34–42, 1973.

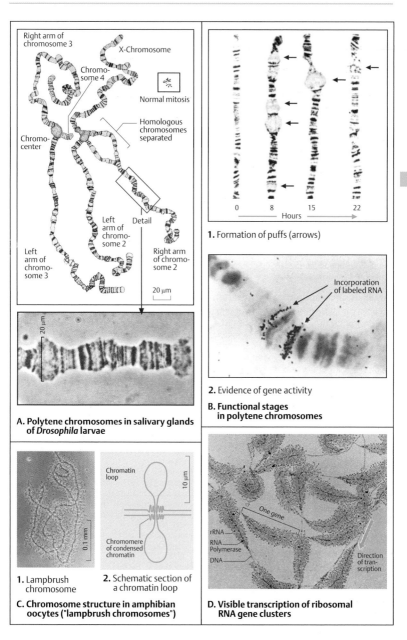

A. Polytene chromosomes in salivary glands of _Drosophila_ larvae

Right arm of chromosome 3

X-Chromosome

Chromosome 4

Normal mitosis

Homologous chromosomes separated

Chromocenter

Left arm of chromosome 2

Detail

Right arm of chromosome 2

Left arm of chromosome 3

20 μm

20 μm

1. Formation of puffs (arrows)

0 8 15 22

Hours

Incorporation of labeled RNA

2. Evidence of gene activity

B. Functional stages in polytene chromosomes

Chromatin loop

10 μm

Chromomere of condensed chromatin

0.1 mm

1. Lampbrush chromosome

2. Schematic section of a chromatin loop

C. Chromosome structure in amphibian oocytes ("lampbrush chromosomes")

One gene

rRNA

RNA Polymerase

DNA

Direction of transcription

D. Visible transcription of ribosomal RNA gene clusters

Chromosome Organization

Chromosomes are visible as separate structures only during mitosis; during interphase they appear as a tangled mass called chromatin. The density of chromatin varies, referred to as heterochromatin and euchromatin (E. Heitz 1928). These relate to the overall activity of genes: euchromatin to active genes, heterochromatin to inactive genes.

A. A histone-free chromosome under the electron microscope

When histone proteins are removed from chromosomes, the DNA chromosomal skeleton becomes visible under the electron microscope (1). A histone-depleted chromosome, with only about 8% of its original protein content, is visible in the center as a dark scaffold surrounded by a halo of DNA (darkly stained threads). A higher magnification (2) shows that DNA is a single continuous thread. (Photographs from Paulson & Laemmli, 1977.)

B. Levels of chromosome organization

From the chromosome to its DNA strand, different levels of organization can be distinguished. The total length of haploid DNA in a dividing human cell is about 1 m. During mitosis, this has to fit into 23 chromosomes of about 3–7 μm each. When a portion of a chromosome arm corresponding to 10% of that chromosome is magnified tenfold, it might be seen to contain about 40 genes, depending on the chromosome segment chosen (eight are shown here, 2). Magnifying a tenth of that section another tenfold (3) would yield a region containing on average 3–4 genes, corresponding to about 1% of the chromosome. A further tenfold magnification shows a single gene (4) with its exon/intron structure. The last level (5) would be that of the nucleotide sequence of the gene and its surrounding DNA. (Figure adapted from Alberts et al., 2002.)

C. Heterochromatin and euchromatin

In 1928, Emil Heitz observed that certain parts of the chromosomes of a moss (*Pellia epiphylla*) remain thickened and deeply stained during interphase, as chromosomes otherwise do only during mitosis. He named these structures *heterochromatin*. Those parts that were less densely stained and became invisible during late telophase and subsequent interphase he called *euchromatin*. Subsequent studies showed that heterochromatin consists of regions with few or no active genes, whereas euchromatin corresponds to regions with active genes. When active genes become located close to the heterochromatin, they usually become inactivated (position effect–variegation). (Figure from Heitz, 1928.)

D. Constitutive heterochromatin at the centromeres (C bands)

The centromeres of eukaryotic chromosomes contain large amounts of repetitive DNA, called α-satellites. They are specific for each chromosome. These sequences are visible in the centromeric region (constitutive heterochromatin). This can be specifically stained (C bands). The distal half of the long arm of the Y chromosome is also C-band-positive. The heterochromatin of the centromeres of chromosomes 1, 9, and 16 and of the long arm of the Y chromosome differs in length in different human individuals (chromosomal polymorphism). (Photograph from Verma & Babu, 1989.)

References

Bickmore WA, Sumner AT: Mammalian chromosome banding: an expression of genome organization. Trends Genet 5: 144–148, 1989.

Brown SW: Heterochromatin. Science 151: 417–425, 1966.

Grasser SM, Laemmli UK: A glimpse at chromosomal order. Trends Genet 3: 16–22, 1987.

Grewal SIS, Moazed D: Heterochromatin and epigenetic control of gene expression. Science 301: 798–802, 2003.

Heitz E: Das Heterochromatin der Moose. I. Jahrb Wiss Bot 69: 762–818, 1928.

Lewin B: Genes VIII. Pearson Education International, 2004.

Manuelidis L: View of metaphase chromosomes. Science 250: 1533–1540, 1990.

Passarge E: Emil Heitz and the concept of heterochromatin: Longitudinal chromosome differentiation was recognized fifty years ago. Am J Hum Genet 31: 106–115, 1979.

Paulson JR, Laemmli UK: The structure of histone-depleted metaphase chromosome. Cell 12: 817–828, 1977.

Pluta AF et al: The centromere: Hub of chromosomal activities. Science 270: 1591–1594, 1995.

Sumner A: Chromosomes: Organization and Function. Blackwell, Malden, MA, 2003.

Verma AS, Babu A: Human Chromosomes. Pergamon Press, New York, 1989.

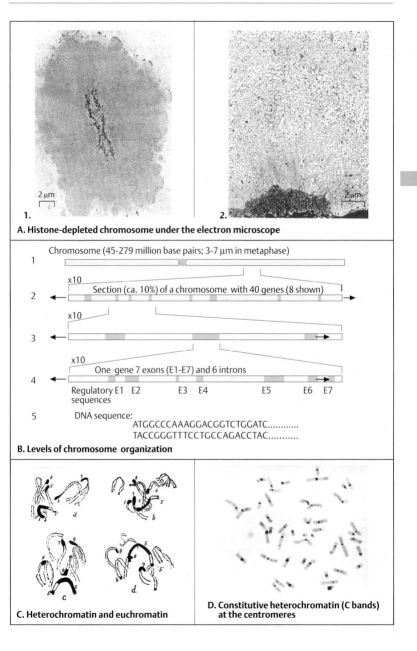

A. Histone-depleted chromosome under the electron microscope

B. Levels of chromosome organization

1 Chromosome (45-279 million base pairs; 3-7 µm in metaphase)

2 x10
   Section (ca. 10%) of a chromosome with 40 genes (8 shown)

3 x10

4 x10
   One gene 7 exons (E1-E7) and 6 introns
   Regulatory E1 E2   E3 E4   E5   E6 E7
   sequences

5   DNA sequence:
                ATGGCCCAAAGGACGGTCTGGATC............
                TACCGGGTTTCCTGCCAGACCTAC...........

C. Heterochromatin and euchromatin

D. Constitutive heterochromatin (C bands) at the centromeres

Functional Elements of Chromosomes

A chromosome requires three types of structures to function properly: (i) centromeric sequences (CEN), (ii) autonomous replicating sequences (ARS, origin of replication), and (iii) telomeric sequences (TEL). The individual contributions of these three types of functional chromosomal elements have been demonstrated in mutant yeast cells (*S. cerevisiae*, baker's yeast). Yeast artificial chromosomes (YACS) must contain all three elements.

A. Basic features of a eukaryotic chromosome

The *centromere* and the *telomeres*, one at each end, are distinct hallmarks of a chromosome. The centromere attaches a chromosome to the spindle at mitosis. Centromeric DNA contains large amounts of repetitive DNA called α-satellite DNA, the most abundant type being long tandem repeats of a 170-bp monomeric sequence. The total length of centromeric DNA ranges from about 300 to 5000 kb. Cloned α-satellite fragments hybridize specifically to individual chromosomes.

The subtelomeric sequences, located proximal to the telomeric sequences, contain sequence homologies shared among subsets of other chromosomes (Martin et al., 2002). Subtelomeric sequences are medically important because rearrangements of these sequences are found in about 5% of patients with mental retardation (De Vries et al., 2003; Knight et al., 1999). Telomeres are described on p. 188.

B. Autonomous replicating sequences

Mutant yeast cells that cannot synthesize the amino acid leucine (Leu) can be transformed with cloned plasmids that contain the gene for leucine synthesis. However, such cells still cannot grow in culture medium lacking leucine because they cannot replicate DNA (**1**). Replication can be restored if ARS sequences are transferred along with the *Leu* gene, because plasmid DNA containing the *Leu* gene and ARS is able to replicate (**2**). However, only about 5–20% of daughter cells receives the plasmid DNA and can grow on Leu⁻ media.

C. Centromeric sequences (CEN)

If in addition to the *Leu* gene and ARS the plasmid contains sequences from the centromere (CEN) of the yeast chromosome, nearly 90% of progeny can grow on Leu⁻ medium (**1**) because normal mitotic segregation takes place (**2**). Thus, CEN sequences are necessary for normal distribution of the chromosomes at mitosis (see p. 116). The centromere sequences contain three elements with a total of about 220 base pairs (bp), which occur in all chromosomes.

D. Telomeric sequences (TEL)

When Leu⁻ yeast cells are transfected with a plasmid that in addition to the *Leu* gene contains ARS and CEN sequences (**1**) but that is linear instead of circular, as in C and D, the cells fail to grow in Leu⁻ medium (**2**). However, if telomere sequences (TEL) are attached to both ends of the plasmid (**3**) before it is incorporated into the cells (**4**), normal growth in Leu⁻ medium takes place (**5**). In this case the linearized plasmid behaves as a normal chromosome. (Figures adapted from Lodish et al., 2004.)

References

Burke DT, Carle GF, Olson MV: Cloning of large segments of exogenous DNA into yeast by means of artificial chromosome vectors. Science 236: 806–812, 1987.

Clarke L, Carbon J: The structure and function of yeast centromeres. Ann Rev Genet 19: 29–55, 1985.

Cleveland DW, Mao Y, Sullivan KI: Centromeres and kinetochores: From epigenetics to mitotic checkpoint signaling. Cell 112: 407–421, 2003.

De Vries BBA et al.: Telomeres: a diagnosis at the end of chromosomes. J Med Genet 40: 385–398, 2003.

Knight SJL, Flint J: Perfect endings: a review of subtelomeric probes and their clinical use. J Med Genet 37: 401–409, 2000.

Martin CL et al: The evolutionary origin of human subtelomeric homologies – or where the end begins. Am J Hum Genet 70: 972–984, 2002.

Miller OJ, Therman E: Human Chromosomes, 4th ed. Springer, New York, 2001.

Schlessinger D: Yeast artificial chromosomes: Tools for mapping and analysis of complex genomes. Trends Genet 6: 248–258, 1990.

Schueler MG et al: Genomic and genetic definition of a functional human centromere. Science 294: 109–115, 2001.

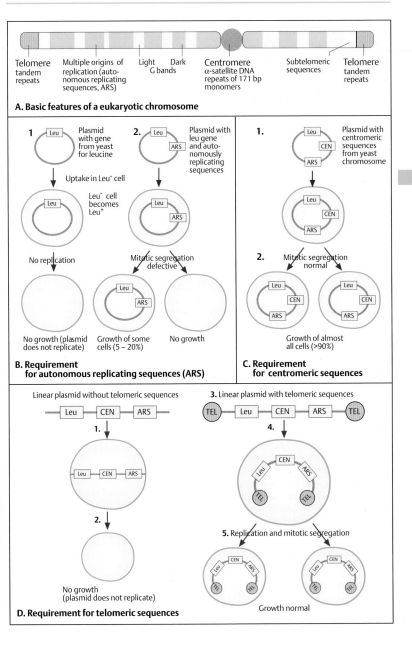

A. Basic features of a eukaryotic chromosome

B. Requirement for autonomous replicating sequences (ARS)

C. Requirement for centromeric sequences

D. Requirement for telomeric sequences

DNA and Nucleosomes

During interphase, DNA and its associated proteins (histones) are tightly packed in chromatin. The overall packing ratio of DNA is about 1000–10 000-fold. This is achieved by its hierarchical organization. The fundamental subunit of chromatin is the nucleosome, first delineated by R.D. Kornberg in 1974. Histone genes are highly conserved in evolution (the amino acid sequences of histone H4 from a pea and a cow differ at only 2 of their 102 positions). (Alberts et al., 2002, p. 210.)

A. The nucleosome, the basic unit of DNA packing

A nucleosome consists of a core of eight histone molecules (octamer), two copies each of H2 A, H2 B, H3, and H4 (**1**), and about 150 bp DNA wrapped around it. The total protein content is 108 kD (28 kD each for H2 A and H2 B, 30 kD for H3, and 22 kD for H4; Lewin, 2004, p. 572). The histone octamer forms a disc-shaped core (the figures in A are highly schematic). About 140–150 (147 bp in humans) base pairs of DNA are wrapped 1.67 times in left-handed turns (a left-handed superhelix, **2**) around the histone core to form a nucleosome about 11 nm in diameter and 6 nm high. The DNA enters and leaves the nucleosome at points close to each other. A fifth type of histone, H1, is located here and attaches to the DNA between two nucleosomes. Each nucleosome is separated from the other by 50–70 bp of linker DNA, which yields a repeat length of 157–240 bp. For transcription and repair the tight association of histones and DNA has to be loosened (see p. 230).

B. Three-dimensional structure of a nucleosome

The ribbon diagram of the nucleosome shown from above, based on the X-ray structure at a high resolution of 2.8 Angström, shows DNA wrapped around its histone core. One strand of DNA is shown in green, the other in brown. The histones are shown in different colors as indicated. (Photograph from Luger et al., 1997, with kind permission by T. J. Richmond.)

C. Chromatin structures

Chromatin occurs in a condensed (tightly folded), a less condensed (partially folded), and an extended, unfolded form. When extracted from cell nuclei in isotonic buffers, most chromatin appears as fibers of about 30 nm diameter. The corresponding electron-microscopic photographs obtained by different techniques show the condensed (folded) chromatin as compact 300–500-Å structures (top), a 250-Å fiber when partially folded (middle), and as a "beads on a string" 100-Å chromatin fiber (bottom). (Figure adapted from Alberts et al. 2002; the electron micrographs are from Thoma, Koller, & Klug, 1979.)

D. Chromatin segments

The chromatin structures shown in part C correspond to the third level of organization, the packing of the 30-nm fiber. This yields an overall packing ratio of 1000 in euchromatin (about the same in mitotic chromosomes) and 10,000-fold in heterochromatin in both interphase and mitosis (Lewin, 2004, p. 571). (Figure adapted from Alberts et al., 2002.)

References

Alberts B et al: Molecular Biology of the Cell, 4th ed. Garland Publishing, New York, 2002.

Dorigo B et al: Nucleosome arrays reveal the two-start organization of the chromatin fiber. Science 306: 1471–1573, 2004.

Kornberg RD, Lorch Y: Twenty-five years of the nucleosome, fundamental particle of the eukaryote chromosome. Cell 98: 285–294, 1999.

Khorasanizadeh S: The nucleosome: From genomic organization to genome regulation. Cell 116: 259–272, 2004.

Lewin B: Genes VIII. Pearson International, 2004.

Lodish H et al: Molecular Cell Biology, 5th ed. WH Freeman & Co, New York, 2004.

Luger K et al: Crystal structure of the nucleosome core particle at 2.8 Å resolution. Nature 389: 251–260, 1997.

Mohd-Sarip A, Verrijzer CP: A higher order of silence. Science 306: 1484–1485, 2004.

Richmond TJ, Dave, CA: The structure of DNA in the nucleosome. Nature 423: 145–150, 2003.

Schalch T et al: X-ray structure of a tetranucleosome and its implications for the chromatin fibre. Nature 436: 138–141, 2005.

Thoma F, Koller T, Klug A: Involvement of histone H1 in the organization of the nucleosome and of the salt-dependent superstructures of chromatin. J Cell Biol 83: 403–427, 1979.

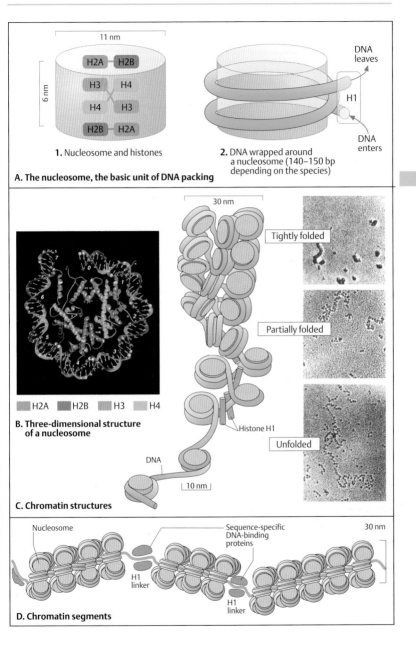

A. The nucleosome, the basic unit of DNA packing

11 nm

6 nm

H2A   H2B

H3   H4

H4   H3

H2B   H2A

1. Nucleosome and histones

DNA leaves

H1

DNA enters

2. DNA wrapped around a nucleosome (140–150 bp depending on the species)

H2A   H2B   H3   H4

B. Three-dimensional structure of a nucleosome

30 nm

Tightly folded

Partially folded

Histone H1

Unfolded

DNA

10 nm

C. Chromatin structures

Nucleosome

Sequence-specific DNA-binding proteins

30 nm

H1 linker

H1 linker

D. Chromatin segments

DNA in Chromosomes

Prior to mitosis, interphase chromosomes are condensed into mitotic chromosomes. This happens in a highly organized manner. The change from interphase to mitotic chromosomes requires a class of proteins called *condensins*. These use energy from ATP hydrolysis to coil each interphase chromosome into a mitotic chromososome. Two of the five subunits of condensins interact with ATP and DNA. The cell cycle protein cdc2 (see cell cycle, p. 124) is required for both interphase and mitotic condensation (Aono et al., 2002).

A. Model for chromatin packing

Chromosomal DNA is folded and packed in an efficient manner. Six successive levels of the hierarchical organization of DNA packing in a metaphase chromosome are schematically shown top to bottom. First, a condensed section loops out of a metaphase chromosome. A higher magnification of the section shows a slightly extended part in a scaffold-associated region with loops of DNA attached to a scaffold. These loops correspond to the 30-nm chromatin fiber of packed nucleosomes shown at the next level below. The 11-nm "beads-on-a-string" form of chromatin follows. A short region of DNA double helix (five turns) marks the molecular level of DNA. (Figure modified from Alberts et al., 2002, and Lodish et al, 2000.)

B. Chromosome territories during interphase

Individual chromosomes occupy particular territories in an interphase nucleus. Recently, T. Cremer and co-workers (Bolzer et al., 2005) reported that small chromosomes are located preferentially towards the center of fibroblast nuclei, whereas large chromosomes are positioned preferentially towards the nuclear rim. This was independent of gene density. Measurements along the optical axes of the chromosome territories of human chromosomes 18 and 19 suggest that the gene-poor chromosome 18 is closer to the top or bottom of the nuclear envelope than chromosome 19. This agrees with their observation that a layer of Alu- and gene-poor chromatin lies close to the nuclear envelope, while chromatin that is rich in Alu sequences and genes was found preferentially in the interior of the nucleus. Probably complex genetic and epigenetic mechanisms act at different levels to establish, maintain, or alter higher-order chromatin arrangements as required for proper nuclear functions.

The numbers in (**1**) on the left indicate different degrees of chromosome decondensation (Monto Carlo relaxation steps 200, 1000, and 400 000; see Bolzer et al., 2005).

(Images from Bolzer et al., 2005; courtesy of Prof. Thomas Cremer, München.)

References

Alberts B et al: Molecular Biology of the Cell, 4th ed. Garland Publishing, New York, 2002.

Aono N et al: Cdn2 has dual roles in mitotic chromosomes. Nature 417: 197–2002, 2002.

Bolzer A et al: Three-dimensional maps of all chromosomes in human male fibroblast nuclei and prometaphase rosettes. PloS Biol, Vol 3, No 5, e157, May 2005.

Cremer T, Cremer C: Chromosome territories, nuclear architecture and gene regulation in mammalian cells. Nature Rev Genet 2: 292–301, 2001.

Gilbert N et al: Chromatin architecture of the human genome: Gene-rich domains are enriched in open chromatin fibers. Cell 118: 555–566, 2004.

Hagstrom KA, Meyer BJ: Condensin and cohesin: more than chromosome compactor and glue. Nature Rev Genet 4: 520–534, 2003.

Lodish H et al: Molecular Cell Biology, 5th ed. WH Freeman, New York, 2004.

Riddihough G: Chromosomes through space and time. Science 301: 779, 2003.

Science Special Issue: The Dynamic Chromosome. Science 301: 717–876, 2003.

Sun HB, Shen J, Yokota H: Size-dependent positioning of human chromosomes in interphase nuclei. Biophys J 79: 184–190, 2000.

Tyler-Smith C, Willard HF: Mammalian chromosome structure. Curr Opin Genet Dev 3: 390–397, 1993.

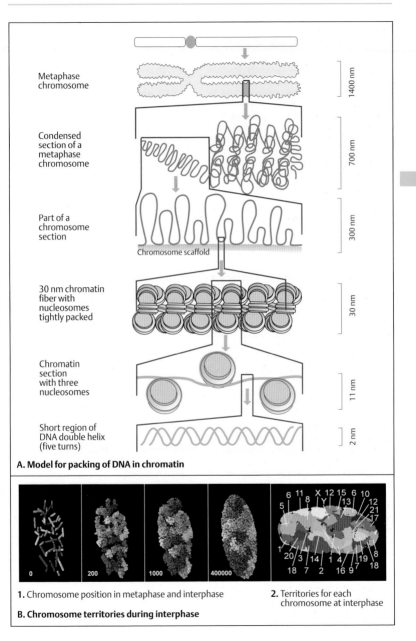

Metaphase
chromosome

1400 nm

Condensed
section of a
metaphase
chromosome

700 nm

Part of a
chromosome
section

300 nm

Chromosome scaffold

30 nm chromatin
fiber with
nucleosomes
tightly packed

30 nm

Chromatin
section
with three
nucleosomes

11 nm

Short region of
DNA double helix
(five turns)

2 nm

A. Model for packing of DNA in chromatin

6 11   X 12 15 6 10
5   8   13   12
                   21
                   17

1
20 3 14   1 4   19 8
18 7 2   16 9   18

1. Chromosome position in metaphase and interphase

2. Territories for each
chromosome at interphase

B. Chromosome territories during interphase

The Telomere

The telomere is a special structure that "seals" the ends of a chromosome. Mammalian telomeric DNA sequences are tandemly repeated along about 60 kb. One strand is G-rich, the other C-rich. All telomeric sequences can be written in the general form $C_n(A/T)_m$, with $n > 1$ and $m = 1–4$ (Lewin, 2004, p. 563). The TTAGGG repeat region in humans is 10–15 kb long, in mice 25–50 kb (Blasco, 2005). At each cell division, somatic cells lose nucleotides from their telomeres. As a result the chromosome ends shorten over time.

A. The replication problem

Since DNA is synthesized in the 5′ to 3′ direction only, the two templates of the parent molecule differ with respect to the continuity of synthesis. From the 3′ to 5′ strand template, new DNA is synthesized as one continuous 5′ to 3′ strand from a single short RNA primer (see DNA replication, p. 50). However, from the 5′ to 3′ template strand, DNA is synthesized in short fragments (the Okazaki fragments), each from a short RNA primer (lagging strand synthesis). Here, 8–12 bases at the end of the lagging strand template cannot be synthesized by DNA polymerase because the required primer cannot be attached. Hence, at each round of replication before cell division, these 8–12 nucleotides will be lost at the chromosome ends.

B. G-rich repetitive sequences

DNA at the telomeres consists of G-rich tandem sequences (5′-TTAGGG-3′ in vertebrates, 5′-TGTGGG-3′ in yeast, 5′-TTGGGG-3′ in protozoa). The G-strand overhang is important for telomeric protection by formation of a duplex loop (see C). Telomerase binds to the G-strand overhang.

C. DNA duplex loop formation

Two features characterize the telomere: (i) telomerase activity to compensate for replication-related loss of nucleotides at the chromosome ends and (ii) formation of a loop of telomeric DNA to stabilize the chromosome ends. Telomerase belongs to a family of modified reverse transcriptases (TERT, telomerase reverse transcriptase) composed of an RNA molecule (about 450 nucleotides) with a well-defined secondary structure and three classes of proteins (Cech, 2004). The RNA nucleotides provide the template for adding nucleotides to the 3′ end of the chromosome. After telomerase has extended the 3′ (G-rich) strand, a new Okazaki fragment can be synthesized at the 5′ strand by DNA polymerase.

Telomeric duplex DNA forms a loop (Griffith et al., 1999), which is mediated by two related proteins, TRF1 and TRF2 (telomeric repeat-binding factor 1 and 2), binding to telomere repeats. The loop is anchored by the insertion of the G-strand overhang (see B) into a proximal segment of duplex telomeric DNA. (Figure adapted from Griffith et al., 1999.)

D. General structure of a telomere

In the terminal 6–10 kb of a chromosome, telomere sequences and telomere-associated sequences can be differentiated (1). The telomere-associated sequences contain autonomously replicating sequences (ARS). The telomere sequences themselves consist of about 250 to 1500 G-rich repeats (~9 kb). They are highly conserved among different species (2). Telomerase activity is essential for survival in protozoans and yeast. In vertebrates, it occurs mainly in germ cells, and no telomerase activity is found in somatic tissues.

Medical relevance

Telomere shortening is a cause for cell senescence (growing old). High telomerase activity has been observed in immortal cells in culture and malignant cells (Li et al, 2003; Hiyama et al, 1995). A telomerase component is defective in a severe disease, dyskeratosis congenita (MIM 305000; Mitchell et al., 1999).

References

Blackburn EH: Telomere states and cell fate. Nature 408: 53–56, 2000.

Blasco MA: Telomeres and human disease: Ageing, cancer and beyond. Nature Rev Genet 6: 611–622, 2005.

Cech TR: Beginning to understand the end of the chromosome. Cell 116: 273–279, 2004.

DeLange T: T loops and the origin of telomeres. Nature Rev Mol Cell Biol 5: 323–329, 2004.

Griffith JD et al: Mammalian telomeres end in a large duplex loop. Cell 97: 503–514, 1999.

Hodes RJ: Telomere length, aging, and somatic cell turnover. J Exp Med 190: 153–156, 1999.

Mitchell JR; Wood E, Collins K: A telomerase component is defective in the human disease dyskeratosis congenital. Nature 402: 551–555, 1999.

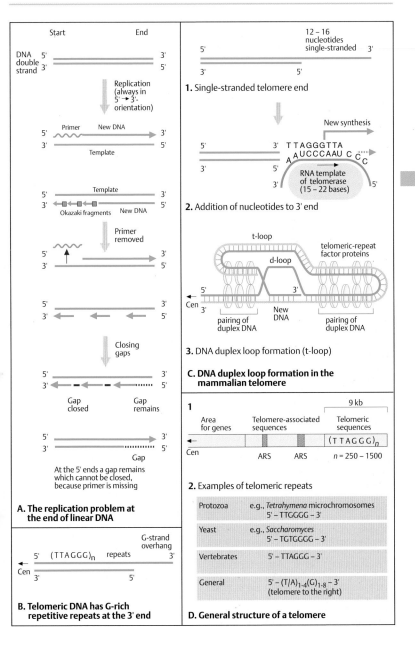

A. The replication problem at the end of linear DNA

Start   End

DNA double strand   5'   3'
                    3'   5'

Replication (always in 5' → 3'-orientation)

5'   Primer   New DNA   3'
3'   Template   5'

5'   Template   3'
3'   5'
Okazaki fragments   New DNA

Primer removed

5'   3'
3'   5'

5'   3'
3'   5'

Closing gaps

5'   3'
3'   5'
Gap closed   Gap remains

5'   3'
3'   5'
Gap

At the 5' ends a gap remains which cannot be closed, because primer is missing

B. Telomeric DNA has G-rich repetitive repeats at the 3' end

G-strand overhang
5'   (TTAGGG)n   repeats   3'
Cen
3'   5'

1. Single-stranded telomere end

12 – 16 nucleotides single-stranded
5'   3'
3'   5'

2. Addition of nucleotides to 3' end

New synthesis
5'   3'   T TAGGGTTA
                    A UCCCAAU C C
3'   5'   C
RNA template of telomerase (15 – 22 bases)
3'   5'

3. DNA duplex loop formation (t-loop)

t-loop
telomeric-repeat factor proteins
d-loop
5'   3'
Cen
3'   5'
New DNA
pairing of duplex DNA   pairing of duplex DNA

C. DNA duplex loop formation in the mammalian telomere

1

9 kb
Area for genes   Telomere-associated sequences   Telomeric sequences
Cen
ARS   ARS $(TTAGGG)_n$
$n = 250 – 1500$

2. Examples of telomeric repeats

Protozoa	e.g., *Tetrahymena* microchromosomes 5' – TTGGGG – 3'
Yeast	e.g., *Saccharomyces* 5' – TGTGGGG – 3'
Vertebrates	5' – TTAGGG – 3'
General	5' – $(T/A)_{1-4}(G)_{1-8}$ – 3' (telomere to the right)

D. General structure of a telomere

The Banding Patterns of Human Chromosomes

When Tjio and Levan and Ford and Hamerton in 1956 independently determined the chromosome number of man to be 46, many of the chromosome pairs could not be distinguished. The complete set of chromosomes of an individual or a species is the *karyotype*, a term introduced by Levitsky in 1924. The chromosomes are arranged in homologous pairs in a *karyogram*. Before 1971, it was not possible to identify each chromosome unambiguously. Since then, several techniques have been developed for inducing specific patterns of light and dark transverse bands along each metaphase chromosome: the *banding patterns*, which can be visualized under the microscope. The most frequently used type of banding is *G-banding*. This name is derived from the Giemsa stain, applied to the chromosomal preparation after it has been treated with the proteolytic enzyme trypsin. Other types of band are Q bands (quinacrine-induced), R bands (reverse of G), C bands (centromeric constitutive heterochromatin), T bands (telomeric), and others (see ISCN, 2005 and table in appendix). The unambiguous identification of all chromosomes is a prerequisite for determining the site on a chromosome at which a given gene locus resides (gene mapping) and for defining the breakpoints of structural chromosomal aberrations.

A. Banding patterns and sizes of human chromosomes 1–12

A band is defined as part of a chromosome that can be clearly distinguished from its adjacent region, which represents another band of opposite staining. This figure schematically shows the G-banding pattern of human chromosomes 1–12 (for chromosomes 13–22, the X, and the Y chromosomem, see next page).

Each chromosome is divided into regions along the short (p) arm and the long (q) arm. The regions are numbered from the centromere to the telomeres. For example, the proximal short arm of chromosome 1 begins (next to the centromere) with region 1, containing a dark and a light band, followed by regions 2 and 3. Within each region, bands are also numbered proximally to distally (towards the telomere). The number of each band is stated directly after the

number of the region. For example, bands 2 and 3 of region 2 are respectively designated 22 and 23; bands 1–6 of region 3, respectively 31, 32, 33, 34, 35, and 36. Each band is designated according to its chromosome, chromosome arm, region, and band. Thus, 1p23 indicates region 2, band 3 of the short arm of chromosome 1. Bands that are visible with increased resolution are indicated by a decimal: if additional bands can be distinguished in 1p23, they are designated 1p23.1, 1p23.2, 1p23.3, etc. This system, first proposed at the Paris Conference of 1971, is outlined in detail in the International System of Chromosome Nomenclature (ISCN, 2005).

In addition, some chromosomes, such as chromosomes 1 and 9, show a secondary constriction adjacent to the centromere. This is a region of polymorphic centromeric heterochromatin. Its length varies among individuals, about 2–3% of individuals having relatively broad centromeric heterochromatin. A pericentric inversion of this region in chromosome 9 occurs in about 1–2% of individuals (see p. 202 about inversions).

References

Bickmore WA, Craig J: Chromosome Bands: Patterns in the Genome. Chapman & Hall, New York, 1997.

Ford CE, Hamerton JL: The chromosomes of man. Nature 178: 1020–1023, 1956.

ISCN 2005: An International System for Human Cytogenetic Nomenclature. Shaffer LG, Tommerup N (eds) Cytogenetics and Cell Genetics, Karger, Basel, 2005.

Miller OJ, Therman E: Human Chromosomes, 4th ed. Springer Verlag, New York, 2001.

Philip AGS, Polani PE: Historical perspectives: Chromosomal abnormalities and clinical syndromes. NeoReviews, Aug 2004; 5: e315–e320 (online at NeoReviews.org).

Tjio JH, Levan A: The chromosome number of man. Hereditas 42: 1–6, 1956.

Traut W: Chromosomen. Klassische und molekulare Cytogenetik. Springer, Heidelberg, 1991.

Verma RS, Babu A: Human Chromosomes. Principles and Techniques. McGraw-Hill, New York, 1995.

Wegner RD, ed.: Diagnostic Cytogenetics. Springer, Berlin, 1999.

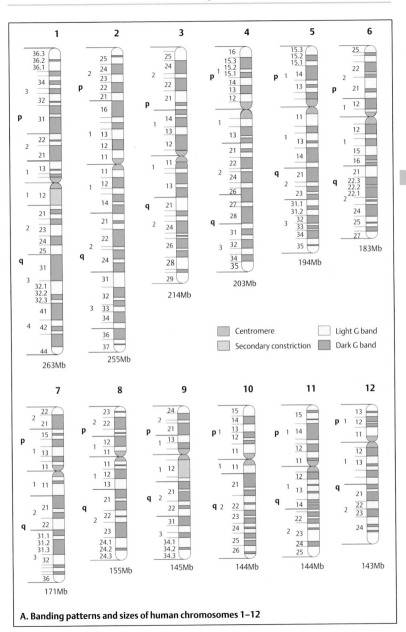

Centromere

Secondary constriction

Light G band

Dark G band

A. Banding patterns and sizes of human chromosomes 1–12

Karyotypes of Man and Mouse

A. Banding patterns of human chromosomes 13–22, X, and Y

The schematic representation of the human karyotype is continued from the previous page.

B. Karyotype of the mouse

The mouse (*Mus musculus*) has a standard karyotype of 20 chromosomes, each with a specific banding pattern (**1**). All chromosomes except the X chromosome are acrocentric, with the centromere at the very end.

Variant strains of mice have a karyotype with metacentric chromosomes. These are the result of centric fusion between two acrocentric chromosomes. This example shows fused chromosome pairs 4 and 2 (4/2), 8 and 3, 7 and 6, 13 and 5, 12 and 10, 14 and 9, 18 and 11, 17 and 16. Chromosome pairs 1, 15, 19 are unchanged, as in the standard karyotype. The fused chromosomes are present in all mice of this particular population. They represent an example of chromosome evolution by fusion. (Photograph in 1 from Traut, 1991; in 2 kindly provided by Dr. H. Winking, Lübeck, Germany.)

Human chromosome nomenclature

An elaborate system of nomenclature has been developed. It is updated from time to time, the last update being from 2005. This is used to designate normal and abnormal chromosome findings. The most important examples are listed in a table in the appendix. For details, the reader is referred to the International System for Human Cytogenetic Nomenclature (ISCN 2005).

References

ISCN 2005: An International System for Human Cytogenetic Nomenclature. Shaffer LG, Tommerup N (eds); S. Karger, Basel, 2005.

Traut W: Chromosomen. Klassische und molekulare Cytogenetik. Springer, Heidelberg, 1991.

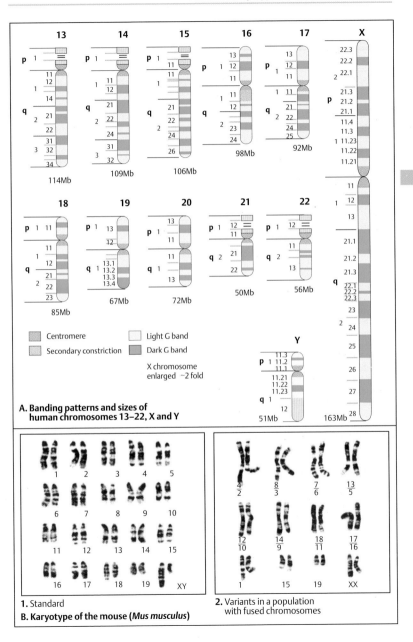

A. Banding patterns and sizes of human chromosomes 13–22, X and Y

Centromere

Secondary constriction

Light G band

Dark G band

X chromosome enlarged ~2 fold

1. Standard

B. Karyotype of the mouse (*Mus musculus*)

2. Variants in a population with fused chromosomes

Preparation of Metaphase Chromosomes for Analysis

For diagnostic purposes, chromosomes are usually analyzed in metaphases prepared from lymphocyte cultures. Metaphase preparations can also be obtained from cultured skin fibroblasts, cultured amniocytes from amniotic fluid, cultured chorionic cells from chorionic villae, or bone marrow cells. Peripheral blood lymphocytes grow in a suspension culture after they have been stimulated to divide by phytohemagglutinin. Their lifespan is limited to a few cell divisions. However, by exposing the culture to Epstein–Barr virus they can be transformed into a lymphoblastoid cell line with permanent growth potential (p. 128).

A. Chromosome analysis from blood

Five principal steps are required: 1, lymphocyte culture, 2, harvest of metaphase chromosomes, 3, chromosome preparation, 4, staining with special dyes that stain chromosomes (and chromatin), and 5, analysis by microscopy, nowadays assisted by computer analysis.

For a lymphocyte cell culture, either peripheral blood is used directly or lymphocytes are isolated from peripheral blood (T lymphocytes). A sample of about 0.5 mL of peripheral blood is needed. Heparin must be added to prevent clotting, which would interfere with proliferation of the cultured cells. The proportion of heparin to blood is about 1 : 20. A lymphocyte culture requires about 72 hours at 37 °C for two cell divisions. Cells reaching mitosis are arrested in metaphase by adding a suitable concentration of a colchicine derivative (colcemide) for two hours prior to harvest. Colcemide interferes with spindle formation and thus arrests mitosis in metaphase, yielding a relative enrichment of cells in metaphase. About 5% of cells will be in mitosis after 72 hours. The culture is then terminated and the cells in metaphase are harvested (1).

At harvest, the culture solution is centrifuged (2). Hypotonic potassium chloride solution (KCl, 0.075 molar) is added for 20 minutes (3), after which a fixative solution of a 3 : 1 mixture of methyl alcohol and glacial acetic acid is added (4). Usually the fixative is changed 4–6 times with subsequent centrifugation. The fixed cells are taken up in a pipette, dropped onto a clean, wet, fat-free glass slide suitable for microscopic analysis, and air-dried (5). The preparation is treated according to the type of bands desired (6), stained (7), and the slide is covered with a cover glass (8).

Suitable metaphases are located under the microscope with about 100-fold magnification and are subsequently examined at about 1250-fold magnification (9). During direct analysis with the microscope, the number of chromosomes as well as the presence or absence of a chromosome and/or recognizable chromosome segments are noted. Since the preparation procedure itself may induce deviations from the normal chromosome number or structure in some cells, more than one cell has to be analyzed. Depending on the purpose of the analysis, between 5 and 100 metaphases (usually 10–20) are examined. Some of the metaphases are photographed under the microscope and subsequently can be cut out from the photograph (karyotyping). In this way a karyogram can be obtained from the photograph of a metaphase. The time needed for a chromosome analysis varies depending on the problem, but is usually 3–4 hours. Analysis and karyotyping time can be considerably shortened by computer procedures. The karyogram can be obtained by computer-assisted analysis.

References

Arakaki DT, Sparkes RS: Microtechnique for culturing leukocytes from whole blood. Cytogenetics 85: 57–60, 1963.

Miller OJ, Therman E: Human Chromosomes, 4th ed. Springer Verlag, New York, 2001.

Moorhead PS, Nowell P, Mellman WJ, Battips DM, Hungerford DA: Chromosome preparations of leucocytes cultured from human peripheral blood. Exp Cell Res 20: 613–616, 1960.

Schwarzacher HG: Preparation of metaphase chromosomes. In: Schwarzacher HG, Wolf U, Passarge E, eds.: Methods in Human Cytogenetics. Springer, Berlin, 1974.

Schwarzacher HG, Wolf U, Passarge E eds.: Methods in Human Cytogenetics. Springer, Berlin, 1974.

Verma RS, Babu A: Human Chromosomes. Manual of Basic Techniques. Pergamon Press, New York, 1989.

Wegner RD ed.: Diagnostic Cytogenetics. Springer, Berlin, 1999.

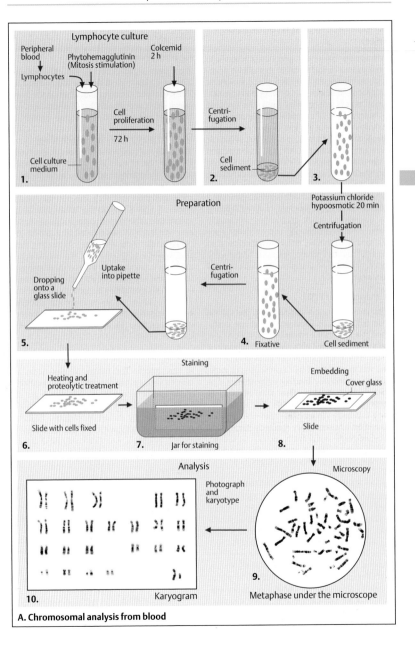

A. Chromosomal analysis from blood

Fluorescence In-Situ Hybridization (FISH)

FISH applies molecular genetic techniques to chromosome preparations in metaphase or interphase nuclei, an approach called *molecular cytogenetics*. The aim is to detect small chromosomal rearrangements that cannot be detected by microscopy. Conventional chromosomal analysis can detect the loss or gain of chromosomal material of 4 million base pairs (4 Mb) or more. In FISH a labeled DNA probe is hybridized *in situ* to single-stranded chromosomal DNA on a microscope slide. Site-specific hybridization results in a signal visualized over the chromosome. A distinction is made between direct and indirect nonisotopic labeling. In *direct labeling* the fluorescent label, a modified nucleotide (often 2′ deoxyuridine 5′ triphosphate) containing a fluorophore, is directly incorporated into DNA. *Indirect labeling* requires labeling the DNA probe with a fluorophore to make the signal visible.

A. Principle of FISH

With indirect nonisotopic labeling, metaphase or interphase cells fixed on a slide are denatured (**1a**) into single-stranded DNA (**2**). A DNA probe (**1b**) is labeled with biotin and hybridized *in situ* to its specific site on the chromosome (**3**). This site is visualized by fluorescence in darkfield microscopy by binding a fluorescent-dye labeled antibody (for biotin this antibody is streptavidin) to the biotin (**4**). This is the primary antibody. To enhance the intensity of fluorescence, a secondary antibody (here, a biotinylated anti-avidin antibody) is attached (**5**). The resulting signal is amplified by attaching additional labeled antibodies (**6**).

B. Example of FISH in metaphase

Here all chromosomes are stained dark blue by DAPI (4′,6-diamidino-2-phenylindole) except five chromosomes carrying a fluorescent signal. Two chromosomes 3 are identified by two red signals. A green fluorescent probe (D354559, Vysis Inc., Downers Grove, Il) has been hybridized to the ends of the short arms (3p). A further green signal is visible over the long arm of one chromosome 16. These three green signals indicate the presence of three chromosomal segments of the long arm of chromosome 3 (*partial trisomy 3p*). Two chromosomes 21 at

the upper right of the metaphase are unspecifically stained.

C. Interphase FISH analysis

Two green signals identify the chromosomes 22. A red signal identifies the long arm (22q). The upper cell (**1**) is normal, whereas the lower one lacks a red signal (**2**). This indicates loss (deletion) of the 22q chromosomal region.

D. FISH analysis of a translocation

This shows a reciprocal translocation between the long arm of a chromosome 8 and the short arm of a chromosome 4 with breakpoints at 8q24 and 4p15.3, yielding two derivative chromosomes, der4 and der8. The normal chromosomes 4 and 8 are in the lower left of the metaphase. The region embracing the breakpoint on 8 at q24 was labeled by a 170-kb YAC (yeast artificial chromosome). As a result, three signals are visible: one over the normal chromosome 8, one over the altered chromosome 8 (der8), and one over the altered chromosome 4 (der4). This identifies a reciprocal translocation. The four chromosomes involved are identified by labeling their centromeres with centromere-specific probes. (Photograph courtesy of H. J. Lüdecke, Essen.)

E. FISH of telomere sequences

In this human metaphase all telomeres are labeled with a probe that hybridizes specifically to the telomere sequences. Two signals are visible at each telomere, one over each chromatid. (Photographs by Robert M. Moyzes, 1991, Los Alamos Laboratory, with kind permission of the author, Scientific American, August 1991, pp. 34–41.)

References

Miller OJ, Therman E: Human Chromosomes, 4th ed. Springer, New York, 2001.

Ried T, Schröck E, Ning Y, Wienberg J: Chromosome painting: a useful art. Hum Mol Genet 7: 1619–1626, 1998.

Speicher MR, Carter NP: The new cytogenetics: Blurring the boundaries with molecular biology. Nature Rev Genet 6: 782–792, 2005.

Strachan T, Read AP: Human Molecular Genetics, 3rd ed. Garland Science, London-New York, 2004.

Online information:
Website Cytogenetics Resource:
  http://www.kumc.edu/gec/prof/cytogene.html.

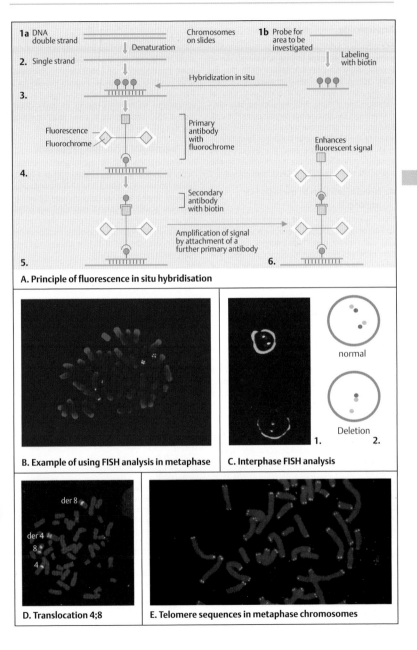

A. Principle of fluorescence in situ hybridisation

B. Example of using FISH analysis in metaphase

C. Interphase FISH analysis

D. Translocation 4;8

E. Telomere sequences in metaphase chromosomes

Aneuploidy

Aneuploidy is a deviation of the normal chromosome number. It leads to loss or gain of one or several individual chromosomes from the diploid set. A loss is called *monosomy*, a gain *trisomy*. Both result from nondisjunction at meiosis (see p. 120). Nondisjunction was discovered 1913 in *Drosophila melanogaster* by C. B. Bridges, who coined the term. It was considered proof of the chromosome theory of heredity by T. H. Morgan and his group. The frequency of nondisjunction in humans is influenced by the age of the mother at the time of conception.

A. Nondisjunction in meiosis I or meiosis II

Nondisjunction may occur during either meiosis I or meiosis II. During meiosis I, one daughter cell will receive two chromosomes instead of one, whereas the other daughter cell receives none (in the figure opposite only one pair of chromosomes is shown). As a result, gametes will be formed carrying either two chromosomes (disomy) or none (nullisomy), instead of one each. If nondisjunction occurs during meiosis II, the first meiotic division (see p. 120) is normal. During the second meiotic division, one daughter cell will receive two chromosomes (becoming disomic), the other none (nullisomic). After fertilization, a disomic gamete gives rise to a trisomic zygote and a nullisomic gamete to a monosomic zygote. In humans the only monosomy to occur in liveborn infants is that of the X chromosome.

B. Common types of aneuploidy

Three states of aneuploidy can be distinguished: (i) trisomy (three chromosomes instead of two in one pair), (ii) monosomy (one chromosome instead of two), (iii) triploidy and tetraploidy (all chromosomes present in triplicate or quadruplicate). Triploidy does not result from nondisjunction at meiosis, but from one of a variety of other processes. Two sperms may have penetrated the egg (dispermia); the egg or sperm may have an unreduced chromosome set as a result of restitution in the first or second meiotic division; or the second polar body may have reunited with the haploid egg nucleus. With dispermy, two of the three sets of chromosomes will be of paternal origin, resulting in either 69,XYY, 69,XXY or 69,XXX. Dispermy is the cause of triploidy in about 66% of cases;

fertilization of a haploid egg by a diploid sperm in 24% (failure at meiosis I); and a diploid egg in 10% (Jacobs et al., 1978). Triploidy is one of the most frequent chromosomal aberrations in man, causing 17% of spontaneous abortions. Only 1 in 10000 triploid zygotes results in a liveborn infant, but with severe congenital malformations invariably leading to early death (see p. 412). Tetraploidy is rarer than triploidy (for details, see Miller and Therman, 2001.)

C. Autosomal trisomies in humans

In humans, only three autosomal trisomies occur in liveborn infants: trisomy 13 at a population frequency of about 1 in 12000 newborns; trisomy 18 at 1 in 6000; and trisomy 21 at 1 in 650 (see p. 412). Beyond a maternal age of 35, the incidence of autosomal trisomy rises, reaching about 10-fold the normal incidence in mothers over the age of 40 (see Miller and Therman, 2001; Harper, 2004.)

D. Additional X or Y chromosome

An additional X or Y chromosome occurs in about 1 in 800 newborns. Although neither is associated with a distinct phenotype, a proportion of individuals with such aberrations are more liable to have impaired speech development, limited learning ability, and behavioral problems (for details, see Harper, 2004.)

References

Bridges CB: Nondisjunction of the sex chromosomes of Drosophila. J Exp Zool 15: 587–606, 1913.

Grant R, McKinlay J, Sutherland GR: Chromosome Anomalies and Genetic Counseling, 3rd ed. Oxford University Press, Oxford, 2004.

Harper P: Practical Genetic Counseling, 6th ed. Edward Arnold, London, 2004.

Hassold T, Jacobs PA: Trisomy in man. Ann Rev Genet 18: 69–97, 1984.

ISCN 2005: An International System for Human Cytogenetic Nomenclature. Shaffer LG, Tommerup N (eds). Karger, Basel, 2005.

Jacobs PA, Hassold T: The origin of numerical chromosome abnormalities. Adv Hum Genet 33: 101–133, 1995.

Jacobs PA et al: The origins of human triploids. Ann Hum Genet 42: 49–57, 1978.

Miller OJ. Therman E: Human Chromosomes, 4th ed. Springer, New York, 2001.

Rooney DE, Czepulski BH (eds): Human Cytogenetics. A Practical Approach, 2nd ed. Oxford University Press, Oxford, 2001.

Schinzel A: Catalogue of Unbalanced Chromosome Aberrations in Man. De Gruyter, Berlin, 2001.

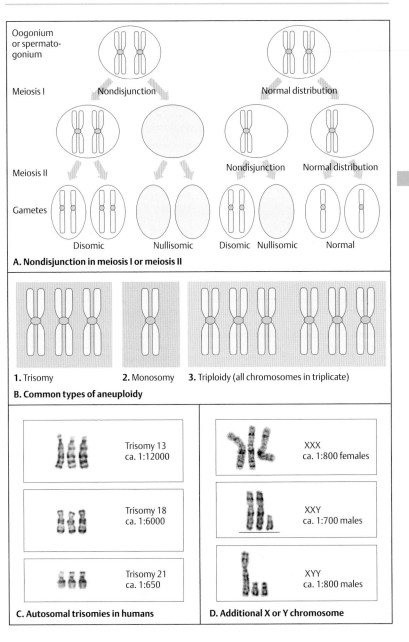

A. Nondisjunction in meiosis I or meiosis II

Oogonium or spermato-gonium

Meiosis I   Nondisjunction   Normal distribution

Meiosis II   Nondisjunction   Normal distribution

Gametes

Disomic   Nullisomic   Disomic   Nullisomic   Normal

B. Common types of aneuploidy

1. Trisomy   **2.** Monosomy   **3.** Triploidy (all chromosomes in triplicate)

C. Autosomal trisomies in humans

Trisomy 13
ca. 1:12000

Trisomy 18
ca. 1:6000

Trisomy 21
ca. 1:650

D. Additional X or Y chromosome

XXX
ca. 1:800 females

XXY
ca. 1:700 males

XYY
ca. 1:800 males

Chromosome Translocation

Translocation refers to a process by which a part of (or a whole) chromosome is moved from one location to another. Usually this is reciprocal, one part being exchanged with another, with no chromosomal material being lost or added. Thus, reciprocal translocations usually have no effect on the phenotype. Translocations occur spontaneously but can be transmitted to offspring over several generations. Occasionally, the breakpoint of a translocation is in a gene and disrupts its function. This is an important cause of tumors originating in the hemato-poietic system.

A special type of translocation, the fusion of two acrocentric chromosomes, is called a Robert-sonian translocation after W.R.B. Robertson, who in 1911 first observed them in insects. He concluded that during evolution, metacentric chromosomes may have arisen by the fusion of acrocentrics.

A. Reciprocal translocation

A reciprocal translocation between two chromosomes is usually balanced and does not carry a risk of disease unless one of the breakpoints interrupts a gene function. Since coding and regulatory regions constitute only 5% of the genome, this is a rare event. However, meiotic segregation in a carrier of a reciprocal translocation sometimes leads to gametes with unbalanced chromosome complements, resulting in a chromosomal disease in the offspring.

During meiosis, the normal homologous chromosomes and the chromosomes involved in the reciprocal translocation pair in meiosis I as usual. Each of the chromosomes not involved in the translocation pairs with its homologous partner that is involved in the translocation. However, the translocation chromosomes can only pair by forming a quadriradial configuration. Disregarding crossing-over and secondary nondisjunction, several outcomes are possible. The simplest is segregation of the two normal chromosomes to one gamete and the two translocation chromosomes to the other (*alternate segregation*). This results in chromosomally balanced gametes.

However, if two neighboring (adjacent) chromosomes are distributed (segregate) to the same gamete, the latter will be chromosomally unbalanced. It will either contain a chromo-

somal segment twice (duplication) or lack it (deficiency). Two types of adjacent segregation can be distinguished. In *adjacent-1* segregation, the gametes receive a normal chromosome and a translocation chromosome with opposite centromeres. Two types of unbalanced gamete result. In the rare *adjacent-2* segregation, homologous centromeres go to the same pole. This produces two generally more extreme types of unbalanced gamete (1 : 3 and 0 : 4 segregation).

B. Centric fusion of acrocentric chromosomes

Robertsonian translocations (centric fusion) involve either a pair of homologous or two nonhomologous acrocentric chromosomes. The outcomes after meiotic segregation differ drastically. In homologous fusion, only disomic and nullisomic gametes are produced. Thus, a zygote will be either trisomic or monosomic, but not balanced. However, fusion of nonhomologous acrocentric chromosomes is much more common. Chromosome 14 and chromosome 21 (**1**) are most frequently involved in centric fusion (about 1 in 1000 newborns). When the long arm of a chromosome 21 (21q) and the long arm of a chromosome 14 (14q) fuse to form a chromosome 14q21q (**2**), the satellite-carrying short arms of both chromosomes are lost, but this is insignificant. In principle, four types of gamete can be formed: normal, balanced, disomic, or nullisomic for chromosome 21 (**3**). After fertilization, the corresponding zygotes contain either only one chromosome 21 (nonviable monosomy 21), a normal chromosome complement, a balanced chromosome complement with the fused chromosome, or three chromosomes 21 (trisomy 21, see p. 412).

References

Harper P: Practical Genetic Counseling, 6th ed. Edward Arnold, London, 2004.

Miller OJ, Therman E: Human Chromosomes, 4th ed. Springer, New York, 2001.

Rooney DE, Czepulski BH (eds): Human Cytogenetics. A Practical Approach, 2nd ed. Oxford University Press, Oxford, 2001.

Schinzel A: Catalogue of Unbalanced Chromosome Aberrations in Man. De Gruyter, Berlin, 2001.

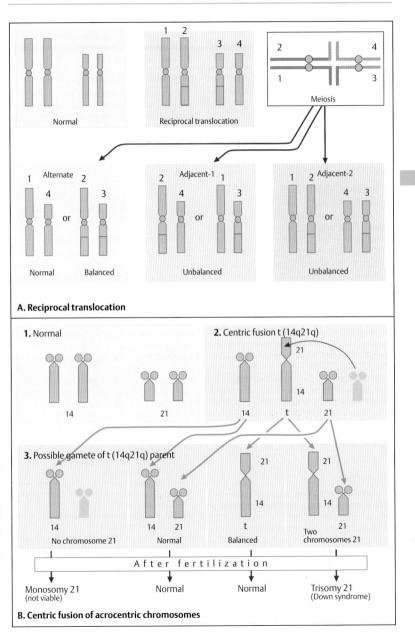

A. Reciprocal translocation

1. Normal

2. Centric fusion t (14q21q)

3. Possible gamete of t (14q21q) parent

No chromosome 21   Normal   Balanced   Two chromosomes 21

After fertilization

Monosomy 21
(not viable)   Normal   Normal   Trisomy 21
(Down syndrome)

B. Centric fusion of acrocentric chromosomes

Structural Chromosomal Aberrations

Structural chromosome abnormalities result from a break or breaks that disrupt the continuity of a chromosome. The principal types of structural chromosomal aberration as observed in human individuals are deletion, duplication, isochromosome, inversion, and the special case of a ring chromosome. Structural chromosomal rearrangements occur with a frequency of about 0.7–2.4 per 1000 mentally retarded individuals. Small supernumerary chromosomes are observed about once in 2500 prenatal diagnoses.

A. Deletion, duplication, isochromosome

Chromosomal deletion (deficiency) arises from a single break with loss of the distal fragment (terminal deletion, **1**) or from two breaks and loss of the intervening segment (interstitial deletion, **2**). In molecular terms, terminal deletions are not terminal: they are capped by a telomeric $(TTAGGG)_n$ repeat. Duplications (**3**) occur mainly as small supernumerary chromosomes. About half of these small chromosomes are derived from chromosome 15, being inverted duplications of the pericentric region [invdup (15)]. These represent one of the most common structural aberrations in man (Schreck et al., 1977).

An isochromosome (**4**) is an inverted duplication. It arises when a normal chromosome divides transversely instead of longitudinally, being then composed of two long arms or of two short arms. In each case, the other arm is missing. The most common isochromosome is that of the long arm of the human X chromosome [i(Xq)].

B. Inversion

An inversion is a 180-degree change in direction of a chromosomal segment caused by a break at two different sites, followed by reunion of the inverted segment. Depending on whether the centromere is involved, a *pericentric* inversion (the centromere lies within the inverted segment) and a *paracentric* inversion can be differentiated.

C. Ring chromosome

A ring chromosome arises after two breaks with the loss of both ends, followed by joining of the two newly resulting ends. Since its distal segments have been lost, a ring chromosome is unbalanced. Ring chromosomes are unstable in mitosis and meiosis (see E).

D. Aneusomy by recombination

When a normal chromosome pairs with its inversion-carrying homolog during meiosis, a loop is created in the region of the inversion (**1**). When the inverted segment is relatively large, crossing-over may occur within this region (**2**). In the daughter cells shown here, one resulting chromosome will contain a duplication of segments A and B and a deficiency of segment F (**3**), whereas the other resulting chromosome lacks segments A and B and has two segments F (**4**). These chromosome segments are not balanced (aneusomy by recombination).

E. A ring chromosome at meiosis

A ring chromosome at mitosis and meiosis is unstable and is frequently lost. If crossing-over occurs during meiosis, a dicentric ring is created. At the following anaphase, the ring breaks at variable locations as the centromeres go to different poles. The daughter cells will receive different parts of the ring chromosome, resulting in deficiency in one cell and duplication in the other. Ring chromosomes are often dicentric. Subsequently a breakage-fusion-bridge cycle (McClintock, 1938) occurs as the chromosome is pulled toward the opposing poles during anaphase. The same happens during mitosis (without crossing-over). Thus, ring chromosomes tend to generate new variants of derivative chromosomes, all imbalanced.

References

Madan K: Paracentric inversions: a review. Hum Genet 96: 503–515, 1995.

Meltzer PS, Guan X-Y, Trent JC: Telomere capture stabilizes chromosome breakage. Nature Genet 4: 252–273, 1993.

Miller OJ, Therman E: Human Chromosomes, 4th ed. Springer, New York, 2001.

Niss R, Passarge E: Derivative chromosomal structures from a ring chromosome 4. Humangenetik 28: 9–23, 1975.

Schreck RR et al: Preferential derivation of abnormal human G-group-like chromosomes. Hum Genet 36: 1–12, 1977.

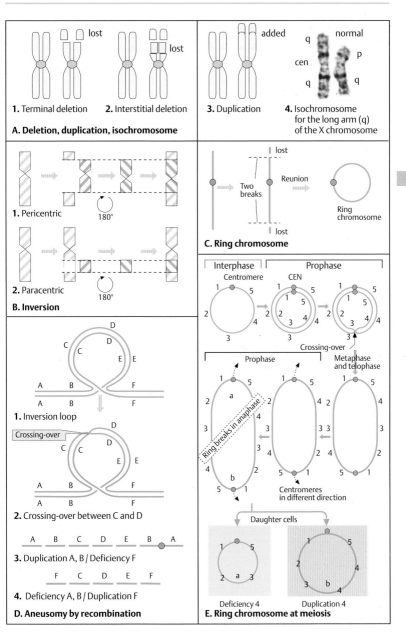

1. Terminal deletion   **2.** Interstitial deletion

A. Deletion, duplication, isochromosome

lost

lost

added

3. Duplication   **4.** Isochromosome
for the long arm (q)
of the X chromosome

q
cen
q

normal
q
p
q

1. Pericentric   180°

2. Paracentric   180°

B. Inversion

C. Ring chromosome

Two
breaks

Reunion

Ring
chromosome

lost

lost

D
C   C   D
E   E
A   B   F
A   B   F

1. Inversion loop

Crossing-over

D
C   C   D
E   E
A   B   F
A   B   F

2. Crossing-over between C and D

A   B   C   D   E   B   A

3. Duplication A, B / Deficiency F

F   C   D   E   F

4. Deficiency A, B / Duplication F

D. Aneusomy by recombination

Interphase

Centromere

Prophase

CEN

Crossing-over

Prophase

Metaphase
and telophase

Ring breaks in anaphase

Centromeres
in different direction

Daughter cells

Deficiency 4   Duplication 4

E. Ring chromosome at meiosis

Multicolor FISH Identification of Chromosomes

Computer-based methods have been developed to identify each chromosome and to detect small rearrangements. When sets of chromosome-specific probes hybridize to an entire chromosome or parts of a chromosome, the result is referred to as "chromosome painting." To increase the number of chromosomes that can be differentiated at one time, a combination of labels (DNA probes that are labeled with more than one type of fluorochrome) and different ratios of labels are applied. The resulting mixed colors are detectable by automated digital image analysis. Two approaches have proved particularly useful: multiplex fluorescence *in-situ* hybridization (M-FISH, Speicher et al., 1996) and spectral karyotyping (SKY, Schröck et al., 1996). Other approaches and modifications exist, e.g., the use of artificially extended DNA or chromatin fibers.

A. Multicolor FISH

Multicolor FISH (M-FISH, Speicher et al., 1996) uses sets of chromosome-specific DNA probes, which are hybridized to denatured metaphase chromosomes. Each probe is contained in a YAC (yeast artificial chromosome) and labeled with a combination of DNA-binding fluorescent dyes specific for each chromosome. Five different fluorophores can produce 24 different colors for image analysis by epifluorescence microscopy using a charge-coupled device (CCD) camera. An image of each chromosome in a pseudocolor is displayed by appropriate computer software. (Photograph courtesy of Drs. Sabine Uhrig and Michael Speicher, Graz, formerly München.)

B. Spectral karyotyping

Spectral karyotyping (SKY, Schröck et al., 1996) combines optical microscopy, Fourier spectroscopy, and CCD imaging. The emission spectra of all points in the sample are measured simultaneously in the visible and near-infrared spectral range and the unique spectral signature of each chromosome allows for automated chromosome classification.

First, all 24 human chromosomes (chromosomes 1–22, X, and Y) are separated by flow-sorting and labeled with at least one, but possibly as many as five fluorochrome combinations (SKY probes). Following DNA denatura-tion, the SKY probe is hybridized to a metaphase chromosome preparation at 37 °C for 24–72 hours.

An interferometer and a CCD camera (SpectraCube) are coupled to an epifluorescence microscope to visualize the SKY probes. All fluorochrome dyes contained in the sample are excited simultaneously within a single exposure. The emitted light is sent through a Sagnac interferometer and focused onto the CCD camera. In this way, an interferogram based on the optical path difference of the divided light beam is measured for each pixel of the image. The resulting unique spectral "signature" of each chromosome based on the raw spectral information is displayed as an RGB (red, green, blue) image of the metaphase and is used for automated chromosome classification. During this process, a discrete false color is assigned to all pixels of the image with identical spectra, revealing structural and numerical chromosome aberrations. Spectral karyotyping has a wide range of diagnostic applications in clinical and cancer cytogenetics. (Photographs courtesy of Professor Evelin Schröck, University of Dresden.)

References

Heiskanen M, Peltonen L, Palotie A: Visual mapping by high resolution FISH. Trends Genet 12: 379–384, 1996.

Lichter P: Multicolor FISHing: what's the catch? Trends Genet 13: 475–479, 1997.

Miller OJ, Therman E: Human Chromosomes, 4th ed. Springer, New York, 2001.

Ried T et al: Chromosome painting: a useful art. Hum Mol Genet 7: 1619–1626, 1998.

Schröck E et al: Multicolor spectral karyotyping of human chromosomes. Science 273: 494–497, 1996.

Speicher MR, Ballard SG, Ward DC: Karyotyping human chromosomes by combinatorial multi-fluor FISH. Nature Genet 12: 368–375, 1996.

Strachan T, Read A: Human Molecular Genetics, 3rd ed. Bios Scientific Publishers, Oxford, 2004.

Uhrig S et al: Multiplex-FISH for pre- and postnatal diagnostic application. Am J Hum Genet 65: 448–462, 1999.

Online information:

NCI and NCBI's SKY/M-FISH and CGH Database 2005:
   http://www.ncbi.nlm.nih.gov/entrez/query.
   fcgi?db = CancerChromosomes
   http://www.ncbi.nlm.nih.gov/projects/sky/

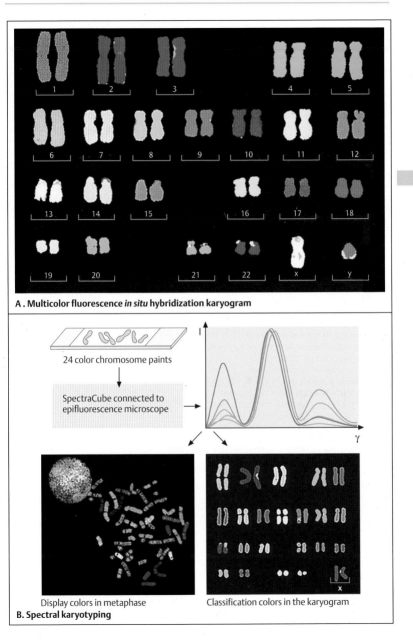

A . Multicolor fluorescence *in situ* hybridization karyogram

24 color chromosome paints

SpectraCube connected to epifluorescence microscope

Display colors in metaphase

Classification colors in the karyogram

B. Spectral karyotyping

Comparative Genomic Hybridization

Comparative genomic hybridization (CGH) detects differences in the copy number of DNA in all chromosomes simultaneously. It combines fluorescence in-situ hybridization and chromosome painting by using probes that hybridize to entire chromosomes. CGH is widely used to investigate tumor cells in metaphase or interphase. In forward chromosome painting, the probe for a particular chromosome is hybridized to a tumor metaphase to identify the chromosome of the tumor to which the probe hybridizes. Modifications of CGH are array-based CGH (aCGH, see p. 260) and multiplex ligation-dependent probe amplification (MLPA, Sellner & Taylor, 2004).

A. Application of CGH

CGH compares DNA isolated from a tumor sample with genomic DNA from a normal individual as control. Here, whole-genomic tumor DNA labeled with a green fluorescent fluorophore (FITC, fluorescein isothiocyanate) is hybridized to normal DNA labeled with rhodamine, which fluoresces red. Hybridization of highly repetitive sequences is suppressed in order to assess the actual copy number in the chromosomal regions where the probes competitively hybridize to the complementary DNA sequences in the target chromosomes. This example shows a metaphase at the top from a patient with von Hippel-Lindau syndrome, a hereditary form of renal cell carcinoma (MIM 193300). At the left the metaphase is labeled green by fluorescein isothiocyanate (FITC), which in this case labels tumor DNA (1). In the middle the same metaphase is labeled red by rhodamine labeling control DNA (2). On the right (3) is an electronically blended composite of the fluorophore labels of green and red. Yellow fluorescence indicates a normal 1-to-1 green-to-red ratio without loss or gain of chromosomal material. Red fluorescence indicates loss in the tumor; green fluorescence indicates gain. The hybridization pattern is shown for all chromosomes in the CGH profile (4). The individual chromosomes are identified by a blue dye that stains AT-rich sequences (DAPI, 4,6-diamidinophenylindol), visualized by fluorescence microscopy. The metaphase is sequentially analyzed with a charge-coupled device (CCD) camera for red, green, and blue (DAPI counterstain) using different filter systems. The blue DAPI fluorescence is converted into a black-and-white image. This is enhanced to result in a banding pattern similar to G-bands for chromosome identification. The CGH profile scans each chromosome for a deviation to the red (first vertical line on the left of each field right to a chromosome) to search for loss or for a deviation to green (third vertical line), which will indicate a gain. A deviation beyond the red line to the left is visible along the long arm of chromosome 4. This indicates chromosomal loss in 4q. A deviation to the green over the long arm of chromosome 10 indicates a gain of chromosomal material in 10q. The centromeres are shown as horizontal gray boxes. (Images courtesy of Drs. Nicole McNeil and Thomas Ried, NIH.)

B. Identification of extra chromosomal material by M-FISH

Here extra chromosomal material, invisible in a standard karyotype (left), is visualized by multicolor fluorescence in-situ hybridization (M-FISH, see previous page). The multiplex FISH karyogram (right) shows a small extra band at the end of the long arm of a chromosome 1 (arrow). This extra band is derived from a chromosome 12. (Photographs kindly provided by Dr. Sabine Uhrig and Dr. Michael Speicher, Graz.)

References

Chudoba I et al: High-resolution multicolor-banding: A new technique for refined FISH analysis of human chromosomes. Cytogenet Cell Genet 84: 156–160, 1999.

Sellner LN, Taylor GR: MLPA and MAPH: New techniques for detection of gene deletions. Hum Mutat 23: 413–419, 2004.

Speicher MR, Carter NP: The new cytogenetics: Blurring the boundaries with molecular biology. Nature Rev Genet 6: 782–292, 2005.

Wong A et al: Detection and calibration of microdeletions and microduplications by array-based comparative genomic hybridization and its applicability to clinical genetic testing. Genet Med 7: 264–271, 2005.

Online information:

CGH Data Base, Department of Molecular Cytogenetics (MCG), Medical Research Institute Tokyo: http://www.cghtmd.jp/cghdatabase/index_e.html

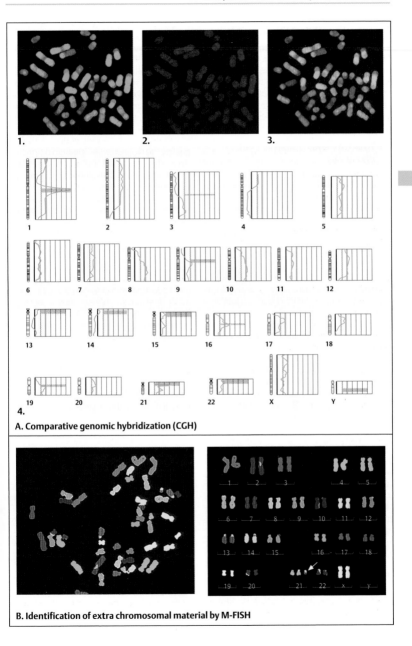

A. Comparative genomic hybridization (CGH)

B. Identification of extra chromosomal material by M-FISH

Ribosomes and Protein Assembly

Ribosomes are large ribonucleoprotein particles that coordinate the interaction of mRNA and tRNAs during protein synthesis. All proteins produced by a cell at any given time are called the *proteome*. The total number of different proteins required for eukaryotes is estimated to be in the range of 90 000. Ribosomes are the products of individual genes (ribosomal genes).

A. Structure and components of ribosomes

Ribosomes are made up of a small and a large subunit. Bacterial ribosomes contain three different types of rRNA molecules and up to 83 proteins. A ribosome of the bacterium *Escherichia coli* has a sedimentation coefficient of 70 S (the sedimentation coefficient is a measure of the rate of sedimentation in an ultracentrifuge of a molecule suspended in a less dense solvent. It is measured in Svedberg units, S. The S values are not additive. A dalton is a unit of atomic or molecular mass). A bacterial cell contains about 20 000 ribosomes, which account for about 25% of its mass.

The prokaryote 70 S ribosome can be dissociated into a smaller subunit of 30 S and a larger subunit of 50 S. The 50 S subunit in turn consists of smaller rRNAs (5 S) and larger rRNAs (23 S), with 120 and ~2900 ribonucleotides, respectively. In addition, 33–35 different proteins are present. The 30 S subunit contains a large 16 S rRNA and 21 proteins. The 30 S subunit is the site where genetic information is decoded. It also has a proofreading mechanism. The 50 S subunit provides peptidyltransferase activity. The entire ribosome has a molecular weight of 2.5 million daltons (MDa).

The eukaryotic ribosome is much larger (80 S, 4.2 MDa), consisting of a 40 S and a 60 S subunit. The 60 S subunit contains 5 S, 5.8 S, and 28 S rRNAs (120, 160, and 4800 bases, respectively) in addition to 50 proteins. The 40 S subunit has 18 S rRNAs (1900 bases) and 33 proteins. Bacterial 30 S and 50 S ribosomal subunit structures are known at 5 Å resolution.

B. From gene to protein

The cell nucleus directs the production of endogenous proteins (protein synthesis). RNA in the nucleus is bound to nuclear RNA-binding proteins for stabilization. The mature RNA is released from the nucleus into the cytoplasm, where it associates with ribosomes.

C. Nucleolus and ribosomes

The nucleolus is a morphologically and functionally specific region in the cell nucleus in which ribosomes are synthesized. In man, the rRNA genes (200 copies per haploid genome) are transcribed by RNA polymerase I to form 45 S rRNA molecules. After the 45 S rRNA precursors have been produced, they are quickly packaged with ribosomal proteins (from the cytoplasm). Before they are transferred from the nucleus to the cytoplasm, they are cleaved to form three of the four rRNA subunits. These are released into the cytoplasm with the separately synthesized 5 S subunits. Here they form functional ribosomes.

Two types of small RNAs have important functions. Small nuclear RNAs (snRNA) are a family of RNA molecules that bind specifically with a small number of nuclear ribonucleoprotein particles (snRNP, pronounced "snurps"). These play important roles in the modification of RNA molecules after transcription (posttranscriptional modification). snRNAs base-pair with pre-mRNA and with each other during the splicing of RNA. Small nucleolar RNA molecules (snoRNA) assist in processing pre-rRNAs and in assembling of ribosomes. (Figures based on Alberts et al., 2002.)

Medical relevance

A variety of chemical compounds occurring naturally as poisons or synthetic products are used for cancer therapy by inhibition of transcription or translation (see table in the appendix).

References

Agalarov SC et al: Structure of the S15, S6, S18-rRNA complex: assembly of the 30S ribosome central domain. Science 288: 107–112, 2000.

Alberts B et al: Molecular Biology of the Cell, 4th ed. Garland Publishing, New York, 2002.

Garrett R: Mechanics of the ribosome. Nature 400: 811–812, 1999.

Lodish H et al: Molecular Cell Biology, 5th ed. WH Freeman, New York, 2004.

Stryer L: Biochemistry, 4th ed. WH Freeman, New York, 1995.

Wimberly BT et al: Structure of the 30 S ribosomal subunit. Nature 407: 327–339, 2000.

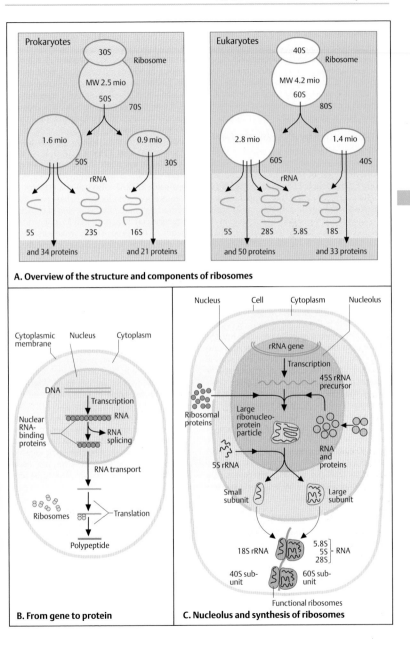

A. Overview of the structure and components of ribosomes

Prokaryotes

30S | Ribosome
MW 2.5 mio
50S | 70S

1.6 mio — 50S
rRNA
5S · 23S
and 34 proteins

0.9 mio — 30S
16S
and 21 proteins

Eukaryotes

40S | Ribosome
MW 4.2 mio
60S | 80S

2.8 mio — 60S
rRNA
5S · 28S · 5.8S
and 50 proteins

1.4 mio — 40S
18S
and 33 proteins

B. From gene to protein

Cytoplasmic membrane | Nucleus | Cytoplasm

DNA
↓ Transcription
RNA
Nuclear RNA-binding proteins
→ RNA splicing

↓ RNA transport

Ribosomes
Translation
Polypeptide

C. Nucleolus and synthesis of ribosomes

Nucleus | Cell | Cytoplasm | Nucleolus

rRNA gene
↓ Transcription
45S rRNA precursor

Ribosomal proteins

Large ribonucleo-protein particle

RNA and proteins

5S rRNA

Small subunit | Large subunit

18S rRNA

5.8S
5S } RNA
28S

40S sub-unit | 60S sub-unit

Functional ribosomes

Transcription

For transcription, the DNA double helix is transiently opened, and the template strand is used to direct synthesis of an RNA strand. This process starts and ends at a defined site. The RNA strand is synthesized from its 5′ to its 3′ end. RNA synthesis is directed by RNA polymerase. Transcription occurs in four stages.

A. Transcription by RNA polymerase II

The first step of transcription is template recognition with binding of RNA polymerase II to a specific sequence of DNA molecules, the promoter (**1**). The double helix then opens, and an initiation complex makes the template strand available for base pairing. Initiation (**2**) begins with synthesis of the first RNA molecules at the initiation complex. RNA polymerase remains at the promoter while it synthesizes the first nine nucleotide bonds. Initiation requires several other proteins, collectively referred to as activators and transcription factors. Elongation (**3**) begins when the enzyme moves along the DNA, thereby extending the RNA chain. As it moves, it unwinds the DNA double helix. Specialized enzymes, DNA helicases, aid in this process. The DNA that has been transcribed rewinds into the double helix behind the polymerase. At termination (**4**), the RNA polymerase is removed from the DNA. At this point, the formation of the unstable primary transcript is completed. Since it is unstable, it is immediately translated in prokaryotes and modified (processed) in eukaryotes (see p. 58). All processes are mediated by the complex interaction of a variety of enzymes (not shown). Transcription is fast, about 40 nucleotides per second (bacterial RNA polymerase), about the same rate as translation (15 amino acids per second), but slower than the rate of replication (800 base pairs per second; Lewin, 2004, p. 243).

B. Polymerase-binding site

The polymerase-binding site defines the starting point of transcription. At the termination site it is closed again (rewinding). Bacterial RNA polymerase binds to a specific region of about 60 base pairs of the DNA.

C. Promoter of transcription

A promoter is a DNA sequence that specifies the site of RNA polymerase binding from which transcription is initiated. The promoter is organized into several regions with sequence homology. In eukaryotes, transcription of protein-coding genes begins at multiple sites, often extending hundreds of base pairs upstream. A specific DNA sequence of about 4–8 base pairs is located about 25–35 base pairs upstream (in the 5′ direction) of the gene. Since this sequence is nearly the same in all organisms, it is called a consensus sequence. One such sequence is called the TATA box, because it contains the sequence TATA (or T)AA(or T)TA/G. It is highly conserved in evolution. In prokaryotes the promoter contains a consensus sequence of six base pairs, TATAAT (also called a Pribnow box after its discoverer), located 10 base pairs above the starting point. Another region of conserved sequences, TTGACA, is located 35 base pairs upstream of the gene. These sequences are referred to as the –10 box and the –35 box, respectively.

D. Transcription units

A transcription unit is the segment of DNA between the sites of initiation and termination of transcription.

E. Determining transcription start sites

A segment of DNA suspected to contain a transcription start site can be analyzed using the nuclease S1 protection assay. Endonuclease S1, an enzyme present in the mold *Aspergillus oryzae,* cleaves single-stranded DNA and RNA, but not double-stranded. Therefore, a DNA fragment that is denatured and mixed with total RNA from cells containing the gene to be analyzed will hybridize with cognate RNA and form a DNA/RNA heteroduplex. Following digestion with S1, all single-stranded DNA will be removed, allowing the relevant DNA to be identified.

References

Alberts B et al: Molecular Biology of the Cell, 4th ed. Garland Publishing, New York, 2002.

Lewin B: Genes VIII. Pearson Education International, 2004.

Lodish H et al: Molecular Cell Biology, 5th ed. WH Freeman & Co, New York, 2004.

Strachan T, Read AP: Human Molecular Genetics, 3rd ed. Garland Science, London-New York, 2004.

Rosenthal N: Regulation of gene expression. N Eng J Med 331: 931–933, 1994.

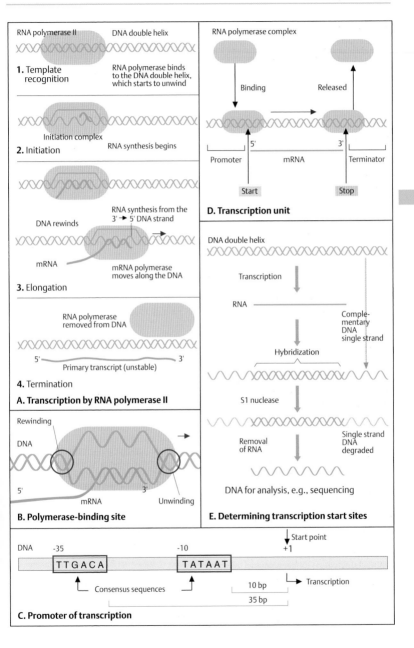

RNA polymerase II   DNA double helix

1. Template
recognition   RNA polymerase binds
                     to the DNA double helix,
                     which starts to unwind

Initiation complex   RNA synthesis begins
2. Initiation

DNA rewinds   RNA synthesis from the
                     3' → 5' DNA strand

mRNA   mRNA polymerase
                     moves along the DNA
3. Elongation

RNA polymerase
removed from DNA

5'   3'
Primary transcript (unstable)
4. Termination

A. Transcription by RNA polymerase II

Rewinding

DNA

5'   3'
mRNA   Unwinding
B. Polymerase-binding site

RNA polymerase complex

Binding   Released

5'   3'
Promoter   mRNA   Terminator

Start   Stop

D. Transcription unit

DNA double helix

Transcription

RNA   Comple-
                             mentary
                             DNA
                             single strand
RNA

Hybridization

S1 nuclease

Removal   Single strand
of RNA   DNA
                     degraded

DNA for analysis, e.g., sequencing

E. Determining transcription start sites

DNA   -35   -10   Start point
                                         +1

TTGACA   TATAAT

Consensus sequences   10 bp   Transcription
                                        35 bp

C. Promoter of transcription

Prokaryotic Repressor and Activator: the *lac* Operon

Bacteria can respond swiftly to changes in environmental conditions. Two basic modes control gene activity in bacteria. One is negative: a repressor protein interferes with RNA polymerase and prevents gene expression. In positive regulation, an activator protein binds to the repressor, thus inducing transcription. The ability to change between repression and activation corresponds to an off/on mode. Two-component regulatory systems control many bacterial responses as a genetic switch. Bacterial genes encoding proteins that function in the same metabolic pathway are usually located next to each other, controlled as a unit called an operon (Jacob & Monod, 1961).

A. Typical bacterial response to environmental change

The bacterium *Escherichia coli* growing in the absence of lactose does not produce enzymes required for lactose breakdown. If lactose is added to the culture medium, enzymes that cleave lactose into galactose and glucose are synthesized within 10 minutes (**1**). This effect is called enzyme induction. In the absence of lactose, enzyme synthesis is blocked by a gene regulatory protein, the lactose (*lac*) repressor (**2**). The *lac* repressor is a small molecule, a tetramer of four identical subunits of 38 kD each with a dual function. It was the first repressor molecule to be isolated (Gilbert & Müller-Hill 1966). Two subunits have a binding site for the operator and two for the inducer. Two alpha helices at the N-terminus fit into the major groove of DNA.

B. The lactose operon in *E. coli*

The bacterial response described above results from a change in transcriptional activity in three genes encoding the three lactose-degrading enzymes, β-galactosidase, β-galactoside permease, and β-galactoside transacetylase. These three genes, *lacZ*, *lacY*, and *lacA*, constitute the lactose (*lac*) operon. Their transcriptional activity is jointly regulated by a control region, which consists of a promoter (P) and an operator (O). The *lac* genes are controlled by negative regulation. Normally they are transcribed unless turned off by the *lac* repressor. This is the product of a regulator gene, *lac i*.

(Figure for *lac* repressor adapted from Koolman & Roehm, 2005.)

C. Control of the *lac* operon

In the absence of lactose, the *lac* repressor binds to the *lac* operator/promoter (**1**). This prevents initiation of RNA polymerase, thus inhibiting transcription. The inducer is β-galactoside, which binds to the repressor. This allows RNA polymerase to bind to the promoter and the three genes to be transcribed (**2**). This process can be reversed rapidly because *lac* mRNA is very unstable, degrading within 3 minutes.

D. The gene-regulatory nucleotide sequence of the *lac* repressor

The *lac* repressor recognizes a specific nucleotide sequence of 21 nucleotides in the operator/promoter region of the *lac* operon. The individual contacts, mediated by hydrogen bonds, ionic bonds, and hydrophobic interactions, are weak, but the 21 contacts allow a specific and strong binding. Such DNA-protein interactions are some of the most specific and tightest molecular interactions known in biology (Alberts et al. 2002, p. 383).

The recognition sequence of the *lac* repressor comprises additional nucleotides at both ends (shown as beige boxes) that have a twofold axis of symmetry. These correspond to the subunits of the repressor. This type of symmetry match is frequent in DNA-protein interactions, as in the palindromic cleavage sites of restriction endonucleases (p. 66).

The difference between the induced and repressed state is about 1000-fold. When the *lac* operator-repressor system is introduced into the mouse genome and connected to a tyrosinase gene, the tyrosinase enzyme is induced because the repressor finds its recognition sequence even in a much larger genome (Lewin, 2004, p. 296). (Figures adapted from Stryer, 1995.)

References

Alberts B et al: Molecular Biology of the Cell, 4th ed. Garland Science, New York, 2002.

Gilbert W, Müller-Hill B: Isolation of the *lac* repressor. Proc Nat Acad Sci 56: 1891–1998, 1966.

Jacob F, Monod J: Genetic regulatory mechanisms in the synthesis of proteins. J Mol Biol 3: 318–356, 1961.

Koolman J, Röhm KH: Color Atlas of Biochemistry, 2nd ed. Thieme, Stuttgart New York, 2005

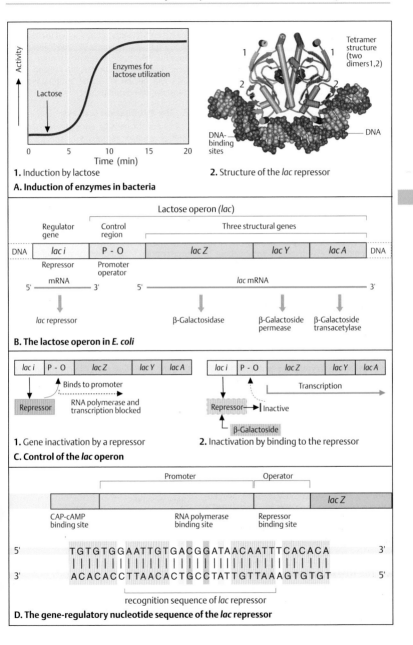

1. Induction by lactose

2. Structure of the *lac* repressor

A. Induction of enzymes in bacteria

Lactose operon *(lac)*

B. The lactose operon in *E. coli*

1. Gene inactivation by a repressor

2. Inactivation by binding to the repressor

C. Control of the *lac* operon

```
5'  TGTGTGGAATTGTGACGGATAACAATTTCACACA  3'
3'  ACACACCTTAACACTGCCTATTGTTAAAGTGTGT  5'
```

recognition sequence of *lac* repressor

D. The gene-regulatory nucleotide sequence of the *lac* repressor

Genetic Control by Alternative RNA Structure

Bacteria can regulate gene expression by premature termination of transcription. A striking example is a mechanism at the mRNA level that terminates transcription in the tryptophan operon of *E. coli*. In the absence of tryptophan, transcription of mRNA for enzymes required for tryptophan synthesis proceeds normally. However, if tryptophan is present, transcription is terminated early by a hairpin-like structure of mRNA formed by pairing of complementary mRNA molecules (attenuation).

A. Regulation of synthesis of the amino acid tryptophan in *E. coli*

Bacteria can synthesize tryptophan when it is not present in the nutrient medium. If tryptophan is added, enzyme activity for tryptophan biosynthesis decreases within about 10 minutes (**1**). Tryptophan is synthesized from chorismate and anthranalite in a metabolic pathway (**2**).

B. Tryptophan biosynthesis

The metabolic pathway of tryptophan synthesis involves four intermediate metabolic products. Chorismate is a common precursor of all three aromatic amino acids, phenylalanine, tyrosine, and tryptophan. Tryptophan synthesis is mediated by three enzymes: anthranilate synthetase, indol-3-glycerol phosphatase, and tryptophan synthetase. In 1964 Yanofsky and four colleagues showed that the protein product and its gene are colinear (see p. 54).

C. The tryptophan operon in *E. coli*

The tryptophan operon in *E. coli* consists of five structural genes, *TrpE, TrpD, TrpC, TrpB*, and *TrpA*. They occur in the order required for the pathway. The *trp* operon also includes the regulatory sequences, a promoter and an operator, a leader sequence (L), and attenuator sequences. The L sequence is the part of mRNA from its 5' end to the start codon. The attenuator sequences are part of the L-sequences.

D. The role of the attenuator

The weakening (attenuation) of the expression of the tryptophan operator is controlled by a sequence of about 100–140 base pairs 3' from the starting point of transcription (tryptophan

mRNA leader). In the presence of tryptophan, the *trp* mRNA leader is interrupted in the region of an attenuator sequence (**1**), and transcription does not take place. When tryptophan is deficient, transcription is delayed, a stop codon UGA will not be read (the attenuation), and transcription continues (**2**).

E. The attenuation process

The key to the attenuation process is two tryptophan residues contained in the *trp* leader peptide. When tryptophan is present (**1**), ribosomes can synthesize the complete leader peptide. The ribosome closely follows the RNA polymerase transcribing the DNA template (not shown). The ribosome has passed region 1 and prevents complementary regions 2 and 3 from forming a hairpin by base pairing. Instead, part of complementary region 3 and region 4 form a stem and a loop, which favors termination of transcription. When tryptophan is deficient (**2**), the ribosome stalls at the two UGG *trp* codons owing to deficiency of tryptophanyl tRNA. This alters the conformation of the mRNA so that regions 2 and 3 pair, region 4 remains single-stranded, and transcription can continue. Attenuation is an example of a tight relationship between transcription and translation. The *trp* mRNA leader region can exist in two alternative base-pair conformations. One allows transcription, the other does not. (Figures adapted from Stryer, 1995.)

References

Alberts B et al: Molecular Biology of the Cell, 4th ed. Garland Publishing Co, New York, 2002.

Bertrand K et al: New features of the regulation of the tryptophan operon. Science 189: 22–26, 1975.

Lewin B: Genes VIII. Pearson International, 2004.

Lodish H et al: Molecular Cell Biology, 5th ed. WH Freeman & Co, New York, 2004.

Stryer L: Biochemistry, 4th ed. WH Freeman & Co, New York, 1995.

Yanofsky C: Attenuation in the control of expression of bacterial operons. Nature 289: 751–758, 1981.

Yanofsky C, Konan KV, Sarsero JP: Some novel transcription attenuation mechanisms used by bacteria. Biochemie 78: 1017–1024, 1996.

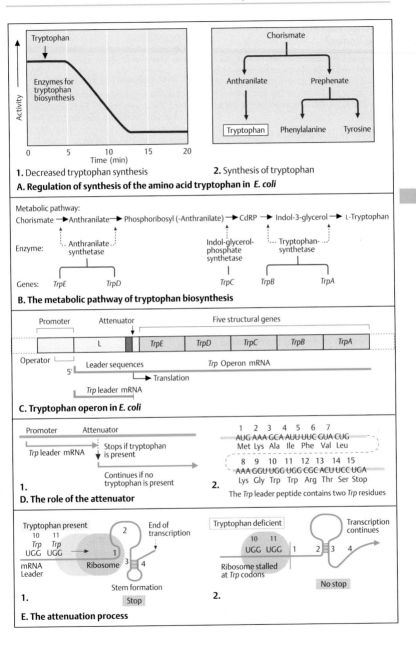

1. Decreased tryptophan synthesis

2. Synthesis of tryptophan

A. Regulation of synthesis of the amino acid tryptophan in *E. coli*

Metabolic pathway:

Chorismate → Anthranilate → Phosphoribosyl (-Anthranilate) → CdRP → Indol-3-glycerol → L-Tryptophan

Enzyme: Anthranilate synthetase   Indol-glycerol-phosphate synthetase   Tryptophan-synthetase

Genes: *TrpE*   *TrpD*   *TrpC*   *TrpB*   *TrpA*

B. The metabolic pathway of tryptophan biosynthesis

Promoter   Attenuator   Five structural genes

Operator | L | *TrpE* | *TrpD* | *TrpC* | *TrpB* | *TrpA*

5′   Leader sequences   *Trp* Operon mRNA

→ Translation

Trp leader mRNA

C. Tryptophan operon in *E. coli*

Promoter   Attenuator

Trp leader mRNA   Stops if tryptophan is present

Continues if no tryptophan is present

1.

```
        1    2    3    4    5    6    7
       AUG AAA GCA AUU UUC GUA CUG
       Met Lys Ala Ile Phe Val Leu

        8    9   10   11   12   13   14   15
       AAA GGU UGG UGG CGC ACU UCC UGA
       Lys Gly Trp Trp Arg Thr Ser Stop
```

The *Trp* leader peptide contains two *Trp* residues

2.

D. The role of the attenuator

1. Tryptophan present

```
  10    11
  Trp   Trp
  UGG   UGG
```

mRNA Leader   Ribosome

Stem formation   Stop

End of transcription

2. Tryptophan deficient

```
  10    11
  UGG   UGG
```

Ribosome stalled at *Trp* codons   No stop

Transcription continues

E. The attenuation process

Basic Mechanisms of Gene Control

Regulatory DNA sequences (promoters and enhancers) bind to specific proteins (transcription factors) and regulate gene activity. The binding of transcription factors at the promoter is the most important mechanism for regulating gene activity.

A. Consensus sequences in the promoter region

Promoters do not tolerate mutational changes and therefore remain constant, even in evolutionarily remote organisms (the sequences are conserved in evolution). In prokaryotes, two important regulatory sequences are 35 and 10 nucleotide base pairs upstream (in the 5′ direction) of the starting point of transcription (Pribnow, 1975). The –10 site (TATA box) is 5′–TATAAT–3′. The –35 sequence is 5′–TATTGACA–3′ (**1**). Mutations at different sites upstream of a gene have different effects, depending on where they are located (**2**). (Figure adapted from Watson et al., Molecular Biology of the Gene, 4th edition, 1987; adapted from M. Rosenberg & D. Court, 1979.)

B. Assembly of general transcription factors

In eukaryotes, the TATA box in the promoter region is about 25–35 bp upstream of the transcription start site (**1**). Several other promoter-proximal elements (different transcription factors, TFs) help to regulate gene activity. General transcription factors associate in an ordered sequence. First, TFIID (transcription factor D for polymerase II) binds to the TATA region (**2**). The TATA box is recognized by a small, 30-kD TATA-binding protein (TBP), which is part of one of the many subunits of TFIID (the bending of the DNA by TBP is not shown here). Then TFIIB binds to the complex (**3**). Subsequently, other transcription factors (TFIIH, followed by TFIIE) and Pol II, escorted by TFIIF, join the complex and ensure that Pol II is attached to the promoter (**4**). Pol II is activated by phosphorylation and transcription can begin (**5**). Other activities of TFIIH involve a helicase and an ATPase. The site of phosphorylation is a polypeptide tail, composed in mammals of 52 repeats of the amino acid sequence YSPTSPS, in which the serine (S) and threonine (T) side chains are phosphorylated.
(The figure, a simplified scheme, is adapted from Alberts et al., 2002, and Lewin, 2004.)

C. RNA polymerase promoters

Eukaryotic cells contain three RNA polymerases (Pol I, Pol II, and Pol III). Each uses a different type of promoter. RNA polymerase II requires a transcription factor complex (TFIID, see B) that binds to a single upstream promoter (**1**). RNA polymerase I has a bipartite promoter, one part 170 to 180 bp upstream (5′ direction) and the other from about 45 bp upstream to 20 bp downstream (3′ direction). The latter is called the core promoter (**2**). Pol I requires two ancillary factors, UPE1 (upstream promoter element 1) and SL1. RNA polymerase III uses either upstream promoters or two internal promoters downstream of the transcription start site (**3**). Three transcription factors are required with internal promoters: TFIIIA (a zinc finger protein, see p. 20), TFIIIB (a TBP and two other proteins), and TFIIIC (a large, more than 500-kD protein; for details see Lewin, 2004, p. 601 ff).
Pol I is located in the nucleolus and synthesizes ribosomal RNA. It accounts for about 50–70% of the relative activity. Pol II and Pol III are located in the nucleoplasm (the part of the nucleus excluding the nucleolus). Pol II represents 20–40 % of cellular activity. It is responsible for the synthesis of heterogeneous nuclear RNA (hnRNA), the precursor of mRNA. Pol III is responsible for the synthesis of tRNAs and other small RNAs. It contributes only minor activity of about 10%. Each of the large eukaryotic RNA polymerases (500 kDa or more) has 8–14 subunits and is more complex than the single prokaryotic RNA polymerase.

References

Alberts B et al: Molecular Biology of the Cell, 4th ed. Garland Publishing Co, New York, 2002.

Lewin B: Genes VIII. Pearson International, 2004.

Lodish H et al: Molecular Cell Biology, 5th ed. WH Freeman, New York, 2004.

Pribnow D: Nucleotide sequence of an RNA polymerase binding site at an early T7 promoter. Proc Nat Acad Sci 72: 784–789, 1975.

Rosenberg M, Court D: Regulatory sequences involved in the promotion and termination of RNA transcription. Ann Rev Genet 13: 319–353, 1979.

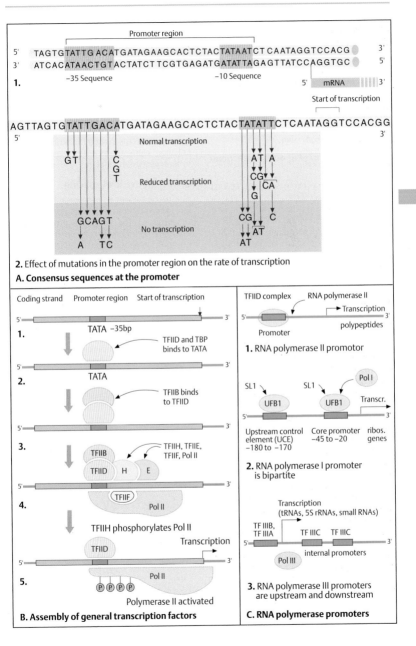

1.

5′ TAGTGTATTGACATGATAGAAGCACTCTACTATAATCTCAATAGGTCCACG 3′
3′ ATCACATAACTGTACTATCTTCGTGAGATGATATTAGAGTTATCCAGGTGC 5′

Promoter region

−35 Sequence   −10 Sequence

5′ mRNA 3′

Start of transcription

2. Effect of mutations in the promoter region on the rate of transcription

AGTTAGTGTATTGACATGATAGAAGCACTCTACTATATTCTCAATAGGTCCACGG

5′   3′

Normal transcription   GT   CGT   AT A

Reduced transcription   CG   CA   G

No transcription   GCAGT A TC   CG AT   C AT

A. Consensus sequences at the promoter

B. Assembly of general transcription factors

Coding strand   Promoter region   Start of transcription

1. TATA −35bp

TFIID and TBP binds to TATA

2. TATA

TFIIB binds to TFIID

3. TFIIB TFIID H E

TFIIH, TFIIE, TFIIF, Pol II

4. TFIIF Pol II

TFIIH phosphorylates Pol II

5. TFIID Transcription

Pol II ℗ ℗ ℗ ℗

Polymerase II activated

C. RNA polymerase promoters

TFIID complex   RNA polymerase II

Transcription

Promoter   polypeptides

1. RNA polymerase II promotor

SL1   SL1   Pol I

UFB1   UFB1   Transcr.

Upstream control element (UCE) −180 to −170   Core promoter −45 to −20   ribos. genes

2. RNA polymerase I promoter is bipartite

Transcription (tRNAs, 5S rRNAs, small RNAs)

TF IIIB, TF IIIA   TF IIIC   TF IIIC

Pol III   internal promoters

3. RNA polymerase III promoters are upstream and downstream

Regulation of Gene Expression in Eukaryotes

The term *gene expression* refers to the entire process of decoding the genetic information of active genes. Genes that are active (expressed) throughout the life of a cell or an organism show *constitutive expression*. Those genes that are transcribed only under certain circumstances, in specific cells or at specific times, show *conditional expression*.

A. Levels of control of eukaryotic gene expression

Schematically, gene expression can be regulated at four distinct levels. The first and by far the most important is primary control of the initiation of transcription. The next level, processing of the transcript to mature mRNA, can be regulated at the level of the primary RNA transcript. Different forms of mRNA are usually obtained from the same gene by alternative splicing (see D). A newly recognized form of gene expression control is RNA interference (see p. 224). Control is possible at the level of translation by mRNA editing (see B). Finally, at the protein level, posttranslational modifications can determine the activity of a protein.

B. RNA editing

RNA editing modifies genetic information at the RNA level. An important example is the apolipoprotein-B gene involved in lipid metabolism (OMIM 107730). It encodes a 512-kD protein of 4536 amino acids. This is synthesized in the liver and secreted into the blood, where it transports lipids. Apo B-48 (250 kD), a functionally related shorter form of the protein with 2152 amino acids is synthesized in the intestine. An intestinal deaminase converts a cytosine in codon 2152 CAA (glutamine) to uracil (UAA). This change results in a stop codon (UAA) and thereby terminates translation at this site.

C. Long-range gene activation by an enhancer

The term enhancer refers to a DNA sequence that stimulates the initiation of transcription (see p. 210). Enhancers act at a distance from the gene. They may be located upstream or downstream on the same DNA strand (*cis*-acting) or on a different DNA strand (*trans*-acting). An enhancer effect is mediated by sequence-specific DNA-binding proteins. One model suggests that DNA forms a loop between an enhancer and the promoter (Blackwood & Kadonga, 1998). An activator protein bound to the enhancer, e.g., a steroid hormone, could then come into contact with the general transcription factor complex at the promoter. The first enhancer to be discovered was a 72-bp tandem repeat near the origin of replication in simian virus 40. When experimentally linked to the β-globin genes (see p. 344), it considerably enhances their transcription (Banerji et al, 1981). Enhancer elements provide tissue-specific or time-dependent regulation.

D. Alternative RNA splicing

Alternative splicing is an important mechanism for generating multiple protein isoforms from a single gene. The resulting proteins differ slightly in their amino acid sequence. This may result in small functional differences. Quite often these differences are restricted to certain tissues, as schematically shown here for the calcitonin gene (OMIM 114130). The primary transcript for the calcitonin gene contains six exons. They are spliced into two different types of mature mRNA. One, calcitonin, consisting of exons 1–4 (excluding exons 5 and 6), is produced in the thyroid. The other, consisting of exons 1, 2, 3, 5, and 6 and excluding exon 4, encodes a calcitonin-like protein in the hypothalamus (calcitonin gene-related product, CGRP).

Alternative splicing clearly represents an evolutionary advantage because it allows for a high degree of functional flexibility (see Gravely, 2001; Modrek & Lee, 2002).

References

Alberts B et al: Molecular Biology of the Cell, 4th ed. Garland Publishing Co, New York, 2002.

Banerji J, Rusconi S, Schaffner S: Expression of a beta-globin in gene is enhanced by remote SV40 DNA sequences. Cell 27: 299–308, 1981.

Blackwood EM, Kadonga JF: Going the distance: A current view of enhancer action. Science 281: 60–63, 1998.

Bulger M, Groudine M: Enhancers. Nature Encyclopedia Hum Genome 2: 290–293, 2003.

Gravely BR: Alternative splicing: increasing diversity in the proteomic world. Trends Genet 17: 100–107, 2001.

Lewin B: Genes VIII, Pearson International, 2004.

Lodish H et al: Molecular Cell Biology, 5th ed. WH Freeman, New York, 2004.

Modrek B, Lee C: A genomic view of alternative splicing. Nature Genet 30: 13–19, 2002.

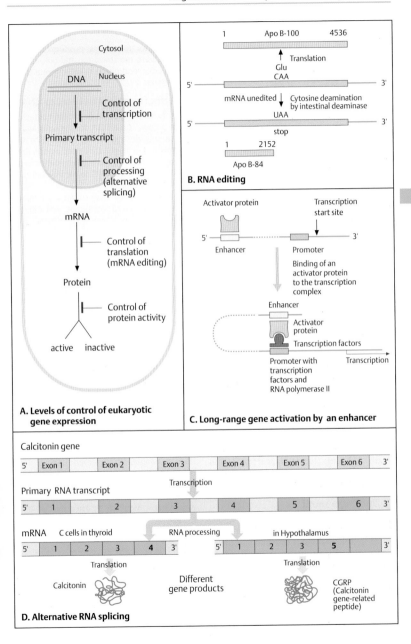

A. Levels of control of eukaryotic gene expression

Cytosol

Nucleus

DNA

Control of transcription

Primary transcript

Control of processing (alternative splicing)

mRNA

Control of translation (mRNA editing)

Protein

Control of protein activity

active   inactive

B. RNA editing

1   Apo B-100   4536

Translation

Glu
CAA

5'   mRNA unedited   3'

Cytosine deamination by intestinal deaminase

UAA

5'   3'

stop

1   2152

Apo B-84

C. Long-range gene activation by an enhancer

Activator protein

Transcription start site

5'   3'

Enhancer   Promoter

Binding of an activator protein to the transcription complex

Enhancer

Activator protein

Transcription factors

Promoter with transcription factors and RNA polymerase II

Transcription

D. Alternative RNA splicing

Calcitonin gene

5' | Exon 1 | Exon 2 | Exon 3 | Exon 4 | Exon 5 | Exon 6 | 3'

Transcription

Primary RNA transcript

5' | 1 | 2 | 3 | 4 | 5 | 6 | 3'

RNA processing

mRNA   C cells in thyroid

5' | 1 | 2 | 3 | 4 | 3'

in Hypothalamus

5' | 1 | 2 | 3 | 5 | 3'

Translation   Translation

Calcitonin   Different gene products   CGRP (Calcitonin gene-related peptide)

DNA-Binding Proteins, I

Regulatory DNA sequences exert their control function by specific interaction with DNA-binding proteins. Regulatory proteins recognize specific DNA sequences by a precise fit of their surface to the DNA double helix.

A. Binding of a regulatory protein to DNA

Gene regulatory proteins can recognize DNA sequence information without affecting the hydrogen bonds within the helix. Each base pair represents a distinctive pattern of hydrogen bond donors (shown in a red rectangle) and hydrogen acceptors (shown in a green rectangle). These proteins bind to the major groove of DNA. A single contact of an asparagine (Asn) of a gene-regulatory protein with a DNA base adenine (A) is shown here. A typical area of surface-to-surface contact involves 10–20 such interactions, resulting in high specificity.
(Figure adapted from Alberts et al., 2002, p. 384.)

B. Interaction of a DNA-binding protein with DNA

The alpha helix of a DNA-binding regulatory protein recognizes specific DNA sequences. Many bacterial repressor proteins are dimeric, so that an α-helix from each dimer can insert itself into two adjacent major grooves of the DNA double helix (recognition or sequence-reading helix). This structural motif is called a helix-turn-helix motif, because two helices lie next to each other. The example shows the tight interaction of the bacteriophage 434 repressor protein with one side of the DNA molecule over a length of 1.5 turns. (Figure adapted from Lodish et al., 2004, p. 463, based on A.K. Aggarwal et al., Science 242: 899, 1988.)

C. Zinc finger motif

Several eukaryotic regulatory proteins harbor regions that fold around a central zinc atom (Zn^{2+}), a structural motif resembling a finger (hence, zinc finger). In the example shown (from a frog protein; M.S. Lee et al., Science 245: 635–637, 1989), the basic zinc finger motif consists of a zinc atom connected to four amino acids of a polypeptide chain. The three-dimensional structure on the right consists of an antiparallel β-sheet (amino acids 1–10), an α-helix (amino acids 12–24), and the zinc connection. Four amino acids, two cysteines at positions 3 and 6 and two histidines at positions 19 and 23, are bonded to the zinc atom and hold the carboxy (COOH) end of the α-helix to one end of the β-sheet.

D. A zinc finger proten binds to DNA

The α-helix of each zinc finger can contact the major groove of the DNA double helix and establish specific and strong interactions over several turns in length. Extraordinary flexibility in gene control has been acquired during evolution through adjustments to the number of interacting zinc fingers.
Zinc finger proteins serve important functions during embryonic development and differentiation. (Figure redrawn from Alberts et al., 2002, p. 386.)

E. Hormone response element

Some DNA-binding proteins act as signal-transmitting molecules. The signal may be a hormone or a growth factor that activates an intracellular receptor. Steroid hormones enter target cells and bind to specific receptor proteins. Three views of the glucocorticoid receptor and its binding to DNA are shown here. The dimeric glucocorticoid receptor consists of two polypeptide chains (*hormone response element*, HRE). Each is stabilized by a zinc ion connected to four cysteine side chains (1). The skeletal model shows the binding of the dimeric protein to the DNA double helix (2). The space-filling model (3) shows how tightly the recognition helix of each dimer of this protein (yellow above, blue below) fits into two neighboring major grooves of DNA (shown in red and green). (Figures adapted from Stryer, 1995, p. 1002, based on B.F. Luisi et al., Nature 352: 497, 1991.)

References

Alberts B et al: Molecular Biology of the Cell, 4th ed. Garland Publishing Co, New York, 2002.

Alberts B et al: Essential Cell Biology. Garland Publishing Co, New York, 1998.

Lodish H et al: Molecular Cell Biology, 5th ed. WH Freeman & Co, New York, 2004.

Stryer L: Biochemistry, 4th ed. WH Freeman & Co, New York, 1995.

Tjian R: Molecular machines that control genes. Sci Am 272: 38–45, 1995.

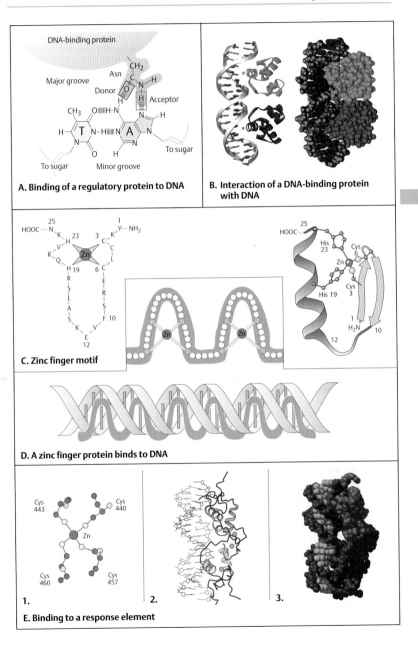

A. Binding of a regulatory protein to DNA

DNA-binding protein

Major groove

Donor

Acceptor

Asn

CH_2

T

A

CH_3

To sugar

Minor groove

To sugar

B. Interaction of a DNA-binding protein with DNA

C. Zinc finger motif

HOOC

25

23

19

Zn

1

3

6

10

12

HOOC

25

His 23

Cys 6

Zn

His 19

Cys 3

12

1

H_2N

10

D. A zinc finger protein binds to DNA

E. Binding to a response element

1.

Cys 443

Cys 440

Zn

Cys 460

Cys 457

2.

3.

DNA-Binding Proteins, II

The DNA-binding domains of eukaryotic transcription activators or repressors can be grouped according to different structural motifs. The specific DNA–protein binding usually involves noncovalent interactions between atoms in an α-helix of the binding domain of the protein and atoms on the outside of the major groove of the DNA double helix. About 2000 transcription factors are encoded by the human genome (Lodish et al., 2004, p. 463). Examples of important classes of DNA-binding domains are (i) *homeodomain proteins* containing a 180-bp sequence that is highly conserved as a *homeobox* in evolution (see part III, genes in embryonic development), (ii) *zinc finger proteins* (see previous page), (iii) *leucine-zipper proteins* (see below), and (iv) *basic helix-loop-helix (bHLH) proteins* (see previous page).

A. Leucine zipper and bHLH protein

Most DNA-binding regulatory proteins have a dimeric structure, which allows a dual function. One part of the molecule recognizes specific DNA sequences, the other stabilizes. One frequently occurring class of proteins has a characteristic structural motif called a leucine zipper. The name is derived from its basic structure. A typical leucine-zipper protein consists of a dimer with a periodic repeat of leucine every seven residues. The leucine residues are aligned along one face of each α-helix and interact with the DNA at adjacent major grooves (1). In basic helix-loop-helix (bHLH) proteins the DNA-binding helices at the N-terminal near the DNA are separated by nonhelical loops (2). (Figure adapted from Lodish et al., 2004, p. 465.)

B. Alternative heterodimeric combinations

Leucine-zipper proteins and basic helix-loop-helix proteins often exist in alternative combinations of dimers consisting of two different monomers. This dramatically increases the number of combinatorial possibilities.
In higher eukaryotes, leucine-zipper proteins often mediate the effect of cyclic adenosine monophosphate (cAMP) on transcription. Genes under this type of control contain a cAMP response element (CRE), a palindromic 8-bp recognition sequence. A protein of 43 kD binds to this target sequence. It is therefore known as the cAMP-response-element-binding protein (CREB). Leucine zipper proteins were first described in 1988 by W.H. Landschulz, P.F. Johnson, and S.L. McKnight (King & Stansfield, 2002). (Figure based on Alberts et al., 2002, p. 389.)

C. Activation by steroid hormone receptor complex binding to an enhancer

Transcriptional enhancers are regulatory regions of DNA that increase the rate of transcription. Their spacing and orientation vary relative to the starting point of transcription. Here an enhancer is activated by binding of a hormone receptor complex to a specific DNA sequence, a hormone response element. This activates the promoter, and transcription begins (active gene). Numerous important genes in mammalian development are regulated by steroids (steroid-responsive transcription).

D. Detection of DNA–protein interactions

The different regulatory proteins bind at specific sites of DNA (transcription-control elements). One common approach for detecting such sites and their cognate protein is the DNase I footprinting assay (1). This is based on the observation that DNA is protected from digestion by nucleases at the protein-binding site, whereas DNA outside the binding site is digested by DNase I. This enzyme cleaves DNA at multiple sites where it is not bound to protein. The protected protein-binding sites of DNA are visible as missing bands ("footprint") after separation of the DNA fragments by gel electrophoresis according to size. The electrophoretic mobility shift assay, or band-shift assay (2), is based on the principle that a DNA-protein complex retards the speed at which a fragment migrates in gel electrophoresis.

References

Alberts B et al: Molecular Biology of the Cell, 4th ed. Garland Publishing Co, New York, 2002.

King RC, Stansfield WD: A Dictionary of Genetics, 6th ed. Oxford University Press, Oxford, 2002.

Lodish H et al: Molecular Cell Biology, 5th ed. WH Freeman & Co, New York, 2004.

Stryer L: Biochemistry, 4th ed. WH Freeman, New York, 1995.

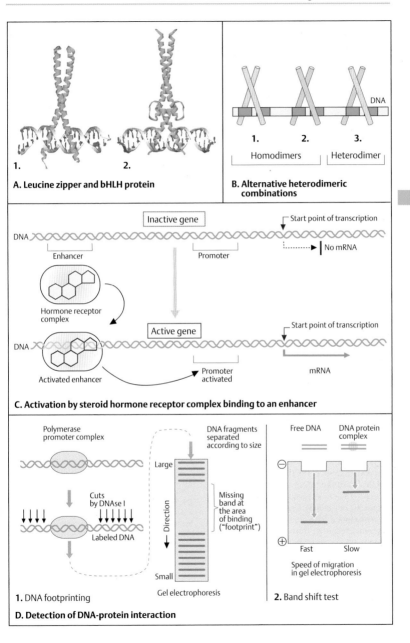

A. Leucine zipper and bHLH protein

1. 2.

B. Alternative heterodimeric combinations

DNA

1. 2. 3.

Homodimers | Heterodimer

C. Activation by steroid hormone receptor complex binding to an enhancer

Inactive gene

Start point of transcription

DNA

Enhancer Promoter

No mRNA

Hormone receptor complex

Active gene

Start point of transcription

DNA

Activated enhancer Promoter activated mRNA

D. Detection of DNA-protein interaction

Polymerase promoter complex

DNA fragments separated according to size

Large

Cuts by DNAse I

Missing band at the area of binding ("footprint")

Direction

Labeled DNA

Small

Gel electrophoresis

1. DNA footprinting

Free DNA DNA protein complex

Fast Slow

Speed of migration in gel electrophoresis

2. Band shift test

RNA Interference (RNAi)

RNA interference (RNAi) is a new biological phenomenon. RNAi selectively blocks transcription. RNAi is induced by short interfering RNA (siRNA). These are short double-stranded RNA molecules (dsRNA) of 21–23 base pairs with a high specificity for the nucleotide sequence of the target molecule, an mRNA.

Similar RNA molecules, called micro-RNAs (miRNAs) can function as antisense regulators of other genes. RNAi is regarded as a natural defense mechanism against endogenous parasites and exogenous pathogenic nucleic acids. Short interfering RNA is a new, important tool for analyzing gene function. The human genome contains about 200–255 genes for micro-RNAs (Lim et al., 2003).

A. Short interfering RNA (siRNA)

Short interfering RNA (siRNA) typically consists of 19 nucleotides of double-stranded RNA with a 2-bp overhang at both ends.

B. RNA-induced silencing complex

In plants and in *Drosophila*, siRNAs are formed by several enzymes with helicase and nuclease activity. An RNA endonuclease cleaves long, double-stranded RNA (dsRNA). The helicase unwinds the dsRNA. This protein complex is known as the RNA-induced silencing complex (RISC).

C. Posttranscriptional gene silencing

The target molecule of posttranscriptional gene inactivation (silencing) is an mRNA (**1**). With energy gained from an ATP to ADP reaction, the helicase activity of the RISC unwinds the short interfering RNA (**2**). The resulting single-stranded segment of the siRNA binds sequence-specifically to the mRNA and silences gene expression (**3**). A specialized ribonuclease III (RNase III) in the RISC cleaves the neighboring single-stranded RNA (red arrows). The mRNA fragments resulting from the degradation are then rapidly degraded by cellular nucleases (**4**).

D. Degradation of dsRNA by a dicer

A (biological) dicer is a complex molecule with endonuclease and helicase activity (RNase III helicase) that can cleave double-stranded RNA (**1**). The dicer complex binds to the dsRNA (**2**). The helicase activity unwinds the dsRNA, and

the RNA endonuclease activity (RNase type III enzyme) cleaves the RNA (**3**). In this way siRNA is formed (**4**).

E. Functional effect of RNAi

RNAi can be used for intentional silencing of a selected gene to assess its normal function. Here RNAi has silenced a gene in the developing worm *C. elegans* (see p. 304). Double-stranded RNAi targeted to a specific gene, based on sequence information, was injected into the gonad of an adult worm (**1**). Its effect is observed in the developing embryo (**2**). The degrading effect of the dsRNA is visualized by fluorescence *in-situ* hybridization of a labeled probe of mRNA from the target gene. The probe hybridizes to the normal, noninjected embryo (purple on the left, 2a) but not to the injected embryo (2b), whose target gene mRNA has been destroyed. (Figures in A–D modified from McManus & Sharp, 2002, and Kitabwalla & Ruprecht, 2002; in E, from Lodish et al., 2004.)

References

Fire A et al: Potent and specific genetic interference by double-stranded RNA in *Caenorhabditis elegans*. Nature 391: 806–811, 1998.

Hannon GJ: RNA interference. Nature 418: 244–251, 2002.

Kitabwalla M, Ruprecht RM: RNA interference – A new weapon against HIV and beyond. New Eng J Med 347: 1364–1367, 2002.

Lim LP et al: Vertebrate microRNA genes. Science 299: 1540, 2003.

Lodish H et al: Molecular Cell Biology, 5th ed. WH Freeman, New York, 2004.

McManus MT, Sharp PA: Gene silencing in mammals by small interfering RNAs. Nature Rev Genet 3: 737–747, 2002.

RNAi. Nature Insight, 16 September 2004, pp 338–378.

RNAi and its application poster. Nature Rev Genet 7: 1, 2006 (Online at www.nature.com/nrg/poster/rnai).

Stevenson M: Therapeutic potential of RNA interference. New Eng J Med 351: 1772–1777, 2004.

Soutchek J et al: Therapeutic silencing of an endogenous gene by systematic administration of modified RNAs. Nature 432: 173–174, 2004.

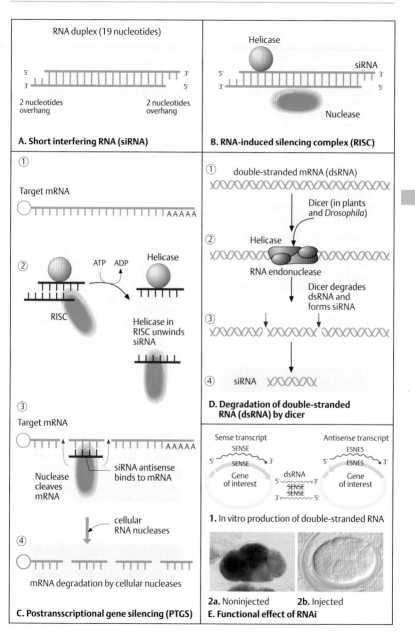

A. Short interfering RNA (siRNA)

RNA duplex (19 nucleotides)

5' — 3'
3' — 5'

2 nucleotides overhang — 2 nucleotides overhang

B. RNA-induced silencing complex (RISC)

Helicase

siRNA

Nuclease

C. Postranscriptional gene silencing (PTGS)

① Target mRNA
AAAAA

② ATP   ADP   Helicase
RISC
Helicase in RISC unwinds siRNA

③ Target mRNA
AAAAA
Nuclease cleaves mRNA
siRNA antisense binds to mRNA

cellular RNA nucleases

④ mRNA degradation by cellular nucleases

D. Degradation of double-stranded RNA (dsRNA) by dicer

① double-stranded mRNA (dsRNA)

Dicer (in plants and *Drosophila*)

② Helicase
RNA endonuclease

Dicer degrades dsRNA and forms siRNA

③

④ siRNA

E. Functional effect of RNAi

Sense transcript
SENSE
5' SENSE 3'
Gene of interest

Antisense transcript
ESNES
5' ESNES 3'
Gene of interest

dsRNA
5' SENSE 3'
3' SENSE 5'

1. In vitro production of double-stranded RNA

2a. Noninjected   **2b.** Injected

Targeted Gene Disruption

Targeted gene disruption refers to experimental inactivation of a gene in order to investigate its function. In a "knockout" animal, usually in mice, the gene under study is inactivated in the germ line by disrupting it (*gene knockout*). The effects can be studied at different embryonic stages and after birth. Ultimately, this knowledge can be utilized to understand the effects of mutations in homologous human genes as seen in human genetic diseases.

A variant of knockout is known as gene *knock-in*. In this case, the targeting construct contains a normal gene that is introduced either in addition to or instead of the gene to be studied. Transgenic animals contain foreign DNA which has been injected during early embryonic stages.

A. Preparation of ES cells with a knockout mutation

The target gene is disrupted (knocked out) in embryonic stem cells (ES) by homologous recombination with an artificially produced nonfunctional allele. The isolation of ES cells with disrupted gene requires positive and negative selection. A bacterial gene conferring resistance to neomycin (*neo^R*) is introduced into the DNA of the artificial allele, partially cloned from the normal target gene (**1,2**). In addition, DNA containing the thymidine kinase gene *(tk$^+$)* from herpes simplex virus is added to the gene replacement construct outside the region of homology (**3**). The selective medium contains the positive and the negative selectable markers neomycin and ganciclovir. Nonrecombinant cells and cells with nonhomologous recombination at random sites cannot grow in this medium. Nonrecombinant cells remain sensitive to neomycin, whereas recombinant cells are resistant (positive selection, not shown). The *thymidine kinase (tk$^+$)* gene confers sensitivity to ganciclovir, a nucleotide analog. Unlike the endogenous mammalian thymidine kinase, the enzyme derived from the herpex simplex virus is able to convert ganciclovir into the monophosphate form. This is modified into the triphosphate form, which inhibits cellular DNA replication. Since nonhomologously recombinant ES cells contain the *tk$^+$* gene at random sites, they are sensitive to ganciclovir and cannot grow in its presence (negative selection, **4**). Only cells that have undergone homologous recombination can survive, because they contain the gene for neomycin resistance (*neo^R*) and do not contain the *tk$^+$* gene (**5**).
(Figure redrawn from Lodish et al., 2004, p. 389.)

B. Knockout mice

In the second phase embryonic stem cells (ES) from a mouse blastocyst are isolated (**1**) after 3.5 days of gestation (of a total of 19.5 days) and transferred to a cell culture grown on a feeder layer of irradiated cells that are unable to divide (**2**). ES cells heterozygous for the knockout mutation are added (**3**). These ES cells are derived from a mouse that is homozygous for a different coat color (e.g. black) from that of the mouse that will develop from the blastocyst (e.g. white). The recombinant ES cells are integrated into the recipient blastocyst (**4**). The early embryos are transplanted into a pseudopregnant mouse (**5**). The offspring that have taken up ES-derived cells are chimeric. They consist of two types of cells, some with and some without the disrupted gene. The transgenic mice can be recognized by black coat color spots on a white (or brown) background (**6**). The chimeric mice are then backcrossed to homozygous white mice (**7**). Black offspring from this mating are heterozygous for the disrupted (mutant) gene (**8**). During further breeding of the heterozygous mice (**9**), some of their offspring, the knockout mice, will be homozygous for the disrupted gene.
(Figure adapted from Alberts et al., 2002.)

References

Alberts B et al: Molecular Biology of the Cell, 4th ed. Garland Publishing Co, New York, 2002.

Capecchi MR: Altering the genome by homologous recombination. Science 244: 1288–1292, 1989.

Capecchi MR: Targeted gene replacement. Sci Am, pp 52–59, March 1994.

Lodish H et al: Molecular Cell Biology, 5th ed. WH Freeman & Co, New York, 2004.

Gordon JW: Genetic transformation of mouse embryos by microinjection of purified DNA. Proc Nat Acad Sci 77: 7380–7384, 1980.

Majzoub JA, Muglia LJ: Knockout mice. Molecular Medicine. New Engl J Med 334: 904–907, 1996.

Strachan T, Read AP: Human Molecular Genetics, 3rd ed. Garland Publishing Co, New York, 2004.

Online information:

Internet Resources for Mammalian Transgenesis: BioMetNet Mouse Knockout Database: www.bioednet.com/db/mkmd. Jackson Laboratory Database: www.jaxmice.jax.org/index.shtml.

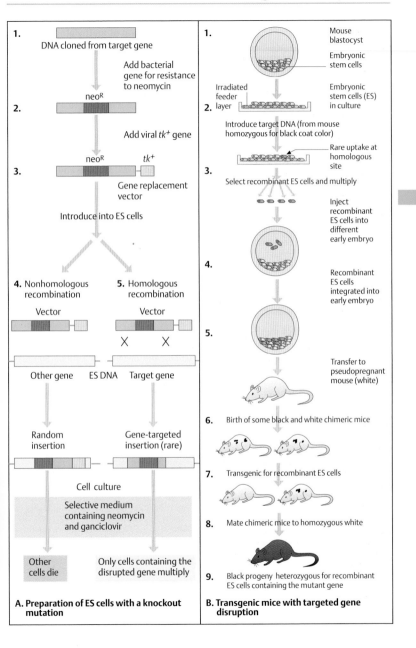

1. DNA cloned from target gene

Add bacterial gene for resistance to neomycin

2. neoR

Add viral *tk*+ gene

3. neoR   *tk*+

Gene replacement vector

Introduce into ES cells

4. Nonhomologous recombination

5. Homologous recombination

Vector

Vector

X   X

Other gene   ES DNA   Target gene

Random insertion

Gene-targeted insertion (rare)

Cell culture

Selective medium containing neomycin and ganciclovir

Other cells die

Only cells containing the disrupted gene multiply

A. Preparation of ES cells with a knockout mutation

1. Mouse blastocyst
Embryonic stem cells

Irradiated feeder layer

Embryonic stem cells (ES) in culture

2.

Introduce target DNA (from mouse homozygous for black coat color)

Rare uptake at homologous site

3. Select recombinant ES cells and multiply

Inject recombinant ES cells into different early embryo

4. Recombinant ES cells integrated into early embryo

5. Transfer to pseudopregnant mouse (white)

6. Birth of some black and white chimeric mice

7. Transgenic for recombinant ES cells

8. Mate chimeric mice to homozygous white

9. Black progeny heterozygous for recombinant ES cells containing the mutant gene

B. Transgenic mice with targeted gene disruption

DNA Methylation

DNA methylation is the addition of methyl groups to specific sites on the DNA, commonly a cytosine. Up to almost 10% of cytosine in higher organisms is methylated. It is present in CG doublets, called CpG islands. They are found at the 5′ end of many genes. DNA methylation is a functionally important epigenetic modification. Epigenetic modifications are heritable changes that influence the expression of certain genes without altering the DNA sequence. The term *epigenetics* was coined by C.H. Waddington in 1939 (Speybroeck, 2002).

The methylation pattern of DNA is functionally important, since altered DNA methylation may result in developmental failure and disease. Mammalian cells contain enzymes that maintain DNA methylation and establish it in the new strand of DNA after replication. These are the DNA methyltransferases (DNMTs) and methyl-cytosine binding proteins (MeCPs) binding to CpG islands. Two types of methyltransferase can be distinguished by their basic functions: maintenance methylation (DNMT1) and *de novo* methylation (DNMT3a and 3b).

A. Maintenance of DNA methylation

This type of methylation adds methyl groups to the newly synthesized DNA strand after replication and cell division. The methylated sites in the parental DNA (**1**) serve as templates for correct methylation of the two new strands after replication (**2**). This ensures that the previous methylation pattern is correctly maintained at the same sites as the parental DNA (**3**). The enzyme responsible for this is Dnmt1 (DNA methylase 1, DNMT1 in humans). Mice deficient in Dnmt3a die within a few weeks after birth as a result of genome-wide demethylation (Okano et al., 1999).

B. *De novo* DNA methylation

Here methyl groups are added at new positions on both strands of DNA. Two genes for different methyltransferases with overlapping functions in global remethylation have been identified: *Dnmt3a* and *Dnmt3b*. Unmethylated DNA (**1**) is methylated by their enzymes (**2**) in a site-specific and tissue-specific manner (**3**). Targeted homozygous disruption of the mouse *Dnmt3a* and *Dnmt3b* genes results in severe developmental defects.

C. Recognition of methylated DNA

Certain restriction enzymes do not cleave DNA when their recognition sequence is methylated (**1**). The enzyme *Hpa*II cleaves DNA only when its recognition sequence 5′-CCGG-3′ is not methylated (**2**). *Msp*I recognizes the same 5′-CCGG-3′ sequence irrespective of methylation and cleaves DNA at this site every time. This difference in cleavage pattern, resulting in DNA fragments of different sizes, serves to distinguish the methylation pattern of the DNA.

D. Human *DNMT3B* gene and mutations

Mutations in the human gene *DNMT3B* encoding type 3B *de novo* methyltransferase cause a distinctive disease called ICF syndrome (immunodeficiency, centromeric chromosomal instability, and facial anomalies, OMIM 242860; Hansen et al., 1999; Xu et al., 1999). The centromeres of chromosomes 1, 9 and 16, where satellite DNA types 2 and 3 are located, are unstable. The human *DNMT3B* gene (**1**) consists of 23 exons spanning 47 kb. Six exons are subject to alternative splicing. The protein (**2**) has 845 amino acids with five DNA methyltransferase motifs (I, IV, V, IX, X) in the C-terminal region. The arrows point to six different mutations. The mutation at position 809 (**3**), a change of A to G in codon 809, i.e., GAC (Asp) to GGC (Gly), involves the replacement of asparagine (Asn) by glycine (Gly). Both parents are heterozygous for this mutation. Multiradiate chromosomes with multiple p and q arms are typical in lymphocytes in ICF syndrome, here derived from chromosomes 1 and 16 shown in R banding (**4**).

(Figure adapted form Xu et al., 1999.)

References

Hansen RS et al: *DNMT3B* DNA methyltransferase gene is mutated in the ICF immunodeficiency syndrome. Proc Natl Acad Sci 96: 14412–14417, 1999.

Okano M et al: DNA methyltransferases Dnmt3a and Dnmt3b are essential for de novo methylation and mammalian development. Cell 99: 247–257, 1999.

Robertson KD: DNA methylation and human disease. Nature Rev Genet 6: 597–610, 2005.

Speybroek L van: From epigenesis to epigenetics. The case of C. H. Waddington. Ann NY Acad Sci 981: 61–81, 2002.

Xu G, Bestor T et al: Chromosome instability and immunodeficiency syndrome caused by mutations in a DNA methyltransferase gene. Nature 402: 187–191, 1999.

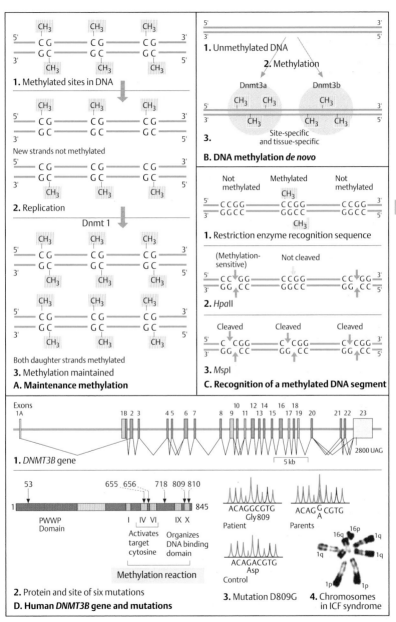

1. Methylated sites in DNA

2. Replication

Dnmt 1

Both daughter strands methylated
3. Methylation maintained
A. Maintenance methylation

1. Unmethylated DNA

2. Methylation

Dnmt3a   Dnmt3b

3. Site-specific and tissue-specific
B. DNA methylation *de novo*

Not methylated   Methylated   Not methylated
1. Restriction enzyme recognition sequence

(Methylation-sensitive)   Not cleaved
2. *Hpa*II

Cleaved   Cleaved   Cleaved
3. *Msp*I
C. Recognition of a methylated DNA segment

Exons

1. *DNMT3B* gene

5 kb

2800 UAG

PWWP Domain

I IV VI   IX X

Activates target cytosine   Organizes DNA binding domain

Methylation reaction

2. Protein and site of six mutations

ACAGGCGTG   ACAG G/A CGTG
Gly 809
Patient   Parents

ACAGACGTG
Asp
Control

3. Mutation D809G   **4. Chromosomes in ICF syndrome**

D. Human *DNMT3B* gene and mutations

Reversible Changes in Chromatin Structure

In euchromatin, genes are accessible to transcription, but not in heterochromatin. The local structure of chromatin (see p. 186) is an *epigenetic* state. This can be changed reversibly by a variety of mechanisms, called chromatin remodeling. Chromatin remodeling is an active, reversible process by which histones from the DNA molecule are displaced to make genes accessible for transcription. The energy required is provided by the hydrolysis of adenosine triphosphate (ATP) in large remodeling complexes that can be classified according to their ATPase subunits (for details see Lewin, 2004, p. 659 ff.).

A. Histone modification

A key event in chromatin remodeling is the modification of the core histones H3 and H4 (see p. 184). Methylation (adding CH_3 groups), acetylation (adding acetyl groups; -NH-CH$_3$), and phosphorylation (adding phosphate groups) are the types of modification that histones H3 (**1**) and H4 (**2**) undergo at certain of the 20 amino acids at the N-terminal ends (tails). The resulting combinations of signals are called the histone code (Turner, 2002). The modifications are mediated by methylases and demethylases, acetylases and deacetylases, and phosphorylation kinases.

Active chromatin is acetylated at the lysine residues of H3 and H4 histones. Inactive chromatin is methylated at the position 9 lysine of H3 and at other lysine residues, and methylated at the cytosines of CpG islands. Modifications can be related to individual functions (see table in the appendix). Lys-9 in H3 can be either methylated or acetylated. Thus, multiple modifications can occur and influence each other.
(Figure based on data in Strachan & Read, 2004, and Lewin, 2004.)

B. Histone acetylation and deacetylation

Acetylation is mediated by histone acetyltransferases (HATs). HAT is part of a large activating complex (**1**). The acetyl groups can be removed in a reversible process, deacetylation (**2**), mediated by deacetylases (HDACs). Two groups of HAT enzymes are distinguished: group A, involved in transcription, and group B, involved in nucleosome assembly (Lewin, 2004, p. 665 ff.).

Deacetylation and methylation may be connected. Two methyl-cytosine binding proteins, MeCP1 and MeCP2, selectively bind to methylated DNA. Transcriptional repression by the methyl-CpG-binding proteins 1 and 2 involve histone deacetylation in a multiprotein complex (Nan et al, 1998). (Figure adapted from Lodish et al., 2004, p. 475.)

C. Chromatin remodeling

Activator proteins can reverse the "gene off" state in heterochromatin. These are specific DNA-binding control elements in chromatin, able to interact with multiprotein complexes. The activator proteins bind to a mediator protein. As a result, chromatin becomes decondensed and the gene assumes the "gene on" state. General transcription factors and RNA polymerase assemble at the promoter and initiate transcription. When repressor proteins bind to control elements, this process is reversed and the initiation of transcription by RNA polymerase is inhibited. The formation of heterochromatin begins with the binding of heterochromatin protein 1 (HP1) to methylated histone 3. A DNA-binding protein, RAP1, recruits other proteins (SIR3/SIR4). These bind to H3/H4 and polymerize along chromatin. (Figure adapted from Lodish et al., 2004, p. 448.)

Medical relevance

Mutations of the *MECP2* gene on Xq28 cause Rett syndrome (OMIM 312750; Amir et al., 1999).

References

Amir RE et al: Rett syndrome is caused by mutations in X-linked *MECP2*, encoding methyl-CpG binding protein 2. Nature Genet 23: 185–188, 1999.

Jaenisch R, Bird A: Epigenetic regulation of gene expression: how the genome integrates intrinsic and environmental signals. Nature Genet Suppl 33: 245–254, 2003.

Lachner M, O'Sullivan RJ, Jenuwein T: An epigenetic road map for histone lysine methylation. J Cell Sci 116: 2117–2124, 2003.

Lewin B: Genes VIII. Pearson International, 2004.

Lodish H et al: Molecular Cell Biology, 5th ed. WH Freeman, New York, 2004.

Strachan T, Read AD: Human Molecular Genetics, 3rd ed. Garland Science, London & New York, 2004.

Turner BM: Cellular memory and the histone code. Cell 111: 285–291, 2002.

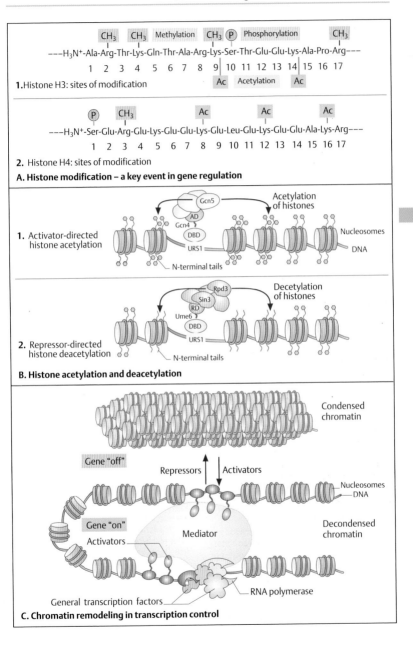

1. Histone H3: sites of modification

2. Histone H4: sites of modification

A. Histone modification – a key event in gene regulation

1. Activator-directed histone acetylation

2. Repressor-directed histone deacetylation

B. Histone acetylation and deacetylation

C. Chromatin remodeling in transcription control

Genomic Imprinting

In eukaryotes, only one allele of certain genes is expressed, while the other is permanently repressed. The state of expression depends on which parent contributes the allele, i.e., whether it is of maternal or paternal origin (parent-specific expression). This is called *genomic imprinting*. Genomic imprinting is an important epigenetic change in mammalian cells. Imprinting is assumed to have evolved in mammals in response to intrauterine competition for resources. Natural selection acts differently on genomes of maternal and paternal origin. A balance between maternal survival and fetal growth is favorable.

A. The importance of two different parental genomes

In mice, different developmental results are observed depending on whether the female pronucleus (2) or the male pronucleus (4) is removed from a diploid zygote (1) before they fuse, instead of leaving both in place (3). If the female pronucleus is replaced by a male pronucleus, an androgenetic zygote results. In this case the zygote initially appears normal. However, if implantation ensues, nearly all androgenotes will fail to complete preimplantation (2). The rare few that reach postimplantation develop abnormally and do not progress beyond the 12-somite stage.

In contrast, when a male pronucleus is replaced by a female pronucleus, a gynogenetic zygote results (4), which differs markedly from the androgenote. Although about 85% of gynogenotes develop normally until preimplantation, the extraembryonic membranes are absent or underdeveloped. As a result the embryo dies at or before the 40-somite stage. (Figure adapted from Sapienza & Hall, 2001.)

B. Requirement for a maternal and a paternal genome

A naturally occurring human androgenetic zygote is a hydatidiform mole (1). This is an abnormal placental formation containing two sets of paternal chromosomes and none from the mother. An embryo does not develop, although implantation takes place. The placental tissues develop many cysts (2). When only maternal chromosomes are present, an ovarian teratoma with many different types of fetal tissue develops (3). No placental tissue is present in this naturally occurring gynogenetic zygote. In triploidy, a relatively frequent fatal human chromosomal disorder (see p. 412), extreme hypoplasia of the placenta and fetus is observed when the additional chromosomal set is of maternal origin (4). (Photographs kindly provided by Professor Helga Rehder, Marburg.)

C. Genomic imprinting is established in early embryonic development

The changes responsible for imprinting occur in early embryogenesis. The imprint pattern typically present in somatic cells (1), is erased in primordial germ cells (2). During the formation of gametes, the imprinting pattern is reset (3). Imprinted chromosomal regions of paternal origin receive the paternal pattern; those of maternal origin receive the maternal pattern. As a result, after fertilization the correct imprint pattern is present in the zygote (4) and is maintained through all subsequent cell divisions.

Medical relevance

Failure to establish the normal pattern of imprinting due to gene rearrangements causes an important, heterogenous group of *imprinting diseases* (see p. 410).

References

Constância M, Kelsey G, Reik W: Resourceful imprinting. Nature 432: 53–57, 2004.

Horsthemke B, Buiting K: Imprinting defects on human chromosome 15. Cytogenet Genome Res 113: 292–299, 2006.

Morrison IM, Reeve AE: Catalogue of imprinted genes and parent-of-origin effects in humans and animals. Hum Mol Genet 7: 1599–1609, 1998.

Reik W, Walter J: Genomic imprinting: parental influence on the genome. Nature Rev Genet 2: 21–32, 2001.

Reik W, Dean W, Walter J: Epigenetic reprogramming in mammalian development. Science 293: 1089–1093, 2001.

Sapienza C, Hall JG: Genetic imprinting in human disease, pp 417–431. In: The Metabolic and Molecular Bases of Inherited Disease, 8th ed. CR Scriver et al (eds), McGraw-Hill, New York, 2001.

Wilkins JF, Haig D: What good is genomic imprinting: the function of parent-specific gene expression. Nature Rev Genet 4: 359–368, 2003.

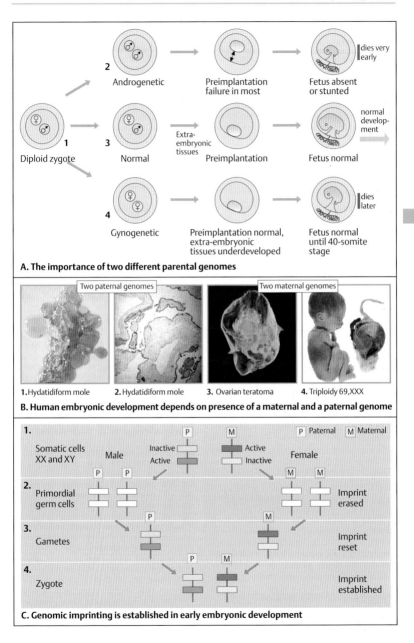

A. The importance of two different parental genomes

1. Hydatidiform mole   2. Hydatidiform mole   3. Ovarian teratoma   4. Triploidy 69,XXX

B. Human embryonic development depends on presence of a maternal and a paternal genome

C. Genomic imprinting is established in early embryonic development

Mammalian X Chromosome Inactivation

In mammals, the genes on one of the two X chromosomes in every cell of females are inactivated. The result is *dosage compensation*, a term introduced 1948 by H.J. Muller. A long 17-kb noncoding RNA molecule called Xist (X-inactivation specific transcript) in early embryonic development coats one of the X chromosomes in cells with two X. This recruits other proteins and results in gene silencing by methylation and histone modifications. Xist is the product of the *Xist* gene (MIM 314670) located at Xq13.2.

A. X chromatin

Small, darkly staining bodies were described by Barr & Bertram in 1949 in the nerve cells of female cats (**1, 3**), but not in males (**2**). Davidson & Smith described similar structures as drumsticks in peripheral blood leukocytes (**4**). X chromatin is visible as a dark density of about 0.8×1.1 µm in the nuclei of oral mucosal cells obtained from a buccal smear or in cultured fibroblasts (**5**) (Figures 1–3 from Barr & Bertram, 1949.).

B. Scheme of X inactivation

The maternal *Xist* is expressed from the morula stage on. It follows random X inactivation involving either the maternal or the paternal X chromosome. The inactivation pattern is stably transmitted to all daughter cells.

C. Mosaic pattern of expression

Mary F. Lyon in 1961 described a mosaic distribution pattern of X-linked coat colors in female mice as a manifestation of X inactivation (**1**). Fingerprints of human females heterozygous for X-linked hypohidrotic ectodermal dysplasia (OMIM 305100) show a mosaic pattern of areas with normal sweat pores, (black points) and areas without sweat pores, as seen in affected males (**2**). In cell cultures from females heterozygous for X-chromosomal HGPRT (hypoxanthine-guanine phosphoribosyltransferase) deficiency (OMIM 308000), the colonies are either HGPRT⁻ or HGPRT⁺ (**3**). (Figure in **1** from Thompson, 1965; in **2** from Passarge & Fries, 1973; in **3** from Migeon, 1971.)

D. X-inactivation profile

Genes in defined regions of the human X chromosome are not activated. An X-inactivation profile (Carrel & Willard, 2005) reveals that 458 (75%) genes are inactivated and 94 (15%) regularly escape inactivation. Surprisingly, 65 genes (10%) are inactivated in some females, but not in others. Thus, 25% of X-linked human genes are not regularly inactivated, and 10% exhibit an interindividual inactivation pattern.
To the left (**a**) of the X chromosome, nine vertical lanes represent 9 rodent/human cell hybrids. Genes expressed on the inactive X chromosome are shown in blue; silenced genes are in yellow. The right (**b**) illustrates the level of expression in the inactive X chromosome. (Figure kindly provided by Dr. Laura Carrel, Hershey Medical Center, Pennsylvania; from Carell & Willard, 2005)

E. Evolutionary strata on the X

The human X chromosome harbors strata (S1-S5) of different evolutionary origin and time (see p. 256).

References

Barr ML, Bertram EG: A morphological distinction between neurones of the male and female, and the behaviour of the nucleolar satellite during accelerated nucleoprotein synthesis. Nature 163: 676–677, 1949.

Davidson WM, Smith DR: A morphological sex difference in the polymorphonuclear neutrophil leukocytes. Brit Med J 2: 6–7, 1954.

Lyon MF: Gene action in the X-chromosome of the mouse (*Mus musculus* L.). Nature 190: 372–373, 1961.

Marberger E, Boccabella, R, Nelson WO: Oral smear as a method of chromosomal sex detection. Proc Soc Exp Biol (NY) 89: 488–489, 1955.

Migeon BR: Studies of skin fibroblasts from 10 families with HGPRT deficiency, with reference to X-chromosomal inactivation. Am J Hum Genet 23: 199–200, 1971.

Okamoto I et al: Epigenetic dynamics of imprinted X inactivation during early mouse development. Science 303: 644–649, 2004.

Passarge E, Fries E: X chromosome inactivation in X-linked hypohidrotic ectodermal dysplasia. Nature New Biology 245: 58–59, 1973.

Thompson MW: Genetic consequences of heteropyknosis of an X chromosome. Canad J Genet Cytol 7: 202–213, 1965.

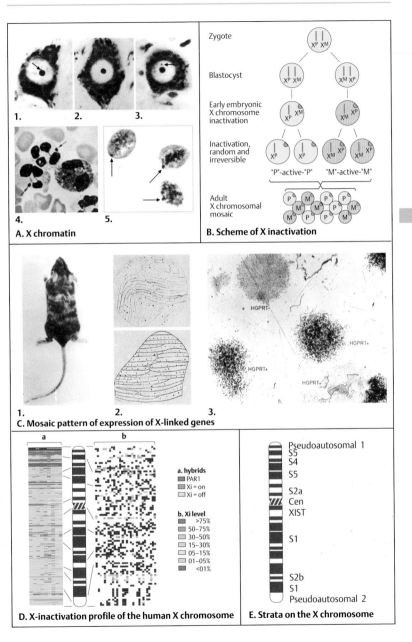

A. X chromatin

Zygote

Blastocyst

Early embryonic
X chromosome
inactivation

Inactivation,
random and
irreversible

"P"-active-"P"   "M"-active-"M"

Adult
X chromosomal
mosaic

B. Scheme of X inactivation

C. Mosaic pattern of expression of X-linked genes

HGPRT−

HGPRT+

HGPRT+

HGPRT−

a   b

a. hybrids
  PAR1
  Xi = on
  Xi = off

b. Xi level
  >75%
  50–75%
  30–50%
  15–30%
  05–15%
  01–05%
  <01%

D. X-inactivation profile of the human X chromosome

Pseudoautosomal 1
S5
S4
S5

S2a
Cen
XIST

S1

S2b
S1
Pseudoautosomal 2

E. Strata on the X chromosome

Genomics

Genomics, the Study of the Organization of Genomes

Genomics is the scientific field dealing with all structural and functional aspects of the genomes of different species. A central goal is to determine entire DNA nucleotide sequences. Related areas are concerned with all molecules involved in transcription and translation and their regulation (the *transcriptome*); all proteins that a cell or an organism is able to produce (the *proteome*); the functional analysis of all genes (*functional genomics*); the evolution of genomes (*comparative genomics*); and the assembly, storage, and management of data (*bioinformatics*). The study of genomes has important implications for medicine and agriculture.

A. Examples of organisms whose genomes have been sequenced

The human genome contains 2.85 billion (2.85 Gb) nucleotides (2 851 300 913), but remarkably few protein-coding genes, about 22 000. This is surprisingly low compared with the number in other, smaller organisms. Other basic features of the human genome are a low average density of genes along a given stretch of DNA; chromosomal segments that result from duplication events during evolution; a high proportion of interspersed, repetitive sequences; evidence of transposition events; and homology with genes and sequences of other organisms. All 24 human chromosomes have been sequenced. Information about the genome of man and other organisms is available online.

Sources of images: 1: A. Dürer, 1507. Adam und Eva. Museo Nacional del Prado, Madrid; 2: J. Weissenbach, 2004; 3: Nature 429:353–355; 4: www.peter-strohmaier.at/mausfolgen.gif; 5: Robert Geisler, Max Planck Institute for Developmental Biology, Germany (www.zf-models.org/); 6: Marco van Kerkhoven (www.kennislink.nl); 7: University of Guelph, Ontario (www.uoguelph.ca/.../Drosophila2.JPG; image source: www.gen.cam.ac.uk/dept/ashburner.html); 8: Dr. Michel Viso (www.desc.med.vu.nl/ NL-Taxi.htm); 9: Winston Laboratory, Department of Genetics, Harvard Medical School (genetics.med.harvard.edu/~winston/); 10: MichiganTech (www.techalive.mtu.edu/meec/); 11: The Swiss-Prot plant proteome annotation program (www.biologie.uni-ulm.de/bio2/knoop/images/arabidopsis.jpg).

References

Homo sapiens: International Human Sequencing Genome Consortium (IHSGC): Initial sequencing and analysis of the human genome. Nature 409: 860–921, 2001.

International Human Genome Sequencing Consortium: Finishing the euchromatic sequence of the human genome. Nature 431: 931–945, 2004.

Nature Webfocus: The Human Genome (free access at www.nature.com/nature/focus/humangenome/index.html).

Venter JG et al: The sequence of the human genome. Science 291: 1304–1351, 2001.

Chimpanzee: Waterston, RH, Lander ES, Watson RK et al.: Initial sequencing of the chimpanzee genome and comparison with the human genome. Nature 437: 69–87, 2005.

Dog: Lindhblad K et al: Genome sequence, comparative analysis, and haplotype structure of the domestic dog. Nature 438: 803–819, 2005.

Mouse: Waterston RH et al.: Initial sequencing and comparative genomics of the mouse genome. Nature 420: 520–562, 2002.

Rat (not shown): Gibbs RA et al: Genome sequence of the Brown Norway rat yields insights into mammalian evolution. Nature 428: 493–519, 2004.

Zebrafish (*Brachydanio rerio*): 1086 Mb of 1730 Mb were sequenced as of May 18, 2006 (www.sanger.ac.uk/).

***Anopheles gambiae*, the vector of the malaria parasite** *Plasmodium falciparum*: Holt RA et al: The genome sequence of the malaria mosquito *Anopheles gambiae*. Science 270: 129–149, 2002. (The parasite has also been sequenced: Gardner MJ et al, Nature 419: 498–519, 2000).

***Drosophila melanogaster*:** Adams MA et al: The genome sequence of *Drosophila melanogaster*. Science 287: 2185–2195, 2000.

***C. elegans*:** CESC (The *C. elegans* Sequencing Consortium): Genome sequence of the nematode *C. elegans*: a platform for investigating biology. Science 282: 2012–2018, 1998.

Yeast: Goffeau A et al: Life with 6000 genes. Science 274: 562–567, 1996.

Bacterium *E. coli*: Blattner FR et al.: The complete genome sequence of Escherichia coli K-12. Science 277: 1453–1474, 1997.

Plant: Arabidopsis Genome Initiative: Analysis of the genome sequence of the flowering plant *Arabidopsis thaliana*. Nature 408: 796–815, 2000.

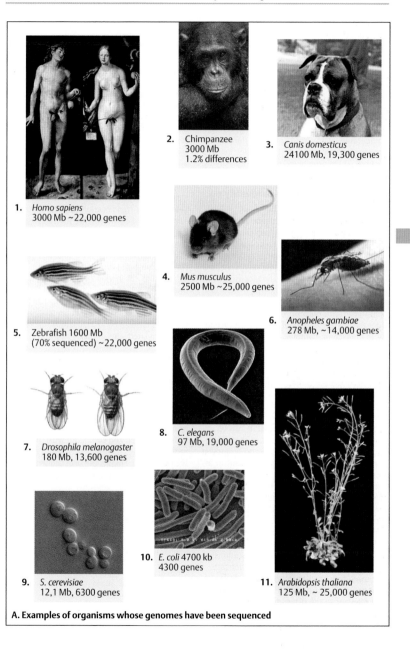

1. *Homo sapiens*
   3000 Mb ~22,000 genes

2. Chimpanzee
   3000 Mb
   1.2% differences

3. *Canis domesticus*
   24100 Mb, 19,300 genes

4. *Mus musculus*
   2500 Mb ~25,000 genes

5. Zebrafish 1600 Mb
   (70% sequenced) ~22,000 genes

6. *Anopheles gambiae*
   278 Mb, ~14,000 genes

7. *Drosophila melanogaster*
   180 Mb, 13,600 genes

8. *C. elegans*
   97 Mb, 19,000 genes

9. *S. cerevisiae*
   12,1 Mb, 6300 genes

10. *E. coli* 4700 kb
    4300 genes

11. *Arabidopsis thaliana*
    125 Mb, ~ 25,000 genes

A. Examples of organisms whose genomes have been sequenced

Gene Identification

The first step in identifying a gene is to determine the exact chromosomal localization of its gene locus. This serves as a starting point for the next level of information, the exon/intron structure of a gene. Partial or complete sequence information will yield insights into its function. Electronic data banks store sequence information about maps of polymorphic marker loci. Comparable data from many different organisms and information from other sources are used. Three principles applied in identifying a gene are outlined here.

A. Different approaches to identifying a disease-related gene

In principle, three approaches have proved useful: (i) *positional cloning*, (ii) *functional cloning*, and (iii) the *candidate gene* approach. The first and crucial step in all approaches is the clinical identification of the disease phenotype, i.e., the clinical diagnosis. The use of the McKusick catalog of human genes and phenotypes (McKusick, 1998) or its online system OMIM is indispensable for this purpose. It is important to consider the likely existence of genetic heterogeneity. In multigenic, complex inheritance it is usually not possible to identify a gene as illustrated here.

Positional cloning starts with information about the chromosomal map position of the gene to be investigated. Commonly this information has been previously obtained by linkage analysis with neighboring loci. When the gene has been identified and isolated, it can be examined for mutations, which can then be related to impaired function. It has to be proved that a presumptive mutation is present in affected individuals only, and not in unaffected family members and normal controls.

Functional cloning requires prior knowledge of the function of the gene. As this information is rarely available at the outset, the utility of this approach is limited. It can be applied when a gene with a known function has been mapped previously and the clinical manifestations of the disease suggest a functional relationship.

The candidate gene approach utilizes independent paths of information. If a gene with a function relevant to the disorder is known and has been mapped, mutations of this gene can be sought in patients. If mutations are present in the candidate gene of patients, this gene is likely to be causally related to the disease.

B. Principal steps in gene identification

To identify a suspected human disease gene, clinical and family data together with blood samples for DNA have to be collected from affected and unaffected individuals. In monogenic diseases, the disorder will follow one of the three modes of inheritance, autosomal recessive, autosomal dominant, or X-chromosomal (**1**). A chromosome region likely to harbor a disease gene can be identified by one of several genetic mapping techniques, such as linkage analysis or physical mapping using a chromosomal structural aberration such as a deletion or a translocation (**2**). The map position is refined by narrowing the region where the gene could be located to about 2–3 Mb (**3**). A contig map of overlapping DNA clones contained in a YAC or BAC (yeast or bacterial artificial chromosome) or a cosmid library is established from the region (**4**). This is further refined by a set of localized polymorphic DNA marker loci previously mapped to this region (**5**). Genes are identified by the presence of open reading frames (ORFs), transcripts, exons, and polyadenylated sites in this region and are then isolated (**6**). Each gene in this region is subjected to mutational analysis (**7**). Genes without mutations in patients are excluded. When a mutation is found in one of the analyzed genes and a polymorphism is excluded, the correct gene has been identified (**8**). Now its exon/intron structure, size, and transcript can be determined. For confirmation, the expression pattern is analyzed and compared with that of homologous genes in other organisms ("zoo blot," see next page). Finally, the DNA sequence of the entire gene can be determined.

References

Brown TA: Genomes, 2nd ed. Bios Scientific Publishers, Oxford, 2002.

McKusick VA: Mendelian Inheritance in Man. Catalog of Human Genes and Genetic Disorders, 12th ed. Johns Hopkins University Press, Baltimore, 1998. Online Version OMIM at www.ncbi.nlm.nih.gov/Omim.

Strachan T, Read AP: Human Molecular Genetics, 3rd ed. Garland Science Publishing, London–New York, 2004.

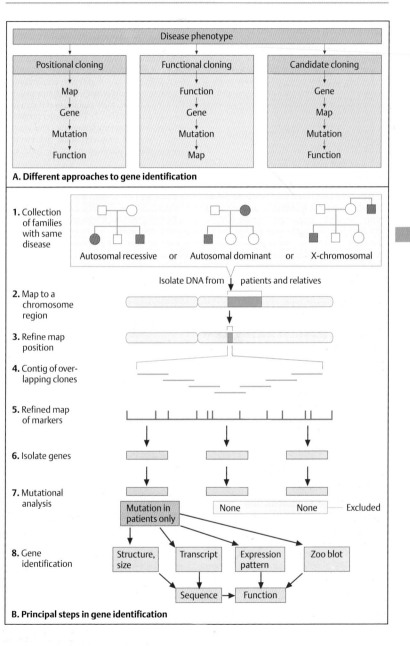

A. Different approaches to gene identification

B. Principal steps in gene identification

Identification of Expressed DNA

Since only 1–2% of DNA corresponds to genes in higher organisms, the search for unknown genes focuses on sequences that are expressed. A number of approaches have been used to identify and analyze a gene of interest and yet avoid the necessity for sequencing long stretches of DNA.

A. Microdissection of a metaphase chromosome

This approach utilizes prior knowledge of the approximate chromosomal map position of a gene to be analyzed. The corresponding chromosomal region is dissected from a metaphase (see red arrows on the right) by micromanipulation. The DNA obtained can subsequently be cloned and analyzed. This method has the advantage that all other chromosomal segments are eliminated. (Photograph kindly provided by Dr. K. Buiting, Essen; from Buiting et al., 1990.)

B. Artificial yeast chromosomes (YACs)

Large fragments (200–300 kb) of foreign DNA can be integrated into yeast cell chromosomes to yield artificial yeast cell chromosomes (YACs, see p. 182) and be replicated in yeast cells. Here six different YACs are analyzed by transverse alternating field electrophoresis (TAFE). Lanes 1, 8, and 9 show size markers. The six YACs are visible as extra bands (marked as yellow dots) among various DNA fragments of different sizes in a gel stained with ethidium bromide (to identify DNA). The variously sized fragments correspond to the naturally occurring yeast chromosomes. Each additional band in lanes 2–7 represents an artificial yeast chromosome. Lane 1 contains the standards for fragment size. (Photograph kindly provided by Drs. K. Buiting and B. Horsthemke, Essen.)

C. Single-strand conformation polymorphism (SSCP)

This procedure involves searching for a rearrangement of a gene by a mutation without the necessity for sequencing large parts or the entire gene. The photograph on the top (**1**) shows five lanes of a polyacrylamide gel electrophoresis performed in different conditions, e.g., of temperature, pH, or other parameters. In lane 4 a difference in the mobility of the DNA fragments is visible. This method is based on the principle that the presence of either a polymorphism or a mutation will alter the three-dimensional spatial arrangement (conformation) of single-stranded DNA. If two DNA fragments differ at one particular site due to a mutation, the two fragments will differ in conformation (single-strand conformation polymorphisms, SSCP). As a result, their mobilities during electrophoresis will differ (**2**). The smaller fragment migrates faster than the larger one. (Photograph of silver-stained polyacrylamide gel electrophoresis kindly provided by Dr. D. Lohmann, Essen.)

D. Exon trapping

In this approach, which was used formerly, a search is made for a gene by looking for an exon. To find this, the DNA fragment is integrated into a vector containing a strong promoter gene and a reporter gene (an expression vector). If the fragment contains an exon from an unidentified gene, it will be expressed with the reporter gene and transcribed into RNA. The RNA becomes spliced, allowing the exons to be recovered from a cDNA copy. The cDNA can be amplified and sequenced. (Figure modified from Davies & Read, 1992.)

E. Zoo blot

A "zoo blot" is a Southern blot that compares genomic DNA from different species. The photograph shows a Southern blot using the same DNA probe on DNA fragments derived from five species. This type of cross-hybridization of DNA across species boundaries ("zoo blot") is an indication that the sequences are used for coding. (Photograph courtesy of K. Buiting, Essen.)

References

Buiting K et al: Microdissection of the Prader–Willi syndrome chromosome region and identification of potential gene sequences. Genomics 6: 521–527, 1990.

Davies KE, Read AP: Molecular Basis of Inherited Disease, 2nd ed. IRL Press, Oxford, 1992.

Lüdecke HJ et al: Cloning defined regions of the human genome by microdissection of banded chromosomes and enzymatic amplification. Nature 338: 348–350, 1989.

Strachan T, Read AP: Human Molecular Genetics, 3rd ed. Garland Science Publishing, London–New York, 2004.

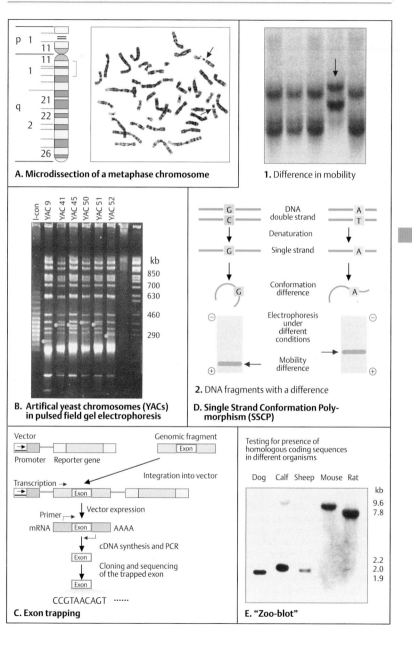

A. Microdissection of a metaphase chromosome

1. Difference in mobility

B. Artifical yeast chromosomes (YACs) in pulsed field gel electrophoresis

kb
850
700
630
460
290

C. Exon trapping

Vector
Promoter   Reporter gene
Genomic fragment
Exon
Transcription →
Exon
Integration into vector
Vector expression
Primer
mRNA   Exon   AAAA
cDNA synthesis and PCR
Exon
Cloning and sequencing of the trapped exon
Exon
CCGTAACAGT ······

D. Single Strand Conformation Polymorphism (SSCP)

G   DNA double strand   A
C   T
Denaturation
G   Single strand   A
Conformation difference
G   A
⊖   Electrophoresis under different conditions   ⊖
Mobility difference
⊕   ⊕

2. DNA fragments with a difference

E. "Zoo-blot"

Testing for presence of homologous coding sequences in different organisms

Dog   Calf   Sheep   Mouse   Rat   kb
9.6
7.8
2.2
2.0
1.9

Approaches to Genome Analysis

Prior to the availability of entire genome sequences, the study of genomes required several approaches that complement each other. Of primary interest is the size of a genome, the number of genes it contains, and their distribution (gene density), function, and evolution. Two basic approaches to sequencing a genome can be distinguished: clone-by-clone sequencing and the so-called shotgun approach. In the former, individual DNA clones of known relation to each other are isolated, arranged in their proper alignment, and sequenced. The shotgun approach breaks the genome into millions of fragments of unknown relation. The individual DNA clones for which prior knowledge of their precise origin is lacking are sequenced. Subsequently, the clones are aligned by high-capacity computers. The two approaches complement each other.

A. Alignment of overlapping DNA clones

A series of overlapping DNA clones (**1**) is assembled to reconstruct the stretch of contiguous genomic DNA from which it is derived (**2**). In the first step a radiolabeled DNA probe (probe A) is hybridized to genomic DNA from a DNA library. This identifies clone 1. Subsequently other probes hybridize to adjacent fragments and establish their order according to which they can be aligned in genomic DNA. This procedure, also called chromosome (DNA) walking, can start from several points and proceed in both directions. A number of human genes were first identified by this approach, e.g., the *CFTR* gene, with mutations causing cystic fibrosis (see p. 284).

B. Range of resolution within the genome

Different levels of resolution can be distinguished: the chromosome (**1**), a series of cloned DNA fragments (**2**), fragments aligned in a contiguous stretch of DNA, called a contig (**3**), landmarks that establish a map (**4**), and finally the sequence (**5**). The landmarks are polymorphic DNA markers that characterize each fragment.

C. STS mapping from a clone library

STS mapping plays a major role in genome mapping. An STS (sequence-tagged site) is a short stretch (60–1000 bp) of a unique DNA nucleotide sequence, An STS has a specific location and can be analyzed by PCR (see p. 60). The relevant information, i.e., the sequence of the oligonucleotide primers used for the PCR reaction and other data, can be stored electronically and does not depend on biological specimens. One can start with a clone library containing DNA fragments in unknown order (**1**). Each end of a chromosomal fragment is characterized by a pattern of restriction sites (see p. 66). The DNA fragments are ordered by determining which ends overlap, then assembling them as a contiguous array of overlapping fragments into a clone contig (**2**). These are linearly arranged. This establishes a map that shows the location and the physical distance of the landmarks, here A, B, C, etc. (**3**). Sequence-tagged sites (STSs) are generated from the two ends of the overlapping clones. This involves sequencing 100–300 bp of DNA (**4**).

D. EST mapping

ESTs (expressed sequence tags) are short DNA sequences obtained from cDNA clones (complementary DNA, see p. 70). Each EST represents part of a gene. Their locations are determined by hybridizing an assembly of different cDNAs (**1**) to genomic DNA (**2**). Thus, the locations of defined sequences of expressed genes can be determined (**3**). These can be mapped to locations on a chromosome to establish an EST map.

References

Brown TA: Genomes. Bios Scientific Publishers, 2nd ed. Oxford, 2002.

Green ED: The human genome project and its impact on the study of human disease, pp 259–298. In: Scriver CR et al (eds) The Metabolic and Molecular Bases of Inherited Disease, 8th ed. McGraw-Hill, New York; 2001.

Strachan T, Read AP: Human Molecular Genetics, 3rd ed. Bios Scientific Publishers, Oxford, 2004.

Online information:
GenBank at *www.ncbi.nlm.nih.gov*.

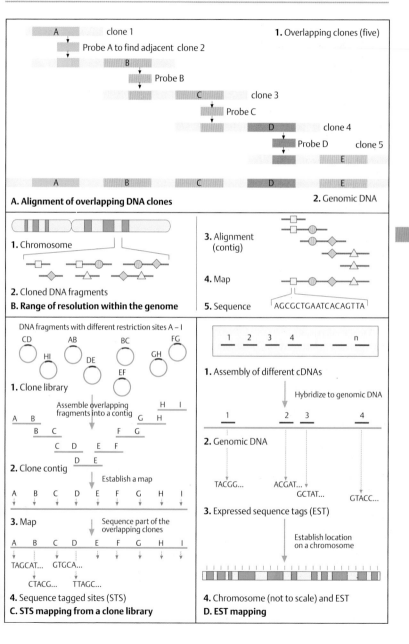

1. Overlapping clones (five)

A clone 1

Probe A to find adjacent clone 2

B

Probe B

C clone 3

Probe C

D clone 4

Probe D clone 5

E

A B C D E

A. Alignment of overlapping DNA clones

2. Genomic DNA

1. Chromosome

2. Cloned DNA fragments

B. Range of resolution within the genome

3. Alignment (contig)

4. Map

5. Sequence AGCGCTGAATCACAGTTA

DNA fragments with different restriction sites A – I

CD AB BC FG

HI DE GH

EF

1. Clone library

Assemble overlapping fragments into a contig

A B H I G H

B C F G

C D E F

D E

2. Clone contig

Establish a map

A B C D E F G H I

3. Map

Sequence part of the overlapping clones

A B C D E F G H I

TAGCAT... GTGCA...

CTACG... TTAGC...

4. Sequence tagged sites (STS)

C. STS mapping from a clone library

1 2 3 4 n

1. Assembly of different cDNAs

Hybridize to genomic DNA

1 2 3 4

2. Genomic DNA

TACGG... ACGAT...

GCTAT... GTACC...

3. Expressed sequence tags (EST)

Establish location on a chromosome

4. Chromosome (not to scale) and EST

D. EST mapping

Genomes of Microorganisms

Bacterial genomes are small, ranging from 500 to 10 000 kb, and are tightly packed, with genes aligned nearly contiguously along the circular chromosome. The coding regions are small (average 1 kb) and contain no introns. Bacteria can be classified according to the sizes and principal features of their genomes. By selectively eliminating genes from a bacterial genome, it could be determined that about 265–350 genes are essential under laboratory conditions (Hutchison et al., 1999).

Owing to their small size, bacterial genomes were the first to be completely sequenced, e.g., *Haemophilus influenzae* in 1995 (Fleischmann et al, 1995). Recent data show that bacterial genomes can contain pseudogenes. About 1 in 20 of the coding regions of *E. coli* K-12 are pseudogenes. The smallest bacterial genome is that of *Mycoplasma genitalium* (580 073 nucleotide base pairs, 483 genes). It is an obligate intracellular pathogen. Many genes encoding proteins for metabolic functions are absent. The limited capacity for metabolism in *M. genitalium* is compensated by the transport of life-supporting molecules from its extracellular environment into the cell.

The smallest free-existing bacterium is *Pelagibacter ubique* (1 308 759 base pairs, 1354 genes; Giovannoni et al., 2005). Its small, compact genome contains the complete biosynthetic pathways for all amino acids, in contrast to all other bacteria.

A. The genome of a small bacteriophage

Phage ΦX174, with ten genes (A-J) contained in 5386 nucleotides in single-stranded DNA, was the very first organism to be sequenced (F. Sanger et al., 1977). The genome of phage ΦX174 is so compact that several genes overlap. (Figure adapted from Sanger et al., 1977.)

B. Overlapping genes in ΦX174

The reading frames of genes A and B, B and C, and D and E partially overlap. The overlapping genes are transcribed in different reading frames: in the start codon ATG of gene E (above the sequences shown), the last two nucleotides (AT) are part of the codon TAT for tyrosine (Tyr) in gene D. Similarly, the stop codon TGA for the E gene is part of codons GTG (valine) and ATG (methionine) in the D gene. Thus, this small genome is used very efficiently.

C. Genome of *Escherichia coli*

This simplified figure shows the essential features of a bacterial genome. Functionally related genes usually cluster in operons (four of many shown). About half of the genes of *E. coli* are in operons.

References

Brown TA: Genomes, 2nd ed. Bios Scientific Publ, Oxford, 2002.

Fleischmann RD et al: Whole-genome random sequencing and assembly of *Haemophilus influenzae* Rd. Science 269: 496–512, 1995.

Fraser CM, Eisen JA, Sulzberg SL: Microbial genome sequencing. Nature 406: 799–803, 2000.

Giovannoni SJ et al: Genome streamlining in a cosmopolitan oceanic bacterium. Science 309: 1242–1245, 2005.

Ochman H, Davalos LM: The nature and dynamics of bacterial genomes. Science 311: 1730–1733, 2006.

Sanger F et al: Nucleotide sequence of the bacteriophage ΦX174 DNA. Nature 265: 687–695, 1977.

Bacterial genome information online: www.tigr.org/tdb/mdb/mdbinprogress.html. (Entrez Genomes and search for sequenced microorganisms at http://www.ncbi.nlm.nih.gov/).

GenomesOnLine Database www.genomesonline.org/.

Parasite Genomes Website: www.ma.ucla.edu/par/.

Size and general contents of bacterial genomes

Main genomic features	Free-living	Facultative pathogen	Pathogen or obligate symbiont
Genome size	large (5–10 Mb)	2–5 Mb	small (0.5–1.5 Mb)
Genome stability	stable or unstable	unstable	stable
Lateral gene transfer	frequent	frequent/rare	rare or none
Number of pseudogenes	few	many	rare
Population size	large	small	small
Pathogenicity factors	absent	present	present

(Data from Ochman & Davalos, 2006) Mb: one million base pairs.

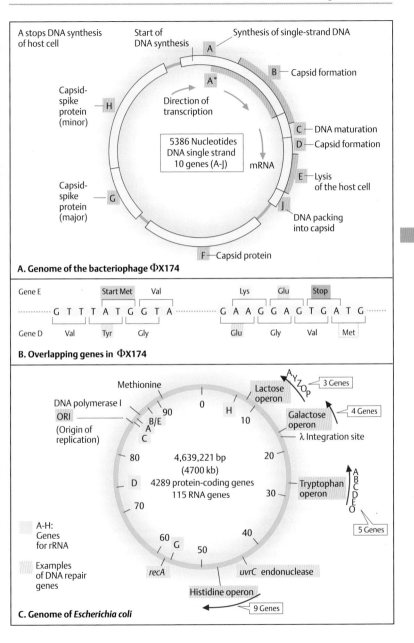

A stops DNA synthesis of host cell

Start of DNA synthesis

Synthesis of single-strand DNA

A

A*

Direction of transcription

B — Capsid formation

Capsid-spike protein (minor) — H

5386 Nucleotides
DNA single strand
10 genes (A-J)

mRNA

C — DNA maturation

D — Capsid formation

E — Lysis of the host cell

Capsid-spike protein (major) — G

J

DNA packing into capsid

F — Capsid protein

A. Genome of the bacteriophage ΦX174

Gene E   Start Met   Val   Lys   Glu   Stop

G T T T A T G G T A ········· G A A G G A G T G A T G ·········

Gene D   Val   Tyr   Gly   Glu   Gly   Val   Met

B. Overlapping genes in ΦX174

Methionine

DNA polymerase I

ORI (Origin of replication)

B/E
A
C

A Y Z O P
Lactose operon   3 Genes

H

Galactose operon   4 Genes

λ Integration site

90
0
10

80

4,639,221 bp
(4700 kb)
4289 protein-coding genes
115 RNA genes

20

D

Tryptophan operon

A B C D E O

70

30

5 Genes

A-H: Genes for rRNA

Examples of DNA repair genes

60
G
50

40

recA

uvrC endonuclease

Histidine operon   9 Genes

C. Genome of *Escherichia coli*

The Complete Sequence of the *Escherichia coli* Genome

Escherichia coli is a regular inhabitant of the gastrointestinal tract of humans and animals. Thus, it is an indicator of fecal contamination. Certain strains of *E. coli* are the most frequent causes of human infections. The report of the complete sequence of the 4 639 221-base pair (4.6 Mb) genome of the *E. coli* K-12 strain in 1997 (Blattner et al., 1997) with 4289 protein-coding genes is presented here as an example of one of many sequenced microorganism genomes. Genome sequence data can be correlated with other information to identify sites that determine gene function, sites that determine infectivity (virulence factors), and sites with homology to other organisms.

A. Overall structure and comparison with other genome sequences

The figure shows a small section of about 80 kb of the genome from the original publication. Base pair numbers 3 310 000 to – 3 345 000 are shown in the first row (top, **1**) and 3 339 000–4 025 000 in the second row (**2**). The top double line shows color-coded genes of *E. coli* encoding a protein on either of the two strands of the DNA double helix. Five other completed genomes are shown for comparison. The CAI (Codon Adaption Index) reflects the preferred codon usage of an organism.

A total of 2657 protein-coding genes with known function (62% of all genes) and 1632 genes (38%) without known function have been identified. The average distance between genes is 118 bp (base pairs). This is much less than in eukaryotic genomes, with about 1 gene per 100 000 bp. The protein-coding genes (87.8% of the genome) can be assigned to 22 functional groups (see gene function color code at the bottom of the figure). Among these are 45 genes with recognized regulatory functions (1.05% of the total); 243 genes for energy metabolism (5.67%); 115 genes for DNA replication, recombination, and repair (2.68%); 255 genes for transcription, RNA synthesis, and metabolism (5.94%); 182 genes for translation (4.24%); 131 genes for amino acid biosynthesis and metabolism (3.06%); and 58 genes for nucleotide biosynthesis and metabolism (1.35%). This is a typical spectrum of basic gene functions in prokaryotes.

Proteins of *E. coli* have homologies with those of other organisms: 1703 proteins are homologous with proteins of *H. influenzae*, and 468 with those of *Mycoplasma genitalium*. Homologies even exist with eukaryotic proteins. Yeast (*S. cerevisiae*) has 5885 proteins matching *E. coli* proteins. (Figure adapted from a small part of the complete map published by Blattner et al., 1997.)

Medical relevance

Certain strains of *E. coli* cause intestinal infections of various types and severity. Distinct pathogenic strains are (i) Shiga-toxin-producing *E. coli* (STEC), with the most prominent strain O157:H7 occurring in undercooked beef, (ii) enterotoxigenic *E. coli* (ETEC, causing traveler's diarrhea), (iii) enteropathic *E. coli* (EPEC), (iv) enteroaggregative *E. coli* (EIEC, causing dysentery), and (v) diffusely adherent *E. coli* (DAEC, causing traveler's and persistent diarrhea; see Russo, 2005).

References

Blattner FR et al.: The complete genome sequence of Escherichia coli K-12. Science 277: 1453–1474, 1997.

Brown TA: Genomes, 2nd ed. Bios Scientific Publ, Oxford, 2002.

Fraser CM, Eisen JA, Sulzberg SL: Microbial genome sequencing. Nature 406: 799–803, 2000.

Hutchison III CA et al: Global transposon mutagenesis and a minimal mycoplasma genome. Science 286: 2165–2169, 1999.

Kayser FH et al: Medizinische Mikrobiologie, 10th ed. Thieme Verlag, Stuttgart–New York, 2001.

Neidhardt FC et al (eds): *Escherichia coli* and *Salmonella*. Cellular and Molecular Biology. ASM Press, Washington, DC, 1996.

Russo TA: Diseases caused by gram-negative enteric bacilli, pp 878–885. In: Kasper DL et al (eds) Harrison's Principles of Internal Medicine, 16th ed. McGraw-Hill, New York, 2005.

Wren BW: Microbial genome analysis: insights into virulence, host adaptation and evolution. Nature Rev Genet 1: 30–39, 2000.

Online information about sequenced microorganisms:

(Entrez Genomes and search for sequenced microorganisms at http://www.ncbi.nlm.nih.gov/).

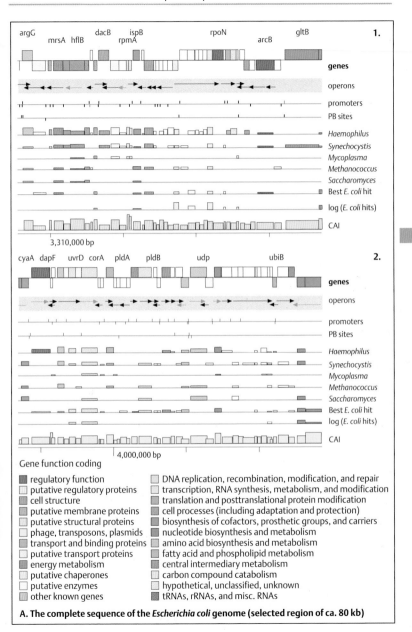

A. The complete sequence of the *Escherichia coli* genome (selected region of ca. 80 kb)

The Genome of a Multiresistant Plasmid

Plasmids are self-replicating double-stranded circular DNA molecules in bacteria separate from the bacterial chromosome. The number of plasmids per bacterial cell varies from a few to thousands, and their sizes range from a few thousand base pairs to more than 100 kb. Plasmids may confer a benefit to the bacterial host cell, because they often contain genes encoding enzymes that inactivate antibiotics. Drug-resistant plasmids pose a major threat to successful antibiotic therapy. Many plasmids also contain transfer genes encoding proteins that form a macromolecular tube, or pilus, through which a copy of plasmid DNA can be transferred to other bacteria. Thus, antibiotic resistance can spread very rapidly. Here a plasmid is presented that is resistant to multiple types of antibiotics. This plasmid is derived from a multiresistant corynebacterium. It is composed of DNA segments derived from bacteria of very different origins, such as bacteria existing in soil, plants, or animals or in humans as pathogens.

A. The multiresistant plasmid pTP10

The plasmid pTP10 has a large genome of 51 409 base pairs (bp). It exists in an opportunistic human pathogen, the Gram-positive *Corynebacterium striatum* strain M82B. This plasmid contains genes encoding proteins that render its host bacterium resistant to 16 antimicrobial agents from six different structural classes (Tauch et al., 2000). When the sequence was published in 2000 this was the largest plasmid to have been sequenced. It contains DNA segments from a plasmid-encoded erythromycin resistance region from the human pathogen *Corynebacterium diphtheriae* (**Em**, shown inside the circular genome diagram), a chromosomal DNA region from *Mycobacterium tuberculosis* containing tetracycline (**Tc**) and oxacillin resistance, a plasmid-encoded chloramphenicol (**Cm**)-resistance region from the soil bacterium *Corynebacterium glutamicum*, and a plasmid-encoded aminoglycoside resistance to kanamycin (**Km**), neomycin, lividomycin, paramomycin, and ribostamycin from the fish pathogen *Pasteurella piscicida*. In addition, the plasmid contains five transposons and four insertion sequences (IS1249, IS1513, IS1250, and IS26) at

eight different sites. Altogether eight genetically distinct DNA segments of different evolutionary origin are present in this plasmid.

B. Genetic map of plasmid pTP10

Plasmid pTP10 has 47 open reading frames (ORFs). These can be assigned to eight different DNA segments. They form a contiguous array of subdivided stretches, which are shown in linear representation as I, II, VIIb, III, VIIa, VIII, IVa, Va, VI, Vb, IVb, and VIIc.

Segment I (shown in green) consists of five ORFs comprising the composite resistance transposon Tn5432. The insertion sequences IS1249b (ORF1) and IS1249a (ORF5) flank the erythromycin-resistance gene regions ermCX (ORF3) and ermLP (ORF4). An identical copy of IS1249 occurs in ORF29 (segment VIII). ORF3, the central region of Tn5432, encodes a 23 S rRNA methyltransferase preceded by a short leader peptide probably involved in the regulation of erythromycin-inducible translational attenuation. This region is virtually identical to the antibiotic resistance gene region (erythromycin, clindamycin) of plasmid pNG2 from *C. diphtheriae* S601.

Segment II (ORFs 6–14), located downstream of Tn5432, contains the tetracycline-resistance genes tetA (ORF6) and tetB (ORF7). This segment (ORFs 6–14) is very similar to ATP-binding cassette (ABC) transporters identified in a mycobacterium (*M. smegmatis*) chromosome. The tandemly arranged genes tetA and tetB also mediate resistance to the (β-lactam antibiotic oxacillin, although it is structurally and functionally unrelated to tetracycline. Presumably this results from TetAB protein heterodimerization and subsequent export of the antibiotics out of the bacterial cell. The other segments have been similarly delineated. (Figures adapted from Tauch et al., 2000; originals kindly provided by Professor Alfred Pühler, University of Bielefeld, Germany.)

References

Kayser FH et al: Medizinische Mikrobiologie, 10th ed. Thieme Verlag Stuttgart–New York, 2001.

Tauch A, Krieft S, Kalinowski J, Pühler A: The 51,409-bp R-plasmid pTP10 from the multiresistant clinical isolate *Corynebacterium striatum* M82B is composed of DNA segments initially identified in soil bacteria and in plant, animal, and human pathogens. Mol Gen Genet 263 : 1–11, 2000.

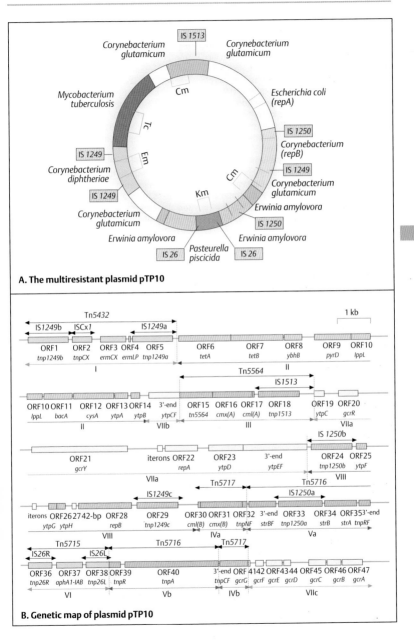

A. The multiresistant plasmid pTP10

B. Genetic map of plasmid pTP10

Architecture of the Human Genome

The human genome is representative of the large, complex genomes of mammals. Outstanding structural features are different types of repetitive, noncoding sequences. Several of these have been integrated into vertebrate genomes during the course of evolution. Each chromosome contains blocks of duplicated sequences (segmental duplications).

A. Types of sequence

The human genome consists of almost 3 billion base pairs (3×10^9 bp or 3000 Mb) per haploid set of chromosomes. Coding DNA in genes (exons) accounts for only 1.2% (34 Mb), and untranslated regions of transcripts account for 21 Mb (0.7%) of the total amount of DNA. (Figure adapted from Brown, 2002; and Strachan & Read, 2004.)

B. Interspersed repetitive DNA

Long interspersed repeat sequences (LINEs, **1**) are mammalian retrotransposons that in contrast to retroviruses lack long terminal repeats (LTRs). They account for 21% of the genome and consist of repetitive sequences up to 6500 bp long that are adenine-rich at their 3′ ends. LINEs encode two open reading frames (ORF1 and 2), which are translated. In addition to a 5′ promoter (P) they have an internal promoter. Approximately 600 000 L1 elements are dispersed throughout the human genome. This can result in genetic disease if one is inserted into a gene (e.g., hemophilia A, see p. 374). LINEs-2 and -3 are inactive because reverse transcription from the 3′ end often fails to proceed to the 5′ end.
Short interspersed repeat sequences (SINEs, **2**) are repetitive segments of about 100–400 bp with tandem duplication of CG-rich segments separated by A-rich segments. They do not encode a protein and are not capable of autonomous insertion. The most abundant type of SINE sequences in humans is the Alu family (Alu sequences) with about 1 200 000 copies (about 6% of the genome). One Alu repeat occurs about once every 3 kb in the human genome. A full-length Alu repeat is a dimer of about 280 bp, 120 bp for each monomer, followed by a short sequence rich in A residues. They are asymmetric: the repeat to the right contains an internal 32-bp sequence, the other not. Alu sequences are specific for primate genomes.

LTR retroposons (**3**) are flanked by long terminal direct repeats (LTRs) containing transcriptional regulatory elements. The autonomous retrotransposons contain *gag* and *pol* genes, which encode proteins required for retrotransposition (see retroviruses, p. 106).
DNA transposons (**4**) resemble bacterial transposons by having terminal inverted repeats and encoding a transposase. At least seven major classes exist, which can be subdivided into familes of independent origins.

C. Segmental duplications

Segmental duplications are blocks of 1–200 kb genomic sequence that are also present at another site in a chromosome (intrachromosomal) or another chromosome (interchromosomal), usually in subtelomeric regions. The three examples show duplications of the X chromosome, chromosome 20, and chromosome 4 shared with blocks from other chromosomes, shown as lines above and below. The human genome contains 1077 duplicated blocks containing 10 310 pairs of genes. When their sequence identity exceeds 95% and their size 10 kb or more, unequal crossing-over may occur. This predisposes to duplication and deletion, leading to *genomic disorders*.

References

Brown TA: Genomes, 2nd ed. Bios Scientific Publ, Oxford, 2002.

Chen J-M et al: A systematic analysis of LINE-1 endonuclease-dependent retropositional events causing human genetic diseases. Hum Genet 117: 411–427, 2005.

Cheng Z et al: A genome-wide comparison of recent chimpanzee and human segmental duplications. Nature 437: 88–93, 2005.

Emmanual BS, Shaikh TH: Segmental duplication: an expanding role in genomic instability and disease. Nature Rev Genet 2: 791–800, 2001.

Kazazian Jr HH: Mobile elements: drivers of genome evolution. Science 303: 1626–1632, 2004.

Kazazian Jr HH: L1 retrotransposons shape the mammalian genome. Science 289:1152–1153, 2000.

Strachan T, Read AP: Human Molecular Genetics, 3rd ed. Bios Scientific Publishers, Oxford, 2004.

Online information:
Nature's Guide to the Human Genome at www.nature.com/nature/focus/humangenome and www.sciencegenomics.org.

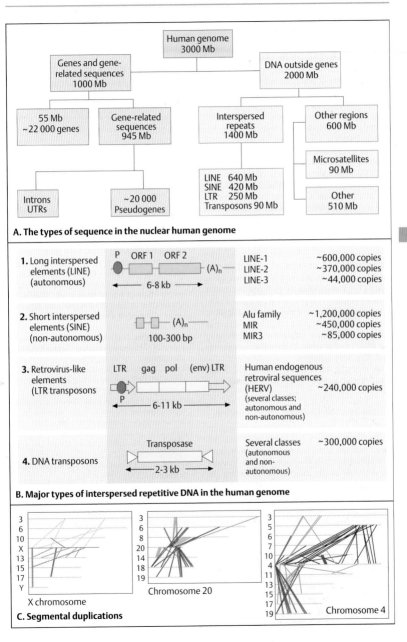

A. The types of sequence in the nuclear human genome

B. Major types of interspersed repetitive DNA in the human genome

C. Segmental duplications

The Human Genome Project

The Human Genome Project (HGP) is an international cooperative effort with the goal of sequencing the euchromatic portion of the human genome. It was launched in 1990, led by five major centers, four in the United States and one in the United Kingdom, joined by groups from France, Germany, Japan, and China. The United States National Human Genome Research Institute (NHGRI) serves as a central agency. A preliminary draft of the human genome sequence was published in 2001 (IHGSC, 2001; Venter et al, 2001) and the completed version in 2004 (IHGSC, 2004). This information is available online (see references). The Human Genome Organization (HUGO), an international organization, is involved in many aspects of the human genome project.

The HGP is concerned with the functions of genes and genomes (*functional genomics*), the entire process of transcription (the *transcriptome*), analysis of all human proteins (*proteome*), the human genome compared with that of other organisms (*comparative genomics*), the development of new techniques for handling the vast amount of data (*bioinformatics*), and epigenetic functions (*epigenome*). In addition to the HGP, similar projects for other organisms have also been initiated.

More recently, the HapMap Project (IHC, 2005) was initiated to investigate common DNA variants.

A. The Human Genome Project online

The speed of progress precludes an up-to-date printed presentation, but access to the various aspects of the human genome project is available on the Internet. The opposite page provides an overview of the main areas, including the HGP itself, the relationship of genes and diseases, gene maps, networks of databases, educational resources, and the genomes of organisms other than man. Several websites provide information beyond that listed in the limited space available here.

Ethical, legal, and social implications

The ethical, legal, and social implications of the Human Genome Project are important. ELSI covers a wide range of issues. These include confidentiality and fairness in the use of individual genetic information, prevention of genetic discrimination, use of genetic methods in clinical diagnostics, conditions for genetic testing, and public and professional education.

Medical relevance

The Human Genome Project has important implications for the theory and practice of medicine. Complete knowledge of human genes will lead to more precise diagnoses, better assessment of genetic risk, and the development of treatments. In particular, certain allelic haplotype combinations that predispose to common, complex disease are likely to be identified (see King, Rotter, Motulsky, 2002).

References

Bailey JA et al: Recent segmental duplications in the human genome. Science 297: 1003–1007, 2002.

Caron H et al: The human transcriptome map: Clustering of highly expressed genes in chromosomal domains. Science 291: 1289–1297, 2001.

Goldstein DB, Cavalleri GL: Genomics: Understanding human diversity. Nature 437: 1241–1242, 2005.

Green ED: The human genome project and its impact on the study of human disease, pp 259–298. In: Scriver CR et al (eds) The Metabolic and Molecular Bases of Inherited Disease, 8th ed. McGraw-Hill, New York, 2001.

ICH: The International HAPMAP consortium: A haplotype map of the human genome. Nature 437: 1299–1220, 2005.

International Human Sequencing Genome Consortium (IHSGC): Initial sequencing and analysis of the human genome. Nature 409: 860–921, 2001.

International Human Genome Sequencing Consortium: Finishing the euchromatic sequence of the human genome. Nature 431: 931–945, 2004.

King R, Rotter J, Motulsky AG (eds): The Genetic Basis of Common Disorders, 2nd ed. Oxford University Press, Oxford, 2002.

Venter JG et al: The sequence of the human genome. Science 291: 1304–1351, 2001.

Online information:

Human Genome Website:
  http://www.ncbi.nlm.nih.gov/.

United States National Human Genome Research Institute (NHGRI: *http://www.nhgri.nih.gov/*).

Various eukaryote genomes: www.sanger.ac.uk/Projects, www.mpg.de/.

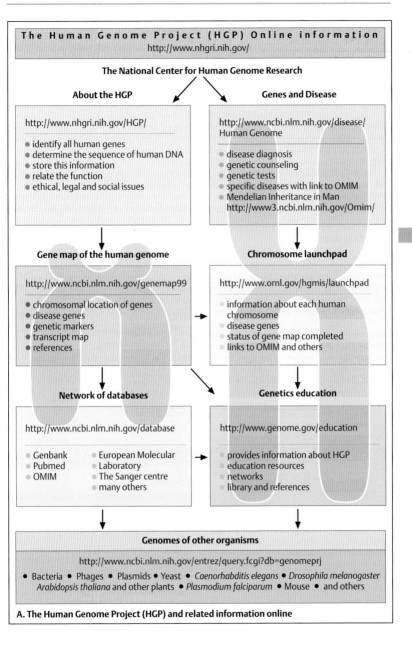

A. The Human Genome Project (HGP) and related information online

Genomic Structure of the Human X and Y Chromosomes

The human X and Y chromosomes have evolved from a pair of ancestral chromosomes during the past 300 million years (Ohno, 1967). While the X chromosome retained many properties of an autosome, the Y chromosome lost most of its genes and became greatly reduced in size. Its genetic function is now limited to inducing male development during embryonic development and to maintaining spermatogenesis in adult males. The two chromosomes undergo pairing and recombination at the distal ends of their short arms in the *pseudoautosomal region* (PAR1), whereas all other regions are exempted from recombination.

A. Genomic structure of the human X chromosome

Functional genes are distributed along the X chromosome as shown by the blue squares (each representing one gene). The approximate locations of nine selected landmark genes and their directions of transcription (arrows) are shown. Most of the short arm (Xp) consists of a region that resulted from the translocation of an ancestral autosome into Xp about 105 MYA (million years ago) (X-added region, XAR). The long arm (Xq) is composed of a region that has been conserved during evolution in mammals (XCR, X-conserved region). Five regions of evolutionary conservation, called evolutionary strata S1-S5, have been identified along the X chromosomes. The human X chromosome contains 1098 genes. The X chromosome has 7.1 genes per million base pairs, one of the lowest gene densities in the human genome (average 10–13). (Figure adapted from Ross et al., 2005.)

B. Genomic structure of the human Y chromosome

The human Y chromosome has a distinct genomic structure comprising five different regions in the euchromatic part: (i) two pseudoautosomal regions at the distal ends of the short (PAR1) and long arms (PAR2), (ii) the Y-specific male determinant (MSY) region of about 35 kb, (iii) about 8.6 Mb (38% of the euchromatic portion) called *X-degenerate*, derived from the ancestral autosome, (iv) 3.4 Mb derived from former X-linked genes by transposition (*X-transposed*), which occurred about 3–4 MYA,

and (v) 10.2 Mb amplified (amplionic) Y-specific sequences, designated as P1-P8, derived from three different processes. These are thought to be derived from former X- and Y-linked genes and to have acquired autosomal male fertility factors by transposition and retroposition. They are termed *amplionic* because they consist of amplified palindromic sequences (amplicons) of various sizes with a marked sequence similarity of 99.9% over long stretches of DNA (ten to hundreds of kilobases). They contain both coding and noncoding genes. Most genes in the amplionic segments are expressed exclusively in testes, presumably being required for spermatogenesis.

Since the male-specific sequences on the Y chromosome do not participate in crossing-over, they are deprived of one mechanism for replacing mutations or structural rearrangements with normal sequences. Gene conversion between these palindromic sequences (Y-Y conversion) presumably serves as a mechanism for restoring normal sequences that have been rendered nonfunctional in one arm of a palindrome. (Figure adapted from Skaletsky et al, 2003.)

C. Homologies between and the X and Y chromosomes

The X and Y chromosomes share regions of homology due to their common evolutionary origin (for details see Ross et al., 2005). (Figure adapted from Ross et al., 2005.)

Medical relevance

Three regions, AZFa, AZFb, and AZFc, in the long arm of the Y chromosome are associated with male infertility when deleted, due to failure to produce viable sperm cells (azoospermia).

References

Jobling MA, Tyler-Smith C: The human Y chromosome: An evolutionary marker comes of age. Nature Rev Genet. 4: 598–612, 2003.

Ohno S: Sex Chromosomes and Sex-linked Genes. Springer, Berlin, 1967.

Repping S et al: High mutation rates have driven extensive structural polymorphisms among human Y chromosomes. Nature Genet 38: 463–467, 2006.

Ross MT et al: The DNA sequence of the human X chromosome. Nature 434: 325–337, 2005.

Skaletsky H et al.: The male-specific region of the human Y chromosome is a mosaic of discrete sequence classes. Nature 423: 825–837, 2003.

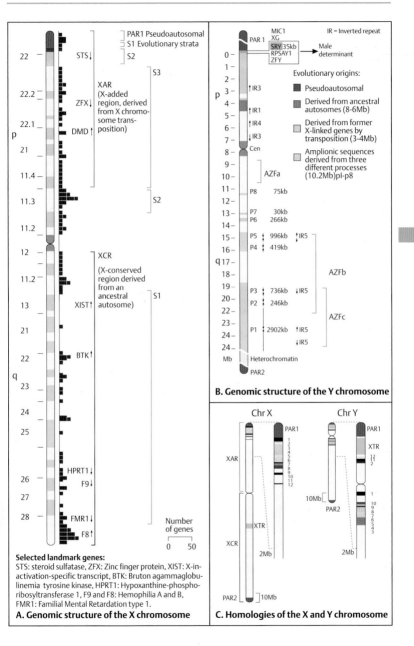

A. Genomic structure of the X chromosome

PAR1 Pseudoautosomal
S1 Evolutionary strata
S2

STS↓

XAR
(X-added
region, derived
from X chromo-
some transposition)

ZFX↓

DMD↑

S3

S2

XCR
(X-conserved
region derived
from an
ancestral
autosome)

XIST↑

S1

BTK↑

HPRT1↓
F9↓

FMR1↓
F8↑

Number
of genes

0   50

Selected landmark genes:
STS: steroid sulfatase, ZFX: Zinc finger protein, XIST: X-in-
activation-specific transcript, BTK: Bruton agammaglobulinemia tyrosine kinase, HPRT1: Hypoxanthine-phospho-
ribosyltransferase 1, F9 and F8: Hemophilia A and B,
FMR1: Familial Mental Retardation type 1.

B. Genomic structure of the Y chromosome

MIC1
XG
SRY 35kb   Male
RPSAY1   determinant
ZFY

IR = Inverted repeat

Evolutionary origins:

■ Pseudoautosomal

■ Derived from ancestral
autosomes (8-6Mb)

□ Derived from former
X-linked genes by
transposition (3-4Mb)

□ Amplionic sequences
derived from three
different processes
(10.2Mb)pl-p8

PAR 1

IR3
IR1
IR4
IR3
Cen

AZFa

P8   75kb
P7   30kb
P6   266kb
P5   996kb ↑IR5
P4   419kb

AZFb

P3   736kb ↓IR5
P2   246kb

AZFc

P1   2902kb ↑IR5
↓IR5

Heterochromatin
PAR2

C. Homologies of the X and Y chromosome

Chr X   Chr Y

PAR1   PAR1
XAR   XTR
XTR   PAR2
XCR
PAR2

Genome Analysis with DNA Microarrays

A microarray or DNA chip is an assembly of oligonucleotides or other DNA probes fixed on a small, fine grid of surfaces. Other specimens such as cDNA clones can also be used. A microarray serves to analyze the expression states of many genes simultaneously. By this means, either mutations or disease-predisposing sequence variations can be identified rapidly and efficiently. The genes may be represented in different ways. One approach is to use cDNA prepared from mRNA (expression screening) or to recognize sequence variations in genes (screening for DNA variation). The advantages of using microarrays are manifold: simultaneous large-scale analysis of thousands of genes at a time, automation, small sample size, and easy handling. Several manufacturers offer highly efficient microarrays that can accomodate 300 000 DNA probes on a small (e.g., 1.28×1.28 cm) high-density glass slide.

Two basic types of DNA microarrays can be distinguished: (i) microarrays of previously prepared DNA clones or PCR products that are attached to the surface, arranged in a grid of high-density arrays in two-dimensional linear coordinates; (ii) microarrays of oligonucleotides synthesized *in situ* on a suitable surface. Both types of DNA arrays can be hybridized to labeled DNA probes in solution. Many variations are being developed (for review, see Nature Genetics Chipping Forecast, 2005).

A. Gene expression profile by cDNA array

This figure shows a microarray of 1500 different cDNAs from the human X chromosome. The cDNAs were obtained from lymphoblastoid cells of a normal male (XY) and a normal female (XX). The cDNAs of the female cells were labeled with the fluorophore Cy3 (red), cDNAs of the male cells with the Cy5 (green). The inactivation of most genes in one of the two X chromosomes in female cells leads to a 1 : 1 ratio of cDNAs from the expressed genes in the male and female X chromosomes. The result is a yellow signal at most sites because the superimposed red fluorescent (female) and green fluorescent (male) signals are present in equal amounts.

At seven sites, marked by a yellow circle, the signal is red. Genes that have escaped inactivation on the inactive female X chromosome (see p. 234) are expressed at a double dosage. (Photograph kindly provided by Drs. G. M. Wieczorek, U. Nuber, and H. H. Ropers, Max-Planck Institute for Molecular Genetics, Berlin.)

B. Gene expression patterns in human cancer cell lines

Microarrays can be used to analyze the pattern of gene expression in cancer cells, as an aid to diagnosis and the monitoring of therapy. The gene expression patterns in 60 cell lines derived from different human cancers are shown. Approximately 8000 genes have been analyzed by this procedure (Ross et al., 2000). A consistent relationship between gene expression patterns and tissue of origin was detectable.

Panel 1 shows the cell-line dendrogram relating the patterns of gene expression to the tissue of origin of the cell lines as derived from 1161 cDNAs in 64 cell lines (**1**). Panel 2 shows a colored microarray representation of the data using Cy3-labeled (red) cDNA reverse-transcribed from mRNA isolated from the cell lines compared with Cy5-labeled (green) cDNA derived from reference mRNA (**2**). Several clusters of red dots in the columns (1161 genes) and the rows (60 cell lines) indicated an increased gene expression. Thus, these regions of the microarray represent genes with altered gene expression patterns in tumor cells. (Figure adapted from Ross, et al., 2000 with kind permission by the authors and Nature Genetics.)

References

Brown TA: Genomes, 2nd ed. Bios Scientific Publ, Oxford, 2002.

Gaasterland T, Bekiranov S: Making the most of microarray data. Nature Genet 24: 204–206, 2000.

Hoheisel JD: Microarray technology: beyond transcript profiling and genotype analysis. Nature Rev Genet 7: 200–210, 2006.

Nature Genetics: The Chipping Forecast III. Nature Genet 37: Supplement, June 2005.

Pinkel D: Cancer cells, chemotherapy and gene clusters. Nature Genet 24: 208–209, 2000.

Ross DT et al: Systematic variation in gene expression patterns in human cancer cell lines. Nature Genet 24: 227–235, 2000.

Strachan T, Read AP: Human Molecular Genetics, 3rd ed. Bios Scientific Publishers, Oxford, 2004.

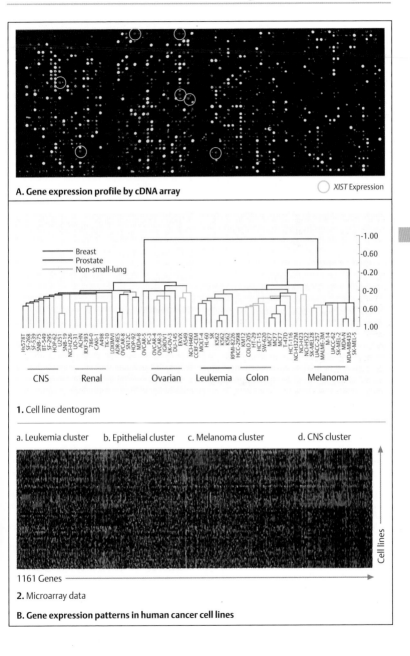

A. Gene expression profile by cDNA array   ◯ *XIST* Expression

Breast
Prostate
Non-small-lung

CNS   Renal   Ovarian   Leukemia   Colon   Melanoma

1. Cell line dentogram

a. Leukemia cluster   b. Epithelial cluster   c. Melanoma cluster   d. CNS cluster

Cell lines

1161 Genes

2. Microarray data

B. Gene expression patterns in human cancer cell lines

Genome Scan and Array CGH

Whole-genome scan and array-comparative genomic hybridization (CGH, see p. 206) represent genetic analyses based on the whole genome. One goal of whole-genome scanning corresponds to linkage analysis (see p. 156). It utilizes polymorphic DNA sequence variants (SNPs, single nucleotide polymorphisms) at multiple loci along all chromosomes to search for linkage with a region that might harbor gene loci involved in predisposing to or causing a disease. A dense map of 1.5 Mio SNPs of the genome is available (Hinds et al., 2005). It is also used for studies that search for an association of certain alleles with a disease-predisposing region. Array-comparative genomic hybridization (aCGH) combines the microarray methodology with genomic hybridization. The principle is illustrated schematically here.

A. Whole-genome scan

A whole-genome scan analyzes all chromosomes (only four are shown here schematically) for linkage of a disease locus with polymorphic markers along each chromosome (multipoint linkage analysis). A peak of the LOD score of nearly 4 over chromosome 2 indicates that linkage between a disease locus and two marker loci (shown in red and blue) may be present. A LOD score of 3 or above indicates a probability ratio for linkage/against linkage of 1000 : 1 or above.

More often, a genome-wide scan aims at detecting an association between polymorphic variants and a region containing susceptibility loci for complex diseases. An association will result in linkage disequilibrium (LD). LD is the result of a preferential segregation of certain alleles with a susceptibility region. Many different computer-based test procedures for analyzing linkage exist (see references).

B. Array-comparative genomic hybridization (aCGH)

Array-comparative genomic hybridization (aCGH) is a combination of microarray techniques (see p. 258) and CGH (see p. 206). In aCGH, many mapped clones in a microarray are used instead of metaphase chromosomes. This greatly increases the detection rate for small deletions or duplications. The genome to be tested and a normal genome as reference are used as probes. They are differently labeled. In the example shown here, the DNA from an individual to be tested is labeled with the fluorophore Cy5, which induces green fluorescence. The DNA from another individual as control is labeled with Cy3, which induces red fluorescence.

If the amount of DNA is the same, a yellow signal will result. However, wherever there is an imbalance, the signal will be red if the amount of DNA is reduced due to a deletion and green in case of duplication, there then being more DNA from the genome being tested. Two sites with this type of imbalance are indicated by arrows. Array CGH is a fast and efficient meaans of detecting deletions and duplications on a genome-wide scale. The resolution for whole-genome arrays has increased to more than 30 000 overlapping clones covering the whole human genome (for details see references). (Figure by A.A. Snijders, Cancer Research Institute San Francisco, D.P. Locke and E.E. Eichler, Department of Genome Sciences, Univ. of Washington, Seattle, kindly provided by Dr. Evan E. Eichler.)

References

Cardon LR, Bell JI: Association study design for complex diseases. Nature Rev Genet 2: 91–99, 2001.

Eichler EE: Widening the spectrum of human genetic variation. Nature Genet 38: 9–11, 2006.

Göring HHH et al: Large upward bias in estimation of locus-specific effects from genomewide scans. Am J Hum Genet 69: 1357–1369, 2001.

Hinds DA et al: Whole-genome patterns of common DNA variation in three human populations. Science 307: 1072–1079, 2005.

Ishkanian AS et al: A tiling resolution DNA microarray with complete coverage of the human genome. Nature Genet 23: 41–46, 2004.

LI J et al: High-resolution human genome scanning using whole-genome BAC arrays. Cold Spring Harbor Symp Quant Biol LXVIII: 323–329, 2003.

Newman TL et al: High-throughput genotyping of intermediate-size variation. Hum Mol Genet 15: 1159–1169, 2006.

Pinkel D et al: High-resolution analysis of DNA copy number variations using comparative genomic hybridization to microarrays. Nature Genet 20: 207–211, 1998.

Thomas DC et al: Recent developments in genomewide association scans: A workshop summary and review. Am J Hum Genet 77: 337–345, 2005.

Vissers LE et al: Identification of disease genes by whole genome CGH arrays. Hum Mol Genet 14: R215–R223, 2005.

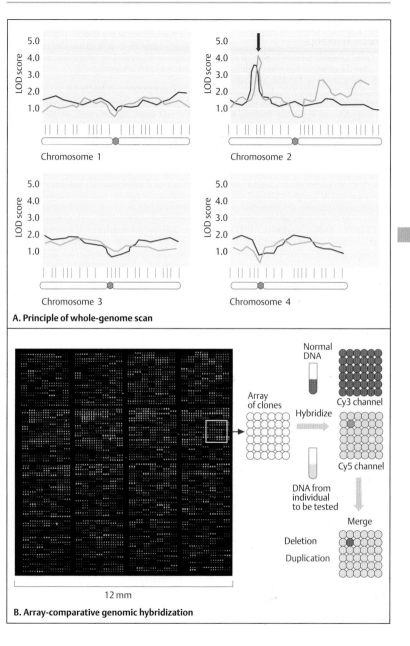

A. Principle of whole-genome scan

B. Array-comparative genomic hybridization

The Dynamic Genome: Mobile Genetic Elements

The term "dynamic genome" refers to observations that the genome of a living organism is not static. Instead, it is flexible and subject to changes. DNA sequences can alter their position within the genome. This unusual phenomenon was first observed in the late 1940s by Barbara McClintock while investigating the genetics of Indian corn (maize, *Zea mays*). She found that certain genes apparently were able to alter their position spontaneously and named them "jumping genes," later *mobile genetic elements*. Today they are known as *transposons* (see transposition, p. 86). Although McClintock's observations were initially met with skepticism, she was awarded the Nobel Prize for this work in 1984 (McClintock, 1984; Fox-Keller, 1983).Transposable genetic elements occur in large numbers in the genomes of most organisms, including man.

A. Stable and unstable mutations

McClintock (1953) determined that certain mutations in maize are unstable. A stable mutation at the C locus causes violet corn kernels (**1**) whereas unstable mutations cause fine pigment spots in individual kernels (variegation, **2**).

B. Effect of mutation and transposition

Normally a gene at the C locus produces a violet pigment of the aleurone in cells of Indian corn (**1**). This gene can be inactivated by insertion of a mobile element *(Ds)* into the gene, resulting in a colorless kernel (**2**). If *Ds* is removed by transposition, C-locus function is restored and small pigmented spots appear (**3**).

C. Insertion and removal of *Ds*

Activator-dissociation (Ac/Ds) is a system of controlling elements in maize. *Ac* is an inherently unstable autonomous element. It can activate another locus, dissociation *(Ds),* and cause a break in the chromosome (**1**). While *Ac* can move independently (*autonomous transposition*), *Ds* can move to another location in the chromosome only under the influence of *Ac* (*nonautonomous transposition*). The *Ac* locus is a 4.6-kb transposon; *Ds* is defective without a transposase gene (see p. 86). The C locus (**2**) is inactivated by the insertion of *Ds*. *Ds* can be re-moved under the influence of *Ac*. This restores normal function at the C locus. If transposition occurs early in development, the pigmented spots are relatively large; if transposition occurs late, the spots are small.

D. Transposons in bacteria

Transposons are classified according to their effect and molecular structure: simple insertion sequences (IS) and the more complex transposons (Tn). A transposon contains additional genes, e.g., for antibiotic resistance in bacteria. Transposition is a special type of recombination by which a DNA segment of about 750 bp to 10 kb is able to move from one position to another, either on the same or on another DNA molecule. The insertion occurs at an integration site (**1**) and requires a break (**2**) with subsequent integration (**3**). The sequences on either side of the integrated segment at the integration site are direct repeats. At both ends, each IS element or transposon carries inverted repeats whose lengths and base sequences are characteristic for different IS and Tn elements. One *E. coli* cell contains on average about ten copies of such sequences. They have also been demonstrated in yeast, *Drosophila*, and other eukaryotic cells. (Photographs from N.V. Fedoroff, 1984.)

References

Fedoroff NV: Transposable genetic elements in maize. Sci Am 250: 65–74, 1984.

Fedoroff NV, Botstein D (eds): The Dynamic Genome: Barbara McClintock's Ideas in the Century of Genetics. Cold Spring Harbor Laboratory Press, New York, 1992.

Fox-Keller E: A Feeling for the Organism: The Life and Work of Barbara McClintock. WH Freeman & Co, San Francisco, 1983.

McClintock B: Introduction of instability at selected loci in maize. Genetics 38: 579–599, 1953.

McClintock B: Controlling genetic elements. Brookhaven Symp Biol 8: 58–74, 1955.

McClintock B: The significance of responses of the genome to challenge. Science 226: 792–801, 1984.

Schwartz RS: Jumping genes. New Engl J Med 332: 941–944, 1995.

Zhang J, Peterson T: A segmental deletion series generated by sister-chromatid transposition of Ac transposable elements in maize. Genetics 171: 333–344, 2005.

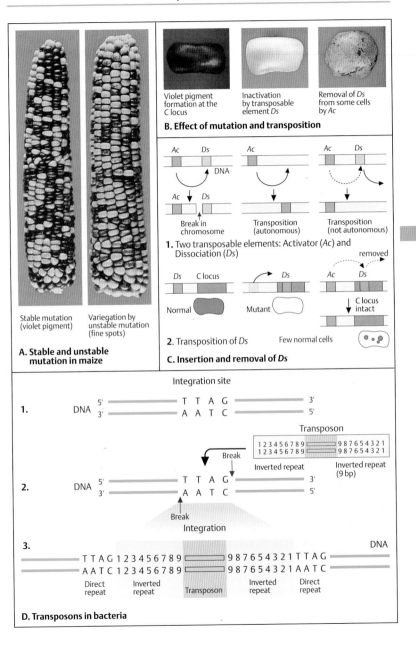

Violet pigment formation at the C locus

Inactivation by transposable element *Ds*

Removal of *Ds* from some cells by *Ac*

B. Effect of mutation and transposition

Ac   *Ds*   *Ac*   *Ac*   *Ds*
DNA

Ac   *Ds*

Break in chromosome

Transposition (autonomous)

Transposition (not autonomous)

1. Two transposable elements: Activator (*Ac*) and Dissociation (*Ds*)

removed

Ds   C locus   *Ds*   *Ac*   *Ds*

Normal   Mutant   C locus intact

Few normal cells

2. Transposition of *Ds*

Stable mutation (violet pigment)

Variegation by unstable mutation (fine spots)

A. Stable and unstable mutation in maize

C. Insertion and removal of *Ds*

Integration site

1. DNA   5′   T T A G   3′
             3′   A A T C   5′

Transposon

1 2 3 4 5 6 7 8 9   9 8 7 6 5 4 3 2 1
1 2 3 4 5 6 7 8 9   9 8 7 6 5 4 3 2 1

Break

Inverted repeat   Inverted repeat (9 bp)

2. DNA   5′   T T A G   3′
             3′   A A T C   5′

Break

Integration

3.   DNA

T T A G 1 2 3 4 5 6 7 8 9   9 8 7 6 5 4 3 2 1 T T A G
A A T C 1 2 3 4 5 6 7 8 9   9 8 7 6 5 4 3 2 1 A A T C

Direct repeat   Inverted repeat   Transposon   Inverted repeat   Direct repeat

D. Transposons in bacteria

Evolution of Genes and Genomes

Genes and genomes existing today are the cumulative result of events that have taken place in the past. The classical theory of evolution as formulated by Charles Darwin in 1859 states that (i) all living organisms today have descended from organisms living in the past; (ii) organisms that lived during earlier times differed from those living today; (iii) the changes were more or less gradual, with only small changes at a time; and (iv) the changes usually led to divergent organisms, with the number of ancestral types of organisms being smaller than the number of types today.

A. Gene evolution by duplication

Studies of the various genomes indicate that different types of duplication must have occurred: of individual genes or parts of genes (exons), subgenomic duplications, and rarely, duplications of the whole genome (Ohno, 1970). Duplication of a gene relieves the selective pressure on that gene. After a duplication event, the gene can accumulate mutations without compromising the original function, provided the duplicated gene has separate regulatory control. (Figure adapted from Strachan and Read, 2004.)

B. Gene evolution by exon shuffling

The exon/intron structure of eukaryotic genes provides great evolutionary versatility. New genes can be created by placing parts of existing genes into a new context, using functional properties in a new combination. This is referred to as exon or domain shuffling (Gilbert, 1987; Kaessmann et al, 2002). (Figure adapted from Strachan and Read, 2004.)

C. Evolution of chromosomes

Evolution also occurs by structural rearrangements of the genome at the chromosomal level. Related species, e.g., mammals, differ in the number of their chromosomes and chromosomal morphology, but not in the number of genes, which often are conserved to a remarkable degree. The human chromosome 2 appears to have evolved from the fusion of two primate chromosomes. The differences in chromosome 3 are much more subtle. The orang-utan chromosome 3 differs from that of man and the other primates by a pericentric inversion. The banding patterns of all primate chromosomes are remarkably similar. This reflects their close evolutionary relationship. (Figure adapted from Yunis and Prakash, 1982.)

D. Molecular phylogenetics and evolutionary tree reconstruction

A phylogenetic tree can be based on different types of evidence: fossils, differences in proteins, immunological data, DNA–DNA hybridization, and DNA sequence similarity. The number of events that have to have taken place to explain the diversity observed today is determined. In the path from an ancestral gene (**1**). Two events are shown schematically. Two categories of homologs are distinguished: *paralogs* and *orthologs* (**2**). Paralogs are homologous genes that have evolved by duplication of an ancestral gene within a species. Orthologs are genes that have evolved by vertical descent from an ancestral gene between different, but related species. The human α- and δ-globin loci are examples of paralogs. The β-globin genes of humans and other mammals are examples of orthologs. The adjective *paralogous* refers to nucleotide sequence comparisons.

References

Brown TA: Genomes, 2nd ed. Bios Scientific Publ, Oxford, 2002.

Eichler EE, Sankoff D: Structural dynamics of eukaryotic chromosome evolution. Science: 301: 793–797, 2003.

Gilbert W: The exon theory of genes. Cold Spring Harbor Symp Quant Biol 46: 151–153, 1987.

Jobling MA, Hurles ME, Tyler-Smith C: Human Evolutionary Genetics. Origins, Peoples, and Disease. GS Garland Science Publishers, New York, 2004.

Klein J, Takahata N: Where Do We Come From? The Molecular Evidence for Human Descent. Springer, Berlin–Heidelberg–New York, 2002.

Ohno S: Evolution by Gene Duplication. Springer Verlag, Heidelberg, 1970.

Strachan T, Read AP: Human Molecular Genetics, 3rd ed. Bios Scientific Publishers, Oxford, 2004.

Yunis JJ, Prakash O: The origin of man: A chromosomal pictorial legacy. Science 215: 1525–1530, 1982.

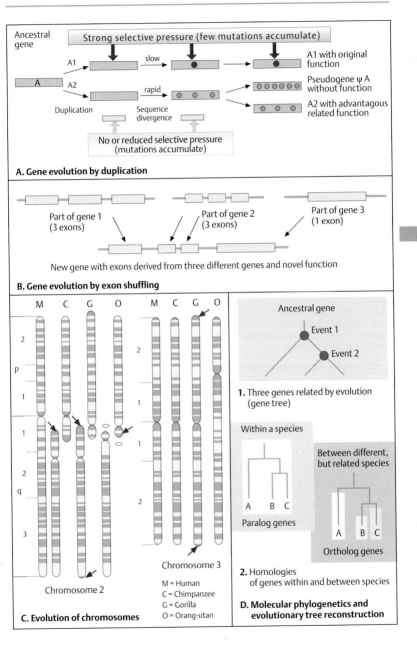

Ancestral gene

Strong selective pressure (few mutations accumulate)

A1 — slow → A1 with original function

A

A2 — rapid → Pseudogene ψ A without function

A2 with advantageous related function

Duplication

Sequence divergence

No or reduced selective pressure (mutations accumulate)

A. Gene evolution by duplication

Part of gene 1 (3 exons)

Part of gene 2 (3 exons)

Part of gene 3 (1 exon)

New gene with exons derived from three different genes and novel function

B. Gene evolution by exon shuffling

M C G O   M C G O

Chromosome 2

Chromosome 3

M = Human
C = Chimpanzee
G = Gorilla
O = Orang-utan

C. Evolution of chromosomes

Ancestral gene

Event 1

Event 2

1. Three genes related by evolution (gene tree)

Within a species

A B C

Paralog genes

Between different, but related species

A   B C

Ortholog genes

2. Homologies of genes within and between species

D. Molecular phylogenetics and evolutionary tree reconstruction

Comparative Genomics

Genomes of different organisms differ or resemble each other according to their evolutionary relationship. The number and types of differences depend on the time that has passed since their common ancestor existed and the frequency of genomic changes and mutations that have occurred since they diverged from the ancestor. If two genomes are sufficiently closely related, the order of genes along each chromosome will have been preserved.

Two areas of comparative genomics have practical applications: plant genomes and the search for human disease genes that have homologs in other organisms (see Brown, 2002, p. 214). For example, wheat, one of the important food plants, has a huge genome of 16000 Mb (5 times larger than the human), whereas rice, perhaps more important, has a genome of 430 Mb, which has recently been sequenced. Comparative genomics has been used to identify genes that contribute to harvest yield, pest resistance, and other plant attributes. The approximately 6200 genes of the yeast genome contain numerous homologs to human disease genes. The same holds true for genes in the genomes of the nematode *C. elegans* and the fruit fly *Drosophila melanogaster*.

A. Homologies of human proteins

The sequence of the human genome has revealed considerable homologies of its proteins with those of other organisms, e.g., 21% with other eukaryotes and prokaryotes. (Figure adapted from IHGSC, 2001.)

B. Chromosome-associated proteins

The chromatin-associated proteins and transcription factors are conserved in evolution. As a consequence, 60% of human, *Drosophila*, and the nematode *C. elegans* chromatin architectures are shared. (Figure adapted from IHGSC, 2001.)

C. Chromosomal segments conserved between man and mouse

The human genome contains 183 defined segments shared by the human and the mouse genomes. The average length of a segment is 15.4 Mb with a range from 24 kb to 90.5 Mb (IHGSC, 2001). Segments containing at least two genes in conserved order between man and mouse are indicated in a color corresponding to the mouse chromosome. (Figure adapted from IHGSC, 2001.)

References

Brown TA: Genomes, 2nd ed. Bios Scientific Publishers, Oxford, 2002.

Elgar G et al: Small is beautiful: comparative genomics with the pufferfish (*Fugu rubripes*). Trends Genet 12: 145–150, 1996.

International Human Sequencing Genome Consortium (IHGSC): Initial sequencing analysis of the human genome. Nature 409: 860–921, 2001.

Rubin GM et al: Comparative genomics of the eukaryotes. Science 287: 2204–2215, 2000.

The Genome of *Homo sapiens*. Cold Spring Harbor Quant Biol LXVIII: 1–512, 2003.

Steinmetz LM et al: Systematic screen for human disease genes in yeast. Nature Genet 31: 400–404, 2002.

Strachan T, Read AP: Human Molecular Genetics, 3rd ed. Garland Science Publishing, London – New York, 2004.

Comparative genomics of four nonvertebrate organisms based on genome sequence data

Organism	Number of genes	Genes arisen by duplication	Gene families (core proteome)	Type of organism
H. influenzae	1709	284	1425	Bacterium
C. cerevisiae	6241	1858	4383	Yeast
C. elegans	18424	8971	9453	Nematode
Drosophila	13601	5536	8065	Insect

(Data from Rubin et al., 2000)

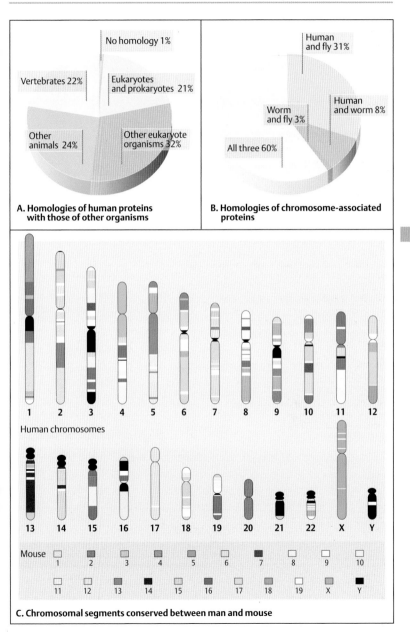

A. Homologies of human proteins with those of other organisms

- No homology 1%
- Eukaryotes and prokaryotes 21%
- Vertebrates 22%
- Other animals 24%
- Other eukaryote organisms 32%

B. Homologies of chromosome-associated proteins

- Human and fly 31%
- Human and worm 8%
- Worm and fly 3%
- All three 60%

Human chromosomes

1 2 3 4 5 6 7 8 9 10 11 12

13 14 15 16 17 18 19 20 21 22 X Y

Mouse 1 2 3 4 5 6 7 8 9 10

11 12 13 14 15 16 17 18 19 X Y

C. Chromosomal segments conserved between man and mouse

Genetics and Medicine

Intracellular Signal Transduction

Multicellular organisms use a broad repertoire of extracellular signaling molecules for communication between and within cells. The specific binding of an extracellular signaling molecule (ligand) to its receptor on a target cell triggers a specific response. This consists of a series of mutually activating or inhibitory molecular events, called a *signal transduction pathway* (or signaling pathway).

A. Main intracellular functions controlling growth

Growth factors are important signaling molecules comprising a large group of secreted proteins (1). Each binds with high specificity to a cell surface receptor protein (2). This activates intracellular signal transduction proteins (3) and initiates a cascade of activations of responsive proteins (often by phosphorylation) that act as second messengers (4). Hormones are small signaling molecules (5) that arrive via the bloodstream. They enter the cell either by diffusion or by binding to a cell surface receptor (6). Some hormones bind to an intranuclear receptor (7). Activated transcription factors (8) together with cofactors initiate transcription (9). Prior to transcription, an elaborate system of DNA damage recognition and repair mechanisms (10) checks DNA integrity (cell cycle control, 11). Cell division proceeds if faults in DNA structure have been repaired; if not, the cell is sacrificed by apoptosis (cell death, 12). (Figure adapted from Lodish et al., 2004.)

B. Receptor tyrosine kinase family

Receptor tyrosine kinases (RTKs) are a major class of cell surface receptors. They consist of a single transmembrane protein with an extracellular N-terminal part, a transmembrane part (TM), and an intracellular C-terminal part. The intracellular part contains the tyrosine kinase domain. RTK ligands are growth factors that control a wide variety of functions involving growth and differentiation. RTK receptors share structural features but differ in function. The extracellular, ligand-binding domains of RTKs contain cysteine-rich regions. In certain RTKs the binding domains resemble immunoglobulin chains (Ig-like domains), known for their ability to bind other molecules (see p. 140).

Medical relevance

Mutations in genes encoding RTKs may result in a proliferative signal in the absence of a growth factor and cause errors in embryonic development and differentiation (congenital malformations) or cancer. RTK mutations cause a group of important human diseases and malformation syndromes. The phenotypes due to the mutations differ according to the particular type of RTK involved and the type of mutation (see table in the appendix).

References

Alberts B et al: Molecular Biology of the Cell, 4th ed. Garland Publishing Co, New York, 2002.

Brivanlou AH, Darnell JE: Signal transduction and the control of gene expression. Science 295: 813–818, 2002.

Cohen jr MM: FGFs/FGFRs and associated disorders, pp 380–400. In: Epstein CJ, Erickson RP, Wynshaw-Boris A (eds) Inborn Errors of Development. The Molecular Basis of Clinical Disorders of Morphogenesis. Oxford University Press, Oxford, 2004.

Lodish H et al: Molecular Cell Biology, 5th ed. WH Freeman & Co, New York, 2004.

Muenke M et al: Fibroblast growth factor receptor-related skeletal disorders: craniosynostosis and dwarfism syndromes, pp. 1029–1038. In: JL Jameson (ed) Principles of Molecular Medicine. Humana Press, Totowa, New Jersey, 1998.

Münke M, Schell U: Fibroblast-growth-factor receptor mutations in human skeletal disorders. Trends Genet 11:308–313, 1995.

Robertson SC, Tynan JA, Donoghue DJ: RTK mutations and human syndromes: when good receptors turn bad. Trends Genet 16: 265–271, 2000.

Tata JR: One hundred years of hormones. A new name sparked multidisciplinary research in endocrinology, which shed light on chemical communication in multicellular organisms. EMBO J Reports 6: 490–496, 2005.

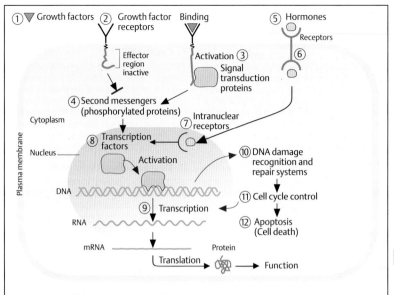

A. Main intracellular functions controlling cell growth

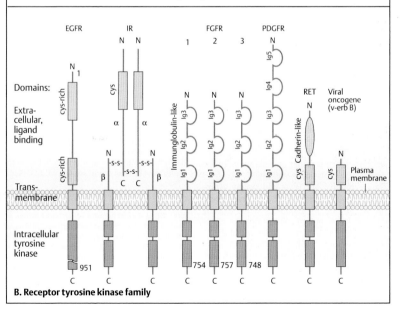

B. Receptor tyrosine kinase family

Signal Transduction Pathways

A signal transduction pathway transmits a signal into the cell by a series of consecutive events that regulate the activity of a gene. A receptor on the surface of a responding cell becomes activated by binding to its ligand (see p. 110). This causes a conformational change in the receptor's structure. A cascade of intracellular activations and inhibitions results, transmitting the signal to the nucleus, where it induces or inhibits the transcription of a target gene. Activation may be achieved by inhibiting an inhibitor.

A. Receptor tyrosine kinase (RTK)

The binding of a ligand activates a receptor tyrosine kinase (RTK, see previous page) by dimerization of two neighboring transmembrane proteins. As a result, intracellular tyrosine residues are phosphorylated. This activates another target protein and further signal transduction events.

B. G protein-coupled receptors

G proteins are trimeric guanine nucleotide-binding proteins with subunits, α, β, and λ. They act as molecular switches between an inactive and a brief active state. In their inactive state, the α subunit is bound to guanosine diphosphate (GDP); in their active state, to guanosine triphosphate (GTP). G protein-coupled receptors (GPCRs) traverse the cell membrane seven times back and forth (therefore sometimes called serpentine receptors). Specific binding of a ligand activates the receptor by releasing the GDP from the α subunit. GDP is replaced by GTP. This change dissociates the trimer into two activated components, the α subunit and a $\beta\lambda$ complex. The $\beta\lambda$ complex thus released interacts with a target protein (effector protein), which is either an enzyme or an ion channel in the plasma membrane.
The α-subunit is a GTPase and rapidly hydrolyzes its bound GTP to GDP. As a result the α subunit and the $\beta\lambda$ complex reunite, forming the inactive G protein. Therefore, the dissociated, active state of the α-subunit and the $\beta\lambda$ complex is short-lived. If the rapid reversal to the inactive state is delayed or impossible due to a toxin or a mutation, normal function is severely impaired.
The α-subunits of mammalian G proteins form a large family of signaling molecules that bind to a wide variety of effector proteins. About 20 mammalian α-subunits, 5 β-subunits, and 12 λ-subunits have been identified.

C. Cyclic AMP

The binding of the ligand to a signal transducing receptor molecule (the "first messenger") causes a brief increase or decrease in the concentration of low-molecular intracellular signaling molecules acting as second messengers. 3′,5′-Cyclic adenosyl monophosphate (cAMP) contains a phosphate group bound to the sugar 5′ and 3′ carbons in a cyclic structure.

D. Degradation of cyclic AMP

cAMP activates protein kinase A (PKA). Phosphodiesterase rapidly degrades cAMP into adenenosine monophosphate (AMP). Active PKA phosphorylates more than 100 signaling proteins and transcription factors.

Medical relevance

Cholera toxin inhibits GTP hydrolysis and induces a permanently active state. In epithelial cells of the gastrointestinal tract this causes a massive efflux of water and chloride ions. Pertussis toxin (whooping cough) inhibits adenyl cyclase (inhibitory G protein, G_i) and prevents the α subunit of the inhibitory G protein from interacting with receptors.
Several endocrine disorders result from mutations of genes encoding G protein-coupled receptors or the G proteins themselves. Gain-of-function and loss-of-function mutations have been observed. See table in the appendix for selected examples.

References

Alberts B et al: Molecular Biology of the Cell, 4th ed. Garland Science, New York, 2002.

Clapham DE: Mutations in G protein-linked receptors: novel insights on disease. Cell 75: 1237–1239, 1993.

Lodish H et al: Molecular cell Biology. 5th ed. WH Freeman, New York, 2004.

Lowe WL et al: Mechanisms of hormone action, pp 419–431. In: JL Jameson (ed) Principles of Molecular Medicine. Humana Press, Totowa, New Jersey, 1998.

Newley SE, Aelst L van: Guanine nucleotide-binding proteins, pp 832–848. In: Epstein CJ, Erickson RP, Wynshaw-Boris A (eds): Inborn Errors of Development. The Molecular Basis of Clinical Disorders of Morphogenesis. Oxford University Press, Oxford, 2004.

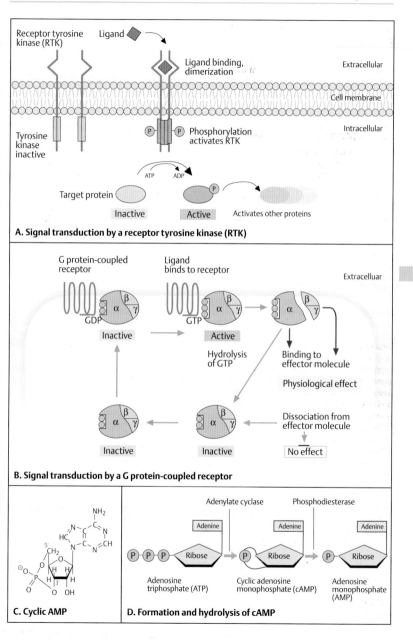

A. Signal transduction by a receptor tyrosine kinase (RTK)

Receptor tyrosine kinase (RTK)
Ligand
Ligand binding, dimerization
Extracellular
Cell membrane
Intracellular
Phosphorylation activates RTK
Tyrosine kinase inactive
ATP   ADP
Target protein
Inactive
Active
Activates other proteins

B. Signal transduction by a G protein-coupled receptor

G protein-coupled receptor
Ligand binds to receptor
Extracellular
GDP
Inactive
GTP
Active
Hydrolysis of GTP
Binding to effector molecule
Physiological effect
Dissociation from effector molecule
No effect
Inactive
Inactive

C. Cyclic AMP

NH_2
Adenosine triphosphate (ATP)

D. Formation and hydrolysis of cAMP

Adenylate cyclase
Phosphodiesterase
Adenine
Adenine
Adenine
Ribose
Ribose
Ribose
Adenosine triphosphate (ATP)
Cyclic adenosine monophosphate (cAMP)
Adenosine monophosphate (AMP)

TGF-β and Wnt/β-Catenin Signaling Pathways

The transforming growth factor-β (TGF-β) and the wingless (Wnt)/β-catenin signal transduction pathways are important examples of signaling systems involved in a variety of developmental processes and cellular functions.

A. The TGF-β signaling pathway

The transforming growth factor-β superfamily includes about 30 structurally related growth and differentiation factors. Their receptors are dimeric transmembrane proteins with intracellular serine/threonine kinase domains. Binding of the ligand induces a heterodimeric complex consisting of two copies each of receptors type II and type I (only one copy each shown). The constitutively phosphorylated receptor type II phosphorylates the GS domain, a conserved glycine/serine-rich sequence. This activates the kinase domain of receptor type I. As a result, members of the Smad family of transcription factors become serine-phosphorylated and form a complex with a nonphosphorylated common Smad (Co-Smad). Smad 4 moves into the nucleus and binds with DNA-binding proteins to initiate transcription.

The TGF-β superfamily has important roles in embryonic development, differentiation, organogenesis, self-renewal and maintenance of stem cells in their undifferentiated stage, selection of cell differentiation lineage, and suppression of carcinogenesis (Mishra et al., 2005). Phylogenetic evidence suggests that this family is one of the oldest signaling pathways, having arisen about 1.3 billion years ago, before the divergence of arthropods and vertebrates (see p. 24). (Figure redrawn from Petryk & O'Connor, 2004.)

B. Wnt/β-catenin signaling pathway

Wingless (*wg*) is a *Drosophila* segment-polarity mutant (see p. 298). The vertebrate ortholog is *Wnt* (derived from wingless and *int-1*, a mouse gene). Its most important function is regulating β-catenin protein levels. β-catenin (in *Drosophila* called armadillo) interacts with other proteins to activate target genes. The receptor for Wnt is a seven-transmembrane receptor named frizzled and a coreceptor, the low-density lipoprotein (LDL, see p. 368)-related receptor protein LRP6. Binding of Wnt to its receptor activates a protein called disheveled (Dsh). Dsh inhibits a protein kinase, glycogen synthase kinase-3 (GSK-3), which normally prevents nuclear accumulation of β-catenin and γ-catenin by proteolysis. Inhibited GSK-3 activity releases β-catenin from the APC protein (adenomatous polyposis coli, see p. 150). β-catenin then migrates into the nucleus, forms a complex with several other proteins (transcription factors LEF and TCF, and others), and regulates numerous target genes. In the absence of a Wnt signal, β-catenin is phosphorylated by GSK-3, APC, and axin (a scaffolding protein) and then degraded (not shown). Of the at least three Wnt-activated pathways downstream of Dsh, the main ("canonical") pathway is schematically shown here. Different Wnt signals have crucial roles in development, e.g., for gastrulation, brain development, limb patterning, and organogenesis. (Figure redrawn from Sheldahl & Moon, 2004.)

Medical relevance

Activated β-catenin activates the proto-oncogene *c-myc*, a stimulator of cell growth and proliferation. Mutations in APC occur in 80% of human cancers of the colon (see p. 150). *Wnt-4* is a potential ovary-determing gene.

References

Brivanlou AH, Darnell JE: Signal transduction and the control of gene expression. Science 295: 813–818, 2002.

Mishra L, Derynck R, Mishra B: Transforming growth factor-β signaling in stem cells and cancer. Science 310: 68–71, 2005.

Moore RT et al: WNT and β-catenin signaling: diseases and therapies. Nature Rev Genet 5: 691–701, 2004.

Nelson WJ, Nusse R: Convergence of Wnt, β-catenin, and cadherin pathways. Science 303: 1483–1487, 2004.

Petryk A, O'Connor MB: The transforming growth factor β (TGF-β) signaling pathway, pp 285–295. In: Epstein CJ, Erickson RP, Wynshaw-Boris A (eds): Inborn Errors of Development. The Molecular Basis of Clinical Disorders of Morphogenesis. Oxford University Press, Oxford, 2004.

Sheldahl LC, Moon RT: The Wnt (Wingless-type) signaling pathway, pp 272–281. In: Epstein CJ, Erickson RP, Wynshaw-Boris A (eds): Inborn Errors of Development. The Molecular Basis of Clinical Disorders of Morphogenesis. Oxford University Press, Oxford, 2004.

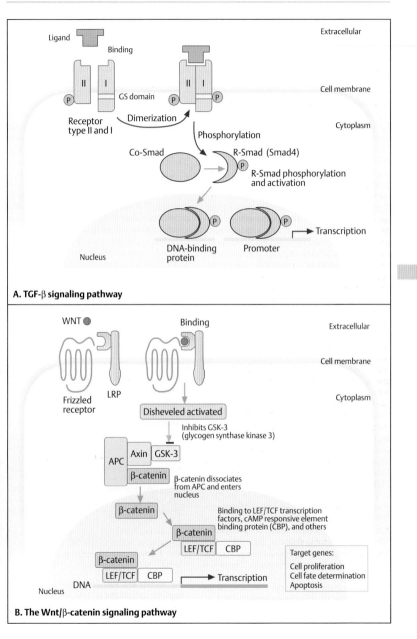

A. TGF-β signaling pathway

B. The Wnt/β-catenin signaling pathway

The Hedgehog and TNF-α Signal Transduction Pathways

A family of secreted proteins (paracrine factors) called hedgehog conveys signals required in many developmental processes, in particular vertebrate limb and neural differentiation. The gene *hedgehog* (*hh*) in *Drosophila* is a segment polarity gene whose protein acts as a transcription factor with both activating and repressing functions in concert with other proteins. Mutant *hh* flies resemble hedgehogs owing to visible spikes on their surfaces. The genomes of vertebrates contain three hedgehog genes, *Sonic hedgehog* (*Shh*), *Desert hedgehog* (*Dhh*), and *Indian hedgehog* (*Ihh*).

Tumor necrosis factors (TNFs) are macrophage-secreted cytokines with cytotoxic effects. TNF-α is involved in two main signaling pathways: (i) kinase and transcription factor activation and (ii) caspase activation in the cell death (apoptosis) pathways (see p. 126). Tumor necrosis factor α (TNF-α) belongs to a large family of 29 TNF receptors (TNFRs) and 18 TNF ligands in humans. TNFR1, TNFR2, and their ligands play a prominent role in the inflammatory response and other immune functions.

A. The hedgehog signaling pathways

More than 12 genes form a network rather than a one-directional pathway. The hedgehog (Hh) signal, secreted as a 45-kD precursor protein, is cleaved into a 20-kD N-terminal fragment and a 25-kD C-terminal fragment. The N-terminal fragment contains an ester-linked hydrophobic cholesterol moiety and an amide-linked palmitoyl moiety critical for signaling activity. Hh is the ligand for a transmembrane receptor protein named patched (Ptch; in humans PTCH). Patched responds to ligand binding by negative control on another transmembrane protein called smoothened (Smo; SMO in humans; names derived from mutant *Drosophila* phenotypes).

Smoothened, a protein with 7 hydrophobic membrane-spanning domains, acts as a hedgehog signal transducer. Without the signal, a microtubule-bound group of proteins (Costal 2, a kinesin-like protein, and Fused, a serine-threonine kinase) are attached to a 155-kD effector protein Ci. Two other proteins, PKA (a protein kinase) and Slimb, cleave Ci into two fragments, one of which represses transcription of hedgehog-responsive genes. Binding of the Hh signal to Ptch suppresses the inhibitory function of Ptch on Smo. This inhibits the action of PKA and Slimb. As a result, Costal 2 and Fused are phosphorylated and activate Ci. Activated Ci moves into the nucleus and initiates transcription together with coactivator CREB-binding protein (CBP) and other factors.

Medical relevance

Mutations and deletions in more than 10 human genes in the hedgehog gene network result in a group of malformation syndromes (table in the appendix).

B. TNF-α signal transduction

The TNF ligands and their receptors are trimeric proteins. The extracellular moiety of TNFR-α consists of 2–8 similar structural motifs of elongated shape stabilized by disulfide bridges.

Upon binding of TNF-α to the receptor, the transcription factor NF-κB (nuclear factor kappa B, first discovered as regulating the κ light chain of B lymphocytes) is activated. This causes rearrangements of the cytosolic domains of the receptor and the recruitment of an intracellular signal protein (RIP, receptor-interacting kinase), adaptor proteins, TRAF2 (TNF-receptor-associated factor 2), and a death-domain protein (TRADD). Signal-induced phosphorylation and degradation of an inhibitory κB (IκB) are the essential features of this pathway. NF-κB proteins control the expression of many genes involved in cell-specific differentiation and in inflammatory and apoptotic responses.

References

Abbas AK, Lichtman AH: Cellular and Molecular Immunology, 5th ed. WB Saunders, Philadelphia, 2005.

Cohen jr MM: The hedghog signaling network. Am J Med Genet 123 A: 5–28, 2003.

Cohen jr MM: The sonic hedgehog pathway, pp 210–228. In: Epstein CJ et al (eds) Inborn Errors of Development. Oxford University Press, Oxford, 2004.

Karim M: Nuclear factor-κB in cancer development and progress. Nature Insight 441: 431–436, 2006.

Lum L, Beachy PA. The Hedgehog response network: sensors, switches, and routers. Science 304: 1755–1759, 2004.

Schneider PL: The tumor necrosis factor signaling pathway, pp 340–358. In: Epstein CJ et al (eds) Inborn Errors of Development. Oxford University Press, Oxford, 2004.

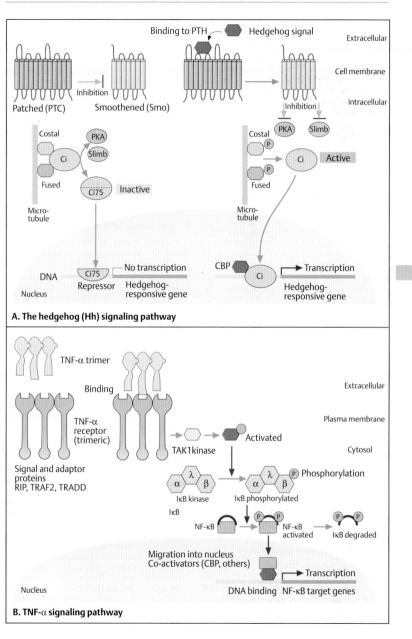

A. The hedgehog (Hh) signaling pathway

B. TNF-α signaling pathway

The Notch/Delta Signaling Pathway

Notch signaling determines cell specification (e.g., neurogenesis, myogenesis, hematopoiesis), embryo patterning, and morphogenesis of various tissues in the developing vertebrate and invertebrate embryo. Notch signals are transmitted by direct interaction of a sending and a receiving cell. A receptor, notch, and different types of ligands, called the DSL group (Delta, Serrate, Lag-2 ligands) mediate the signal. Notch activation involves regulated proteolytic cleavage at three sites of the receptor.

A. Transmission of a notch signal

Both the ligand and the receptor are single-pass transmembrane proteins, and both require proteolytic cleavage (not shown for the ligand in the schematic figure). The first proteolytic cleavage occurs in the *trans* Golgi network by a furinlike convertase. Notch is O-glycosylated by a glycosyltranferase, fringe, which adds a single fucose to some serines, threonines, and hydroxylysines. Binding to its ligand induces the next two proteolytic cleavages. The notch tail migrates to the nucleus and binds to the major effector of notch signaling, the CSL protein (CBF1 in mammals, Suppression of hairless in flies and frogs, and Lag-2 in the nematode *C. elegans*). Binding with other gene regulatory proteins induces transcription of notch-responsive target genes. (Figure based on Alberts et al., 2002, p. 894.)

B. Lateral inhibition by notch signaling

A characteristic function of notch signaling is lateral inhibition in nerve cell development of *Drosophila*. Nerve cells arise as isolated single cells within a sheet of epithelial precursor cells. Cells determined to develop into nerve cells use the notch signal pathways to inhibit the same cell fate in their neighboring cells. If this fails, a fatal excess of nerve cells causes embryonic death. (Figure based on Alberts et al., 2002, p. 894.)

C. The *Drosophila* mutant notch

Notch is a mutant phenotype in *Drosophila melanogaster*, described by T.H. Morgan in 1919. Female heterozygotes exhibit a notch of varied size in the distal part of their wings. Hemizygous males die during embryonic development from hypertrophy of the nervous system.

D. Notch family of receptors

Notch receptors constitute a large family of 50 surface proteins in mammals. The prototype is the *Drosophila* notch (dNotch), a 300-kD protein consisting of 36 epidermal growth factor receptor (EGFR)-like tandem repeats in the extracellular domain. Other features are cysteine (cys)-rich repeats, LNR (Lin-12 Notch-Related region), and a signal peptide (SP) at the N-terminal extracellular part. The intracellular domain contains six ankyrin repeats (ANK), flanked by two nuclear localization signals (NLS). Humans have four slightly different types of notch receptors, the hNotch 1–4 (the smaller *C. elegans* notch receptors Lin-2 and Glp-1 are not shown).

E. DSL family of notch ligands

The notch ligands likewise comprise a large family of cell surface proteins, classified either as delta or serrate ligands. They contain multiple EGFR-like repeats in the extracellular domains, an N-terminal DSL binding motiv, and a signal peptide (SP). (Figures in D and E adapted from Miyamoto & Weinmeister, 2004.)

Medical relevance

A wide variety of human diseases result from disrupted notch signaling e.g., Alagille syndrome (MIM 118450), due to mutations in the gene encoding jagged1 *or NOTCH2*, spondylocostal dysplasia (MIM 122600, 271529), due to mutations in the *deltalike 3* (*DLL3*) gene, and aortic valve disease (MIM 109730), due to mutations in *hNotch1* (Garg et al., 2005).

References

Alberts B et al: Molecular Biology of the Cell, 4th ed. p. 736. Garland Publishing Co, New York, 2002.

Lodish H et al: Molecular Cell Biology, 5th ed. WH Freeman & Co, New York, 2004.

McDaniel R et al.: *NOTCH2* mutations cause Alagille syndrome, a heterogeneous disorder of the notch signaling pathway. Am J Hum Genet 79: July 2006 in press.

Miyamoto A, Weinmaster G: The notch signaling pathway, pp 447–460. In: Epstein CJ et al (eds) Inborn Errors of Development. The Molecular Basis of Clinical Disorders of Morphogenesis. Oxford University Press, Oxford, 2004.

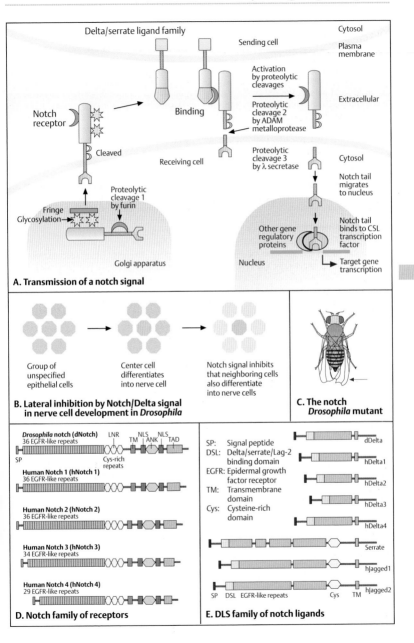

A. Transmission of a notch signal

Delta/serrate ligand family

Cytosol

Sending cell

Plasma membrane

Activation by proteolytic cleavages

Proteolytic cleavage 2 by ADAM metalloprotease

Extracellular

Binding

Notch receptor

Cleaved

Receiving cell

Proteolytic cleavage 3 by λ secretase

Cytosol

Notch tail migrates to nucleus

Proteolytic cleavage 1 by furin

Fringe Glycosylation

Other gene regulatory proteins

Notch tail binds to CSL transcription factor

Golgi apparatus

Nucleus

Target gene transcription

B. Lateral inhibition by Notch/Delta signal in nerve cell development in *Drosophila*

Group of unspecified epithelial cells

Center cell differentiates into nerve cell

Notch signal inhibits that neighboring cells also differentiate into nerve cells

C. The notch *Drosophila* mutant

D. Notch family of receptors

Drosophila notch (dNotch)
36 EGFR-like repeats
SP
LNR
TM NLS ANK NLS TAD
Cys-rich repeats

Human Notch 1 (hNotch 1)
36 EGFR-like repeats

Human Notch 2 (hNotch 2)
36 EGFR-like repeats

Human Notch 3 (hNotch 3)
34 EGFR-like repeats

Human Notch 4 (hNotch 4)
29 EGFR-like repeats

E. DLS family of notch ligands

SP: Signal peptide
DSL: Delta/serrate/Lag-2 binding domain
EGFR: Epidermal growth factor receptor
TM: Transmembrane domain
Cys: Cysteine-rich domain

dDelta
hDelta1
hDelta2
hDelta3
hDelta4
Serrate
hJagged1
hJagged2

SP DSL EGFR-like repeats Cys TM

Neurotransmitter Receptors and Ion Channels

The signal along a nerve cell (neuron) is transmitted by an electric impulse, the action potential. Ion channels regulate the difference in voltage inside and outside the cell (voltage-gated ion channels). Chemicals called neurotransmitters are secreted at the axon terminus when an action potential arrives. These bind to receptors on neighboring cells and alter their membrane potentials. Most neurotransmitters are amino acids (glycine, glutamate, histamine) or are derived from amino acids (dopamine, norepinephrine, epinephrine, serotonin, γ-aminobutyric acid [GABA]), except for the neurotransmitter acetylcholine.

A. Neurotransmitter receptors

Acetylcholine is a neurotransmitter at the neuromuscular junctions, between neurons and muscle cells. Two genetically and functionally different types of acetylcholine receptors are defined by their pharmacological response to either nicotine or muscarine, based on different receptors. The *nicotine-sensitive* acetylcholine receptor is an ion channel for potassium and sodium, regulated by acetylcholine as ligand (a ligand-gated ion channel). It is a pentameric protein consisting of five subunits: two α, one β, one γ, and one δ (**1**). Two acetylcholine molecules bind to the interface of the αδ and αγ subunits (red arrows). Each subunit consists of four transmembrane domains (**2**) and is encoded by a separate gene (**3**). These genes have similar structures and nucleotide base sequences. The *muscarine-sensitive* type of acetylcholine receptor is a seven-α-helix transmembrane protein (**4**). Along the protein different domains can be distinguished according to location and the relative proportion of hydrophilic and hydrophobic amino acids in the intra- and extracellular domains (**5**), corresponding to the exon structure of the gene (**6**). (Figure based on Watson et al., 1992.)

B. Voltage-gated ion channels

Ion channels are transmembrane proteins that efficiently transport inorganic ions into or out of a cell, up to 100 million ions per second. They transport only passively ("downhill"). The ions, mainly Na^+, K^+, Ca^{2+}, or Cl^- diffuse rapidly down an electrochemical gradient across the lipid bilayer. Ion channels are selective for a particular ion. Each ion channel is *gated*: opened briefly in response to a change in voltage across the membrane (*voltage-gated*), to binding of a ligand (*ligand-gated*), or to an ion (*ion-gated*). The potassium (K^+) channel is composed of four identical subunits (only one shown here). Each subunit has six membrane-spanning α-helices (**1**) consisting of 600–700 amino acids. The amino terminal (NH_2) and the carboxy (COOH) terminal of each subunit are located in the cytosol. The N-terminus has a globular shape that is essential for inactivation of the open channel. The fourth transmembrane domain is the voltage-sensing part.

The sodium (Na^+) channel (**2**) is organized as a monomer of four transmembrane domains (I–IV). The subunits of ion channels form narrow pores, which can be opened (**3**) and closed (**4**) in response to an action potential. The structure of calcium (Ca^{++}) ion channels is similar to that of Na^+ channels. (Figure adapted from Watson et al., 1992.)

Medial relevance

Mutations in genes encoding neurotransmitters or their receptors cause a wide variety of neuromuscular and cardiac diseases (see Drachman, 2005).

References

Drachman DB: Myasthenia gravis and other diseases of the neuromuscular junction, pp. 2518–2523. In: Kasper DL et al, editors: Harrison's Principles of Internal Medicine. 16th ed. McGraw-Hill, New York, 2005.

Hauser SL, Beal, MF: Neurobiology of disease, p. 2339–2344. In: Kasper DL et al, editors: Harrison's Principles of Internal Medicine. 16th ed. McGraw-Hill, New York, 2005.

Jiang Y, Ruta V, Chen J, Lee A, MacKinnon R: The principle of gating charge movement in a voltage-dependent K^+ channel. Nature 423: 33–41, 2003.

Lee S-Y et al: Structure of the KvAP voltage-dependent K^+ channel and its dependence on the lipid membrane. Proc Natl Acad Sci USA 102: 15441–15446, 2005.

Long SB, Campbell EB, MacKinnon R: Voltage sensor of Kv1.2: Structural basis of electromechanical coupling. Science 309: 903–908, 2005.

MacKinnon R: Structural biology. Voltage sensor meets lipid membrane. Science 306: 1304–1305, 2004.

Watson JD, Gilman M, Witkowski J, Zoller M: Recombinant DNA, 2nd ed. WH Freeman, Scientific American Books, New York, 1992.

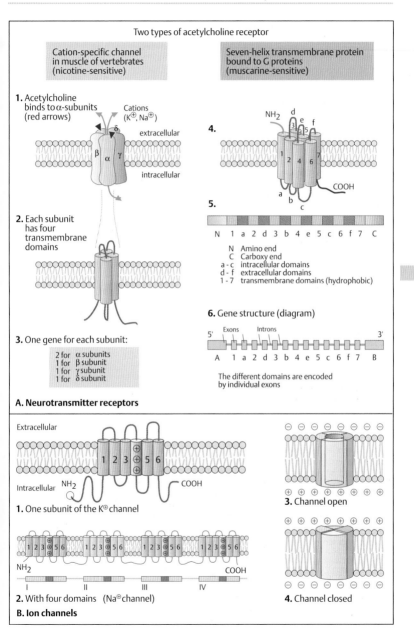

Two types of acetylcholine receptor

Cation-specific channel in muscle of vertebrates (nicotine-sensitive)

Seven-helix transmembrane protein bound to G proteins (muscarine-sensitive)

1. Acetylcholine binds to α-subunits (red arrows)

Cations ($K^{\oplus}$, $Na^{\oplus}$)

extracellular

intracellular

2. Each subunit has four transmembrane domains

3. One gene for each subunit:

2 for	α subunits
1 for	β subunit
1 for	γ subunit
1 for	δ subunit

4.

NH₂

COOH

5.

N 1 a 2 d 3 b 4 e 5 c 6 f 7 C

N   Amino end
C   Carboxy end
a - c   intracellular domains
d - f   extracellular domains
1 - 7   transmembrane domains (hydrophobic)

6. Gene structure (diagram)

5'   Exons   Introns   3'

A 1 a 2 d 3 b 4 e 5 c 6 f 7 B

The different domains are encoded by individual exons

A. Neurotransmitter receptors

Extracellular

Intracellular   NH₂   COOH

1. One subunit of the $K^{\oplus}$ channel

NH₂   COOH

I   II   III   IV

2. With four domains ($Na^{\oplus}$ channel)

3. Channel open

4. Channel closed

B. Ion channels

Genetic Defects in Ion Channels: LQT Syndromes

Mutations in genes encoding ion channels cause more than 30 different genetic disorders. Ion channels play a prominent physiological role in the functioning of cardiac and striated muscle, neuromuscular junctions, and neurons of the central nervous system, inner ear, and retina. A striking example of an ion channel disorder is the long QT (LQT) syndrome. LQT syndrome comprises eight genetically different disorders, LQT syndrome types 1–8 (see table in the appendix and OMIM 192500) resulting from mutations in genes encoding different types of ion channels required in the cardiac conduction system. They are an important cause of cardiac arrhythmia leading to cardiac arrest and sudden death. It is important to distinguish the different types, because the choice of medication differs.

A. Long QT syndrome, a genetic cardiac arrhythmia

LQT syndrome is characterized by a prolonged QT interval in the electrocardiogram (more than 460 ms, corrected for heart rate), sudden attacks of missed heart beats (syncopes) or series of rapid heart beats (*torsade de pointes*), and an increased risk of sudden death from ventricular fibrillation in children and young adults.

B. Different molecular types of long QT syndrome

Prolongation of the QT interval in the electrocardiogram results from an increase in the duration of the cardiac action potential (**1**). The normal potential lasts about 300 ms (phases 1 and 2). The resting membrane potential (phase 3) is reached by progressive repolarization of the cell. In phase 0, the cell is quickly depolarized by activated sodium currents following an excitatory stimulus. LQT1 (**2**, caused by mutations in the *KCNQ1* gene encoding a K+ channel) accounts for about one-third of patients with long QT syndrome. LQT2 (**3**) results from mutations in the *KCNH2* gene (formerly called *HERG*). *KCNH2* encodes a 1159-amino-acid transmembrane protein of the other major potassium channel that participates in phase 3 repolarization. LQT3 (**4**) is caused by defects in the gene encoding a sodium channel protein consisting of four sub-

units (I–IV), each containing six transmembrane domains and a number of phosphate-binding sites.

Syndromic forms of LQT are associated with hearing defects and include the autosomal dominant Romano–Ward syndrome (caused by mutations in *KVLQT1*, the LQT1 gene). Homozygosity for defects of *KVLQT1* or *KCNE1* (LQT5 gene) causes a form of long QT syndrome associated with deafness, the autosomal recessive Jervell and Lange–Nielsen syndrome. (Figure adapted from Ackerman and Clapham, 1997.) (See table in the appendix.)

References

Ackerman AJ: Cardiac channelopathies: it's in the genes. Nature Med 10: 463–464, 2004.

Ackerman MJ, Clapham DD: Ion channels—basic science and clinical disease. New Eng J Med 336: 1575–1586, 1997.

Keating MT, Sanguinetti MC: Molecular and cellular mechanisms of cardiac arrthythmias. Cell 104: 569–580, 2004.

Marks AR: Arrhythmias of the heart: beyond ion channels. Nature Med 9: 263–264, 2003.

Modell SM, Lehmann MH: The long QT syndrome family of cardiac ion channelopathies: A HuGE review. Genet Med 8: 143–155, 2006.

Mohler PJ et al: Ankyrin-B mutation causes type 4 long-QT cardiac arrhythmia and sudden cardiac death. Nature 421: 634–639, 2003.

Schulze-Bahr E et al: *KCNE1* mutations cause Jervell and Lange–Nielsen syndrome. Nature Genet 17: 267–268, 1997.

Schulze-Bahr E et al: The long-QT syndrome. Current status of molecular mechanisms. Z Kardiol 88: 245–254, 1999.

Splawski I et al: Ca (V)1.2 calcium channel dysfunction causes a multisystem disorder including arrhythmia and autism. Cell 119: 19–31, 2004.

Splawski I, Timothy K et al: Severe arrhythmia disorder caused by cardiac L-type calcium channel mutations. Proc Nat Acad Sci 102: 8089–8096, 2005.

Viskin S: Long QT syndromes and torsades de pointes. Lancet 354: 1625–1633, 1999.

Online information:

Gene Connection for the Heart (Europ. Soc. Cardiol.) at www.pc4.fsm.it:81/cardmoc

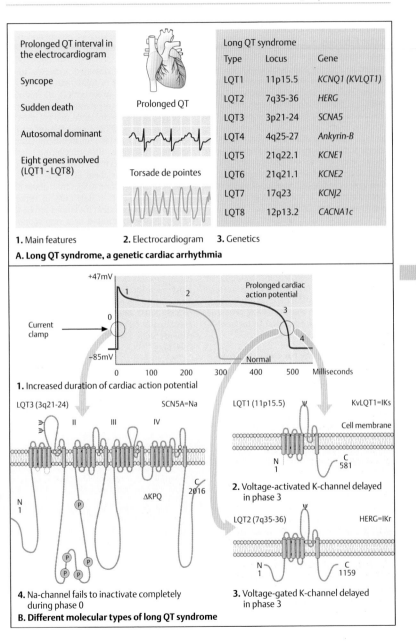

Prolonged QT interval in the electrocardiogram

Syncope

Sudden death

Autosomal dominant

Eight genes involved (LQT1 - LQT8)

Prolonged QT

Torsade de pointes

1. Main features

2. Electrocardiogram

Long QT syndrome

Type	Locus	Gene
LQT1	11p15.5	KCNQ1 (KVLQT1)
LQT2	7q35-36	HERG
LQT3	3p21-24	SCN5A
LQT4	4q25-27	Ankyrin-B
LQT5	21q22.1	KCNE1
LQT6	21q21.1	KCNE2
LQT7	17q23	KCNJ2
LQT8	12p13.2	CACNA1c

3. Genetics

A. Long QT syndrome, a genetic cardiac arrhythmia

+47mV

Current clamp

Prolonged cardiac action potential

−85mV

Normal

0   100   200   300   400   500   Milliseconds

1. Increased duration of cardiac action potential

LQT3 (3q21-24)   SCN5A=Na

ΔKPQ

N
1

C
2016

LQT1 (11p15.5)   KvLQT1=IKs

Cell membrane

N
1

C
581

2. Voltage-activated K-channel delayed in phase 3

LQT2 (7q35-36)   HERG=IKr

N
1

C
1159

3. Voltage-gated K-channel delayed in phase 3

4. Na-channel fails to inactivate completely during phase 0

B. Different molecular types of long QT syndrome

Chloride Channel Defects: Cystic Fibrosis

Cystic fibrosis (CF, mucoviscidosis, MIM 219700) is a highly variable multisystem disorder due to mutations in the cystic fibrosis transmembrane conduction regulator gene *(CFTR)*. It is one of the most frequent autosomal recessive hereditary diseases in populations of European origin (about 1 in 2500 newborns). The high frequency of heterozygotes (1:25) is thought to result from a selective advantage in heterozygotes due to reduced liability to epidemic diarrhea (cholera). Polymorphisms in other genes modify the severity of pulmonary manifestation, e.g., in the 5′ end of the *TGFβ1*-gene (Drumm et al., 2005).

A. Cystic fibrosis: clinical aspects

The disease primarily affects the bronchial system and the gastrointestinal tract. Viscous mucus formation leads to frequent, recurrent lung and bronchial infections, resulting in chronic pulmonary insufficiency. The average life expectancy in typical CF is about 30 years. The disease may take a less severe, almost mild course. Certain mutations occur in males with bilateral absence of the vas deferens.

B. Finding the gene for cystic fibrosis

The gene encodes a chloride channel gene (cystic fibrosis transmembrane conduction regulator, *CFTR*). It was one of the first human genetic diseases identified by positional cloning. Linkage analysis had localized the gene to chromosome 7 at q31. A long-range restriction map of about 1500 kb was narrowed down to 250 kb, and the gene was identified and characterized in a typical manner.

C. The *CFTR* gene and its protein

The large *CFTR* gene spans over 250 kb of genomic DNA, organized into 27 exons (exons 6 and 14 are respectively numbered 6a/6b and 14a/14b), which encode a 6.5-kb transcript with several alternatively spliced forms of mRNA. The protein of 1480 amino acids is a membrane-bound chloride ion channel regulator with five functional domains as shown. The nucleotide-binding domain 1 (NBD1) confers cAMP-regulated chloride channel activity. The most common mutation is a deletion of a phenylalanine codon in position 508 (ΔF508).

The R domain contains putative sites for protein kinase A and protein kinase C phosphorylation. *CFTR* is widely expressed in epithelial cells. The approximately 1300 mutations observed in the *CFTR* gene are classified according to (i) abolished synthesis of full-length protein, (ii) block in protein processing, (iii) reduced chloride channel regulation, (iv) reduced chloride channel conductance, and (v) reduced amount of normal CFTR protein. The underlying genetic defects include missense mutations, nonsense mutations, RNA splicing mutations, and deletions. The most frequent mutation, ΔF508, accounts for about 66% of patients. In Europe, its contribution ranges from about 88% in the north to 50% in Italy (30% in Turkey). Other relatively frequent mutations are G542X (2.4%; glycine substituted by a stop codon at position 542), G551D (1.6%), N1303K (1.3%), and W1282X (1.2%). In Ashkenazi Jewish populations, G542X accounts for 12% and G551D for 3% of all patients. The type of mutation to some extent predicts the severity of the disease. The most severe forms are associated with homozygosity for ΔF508 and compound heterozygotes ΔF508/G551D and ΔF508/G542X. A polymorphic variant (5T) is found in about 40–50% of patients with CBAVD or disseminated bronchiectasis.

References

Bobadilla JL et al: Cystic fibrosis: A worldwide analysis of *CFTR* mutations–correlation with incidence data and application to screening. Hum Mutat 19: 575–606, 2002.

Chillon M et al.: Mutations in the cystic fibrosis gene in patients with congenital absence of the vas deferens. New Engl J Med 332:1475–1480, 1995.

Collins FS: Cystic fibrosis: molecular biology and therapeutic implications. Science 256: 774–779, 1992.

Drumm ML et al: Genetic modifiers of lung disease in cystic fibrosis. New Engl J Med 353: 1443–1453, 2005.

Rosenstein BJ, Zeitline PC: Cystic fibrosis. Lancet 351:277–282, 1998.

Tsui LC: The spectrum of cystic fibrosis mutations. Trends Genet 8:392–398, 1992.

Welsh MJ et al: Cystic fibrosis, pp 5121–5188. In: Scriver CR et al (eds) The Metabolic and Molecular Bases of Inherited Disease, 8th ed. McGraw-Hill, New York, 2001.

Online information.
Toronto Hospital for Sick Children at http://www.genet.sickkids.on.ca/cftr

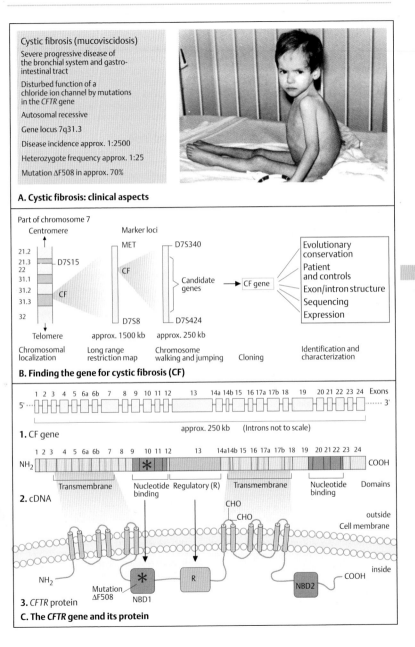

Cystic fibrosis (mucoviscidosis)

Severe progressive disease of the bronchial system and gastro-intestinal tract

Disturbed function of a chloride ion channel by mutations in the *CFTR* gene

Autosomal recessive

Gene locus 7q31.3

Disease incidence approx. 1:2500

Heterozygote frequency approx. 1:25

Mutation ΔF508 in approx. 70%

A. Cystic fibrosis: clinical aspects

Part of chromosome 7

Centromere

Marker loci

MET   D7S340

D7S15

CF

CF

Candidate genes → CF gene

D7S8   D7S424

Telomere   approx. 1500 kb   approx. 250 kb

Evolutionary conservation
Patient and controls
Exon/intron structure
Sequencing
Expression

| Chromosomal localization | Long range restriction map | Chromosome walking and jumping | Cloning | Identification and characterization |

B. Finding the gene for cystic fibrosis (CF)

1 2 3   4 5 6a 6b   7   8 9 10 11 12   13   14a 14b 15 16 17a 17b 18   19   20 21 22 23 24   Exons

5' ···   ···· 3'

approx. 250 kb   (Introns not to scale)

1. CF gene

1 2 3   4 5 6a 6b   7   8 9 10 11 12   13   14a14b 15 16 17a 17b 18   19   20 21 22 23 24

NH₂   *   COOH

Transmembrane   Nucleotide binding   Regulatory (R)   Transmembrane   Nucleotide binding   Domains

2. cDNA

CHO
CHO   outside
Cell membrane
inside

NH₂   *   R   NBD2   COOH

Mutation
ΔF508   NBD1

3. *CFTR* protein

C. The *CFTR* gene and its protein

Rhodopsin, a Photoreceptor

Two types of photoreceptor in specialized cells of the retina provide color vision and light perception in the dark. The human retina has about 6 million cone cells (*cones*) for color vision and about 110 million rod cells (*rods*) that operate in weak light. The photoreceptor responsible for weak light is *rhodopsin*. It is a G protein-coupled receptor activated by light. Only rods contain the trimeric G protein coupled to rhodopsin.

A. Rod cells

A rod cell is a highly specialized cell containing the photoreceptor rhodopsin. The outer segment of a rod contains about 1000 discs with about 4×10^7 molecules of rhodopsin. The approximately 16-nm thick discs are folded by the protein peripherin. At the other end of a rod is an inner segment with the cell nucleus, endoplasmic reticulum, Golgi apparatus, and mitochondria. Each rod has a synapse. From here, a signal is transmitted to the optic nerve and from there to the visual cortex of the brain. The membrane potential of a rod cell in the dark is about – 30 mV, less than the resting potential of – 60 to – 90 mV of a typical nerve cell. This keeps the plasma membrane depolarized with open nonselective ion channels. Hence, neurotransmitters are constantly released. Absorption of light closes the channels and elicits a signal.

B. Photo excitation

In 1958, George Wald and co-workers discovered that light isomerizes 11-*cis*-retinal (**1**) into all-*trans*-retinal (**2**). This ultrafast structural change, in femtoseconds, is so great that it triggers a reliable and reproducible nerve impulse. The absorption spectrum of rhodopsin (**3**) corresponds to the spectrum of sunlight, with an optimum at a wavelength of 500 nm. In the dark, all-*trans*-retinal is converted back into 11-*cis*-retinal. All-*trans*-retinal essentially does not exist in the dark ($\sim$ 1 molecule/1000 years). Although vertebrates, arthropods, and mollusks have anatomically different types of eyes, all three phyla use 11-*cis*-retinal for photoactivation.

C. Light cascade

Photoactivated rhodopsin triggers a series of enzymatic reactions (light cascade). First, the photoactivated rhodopsin activates a G protein–coupled receptor protein, transducin (G_t). Activated transducin is bound to guanosyl-triphosphate (GTP; see p. 272). GTP activates cyclic guanosylmonophosphate phosphodiesterase (cGMP) by binding to its inhibitory γ subunit. This converts cGMP to GMP and closes the cGMP-gated ion channels. The hyperpolarization of the membrane elicits a signal, which is transmitted to the brain via the optic nerve.

D. Rhodopsin

Rhodopsin has a typical seven-helix transmembrane protein structure with binding sites for functionally important molecules such as transducin, rhodopsin kinase, and arrestin on the cytosolic side. The binding site for the light-absorbing 11-*cis*-retinal is lysine in position 296 of the seventh transmembrane domain. X-ray crystallographic studies have shown how the binding of GTP leads to dissociation of G_α from $G_{\beta\gamma}$ (see Lodish et al., 2004, p. 559).

E. cGMP as transmitter of light signals

Cyclic guanosine monophosphate (cGMP) is the second messenger in the light signal transduction system. Rod outer segments normally contain a high concentration of cGMP (ca. 0.07 mM). Light absorption by rhodopsin activates cGMP phosphodiesterase. This hydrolyzes cGMP to 5′-GMP and decreases the cGMP concentration. The high concentration of cGMP in the dark keeps cGMP-gated cation channels open. On exposure to light the channels are closed, the membrane becomes hyperpolarized, and a nerve impulse is triggered. The rod cells are exquisitely sensitive to light. Even a single photon can produce a measurable response by a decrease of the membrane potential by about 1 mV. Humans can detect a flash of about 5 photons (see Lodish et al., 2005, p. 557). (Figures adapted from Stryer, 1995.)

References

Kukura P et al: Structural observation of the primary isomerazation in vision with femtosecond-stimulated Raman. Science 310: 1006–1009, 2005.

Lodish H et al: Molecular Cell Biology, 5th ed. WH Freeman, New York, 2004.

Stryer L: Biochemistry, 4th ed. WH Freeman, New York, 1995.

Stryer L: Molecular basis of visual excitation. Cold Spring Harbor Symp Quant Biol. 53: 28–294, 1988.

Wald G: The molecular basis of visual excitation. Nature 219: 800–807, 1968.

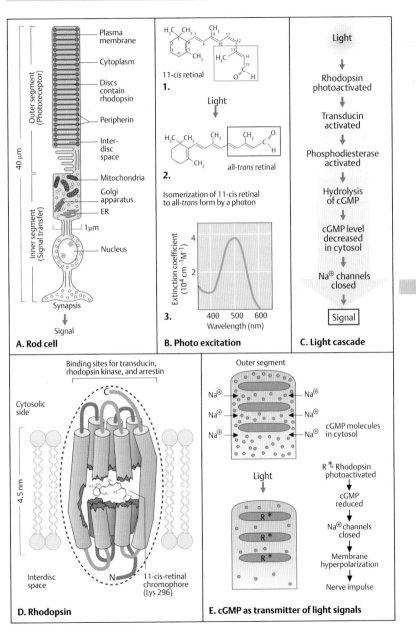

A. Rod cell

- Plasma membrane
- Cytoplasm
- Discs contain rhodopsin
- Peripherin
- Inter-disc space
- Mitochondria
- Golgi apparatus
- ER
- Nucleus
- Synapsis

Outer segment (Photoreceptor)

Inner segment (Signal transfer)

40 μm

1 μm

Signal

B. Photo excitation

11-cis retinal

1.

Light

all-trans retinal

2.

Isomerization of 11-cis retinal to all-trans form by a photon

3.

Extinction coefficient (10^4 cm^{-1}M^{-1})

Wavelength (nm)

C. Light cascade

Light
↓
Rhodopsin photoactivated
↓
Transducin activated
↓
Phosphodiesterase activated
↓
Hydrolysis of cGMP
↓
cGMP level decreased in cytosol
↓
Na$^⊕$ channels closed
↓
Signal

D. Rhodopsin

Binding sites for transducin, rhodopsin kinase, and arrestin

Cytosolic side

4.5 nm

Interdisc space

N

C

11-cis-retinal chromophore (Lys 296)

E. cGMP as transmitter of light signals

Outer segment

Na$^⊕$

cGMP molecules in cytosol

Light

R* Rhodopsin photoactivated
↓
cGMP reduced
↓
Na$^⊕$ channels closed
↓
Membrane hyperpolarization
↓
Nerve impulse

Pigmentary Retinal Degeneration

Pigmentary degeneration of the retina (retinitis pigmentosa, RP) is a large group of eye diseases (MIM 180100, 180102, 312600). Mutations at about 100 gene loci are involved in the causes of retinal degeneration. The different forms of RP occur as autosomal dominant (more than 10 loci), autosomal recessive (more than 22 loci), or X-chromosomal traits (at least 3 loci). Mutations in other genes encoding proteins of the light cascade, such as peripherin, cAMP phosphodiesterase, and others, may also cause retinitis pigmentosa. Retinal degeneration may occur alone (nonsyndromic) or as part of a systemic disorder involving other organ systems (syndromic forms).

A. Retinitis pigmentosa

Affected individuals first lose midperipheral vision; then the visual field is progressively reduced to a small island of central vision (tunnel vision). This is associated with the inability to perceive weak light (night blindness). The fundus of the eye shows thin retinal vessels, a pale, waxy yellow optic nerve, and multiple areas of irregular hyperpigmentation and depigmentation. (Photograph kindly provided by Professor E. Zrenner, Tübingen.)

B. The first mutation in rhodopsin

A typical mutation in the rhodopsin gene is a transversion of cytosine (C) to adenine (A) in codon 23 of exon 1 (Dryja et al., 1990). This exchanges a proline for a histidine in codon 23 (CCC for proline to CAC for histidine, designated as mutation P33 H. The partial DNA sequence of the mutant shows an additional band representing the adenine in codon 23. Proline in position 23 is highly conserved in evolution and occurs in more than ten related G protein receptors.

C. Mutations in rhodopsin

The gene for human rhodopsin (*RHO*) is on the long arm of chromosome 3 in region 2, band 1.4 (3 q21.4). Rhodopsin has 348 amino acid residues, of which 38 are invariant in vertebrates. This figure shows the distribution of selected mutations in the rhodopsin molecule. Most are missense mutations leading to an exchange of one amino acid by another. Rearrangements involving a few base pairs cause small deletions.

More than 100 different mutations are known for autosomal dominant inherited RP. Only a small number of mutations have been recognized to cause autosomal recessive RP.

D. Simplified diagnosis of a mutation

This is an example of a family with 13 affected individuals in three generations (1). Affected individuals are shown in dark, unaffected in white; males as squares, females as circles. The mutation is the same (P23 H) as shown in part B. The information that the mutation allows the normal and the mutant allele to be distinguished by using allele-specific oligonucleotides and the polymerase chain reaction (2). The oligonucleotide corresponding to the mutant allele has the sequence 3′-CATGAGCTT-CACCGACGCA-5′ (the mutation is the presence of adenine [A, underlined] instead of cytosine; compare with the sequence of codons 26 to 21 in part B). This mutant allele-specific oligomer hybridizes with the mutant allele only, not with the normal allele. Thus, all affected individuals shown in the pedigree exhibit a hybridization signal, whereas the unaffected individuals do not (no signal for II-2, II-12, and III-4 because they were not examined). (Data of Dryja et al., 1990.)

References

Dryja TP: Retinitis pigmentosa, pp 5903–5933. In: CR Scriver et al (eds) The Metabolic and Molecular Bases of Inherited Disease, 8th ed. McGraw-Hill, New York, 2001.

Dryja TP et al: A point mutation of the rhodopsin gene in one form of retinitis pigmentosa. Nature 343: 364–366, 1990.

Phelan JK, Bok D: A brief review of retinitis pigmentosa and the identified retinitis pigmentosa genes. Mol Vis 6: 116–124, 2000.

Rattner A, Sun H, Nathans J: Molecular genetics of human retinal diseases. Ann Rev Genet 33: 89–131, 1999.

Rivolta C et al: Retinitis pigmenosa and allied diseases: numerous diseases, genes, and inheritance patterns. Hum Mol Gent 11: 1219–1227, 2002.

Wright AF: New insights into eye disease. Trends Genet 8: 85–91, 1992.

Online information:

Retinal Information Network: www.sph.uth.tmc.edu/Retnet

Retina International: www.retina-international.com/sci-news/database.htm

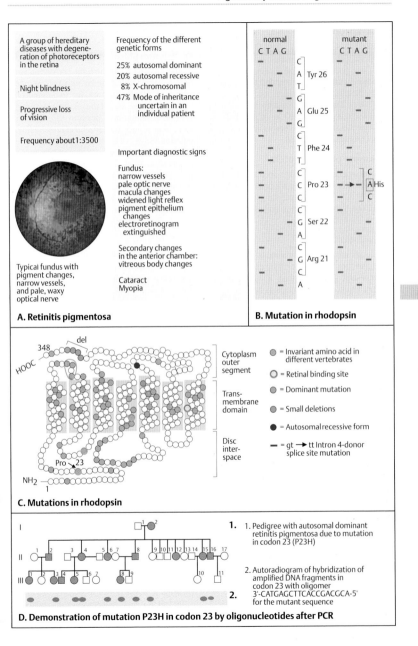

A. Retinitis pigmentosa

A group of hereditary diseases with degeneration of the photoreceptors in the retina

Night blindness

Progressive loss of vision

Frequency about 1:3500

Typical fundus with pigment changes, narrow vessels, and pale, waxy optical nerve

Frequency of the different genetic forms

25% autosomal dominant
20% autosomal recessive
8% X-chromosomal
47% Mode of inheritance uncertain in an individual patient

Important diagnostic signs

Fundus:
narrow vessels
pale optic nerve
macula changes
widened light reflex
pigment epithelium changes
electroretinogram extinguished

Secondary changes in the anterior chamber:
vitreous body changes

Cataract
Myopia

B. Mutation in rhodopsin

normal   mutant
C T A G   C T A G

C
A Tyr 26
T
G
A Glu 25
G
C
T Phe 24
T
C
C Pro 23 → C A His C
C
C
G Ser 22
A
C
G Arg 21
C
A

C. Mutations in rhodopsin

348 del
HOOC

Cytoplasm outer segment

Trans-membrane domain

Disc inter-space

Pro 23
NH2 1

○ = Invariant amino acid in different vertebrates
○ = Retinal binding site
● = Dominant mutation
● = Small deletions
● = Autosomal recessive form
— = gt → tt Intron 4-donor splice site mutation

D. Demonstration of mutation P23H in codon 23 by oligonucleotides after PCR

1.
I
II
III

2.

1. Pedigree with autosomal dominant retinitis pigmentosa due to mutation in codon 23 (P23H)

2. Autoradiogram of hybridization of amplified DNA fragments in codon 23 with oligomer 3'-CATGAGCTTCACCGACGCA-5' for the mutant sequence

Color Vision

In 1802, Thomas Young suggested that color vision in humans is trichromatic: red, green, and blue. Cone cells in the retina contain three types of photoreceptors, which respond to long-wave (red), middle-wave (green) or short-wave (blue) light. The receptor for blue is encoded by an autosomal gene located at 7q31.3-q32; the genes for red and green receptors are located on the X chromosome at Xq28.

A. Human photoreceptors

The overlapping absorption spectra of the three photoreceptors for color have peaks at 420 nm wavelength for blue, 530 nm for green, and 560 nm for red. Subtle variations in color perception exist in the red region.

B. Evolution

The genes for the visual pigment photoreceptors have arisen in evolution by duplication of an ancestral gene encoding a light-sensitive protein. About 800 million years ago, an ancestral visual pigment diverged by a duplication event into the rod pigment rhodopsin and another, not yet differentiated cone pigment. A short-wavelength-responsive (blue) pigment gene and a single mid-wavelength (green-red) gene diverged about 500 million years ago following another duplication event. The rhodopsin–transducin pair was present in vertebrates that lived about 700 million years ago. About 30–40 million years ago, after the separation of the New World monkeys from the Old World monkeys, the single green-red pigment gene duplicated. The two copies then evolved into the green (middle-wave length) and the red (long-wave length) pigment genes. This provided the Old World monkeys and humans with tricolor vision, whereas New World monkeys have dichromatic (blue and green-red) vision.

C. Structural similarity

The four photopigments have identical structures, as heptahelical (seven helices) transmembrane proteins, and similar sequences of amino acids. J. Nathans and co-workers sequenced the genes for color photoreceptors in 1986. The percentage of sequence identity of the gene products is shown here. Open circles indicate variant amino acids present in all; dark circles indicate differences in amino acid sequences. (Figure adapted from Nathans et al., 1986.)

D. Polymorphism in the photoreceptor

Subtle differences in red light perception were detected by Motulsky and co-workers (Winderickx et al., 1992). They are due to polymorphic variants at three sites of the receptor (**1**): serine/alanine at amino acid position 180, isoleucin/threonine at position 230, and alanine/serine at position 233 (**2**). Of the male population studied, 60% had serine at position. Alanine was present at positions 180 in 36%. The distribution of red light perception exhibited by the color-mixing test procedure of Raleigh showed a difference for the red/red and green mix distribution, depending on whether serine or alanine was present (**3**). (Figure adapted from Winderickx et al., 1992.)

E. Defects in color vision

Inherited defects in color vision in humans have been observed for more than 200 years and are mentioned in the Talmud. X-linked red-green color vision defects (MIM 303800/303900) affect about 8% of males. This high frequency can be explained by the tandem arrangement of the genes for red and green pigment (**1**).

Owing to the sequence similarity, unequal crossing-over occurs frequently in the intergenic region (about 15 kb), which results in deletion/duplication and different degrees of red-green vision defects (**2**).

Other vision defects involve the blue pigment (tritanopia, MIM 190900) or cause complete loss of color vision, achromatopsia (MIM 216900).

References

Motulsky AG, Deeb SS: Color vision and its genetic defects, pp 5955–5976. In: CR Scriver et al (eds) The Metabolic and Molecular Bases of Inherited Disease, 8th ed. McGraw-Hill, New York, 2001.

Nathans J, Thomas D, Hogness DS: Molecular genetics of human color vision: the genes encoding blue, green, and red pigments. Science 232:193–202, 1986.

Winderickx J et al: Polymorphism in red photopigment underlies variation in colour matching. Nature 356: 431–433, 1992.

Wissinger B, Sharpe LT: New aspects of an old theme: The genetic basis of human color vision. Am J Hum Genet 63: 1257–1262, 1998.

Young T: On the theory of light and colours. Phil Trans Royal Soc London 92: 12–48, 1802.

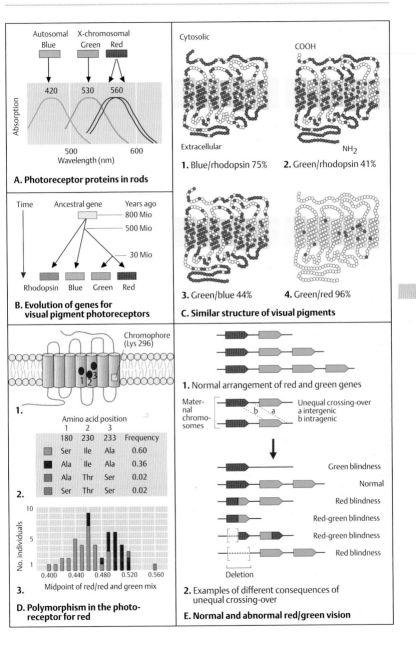

A. Photoreceptor proteins in rods

420   530   560

Wavelength (nm)

B. Evolution of genes for visual pigment photoreceptors

Time   Ancestral gene   Years ago

800 Mio
500 Mio
30 Mio

Rhodopsin   Blue   Green   Red

Cytosolic
COOH
Extracellular   NH₂

1. Blue/rhodopsin 75%   **2.** Green/rhodopsin 41%

3. Green/blue 44%   **4.** Green/red 96%

C. Similar structure of visual pigments

Chromophore (Lys 296)

1.

Amino acid position

	1	2	3	Frequency
	180	230	233	
	Ser	Ile	Ala	0.60
	Ala	Ile	Ala	0.36
	Ala	Thr	Ser	0.02
	Ser	Thr	Ser	0.02

2.

No. individuals

0.400 0.440 0.480 0.520 0.560

3. Midpoint of red/red and green mix

D. Polymorphism in the photoreceptor for red

1. Normal arrangement of red and green genes

Maternal chromosomes

Unequal crossing-over
a intergenic
b intragenic

b   a

Green blindness
Normal
Red blindness
Red-green blindness
Red-green blindness
Red blindness

Deletion

2. Examples of different consequences of unequal crossing-over

E. Normal and abnormal red/green vision

Auditory System

Acoustic signals are essential for communication and proper response to environmental situations. Normal hearing is orchestrated by a diverse ensemble of proteins acting in concert. Specialized sensory cells in the inner ear process the incoming sound waves and convert them into electrical signals. About 1 in 1000 individuals is affected by severe hearing impairment or deafness at birth or in early childhood (prelingual deafness, before development of speech). In 30% of the prelingual forms other organ systems are involved in addition to the ear (syndromic forms with hundreds of different types, see Petit et al., 2001 a). About 95% of deaf children are born to parents with normal hearing. Genetic early childhood deafness is due mainly to single gene mutations.

A. The main components of the ear

Sound waves induce vibrations in the tympanic membrane. They are transmitted from the middle ear to the inner ear by a chain of three small movable bones (malleus, incus, and stapes). In the cochlea, the organ of Corti, the acoustic signals are amplified and processed (auditory pathway). In addition, the inner ear harbors the vestibule with the three semicircular canals, utricle, and saccule, where the sense of equilibrium is regulated.

B. The cochlea

The cochlea is a snail-shaped structure that contains three fluid-filled canals, the scala vestibuli, the scala media, and the scala tympani. Between the endolymph of the scala media and the perilymph of the scala vestibuli and scala tympani, there is an electrical potential difference of about $+85\,mV$. Potassium ions (K^+) secreted from the stria vascularis of the scala media into the endolymph are recycled through supporting cells by K^+ channels and gap junctions (e.g., connexin 26 [Cx26]). The cochlea contains two types of sensory cell: one row of inner hair cells and three rows of outer hair cells. Sound-induced vibrations of the tectorial membrane deflect the stereocilia, open mechanosensitive channels, and cause an influx of K^+ ions. The altered membrane potential elicits an impulse at the acoustic nerve, which is transmitted to the auditory cortex of the brain. (Figure adapted from Willems, 2000, and Kubisch, 2005.)

C. Outer hair cells

The stereocilia of about 50 000 outer hair cells (1) are arranged in arrays resembling an organ pipe. Their tips are connected by tip links containing myosin and the cell adhesion molecule cadherin-23. The regular arrangement of the outer hair cells (2) is disrupted by mutations affecting the cytoskeleton (3). (Figures in 2 and 3 adapted from Self et al., 1998, and Kubisch, 2005.)

D. Congenital deafness

The more than 100 genes involved in hereditary deafness are localized on nearly every chromosome. About 75–85% of mutations are autosomal recessive (gene loci designated as DFNB1 to DFNB55 in 2005), 15% autosomal dominant (DFNA1-A48), and 1–2% X-chromosomal (DFN 1–3). The more than 40 identified genes encode a variety of proteins of the acoustic pathway (table on deafness in the appendix). The most frequent mutations are in Cx26 (DFNA3, DFN B1), in 20–50%. Individuals carrying certain mtDNA mutations are highly sensitive to antibiotics of the aminoglycoside class (e.g., streptomycin).

References

Kubisch C: Genetische Grundlagen nichtsyndromaler Hörstörungen. Dtsch Ärztebl 102 A: 2946–2952, 2005.

Petersen MB, Willems PJ: Non-syndromic, autosomal recessive deafness. Clin Genet 69: 371–392, 2006.

Petit C et al: Hereditary hearing loss, pp 6281–6328. In: Scriver CR, Beaudet AL, Sly WS, Valle D (eds) The Metabolic and Molecular Bases of Inherited Disease, 8th ed. McGraw-Hill, New York, 2001 a.

Petit C, Levilliers J, Hardelin JP: Molecular genetics of hearing loss. Ann Rev Genet 35: 589–646, 2001 b.

Smith RJH, Bale JF jr, White KR: Sensorineural hearing loss in children. Lancet 365: 879–890, 2005.

Toriello H, Reardon W, Gorlin RJ (eds) Hereditary Hearing Loss and its Syndromes, 2nd ed. Oxford University Press, Oxford, 2004.

Willems PJ: Genetic causes of hearing loss. New Engl J Med 342: 1101–1109, 2000.

Online information:
http://deafness.about.com/od/medicalcauses/a/genetics.htm
http://webhost.ua.ac.be/hhh/
http://www.iurc.montp.inserm.fr/cric/audition/english/start2.htm

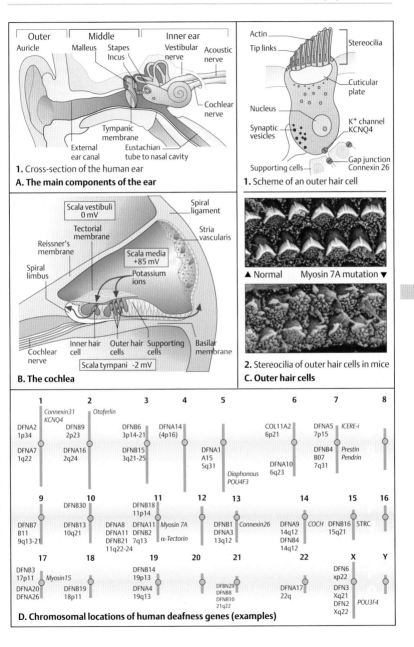

A. The main components of the ear

1. Cross-section of the human ear

Outer — Auricle

Middle — Malleus, Stapes, Incus

Inner ear — Vestibular nerve, Acoustic nerve, Cochlear nerve

Tympanic membrane, External ear canal, Eustachian tube to nasal cavity

1. Scheme of an outer hair cell

Actin, Tip links, Stereocilia, Cuticular plate, Nucleus, K⁺ channel KCNQ4, Synaptic vesicles, Supporting cells, Gap junction Connexin 26

B. The cochlea

Scala vestibuli 0 mV, Tectorial membrane, Reissner's membrane, Spiral limbus, Scala media +85 mV, Potassium ions, Spiral ligament, Stria vascularis, Inner hair cell, Outer hair cells, Supporting cells, Basilar membrane, Cochlear nerve, Scala tympani -2 mV

C. Outer hair cells

▲ Normal   Myosin 7A mutation ▼

2. Stereocilia of outer hair cells in mice

D. Chromosomal locations of human deafness genes (examples)

Odorant Receptors

Vertebrates are able to differentiate thousands of individual odors by means of specific receptors on the cilia of olfactory neurons (odorant receptors, ORs). OR genes have arisen in evolution by many duplication events. Genes of the OR family form the largest family of genes known in mammals, accounting for about 3–4% of all genes. The mammalian genome contains about 1000 OR genes, fish about 100. The rat genome contains 1866 ORs in 113 locations, consisting of 65 multigene clusters and 44 single genes. In humans about 60% are pseudogenes.

A. Sensory olfactory nerve cells

The peripheral olfactory neuroepithelium of the nasal mucous membrane consists of three cell types: olfactory sensory neurons with axons leading to the olfactory bulb, supporting cells, and basal cells. The latter serve as stem cells that replace olfactory sensory neurons. Each olfactory neuron is bipolar, with olfactory cilia in the lumen of the nasal mucous membrane and a projection to the olfactory bulb. From there odorant-induced signals are transmitted via the olfactory nerve to the brain.

B. Odorant-specific receptor

The odorant-specific receptor is a GTP-binding protein with a specific stimulatory α-subunit, the G_{olf}. Binding of the odorant ligand to the receptor activates G_{olf}, which in turn activates adenylate cyclase. The increase in cAMP (cyclic $3',5'$-adenosine monophosphate) opens a cAMP-gated ion channel, which depolarizes the cell membrane and elicits a nerve signal. Each receptor in the cilia of the olfactory neurons binds specifically to one odorant ligand only. Thus, signal amplification in olfaction is fundamentally different from phototransduction.

C. The olfactory receptor protein

The odorant receptor is a typical seven transmembrane G protein-coupled protein. Unlike rhodopsin, the OR proteins contains many variable amino acids, especially in the fourth and fifth transmembrane domains, which is likely to be related to their function (Buck & Axel, 1991).

D. Exclusive gene expression

Only one allele of an OR gene is expressed and each gene is expressed in a few olfactory neurons only. Receptor-specific probes recognize only very few neurons in the olfactory epithelium of the catfish *(Ictalurus punctatus)*: probe 202 hybridizes to two neurons (two black dots, **1**); probe 32 hybridizes to one neuron (**2**). Olfactory neurons are randomly distributed in the periphery, but their axons project to defined locations in the olfactory bulb. Odors are distinguished in the brain according to which neurons are stimulated. (Figure adapted from Ngai et al., 1993.)

E. Subfamilies within the OR family

Olfactory receptor genes form a large family of related genes. Amino acid sequences derived from partial nucleotide sequences of cDNA clones (F2–F24) by Buck & Axel (1991) are variable, especially in transmembrane domains III and IV (**1**). Within a subfamily, considerable amino acid sequence identity is present. For example, families F12 and F13 differ in only 4 of 44 positions, corresponding to 91% sequence identity (**2**). This reflects the ability to distinguish a great number of slightly different odorants. (Figures adapted from Buck & Axel, 1991, and Ngai et al., 1993.)

Medical relevance

Olfactory dysfunction affects about 1% of the population below the age of 60 (Lalwani & Snow, 2005). Anosmia (lack of ability to smell) is associated with hypogonadotropic hypogonadism due to gonadotropin-releasing hormone deficiency in the Kallmann syndrome, due to X-chromosomal (MIM 308700), autosomal dominant (MIM 147950), and autosomal recessive (MIM 244200) mutations.

References

Buck L, Axel R: A novel multigene family may encode odorant receptors: a molecular basis for odor recognition. Cell 65:175–187, 1991.

Emes RD et al: Evolution and comparative genomics of odorant- and pheromone-associated genes in rodents. Genome Res 14: 591–602, 2004.

Lalwani AK, Snow JB: Disorders of smell, taste, and hearing, pp 176–185. In: Kasper DL et al (eds) Harrison's Principles of Internal Medicine, 16th ed. McGraw-Hill, New York, 2005.

Ngai J et al: The family of genes encoding odorant receptors in the channel catfish. Cell 72: 657–666, 1993.

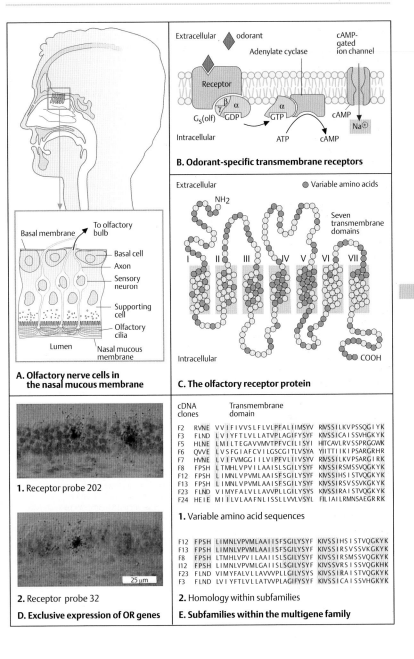

B. Odorant-specific transmembrane receptors

Extracellular — odorant — Adenylate cyclase — cAMP-gated ion channel

Receptor

G_s(olf) / GDP — β / γ / α — α / GTP — cAMP — Na⊕

Intracellular — ATP — cAMP

A. Olfactory nerve cells in the nasal mucous membrane

Basal membrane — To olfactory bulb — Basal cell — Axon — Sensory neuron — Supporting cell — Olfactory cilia — Lumen — Nasal mucous membrane

C. The olfactory receptor protein

Extracellular — ● Variable amino acids

NH₂ — Seven transmembrane domains

I II III IV V VI VII

Intracellular — COOH

D. Exclusive expression of OR genes

1. Receptor probe 202

2. Receptor probe 32 — 25 μm

E. Subfamilies within the multigene family

cDNA clones	Transmembrane domain		
F2	RVNE	VVIFIVVSLFLVLPFALIIMSYV	RIVSSILKVPSSQGIYK
F3	FLND	LVIYFTLVLLATVPLAGIFYSF	KIVSSICAISSVHGKYK
F5	HLNE	LMILTEGAVVMVTPFVCILISYI	HITCAVLRVSSPRGGWK
F6	QVVE	LVSFGIAFCVILGSCGITLVSYA	YIITTIIKIPSARGRHR
F7	HVNE	LVIFVMGGIILVIPFVLIIVSYV	RIVSSILKVPSARGIRK
F8	FPSH	LTMHLVPVILAAISLSGILYSYF	KIVSSIRSMSSVQGKYK
F12	FPSH	LIMNLVPVMLAAISFSGILYSYF	KIVSSIHSISTVQGKYK
F13	FPSH	LIMNLVPVMLAAISFSGILYSYF	KIVSSIRSVSSVKGKYK
F23	FLND	VIMYFALVLLAVVPLLGILYSYS	KIVSSIRAISTVQGKYK
F24	HEIE	MIILVLAAFNLISSLLVVLVSYL	FILIAILRMNSAEGRRK

1. Variable amino acid sequences

F12	FPSH	LIMNLVPVMLAAIISFSGILYSYF	KIVSSIHSISTVQGKYK
F13	FPSH	LIMNLVPVMLAAIISFSGILYSYF	KIVSSIRSVSSVKGKYK
F8	FPSH	LTMHLVPVILAAIISLSGILYSYF	KIVSSIRSMSSVQGKYK
I12	FPSH	LIMNLVPVMLGAIISLSGILYSYF	KIVSSVRSISSVQGKHK
F23	FLND	VIMYFALVLLAVVPLLGILYSYS	KIVSSIRAISTVQGKYK
F3	FLND	LVIYFTLVLLATVVPLAGIFYSF	KIVSSICAISSVHGKYK

2. Homology within subfamilies

Mammalian Taste Receptors

The ability to distinguish different chemical substances by taste benefits feeding behavior by helping the organism select favorable and avoid noxious substances. A sweet taste will indicate desirable carbohydrate content, whereas a bitter taste is associated with toxic substances such as alkaloids, cyanides, or certain aromatic compounds. Humans can distinguish five types of taste: sweet, sour, salt, bitter, and umani (the taste of monosodium glutamate, present in Asian food). Hydrogen ions (H^+) of acids are responsible for sour taste. The salty taste results from direct influx of sodium (Na^+) ions from water-soluble salts. In contrast, bitter, sweet, and umani tastes are mediated via G protein-coupled receptor (GPCR) signaling pathway systems. Especially bitter is perceived at very low concentrations.

A. Mammalian chemosensory epithelia

The oral and nasal cavities of mammals contain three distinct chemosensory epithelia: (i) the main olfactory epithelium (MOE) containing sensory cells with odorant receptors in the nose (see previous page), (ii) the taste sensory epithelium of the taste buds of the tongue, soft palate, and epiglottis, and (iii) the vomeronasal organ (VOM, also called Jacobson's organ). The latter is a tubular structure in the nasal septum containing sensory cells with pheromone receptors. The main olfactory bulb (MOB) relays signals from the MOE to the olfactory cortex of the brain. The accessory olfactory bulb (AOB) relays signals from the VOM to areas of the amygdala and hypothalamus.

B. Chemosensory signal transduction

The mammalian chemosensory receptor cells belong to one of three classes: (i) the olfactory system (see previous page), (ii) the sensory taste system, and (iii) the vomeronasal system, which detects signals arriving as pheromones. Each neuron of the main olfactory sensory system (**1**) sends an axon to specific glomeruli (mitral cells), and from there to the main olfactory bulb and the olfactory nerve. The odorant receptor (OR) gene family comprises about 1000 members, each encoding a seven-transmembrane cyclic nucleotide-gated channel with distinct odorant specificity (G-olfactory proteins, G_{olf}). The bitter taste sensory

system (**2**) connects axonal projections of receptor cells in the taste sensory epithelium of the taste buds to gustatory nuclei in the brain stem. Two families of taste receptors exist, the TIRs (two genes) and T2 Rs (50–80 genes of the gustducin class). The putative mammalian pheromone receptors (V1 Rs and V2 Rs), located in the vomeronasal organ, form two families, encoded by 30–50 and over 100 genes, respectively (**3**).

C. Taste receptor gene family

The figure shows marked sequence differences between the second (TM2) and third (TM3) transmembrane domains of the predicted amino acid sequences of 23 different T2 receptors (T2 Rs) of human (h), rat (r), and mouse (m) origin. Many sequences have been conserved between TM1 and TM2. Underlying dark blue indicates identity in at least half of the aligned sequences; light blue represents conserved substitutions; and the remaining are divergent regions. The T2 R genes cluster on a few chromosomes, human chromosomes 5, 7, and 12 and mouse chromosomes 6 and 15.

D. Expression pattern

Unlike receptor cells of the olfactory system, individual taste receptor cells express multiple receptors (T2 R). Up to ten T2 R probes hybridize to just a few cells, shown as darkened areas (**1**). Double-label fluorescence *in-situ* hybridization shows that different receptor genes are expressed in the same taste receptor cell, shown in green for T2 R-7 (**2**) and in red for T2 R-3 (**3**). The T2 Rs confer high sensitivity for bitter substances at low concentrations. (Figures adapted from Dulac, 2000, and Adler et al., 2000.)

References

Adler E et al: A novel family of mammalian taste receptors. Cell 100: 693–702, 2000.

Buck LB: The molecular architecture of odor and pheromone sensing in mammals. Cell 100: 611–618, 2000.

Chandrashekar J et al: T2 Rs function as bitter taste receptors. Cell 100: 703–711, 2000.

Dulac C: The physiology of taste, vintage 2000. Cell 100: 607–610, 2000.

Emes RD et al: Evolution and comparative genomics of odorant- and pheromone-associated genes in rodents. Genome Res 14: 591–602, 2004.

Malnic B et al: Combinatorial receptor codes for odors. Cell 96: 713–723, 1999.

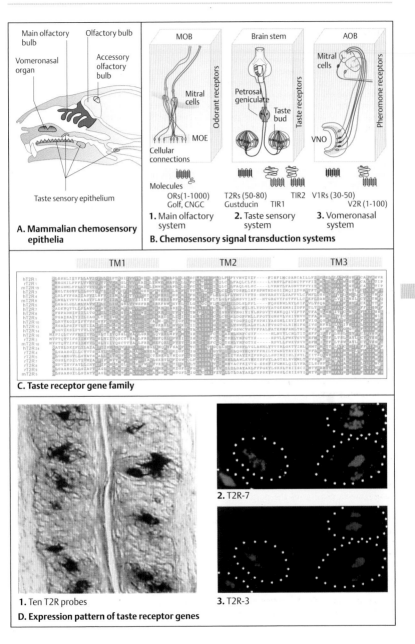

A. Mammalian chemosensory epithelia

B. Chemosensory signal transduction systems

1. Main olfactory system
2. Taste sensory system
3. Vomeronasal system

C. Taste receptor gene family

D. Expression pattern of taste receptor genes

1. Ten T2R probes
2. T2R-7
3. T2R-3

Embryonic Development in *Drosophila*

Different genes determine the early development of the *Drosophila* embryo. They are expressed at defined stages of development and determine the body plan pattern of the growing embryo. First, anterior–osterior and dorsal–ventral polarity is determined, then the segmental pattern of the embryo, and finally the head, the thorax with legs and antennae, and the abdomen. Homologous genes direct the developmental pattern of the mammalian embryo.

A. Life cycle of the fruit fly

The development from a fertilized egg (about 0.5×0.15 mm) to the adult fly (2 mm) takes 9 days. Nine nuclear divisions every 8 minutes without cell division leave the nuclei in a syncytium. Following the ninth nuclear division, 90 minutes after fertilization, the nuclei migrate to the periphery and form a *syncytial blastoderm*. After another four nuclear divisions, plasma membranes grow from the periphery and enclose each nucleus. This establishes the *cellular blastoderm* of about 6000 cells. Up to this stage, the embryo depends largely on maternal mRNA and proteins present before fertilization. The embryo passes through three defined larval stages before forming a pupa cocoon. After 5 days of metamorphosis, an adult fly emerges. (Figure adapted from Carolina Biological Supply Company.)

B. Segmental organization

The adult fly is organized into 14 segments: three segments (C1–3) form the head, three the thorax (T1–3), and eight (A1–8) the abdomen. Each segment has an anterior and a posterior compartment. Each of the initially formed 14 parasegments consists of the posterior compartment of the preceding segment and the anterior compartment of the following segment.

C. Embryonic developmental genes

The genes determining *Drosophila* embryonic development are classified according to their role in development. This is recognized in embryonic fatal mutations produced by inducing mutations in male flies with ethylmethane sulfonate (EMS). Males carrying random mutations are mated with females heterozygous for various mutations. One quarter of the offspring of parents with the same mutations will be homozygous for a mutation and can be analyzed for its phenotypic effects in the embryo (genetic screen). The mutations have colorful names derived from the appearance of the mutant embryo, often applied in the untranslated language of the discoverers.

Three classes of developmental genes function in a hierarchical order: (i) *egg-polarity* (*maternal effect*) genes defining anterior–posterior and dorsal–ventral polarity in the egg and early embryo, (ii) genes determining the cell fate to induce segmentation (*gap* genes and *pair-rule* genes), and (iii) genes determining the structure of body parts after the segmental boundaries are established (*homeotic selector* genes, see next page). The normal embryo shows the region for the head, three segments for the thorax, and eight for the abdomen (**1**). The mutation *bicoid* in an egg-polarity (maternal effect) gene results in lack of anterior parts (**2**). Mutations in one of about nine *gap* genes induce irregular segmentation, as in *Krüppel* (**3**) and *Knirps* (**4**). Characteristic mutant phenotypes of the eight *pair-rule* genes are *even-skipped* (**5**) and *fushi tarazu* (**6**) with loss of portions of alternate segments (*fushi tarazu is* Japanese for too few segments). More than 10 *segment polarity* genes determine the anterior–posterior polarity of each segment, e.g. the mutant *gooseberry* (**7**). Segment polarity genes encode proteins in the Wingless and Hedgehog signal transduction pathways (see p. 274). *Homeotic selector* genes (**8**) determine the ultimate fate of each segment. In the mutant *antennapedia (Ant)*, the antenna is replaced by a leg (homeotic leg).

References

Gilbert SF: Developmental Biology 7th ed. Sinauer, Sunderland, Massachusetts, 2003.

Lawrence PA: The Making of a Fly. The Genetics of Animal Design. Blackwell Scientific, Oxford, 1992.

Nüsslein-Volhard C, Wieschaus E: Mutations affecting segment number and polarity in *Drosophila*. Nature 287: 795–801, 1980.

Nüsslein-Volhard C, Frohnhöfer HG, Lehmann R: Determination of anterior–posterior polarity in *Drosophila*. Science 238: 1675–1681, 1987.

Wolpert L et al: Principles of Development. Current Biology & Oxford University Press, Oxford, 1998.

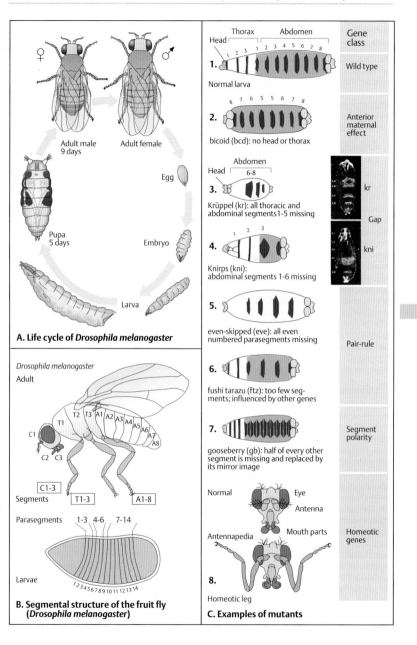

A. Life cycle of *Drosophila melanogaster*

Adult male
9 days

Adult female

Egg

Embryo

Larva

Pupa
5 days

**B. Segmental structure of the fruit fly
(*Drosophila melanogaster*)**

Drosophila melanogaster
Adult

C1

C2 C3

T1 T2 T3 A1 A2 A3 A4 A5 A6 A7 A8

Segments   C1-3   T1-3   A1-8

Parasegments   1-3   4-6   7-14

Larvae   1 2 3 4 5 6 7 8 9 10 11 12 13 14

C. Examples of mutants

		Gene class
Head / Thorax / Abdomen	**1.** Normal larva	Wild type
2. bicoid (bcd): no head or thorax		Anterior maternal effect
3. Krüppel (kr): all thoracic and abdominal segments 1-5 missing		kr — Gap
4. Knirps (kni): abdominal segments 1-6 missing		kni
5. even-skipped (eve): all even numbered parasegments missing		Pair-rule
6. fushi tarazu (ftz): too few segments; influenced by other genes		
7. gooseberry (gb): half of every other segment is missing and replaced by its mirror image		Segment polarity
8. Normal / Antennapedia / Homeotic leg		Homeotic genes

Thorax   Abdomen

Head

1. Normal larva — Wild type

8 7 6 5 5 6 7 8

2. bicoid (bcd): no head or thorax — Anterior maternal effect

Abdomen

Head   6-8

3. Krüppel (kr): all thoracic and abdominal segments 1-5 missing

kr
Gap

4. Knirps (kni): abdominal segments 1-6 missing

kni

5. even-skipped (eve): all even numbered parasegments missing — Pair-rule

6. fushi tarazu (ftz): too few segments; influenced by other genes

7. gooseberry (gb): half of every other segment is missing and replaced by its mirror image — Segment polarity

Normal   Eye
Antenna

Antennapedia   Mouth parts

8.
Homeotic leg — Homeotic genes

Homeotic Selector Genes (Hox genes)

The five principal pattern-determining gene systems in *Drosophila* development act in hierarchical order: (i) *egg-polarity* genes, (ii) *gap* genes, (iii) the *pair-rule* genes, (iv) *segment-polarity* genes, and (v) homeotic selector (*Hox*) genes. Vertebrate *Hox* genes are homologous to homeotic *Drosophila* genes.

A. Regulatory hierarchy of pattern-determining genes

The anterior and posterior system (bicoid, anterior and nanos, posterior) is generated from localized mRNA molecules of maternal origin. The proteins diffuse in a decreasing gradient from the anterior and the posterior pole, respectively. The dorso–ventral axis is determined by a transmembrane receptor called *Toll*. A fourth signal is generated at both ends, a transmembrane tyrosine kinase receptor called *Torso*. The three most important *gap* genes, *Krüppel*, *Hunchback*, and *Knirps*, determine local patterns. *Gap* genes induce the *pair rule* genes, such as *even-skipped* and *fushi-tarazu*. The *segment polarity* genes, expressed in parasegments, determine the correct anterior–posterior orientation of each individual segment. *Homeotic selector* genes determine the development of antennae, wings, legs, and other structures.

B. Homeotic selector genes

Homeotic selector genes form a complex called *Hox* genes (*HOX* in man). Each gene is assigned to the antennapedia cluster or the bithorax/ultrabithorax (*btx/ubx*) cluster. Five genes of the Antp complex are labial (*lb*), proboscis (*pb*), deformed (*dfd*), sex comb reduced (*scr*), and antennapedia (*antp*). *Drosophila* has one set of *Hox* genes, mammals have four. The mammalian *Hox* genes are not derived directly from the *Drosophila* genes. Instead, both share a single ancestral homeo selector gene complex. During the evolution of mammals it was duplicated twice, resulting in 39 genes occurring in four clusters, A–D, in man and mouse. In the process, some genes were lost, while a few genes were added. *HOX* genes are responsible for pattern formation along the body axis and are involved in mammalian limb development. They are sequentially expressed in anterior–posterior orientation from the 3′ end of the

cluster to the 5′ terminus, corresponding to their function. The human chromosomal locations are 7 p15 (*HOXA*), 17 q21-q22 (*B*), 12 q13 (*C*), and 2 q31-q32 (*D*). (Figure adapted from Alberts et al., 2002.)

C. The bithorax mutation

Mutations in the *bithorax* complex (*Bx-t*) induce the development of an additional thoracic segment with completely developed wings (Bridges in 1915). E.B. Lewis recognized in 1978 that the *bithorax* genes have evolved from a small number of ancestral genes by duplication and subsequent specialization into specific functions. (Photograph from Lawrence, 1992, after E. B. Lewis.)

D. The homeobox

Homeotic selector (*Hox*) genes encode gene regulatory proteins, e.g., transcription factors, and are highly conserved in evolution. The *antennapedia* gene contains a 180-bp conserved nucleotide sequence, the homeobox, from which the term Hox is derived. In the protein the corresponding 60 conserved amino acids constitute the homeodomain with four DNA-binding domains (I–IV), a common motif in transcription factors. Their expression depends on segment polarity genes.

Medical relevance

At least 20 human *HOX* genes (MIM 142950 ff) are involved in various disorders. Mutations in *HOXD13* cause synpolydactyly (MIM 142989).

References

Alberts B et al: Molecular Biology of the Cell, 4th ed. Garland Publishing Co., New York, 2002.

Garcia-Fernàndez J: The genesis and evolution of homeobox gene clusters. Nature Rev Genet 6: 881–892, 2005.

Gehring WJ et al: The structure of the homeodomain and its functional implications. Trends Genet 6: 323–329, 1990.

Krumlauf R: Hox genes in vertebrate development. Cell 78 : 191–201, 1994.

Lawrence PA: The Making of a Fly. The Genetics of Animal Design. Blackwell Scientific, Oxford, 1992.

Mark M, Rijli FM, Chambon P: Homebox genes in embryogenesis and pathogenesis. Pediat Res 42: 421–429, 1997.

Scott MP: A rational nomenclature for vertebrate homeobox (HOX) genes. Nucleic Acids Res 21: 1687–1688, 1993.

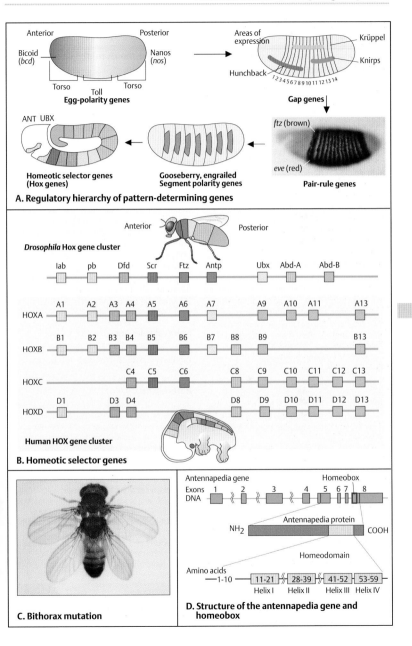

A. Regulatory hierarchy of pattern-determining genes

Anterior   Posterior

Bicoid
(*bcd*)   Nanos
                                 (*nos*)

Torso   Toll   Torso
Egg-polarity genes

Areas of
expression   Krüppel

Hunchback   Knirps
          1 2 3 4 5 6 7 8 9 10 11 12 13 14
                Gap genes

ftz (brown)

eve (red)

Pair-rule genes

ANT UBX

**Homeotic selector genes
(Hox genes)**

Gooseberry, engrailed
Segment polarity genes

B. Homeotic selector genes

Drosophila Hox gene cluster

Anterior   Posterior

Iab   pb   Dfd Scr Ftz Antp   Ubx Abd-A   Abd-B

        A1   A2   A3 A4 A5 A6   A7   A9   A10 A11   A13
HOXA

        B1   B2   B3 B4 B5 B6   B7 B8 B9   B13
HOXB

                      C4 C5 C6   C8   C9 C10 C11 C12 C13
HOXC

        D1   D3 D4   D8 D9 D10 D11 D12 D13
HOXD

Human HOX gene cluster

C. Bithorax mutation

**D. Structure of the antennapedia gene and
homeobox**

Antennapedia gene   Homeobox
Exons   1   2   3   4 5 6 7   8
DNA

                    Antennapedia protein
NH$_2$ COOH

                    Homeodomain

Amino acids
    1-10   11-21   28-39   41-52   53-59
          Helix I Helix II Helix III Helix IV

Genetics in a Translucent Vertebrate Embryo: Zebrafish

Mutations in the zebrafish *(Danio rerio)*, the first vertebrate studied by systematic genetic analysis, have revealed the roles of more than 1000 genes in early development (see Zebrafish issue: *Development* (123: 1–481, December 1996).

A. Embryonic stages

In the optically clear embryo (pharyngula period), the main parts of the brain (forebrain, midbrain, and hindbrain) and the neural tube, somites, and floor plate are discernible 29 hours after fertilization. At 48 hours (hatching period), pigmentation begins, and the fins, eyes, brain, heart, and other structures become visible. At five days (swimming larva), the outline of a fish becomes apparent.

B. Induced mutagenesis

Random mutations induced in male parents are studied in thousands of offspring at various embryonic stages, an approach called genetic screening. Zebrafish adult males are exposed to 3 mM ethylnitrosourea (ENU) in an aqueous solution. The mutagenized males are crossed with wild-type females (P), resulting in the first generation (F1) being heterozygous for mutations (m). Breeding the next generation (F2) results in 50% carrying at least one mutation. Random matings of parents heterozygous for the same mutation result in 25% homozygous mutant offspring. Two examples of mutations, affecting skeletal development and the brain, are described below.

C. Skeletal phenotype of the *fused somites (fss)* mutation

In wild-type fish, the somite anlagen result in the formation of a normal segmental pattern of the vertebrae and muscles of the trunk and the tail (**1**). Five mutants with abnormal somite boundaries have been identified (van Eeden et al., 1996). Four show essentially the same phenotype, with posterior somite defects and neuronal hyperplasia, while the fifth mutant, called *fused somites (fss)*, completely lacks somite formation along the entire anterior–posterior axis (**2**). Irregularly shaped spines grow ectopically at the wrong sites. The *fss* gene encodes a T-box transcription factor, Tbx24

(MIM 607044), required for presomitic mesoderm maturation (Nikaido et al., 2002).

D. The *no isthmus* midbrain mutation

This mutation, in the *noi* gene, is an example of the more than 60 distinct mutant phenotypes affecting the central nervous system and spinal cord in zebrafish (Haffter et al., 1996; Brand et al., 1996). Mutant *noi* embryos lack a conspicuous constriction at the boundary between midbrain and hindbrain (Brand et al., 1996). Normal 28-h embryos (*wt*, wild-type, **1**) show strong expression of the segment-polarity gene *engrailed* (*eng*, see p. 135) between the midbrain and hindbrain, whereas mutant *noi* embryos do not (**2**). Normal eight-somite-stage embryos double stained for eng and krx20 RNA, a marker for rhombomeres 3 and 5 in this region, express *eng* and *krox20* at the midbrain–hindbrain boundary (**3**), whereas *noi* mutants do not express *eng*. The *noi* mutation also eliminates expression of wingless (Wnt1) protein in the posterior tectum at the border of the two brain regions in the 20-somite-stage embryo (**5**, **6**). (Figures adapted from Haffter et al., van Eeden et al., and Brand et al., 1996.)

Medical relevance

Many human genes carrying disease-causing mutations have a homolog in the zebrafish genome.

References

Brand M et al: Mutations in zebrafish genes affecting the formation of the boundary between midbrain and hindbrain. Development 123: 179–190, 1996.

Dodd A et al: Zebrafish: bridging the gap between development and disease. Hum Mol Genet 9: 2443–2449, 2000.

Eeden FJM van et al: Mutations affecting somite formation and patterning in the zebrafish, *Danio rerio.* Development 123: 153–164, 1996.

Haffter P et al: The identification of genes with unique and essential functions in the development of the zebrafish, *Danio rerio.* Development 123: 1–36, 1996.

Nikaido M et al: Tbx24, encoding a T-box protein, is mutated in the zebrafish somite-segmentation mutant fused somites. Nature Genet 31: 195–199, 2002.

Online information:
The Zebrafish Information Network
    www.zfish.uoregon.edu/ and
    www.sanger.ac.uk/ Projects/

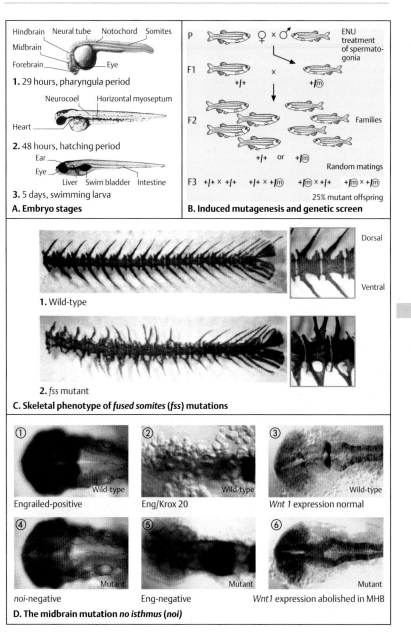

A. Embryo stages

Hindbrain Neural tube Notochord Somites
Midbrain
Forebrain Eye

1. 29 hours, pharyngula period

Neurocoel Horizontal myoseptum
Heart

2. 48 hours, hatching period

Ear
Eye
Liver Swim bladder Intestine

3. 5 days, swimming larva

B. Induced mutagenesis and genetic screen

P   ♀ × ♂   ENU treatment of spermatogonia

F1   +/+   +/⑩

F2   Families

+/+   or   +/⑩

Random matings

F3   +/+ × +/+   +/+ × +/⑩   +/⑩ × +/+   +/⑩ × +/⑩

25% mutant offspring

C. Skeletal phenotype of *fused somites* (*fss*) mutations

Dorsal
Ventral

1. Wild-type

2. *fss* mutant

D. The midbrain mutation *no isthmus* (*noi*)

① Engrailed-positive — Wild-type
② Eng/Krox 20 — Wild-type
③ *Wnt 1* expression normal — Wild-type
④ *noi*-negative — Mutant
⑤ Eng-negative — Mutant
⑥ *Wnt1* expression abolished in MHB — Mutant

Cell Lineage in a Nematode, *C. elegans*

Caenorhabditis elegans (*C. elegans*) is a small organism consisting of a precise number of somatic cells, each of which can be traced back to a founder cell. Sydney Brenner introduced this tiny worm as a model organism in 1965. Systematic genetic analysis of many mutant phenotypes has yielded important insights into the interaction of genetic, anatomical, and physiological traits in development. The complete cell lineage has been established in this organism. Each cell of the adult can be traced back to the cell from which it was derived.

The 97-Mb genome of *C. elegans* contains about 19000 protein-coding genes and over 1000 genes encoding untranslated RNA (CSC, 1998). About 32% of the coding sequences are homologous to sequences in man, and about 70% of known human proteins have homologies in *C. elegans*. The largest group of genes encodes transmembrane receptors (790), in particular chemoreceptors, zinc finger transcription factors (480), and proteins with protein-kinase domains. Interfering RNA (RNAi, see p. 224) has important functions in *C. elegans.*

A. Caenorhabditis elegans

C. elegans is a about 1 mm long, transparent worm with a life cycle of about 3 days. Its basic structure is a bilaterally symmetric elongated body of nerves, muscles, skin, and intestines. It exists as one of two sexes, hermaphrodite or male. Hermaphrodites produce eggs and sperm and can reproduce by self-fertilization. The adult hermaphrodite worm has 959 somatic cell nuclei; the adult male worm, 1031. In addition, there are 1000–2000 germ cells. (Figure adapted from Wood, 1988, after Sulston & Horvitz, 1977.)

B. Origin of individual cells

All tissues arise from six founder cells. At each cell division, genetically established rules determine the fate of the two daughter cells. Differentiated cells are derived from more than one founder cell, except for cells of the intestine and gonad. The gut is formed from a single founder (E) cell at the 8-cell stage and the gonad from a different founder cell (P_4) at the 16-cell stage. Of the 959 adult cells, 302 are nerve cells.

C. Developmental control genes

Many genes directing development have been identified by analysis of the mutations induced by ethyl methanesulfonate. The principal types of mutation cause cells to differentiate into incorrect cell types (e.g., Z instead of B); other cells divide too early or too late (division mutants).

D. Apoptosis in *C. elegans*

Programmed cell death (apoptosis) is a normal part of vertebrate and invertebrate development (see p. 126). During embryonic development of *C. elegans,* 131 of 947 nongonadal cells of the adult hermaphrodite undergo apoptosis at a defined time and branching point during different stages of development (**1**). Mutations in the *ced-9* gene induce apoptosis. Normally, *ced-9* suppresses apoptosis, and *ced-3* and *ced-4* are proapoptotic genes. Apoptosis does not occur in *ced-9/ced-3* double mutants because *ced-9* is upstream of *ced-3* in the apoptosis pathway. The photographs (**2**) show the death of a cell (cell p11.aap) over a time span of about 40 minutes. (Photographs from Wood, 1988, after Sulston & Horvitz, 1977.)

Medical relevance

The human *BCL-2* gene, a homolog of *ced-9*, encodes an inner mitochondrial membrane protein that inhibits apoptosis in pro-B lymphocytes. Disruption of this gene, located on human chromosome 18, causes follicular lymphoma, a B-cell tumor (MIM 151430).

References

Brenner S: The genetics of *Caenorhabditis elegans.* Genetics 77: 71–94, 1974.

CSC – The *C. elegans* Sequencing Consortium: Genome sequence for the nematode *C. elegans*: A platform for investigating biology. Science 282: 2012–2018, 1998.

Culetto E, Sattelle DB: A role for *Caenorhabditis elegans* in understanding the function and interactions of human disease genes. Hum Mol Genet 9: 869–878, 2000.

Jorgensen EM, Mango SE: The art and design of genetic screens: *C. elegans*. Nature Rev Genet 3: 356–369, 2002.

Wood WB and the Community of C. elegans Researchers: The Nematode *Caenorhabditis elegans.* Monograph 17, Cold Spring Harbor, New York, 1998.

Online information:
Wormbase at www.wormbase.org

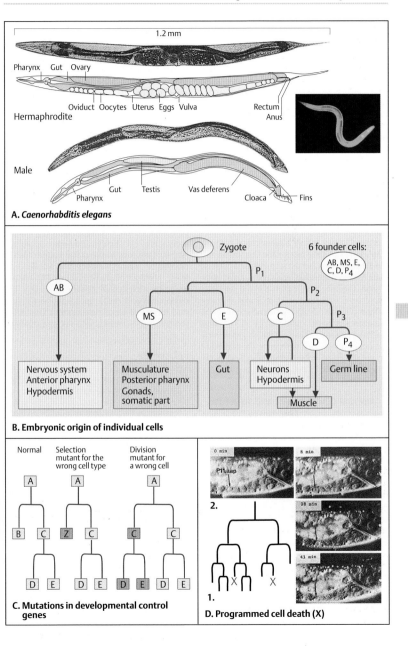

A. *Caenorhabditis elegans*

1.2 mm

Pharynx Gut Ovary

Oviduct Oocytes Uterus Eggs Vulva

Rectum
Anus

Hermaphrodite

Male

Gut   Testis   Vas deferens   Cloaca   Fins
Pharynx

Zygote

6 founder cells:
AB, MS, E, C, D, P₄

P₁

AB

MS   E   C

P₂

P₃

D   P₄

| Nervous system Anterior pharynx Hypodermis | Musculature Posterior pharynx Gonads, somatic part | Gut | Neurons Hypodermis | Germ line |

Muscle

B. Embryonic origin of individual cells

Normal

Selection mutant for the wrong cell type

Division mutant for a wrong cell

A   A   A

B   C   Z   C   C   C

D   E   D   E   D   E   D   E

C. Mutations in developmental control genes

0 min
P11aap

8 min

2.

28 min

X   X   X

41 min

1.

D. Programmed cell death (X)

Developmental Genes in a Plant, *Arabidopsis*

About 1.6 billion years separate animals from plants in evolutionary history. Derived from a common unicellular eukaryotic ancestor, they have evolved into completely different multicellular organisms. Yet their early developmental stages still are comparable. According to fossil evidence, flowering plants arose only 125 million years ago (compared with 350 million years for vertebrates). Systematic genetic analysis of the small flowering plant *Arabidopsis thaliana* (common wall cress) has provided insights into the principles of plant development. *Arabodopsis* is the first plant to be sequenced (AGI, 2000). Its small, compact genome (115 Mb DNA) contains about 25 498 genes (4–5 genes per 1 kb). The early developmental stages follow a hierarchical pattern. Four classes of gene induce the four organs of the flower, sepals, petals, stamens, and carpels. Class A gene determine sepals; class A and class B genes together petals; class B and C stamens; and class C carpels. The class A and C genes mutually inhibit one another. Genetic screens of mutagenized seed (induced by 0.3 % ethyl methane sulfonate) have identified genes with numerous alleles that determine the organization of the plant embryo along an apical–basal longitudinal axis, its radial pattern, and its form (Mayer et al., 1991).

A. Normal development and structure

The basic structural plan follows axial and radial patterns superimposed on each other. Before the seedling is formed, an octant stage, a globular stage, and a so-called heart stage can be distinguished. The regions A, C, and B of the octant stage correspond to the regions A, C, and B of the subsequent heart stage. Region A forms the cotyledon and the meristem, C forms the hypocotyl region, and B forms the root. The seedling consists of a set of seven identifiable structures including vessels (v), external epidermis (e), short meristem (s), cotyledons (c), with hypocotyl (h), ground tissue (g), and root primordium (r) at the bottom. The organizational pattern is predetermined at the heart stage. (Figure adapted from Mayer et al., 1991.)

B. Mutant phenotypes

Using complementation analysis, Mayer et al. (1991) determined mutant phenotypes in three areas of the plant, affecting the apical–basal pattern, the radial pattern, and the shape. Apical–basal mutations involve one of several genes, each leading to a characteristic phenotype: apical deletion to *Gurke*, central deletion to *Fackel,* basal deletion to *monopteros,* and terminal deletion to *Gnom.* (Figure adapted from Mayer et al., 1991.)

C. Wild type

The normal structure of embryonic *Arabidopsis* results from two basic processes: formation of patterns (apical–basal and radial orientation) and morphogenesis through different cell forms and regional differences in cell division.

D. Phenotypes of embryonic mutants

The four mutant phenotypes in the apical–basal pattern are *Gurke* (9 alleles), *Fackel* (5 alleles), monopteros (11 alleles), and *Gnom* (15 alleles) (see **B**). Deletions in the radial pattern lead to phenotypes *Keule* (9 alleles) and *Knolle* (2 alleles). Mutants of shape are *Fass* (12 alleles), *Knopf* (6 alleles), and *Mickey* (8 alleles). The *monopteros* gene *(ml)* is apparently very important for apical–basal development (Berleth & Jürgens, 1993). (Photographs in C and D: Mayer et al., 1991.)

References

AGI – The Arabidopsis Genome Initiative: Analysis of the genome sequence of the flowering plant *Arabidopsis thaliana.* Nature 408: 796–815, 2000.

Berleth T, Jürgens G: The role of the monopteros gene in organising the basal body region of the *Arabidopsis* embryo. Development 118: 575–587, 1993.

Friml J et al: Efflux-dependent auxin gradients establish the apical-basal axis of *Arabidopsis.* Nature 426: 147–153, 2003.

Mayer U et al: Mutations affecting body organization in the *Arabidopsis* embryo. Nature 353: 402–407, 1991.

Pelaz S et al: B and C floral organ identity functions require *SEPALLATA* MADS-box genes. Nature 405: 200–203, 2000.

Pennisi E: Plant Genomics: *Arabidopsis* comes of age. Science 290, 32–35, 2000.

Sommerville C, Koorneef M: A fortunate choice: the history of *Arabidopsis* as a model plant. Nature Rev Genet 3: 883–889, 2002.

Sommerville S: Plant functional genomics. Science 285: 380–383, 1999.

Online Information:
Arabidopsis Information Resource
www.arabidopsis.org/

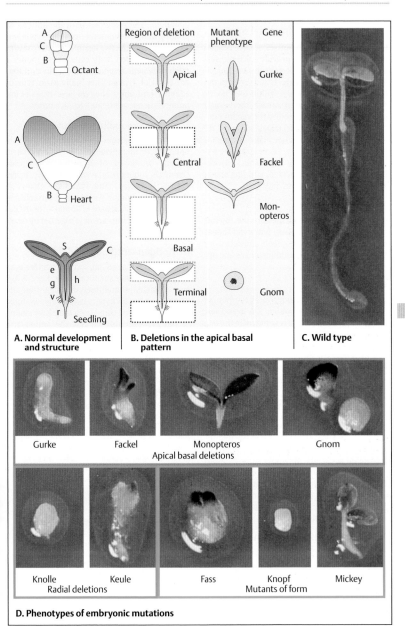

A. Normal development and structure

Octant

A
C
B

Heart

A
C
B

Seedling

S
e
g
h
v
r
C

B. Deletions in the apical basal pattern

Region of deletion	Mutant phenotype	Gene
Apical		Gurke
Central		Fackel
Basal		Mon-opteros
Terminal		Gnom

C. Wild type

D. Phenotypes of embryonic mutations

Gurke   Fackel   Monopteros   Gnom
Apical basal deletions

Knolle   Keule   Fass   Knopf   Mickey
Radial deletions   Mutants of form

Components of the Immune System

The immune system comprises organs, tissues, cells, and molecules that provide a defense system against invading microorganisms and viruses. *Innate immunity* detects and destroys microorganisms immediately, within four hours. Infectious organisms that breach these early lines of defense are repulsed by the *adaptive immune response*. Three principal methods are used by the host to inactivate and eliminate invading foreign molecules: (i) neutralizing extracellular pathogens by antibodies, (ii) destroying a cell that is infected, and (iii) killing bacteria directly by macrophages.

A. Lymphatic organs

The primary lymphoid tissues are the thymus and bone marrow. Secondary lymphoid tissues are the lymph nodes, the spleen, and accessory lymphoid tissue (ALT) including tonsils and appendix.

B. Lymphocytes

The about 2×10^{12} lymphocytes in the human body equal in mass the brain or the liver. Their role in adaptive immunity was shown in the late 1950s by irradiation experiments. A mouse irradiated above a certain dose was no longer able to mount an immune response. The immune response could be restored by lymphocytes from an unirradiated mouse.

C. T cells and B cells

Two functionally different types of lymphocyte exist, T lymphocytes and B lymphocytes. Immature T lymphocytes differentiate in the thymus during embryonic and fetal development (thus designated as T cells). B lymphocytes differentiate in the bone marrow in mammals and in the bursa of Fabricius in birds (thus designated as B cells). Further maturation and differentiation take place in the lymph nodes (T cells) and in the spleen (B cells).

D. Cellular and humoral immune response

The first phase of the immune response induced by an antigen (e.g., a bacterium, virus, fungus, or foreign protein) is rapid proliferation of B cells (*humoral immune response*). Mature B cells develop into plasma cells, which secrete effector molecules, the antibodies (immuno-globulins). These interact with the antigen by binding to it. The humoral immune response is rapid, but is ineffective against microorganisms that have invaded body cells. These induce a *cellular immune response*, carried out by different types of T cells. This is the main type of adaptive immunity. The two basic types of immune response are related to each other.

E. Immunoglobulin molecules

The basic structural motif of an antibody molecule (immunoglobulin, Ig) is a Y-shaped protein composed of different polypeptide chains. A common type of Ig has two heavy chains (H chains) and two light chains (L chains). Both contain regions with variable and with constant sequences of amino acids. At defined sites the chains are held together by disulfide bonds.

F. Antigen–antibody binding

The structure that an antibody recognizes is an antigenic determinant or epitope. Here a foreign molecule, the antigen, is recognized and firmly bound to six hypervariable regions (three from the light and three from the heavy chain). In the hypervariable regions the amino acid sequences differ from one molecule to the next. As a result the antibodies can bind a wide spectrum of different antigenic molecules.
(Figures adapted from Alberts et al., 2002.)

References

Abbas AK, Lichtman A: Cellular and Molecular Immunology, 5th ed. WB Saunders Company, Philadelphia, 2005.

Alberts B et al: Molecular Biology of the Cell, 4th ed. Garland Publ, New York, 2002.

Burmester G-R, Pezzutto A: Color Atlas of Immunology. Thieme Medical Publishers, Stuttgart–New York, 2003.

Haynes BF, Fauci AS: Introduction to the immune system, pp 1907–1930. In: Kasper DL et al (eds) Harrison's Principles of Internal Medicine, 16th ed. McGraw-Hill, New York, 2005.

Janeway CA, Travers P, Walport M, Shlomchik MJ: Immunobiology, 6th ed. The Immune System in Health and Disease. Garland Science, New York–London, 2005.

Nossal GJ: The double helix and immunology. Nature 421: 440–444, 2003.

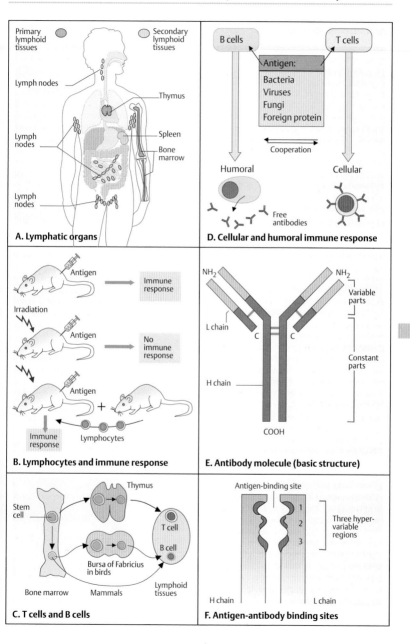

A. Lymphatic organs

Primary lymphoid tissues

Secondary lymphoid tissues

Lymph nodes

Thymus

Lymph nodes

Spleen

Bone marrow

Lymph nodes

D. Cellular and humoral immune response

B cells

T cells

Antigen:
Bacteria
Viruses
Fungi
Foreign protein

Cooperation

Humoral

Cellular

Free antibodies

B. Lymphocytes and immune response

Antigen

Immune response

Irradiation

Antigen

No immune response

Antigen

Immune response

Lymphocytes

E. Antibody molecule (basic structure)

NH_2

NH_2

Variable parts

L chain

C

C

Constant parts

H chain

COOH

C. T cells and B cells

Thymus

Stem cell

T cell

B cell

Bursa of Fabricius in birds

Bone marrow

Mammals

Lymphoid tissues

F. Antigen-antibody binding sites

Antigen-binding site

1
2
3

Three hyper-variable regions

H chain

L chain

Immunoglobulin Molecules

Immunoglobulins are the effector molecules of the immune system. They exist in two basic forms: as membrane-bound cell surface receptor molecules or as free antibodies, each in a vast array of variants. An important feature is the binding site for an antigen, a foreign molecule. The binding site contains regions that vary in their amino acid sequence among individual immunoglobulins and enable each molecule to specifically bind a particular epitope on a foreign protein. A vast number of different effector molecules provide spectacular diversity. Although they differ in details of their structure and function, they share a relatively simple basic pattern, which is derived from a common ancestral molecule.

A. Immunoglobulin G (IgG)

Immunoglobulin G is the prototype of secreted antibody molecules in humoral immunity. The molecule has two H chains and two L chains, held together by disulfide bonds. Each H chain has three constant regions (domains C_H1, C_H2, and C_H3) and one variable region (V_H) with a total of 440 amino acids (110 in the V region). Each L chain (214 amino acids) has one constant (C_L) and one variable (V_L) domain, also with 110 amino acids. The variable regions of both chains contain the antigen-binding site with three hypervariable regions called complementarity determining regions (CDRs), where the actual physical contact with foreign epitopes takes place. A region called hinge joins the constant region 1 (C_H1) and constant region 2 (C_H2) of the heavy chains. This allows considerable flexibility of the molecule. The two H chains, and the H and the L chains are held together by interchain disulfide bonds (–S–S–). Furthermore, there are intrachain disulfide bonds within each of the polypeptide chains. The L chains are of one of two types, kappa (κ) or lambda (λ).

Several types of immunoglobulin exist, IgA (C_α), IgD (C_δ), and IgE (C_ε), which differ from each other in the constant part of the H chain. The largest secreted immunoglobulin, IgM, exists as a pentamer of five Ig molecules. The different types of H chains are referred to as isotypes.

Two proteolytic enzymes cleave Ig molecules into characteristic fragments (**1**). Papain cleaves at the amino terminal disulfide bond (–S–S–) to produce three pieces, two Fab fragments (fragment antigen binding) and one Fc fragment (fragment crystallizable). Pepsin cleaves Ig into one F(ab')$_2$ fragment with an intact disulfide bridge and an Fc fragment (pFc') with several small pieces.

B. Structure of an immunoglobulin

Three globular domains of similar size form the Y-shaped antibody molecule (immunoglobulin, Ig). The three regions, called domains, are connected by a flexible tether, called hinge. The two heavy (H) chains are shown in yellow and blue, the two light (L) chains in red. The two antigen-binding sites are the two ends of the Y. (Figure adapted from Janeway et al., 2005.)

C. Genes encoding different polypeptides of an immunoglobulin

Each immunoglobulin and receptor molecule is encoded by different DNA sequences, which belong to a multigene family. The genes for the H chain are located on chromosome 14 at q32 in humans and chromosome 14 in mice. The genes for the κ light chain are located on chromosome 2 at p12 in humans and chromosome 6 in mice, and for the λ light chain on chromosome 22 at q11 in humans and on chromosome 16 in mice.

References

Abbas AK, Lichtman A: Cellular and Molecular Immunology, 5th ed. WB Saunders Company, Philadelphia, 2005.

Alberts B et al: Molecular Biology of the Cell, 4th ed. Garland Publ, New York, 2002.

Delves PJ, Roitt IM: The immune system. Two parts. New Engl J Med 343: 37–49 and 108–117, 2000.

Haynes BF, Fauci AS: Introduction to the immune system, pp 1907–1930. In: Kasper DL et al (eds) Harrison's Principles of Internal Medicine, 16th ed. McGraw-Hill, New York, 2005.

Janeway CA, Travers P, Walport M, Shlomchik MJ: Immunobiology, 6th ed. The Immune System in Health and Disease. Garland Science, New York–London, 2005.

Nossal GJ: The double helix and immunology. Nature 421: 440–444, 2003.

Strominger JL: Developmental biology of T cell receptors. Science 244: 943–950, 1989.

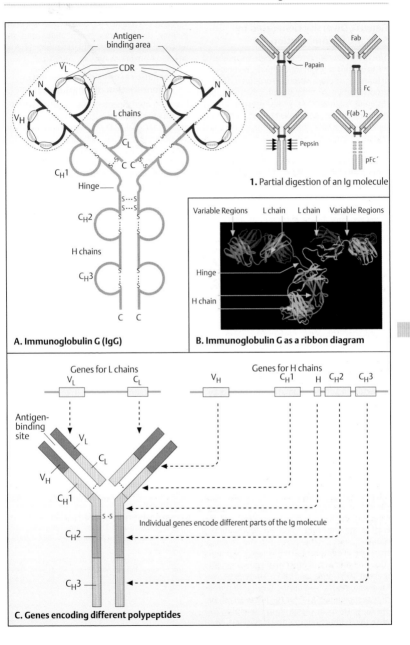

1. Partial digestion of an Ig molecule

A. Immunoglobulin G (IgG)

B. Immunoglobulin G as a ribbon diagram

Individual genes encode different parts of the Ig molecule

C. Genes encoding different polypeptides

Genetic Diversity Generated by Somatic Recombination

The vast array of different antigens encountered by an organism is met by an equally diverse repertoire of antigen-specific receptors. During differentiation of both B and T lymphocytes, different segments of genes present in the germline are randomly selected and arranged in a new combination that is specific for each cell and its progeny. This generates a great diversity of cells, each expressing a different antigen-binding molecule. Each cell expresses only one particular type of receptor. This is called *allelic exclusion*. The main mechanism is somatic recombination during the differentiation of B cells and T cells. For each domain of an Ig molecule a number of selectable sequences are present in genomic DNA. In lymphocyte DNA, these are combined in a variety of combinations for each molecule.

During differentiation of both B and T lymphocytes, diversity is achieved by a process called somatic recombination. Functional genes are produced by gene rearrangements of different elements (V, D, J) from the germline DNA. For each element, a number of selectable sequences are present. These are combined in a variety of combinations for each gene, encoding one of the polypeptide chains of the T cell or B cell receptor. This generates a great diversity of cells each expressing a different antigen-binding molecule. In addition, somatic mutations occur in the hypervariable regions, leading to further genetic differences.

A. Organization of immunoglobulin loci in the human genome

Both B cells and T cells pass through a series of defined steps of differentiation to reach their final stage. During this process, somatic recombination occurs within the gene loci encoding the two L chains and the H chain. The corresponding loci consist of coding sequences for each of the following segments: V (variable), J (joining), and C (constant). For the H chain, there are in addition 25 D (diversity) segments (D_H). The number of functional gene segments is about 30 V_λ, 40 V_κ, and 50 V_H. These gene segments are rearranged during B cell differentiation as shown in part B for the heavy (H) chain. The C_H genes form a large cluster spanning 200 kb in the 3′ direction from the J segments. Each

segment has a leader or signal sequence (L) for guiding the emerging polypeptides into the lumen of the endoplasmatic reticulum. (Figure adapted from Janeway et al., 2005.)

B. Somatic recombination during differentiation of lymphocytes

Here the rearrangements by somatic recombination are shown for the locus of the immunoglobulin H chain (**1**). The first rearrangement joins D and J segments in lymphocyte DNA (D–J joining). The next rearrangement brings one of the V_H genes together with the DJ segments joined previously, resulting in V–D–J joining (**3**). The result is a transcription unit consisting of one V_H, one D_H, one J, and one C gene arranged in this order in the 5′ to 3′ direction (**4**). Each C segment consists of different exons, corresponding to the domains of the complete C region and different isotypes (C_λ, C_δ), etc. The primary transcript is spliced to yield an mRNA consisting of one V, one D, one J, and one C segment (**5**). This is translated into a polypeptide corresponding to the complete H chain (**6**). Following posttranslational modifications such as glycosylation, the final H chain is produced (**7**).

The result is an array of cells each with a unique combination of molecules in the antigen-binding site. This provides each molecule with an antigen-binding specificity that differs from that of all other cells. The L chains and the genes encoding the T-cell receptor (see p. 316) are formed in a similar manner. In contrast to the H chains, the L chains have no diversity (D) genes, so that a J and a V gene are directly joined by somatic recombination during DNA rearrangement in the lymphocytes. (Figure adapted from Abbas & Lichtman, 1997.)

References

Abbas AK, Lichtman AH, Pober JS: Cellular and Molecular Immunology, 3rd ed. WB Saunders, Philadelphia, 2005.

Janeway CA, Travers P, Walport M, Shlomchik MJ: Immunobiology, 6th ed. The Immune System in Health and Disease. Garland Science, New York–London, 2005.

Schwartz, RS: Diversity of the immune repertoire and immunoregulation. New Eng J Med 348: 1017–1026, 2003.

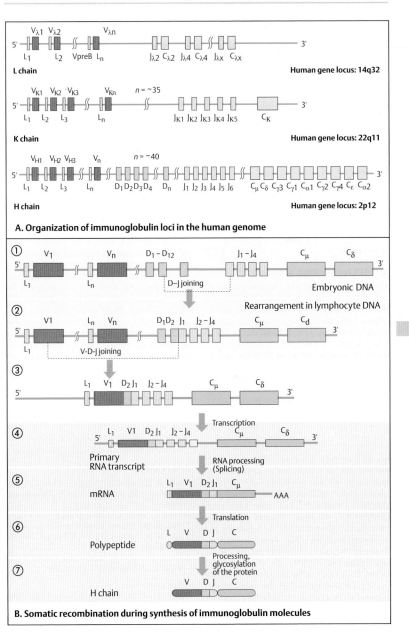

A. Organization of immunoglobulin loci in the human genome

B. Somatic recombination during synthesis of immunoglobulin molecules

Mechanisms in Immunoglobulin Gene Rearrangement

The rearrangement of genes for immunoglobulin molecules during differentiation must assure that a V gene segment joins a D or J segment, and not another V. Conserved noncoding DNA sequences located adjacent to the point of recombination guide this process. The recombinations of V, D, and J segments are carried out by lymphocyte-specific DNA-modifying enzymes called V(D)J recombinase. They mediate a special kind of non-homologous recombination encoded by *RAG1* and *RAG2* genes.

The lymphoid-specific genes *RAG1* and *RAG2* (recombination-activating genes) are expressed in pre-B cells and immature T cells. The other enzymes involved are DNA-modifying proteins required for double-stranded DNA repair, DNA bending, and ligation of the ends of broken DNA.

A. DNA recognition sequences

Recombination takes place between segments located on the same chromosome. A gene segment flanked by a recombination signal sequence (RSS) with a 12-bp spacer can only be joined to one flanked by a 23-bp spacer. This is referred to as the 12/23 rule. As a consequence, a D segment with a 12-bp spacer on both sides must be joined to a heavy chain J segment. Likewise a heavy chain V segment can be joined only to a D segment, but not to a J segment, because V and J segments are both flanked by 23-bp spacers.

The recognition sequences are located in adjacent noncoding DNA segments at the 3′ end of each V exon (variable region) and at the 5′ end of each J segment. The D segments are flanked on both sides by recognition sequences. Recognition sequences are noncoding, but highly conserved DNA segments of seven base pairs (CACAGTG, *heptamer*) and nine base pairs (ACAAAAACC, *nonamer*). Spacers of 23 base pairs (bp) or 12 bp separate the heptamer and nonamer sequences. The heptamer–spacer–nonamer sequences form the recombination signal sequences (RSS).

When an H chain is formed, nonhomologous pairing of the heptamer of a D segment and of a J segment occurs. Lymphocyte-specific recombinase enzymes, RAG-1 and RAG-2, recognize DNA sequences located at the 3′ end of each V exon and the 5′ end of each J exon. The RAG-1

and RAG-2 enzymes cut both strands of DNA at the recognition sites. The RGA proteins align the recognition site sequences and the endonuclease of the RAG complex cuts both strands of the DNA at the 5′ end. This creates a DNA hairpin in the gene segment-coding region. The D and J segments are then joined (D–J joining) by means of recombination: the spacer of 12 or 23 base pairs and all of the intervening DNA form a loop. This is excised, and the D and J segments are joined by the nonhomologous end-joining machinery. By pairing and recombination of the recognition sequences at the 5′ end of a DJ segment and the recognition sequence at the 3′ end of a V gene, a V segment is joined to the DJ segment. Diversity of T-cell receptors (see p. 316) is generated in the same manner.

B. Genetic diversity

The total diversity, of about 10^{18} possible combinations for all types of immunoglobulins and T-cell receptor genes, is the result of different mechanisms. To begin with, different numbers of variable DNA segments are available for different chains (250–1000 for the H chain, 250 for the L chains, 75 for the α chain of the T-cell receptor TCRα, etc.). The different D and J segments also multiply the number of possible combinations. Finally, DNA sequence changes (somatic mutations) occur regularly in the hypervariable regions, further increasing the total number of possible combinations.

Medical relevance

Mutations in the *RAG1* and *RAG2* genes (MIM 179615/16) cause severe combined immune deficiency (SCID) as well as the Omenn syndrome, due to defective V(D)J recombination.

References

Abbas AK, Lichtman AH: Cellular and Molecular Immunology, 5th ed. WB Saunders, Philadelphia, 2005.

Agrawal A, Schaz DG: RAG1 and RAG2 form a stable postcleavage synaptic complex with DNA containing signal end in V(D)J recombination. Cell 89: 43–53, 1997.

Schwartz K et al: RAG mutations in human B cell-negative SCID. Science 274: 97–99, 1996.

Online information:

Undergraduate Immunology Class at Davidson College, Davidson, NC 28035 at

www.bio.davidson.edu/Courses/Immunology/ Students/Spring2003/Beaghan/mfip.html

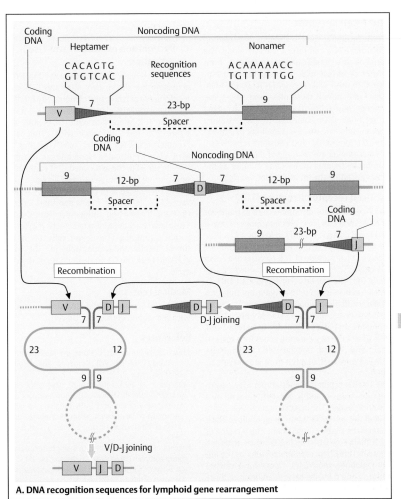

A. DNA recognition sequences for lymphoid gene rearrangement

Mechanism	Immunoglobulin		TCRαβ		TCRγδ	
	H chain	L chains	α	β	γ	δ
Variable domain	250 – 1000	250	75	25	7	10
Number of D segments	12	0	0	2	0	2
Number of J segments	4	4	50	12	2	2
Variable segment combinations	65 000 – 250 000		1825		70	
Total diversity	10^{11}		10^{16}		10^{18}	

B. Genetic diversity in immunoglobulin and T-cell receptor genes

The T-Cell Receptor

T-cells recognize foreign antigens displayed on the surface of a cell from their own body. The antigens are small peptides derived from viruses or intracellular bacteria. They are presented to the T-cell receptor (TCR) by the MHC class I and class II molecules (next page). The genes encoding the α/β and δ TCRs are arranged in germline DNA according to the segments they encode, i.e., variable segments (V), diversity (D), joining (J), and constant regions (C). Gene segments are rearranged during maturation in the thymus in the same manner as the genes encoding the immunoglobulins (Ig).

A. T-cell receptor structure (TCR)

The T-cell receptor resembles the Fab fragment of an Ig molecule. It is a heterodimer of one α and one β chain covalently linked by a disulfide bridge and expressed as an integral membrane protein (**1**). A subtype of TCR consists of a λ and a δ chain. The 3-dimensional structure (**2**) reveals that the hypervariable regions CDR1, 2, and 3 can be aligned with antigen-binding sites of the antibodies. Genomic DNA contains genes for 50–70 V segments, two D segments, 12–60 J segments, and two C segments. In a given T cell, only one of the two parental genes for a given α chain and β chain is expressed (allelic exclusion). (Figure in 2 adapted from Janeway et al., 2005.)

The T-cell receptor is a heterodimer of two polypeptide chains α and β, covalently bound by a disulfide bridge. The basic structure is similar to that of the cell surface immunoglobulins. The β chain is the slightly larger chain. The V region of each chain consists of 102–109 amino acids and contains three hypervariable regions, as the immunoglobulin molecules. Genes for T-cell receptor γ and δ chains exist in addition to those for α and β chains.

B. Interaction of TCR and MHC

The TCR interacts with an antigen-presenting cell carrying an MHC molecule, either an HLA class I or class II (**1**). The 3-dimensional structure (**2**) shows the tight connection between the interacting molecules. T cells able to destroy an infected cell (cytolytic T lymphocytes, CTLs, or "killer cells") recognize their antigen on MHC class I molecules, while "helper cell" T-cells specifically bind to MHC class II molecules. CD8 cells are restricted to MHC class I molecules; CD4 to MHC class II.

C. Recognition of antigen and T-cell activation

A few of the many molecules involved in T-cell activation are shown here. CD4 and CD8 molecules serve as restriction elements. By binding directly to the MHC class II molecule, CD4 stabilizes the TCR interaction with the peptide antigen. CD8 takes this role with cytolytic T lymphocytes, CTLs ("killer cells") binding to MHC class I molecules. Upon antigen recognition by the TCR, the associated CD3 complex is phosphorylated. Complete T cell activation requires the engagement of costimulatory receptors (CD28, LFA-1) on T cells and their ligands (B-7 and ICAM-1) on antigen-presenting cells. This signal transduction activates the *IL2* (Interleukin 2) gene. IL2 is the main T-cell growth factor of T-cells and responsible for the progression from the G1 to the S phase of the cell cycle.

Medical relevance

The gp120 protein of the HIV virus interacts with the second domain of CD4 cells.

References

Abbas AK, Lichtman AH: Cellular and Molecular Immunology, 5th ed. WB Saunders, Philadelphia, 2005.

Amadou C et al: Localization of new genes and markers to the distal part of the human major histocompatibility complex (MHC) region and comparison with the mouse: new insights into the evolution of mammalian genomes. Genomics 26: 9–20, 1995.

Fugger L et al: The role of human major histocompatibility complex (HLA) genes in disease, pp 311–341. In: Scriver CR et al (eds) The Metabolic and Molecular Bases of Inherited Disease, 8th ed. McGraw-Hill, New York, 2001.

Janeway CA, Travers P, Walport M, Shlomchik MJ: Immunobiology, The Immune System in Health and Disease, 6th ed. Garland Science, New York–London, 2005.

Jiang H, Chess L: Regulation of immune responses by T cells. New Engl J Med 354: 1166–1176, 2006.

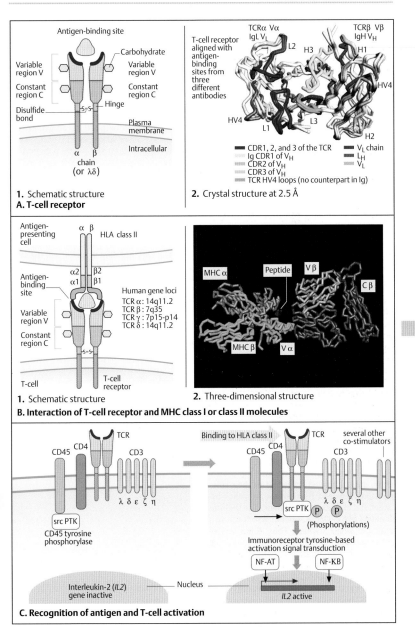

Antigen-binding site

Carbohydrate

Variable region V
Variable region V

Constant region C
Constant region C

Hinge

Disulfide bond

Plasma membrane

Intracellular

α   β
chain
(or λδ)

1. Schematic structure

A. T-cell receptor

T-cell receptor aligned with antigen-binding sites from three different antibodies

TCRα Vα
IgL V_L
L2
H3   H1

TCRβ Vβ
IgH V_H

HV4

HV4
L1   L3

H2

■ CDR1, 2, and 3 of the TCR   ■ V_L chain
Ig CDR1 of V_H   L_H
CDR2 of V_H   V_L
CDR3 of V_H
TCR HV4 loops (no counterpart in Ig)

2. Crystal structure at 2.5 Å

Antigen-presenting cell
α   β
HLA class II

α2 β2
α1 β1

Antigen-binding site

Variable region V

Constant region C

Human gene loci
TCR α : 14q11.2
TCR β : 7q35
TCR γ : 7p15-p14
TCR δ : 14q11.2

T-cell
T-cell receptor

1. Schematic structure

MHC α   Peptide   V β

C β

MHC β   V α

2. Three-dimensional structure

B. Interaction of T-cell receptor and MHC class I or class II molecules

CD45   CD4   TCR   CD3

src PTK
CD45 tyrosine phosphorylase

λ δ ε ζ η

Binding to HLA class II
CD45 CD4   TCR   several other co-stimulators
CD3

λ δ ε ζ η

src PTK (P) (P)
(Phosphorylations)

Immunoreceptor tyrosine-based activation signal transduction

NF-AT   NF-KB

Interleukin-2 (*IL2*) gene inactive

Nucleus

IL2 active

C. Recognition of antigen and T-cell activation

Genes of the MHC Region

The major histocompatibility complex (MHC) is a large chromosomal region containing more than 400 genes along nearly 4000 kb of DNA with the highest density of genes in the human genome (6 per 100 kb, average 1). It is the most polymorphic region with up to 30–60 alleles at many loci, with more than 400 alleles at HLA-A, more than 700 at HLA-B, and more than 500 at HLA-DRB1. Each MHC molecule is composed of two polypeptide chains and is expressed on the surface of various cells. This can be demonstrated serologically by a binding assay or by a cytotoxic cellular reaction in a mixed lymphocyte test.

A. Genomic organization of the MHC complex

In man, the MHC region is located on the short arm of human chromosome 6 (6 p21.3); in the mouse, on chromosome 17. The genes are grouped into three classes, I–III. Class I and class II belong to the HLA system (human leukocyte antigen); in mice called the H2 system. Class I in humans consists of HLA-A, HLA-B, HLA-C, and several other loci (HLA-E to -J). Mouse class I genes form two groups, D and L at the 3′ end, and K at the 5′ end. Human class II MHC molecules are grouped into DP, DQ, and DR (in mouse, I-A and I-E; letter I, not the Roman numeral). Class III includes genes that are not directly involved in the immune system as well as genes involved in the immune response, such as tumor necrosis factor (TFN), lymphotoxin, and others.

B. Gene loci in the MHC

The subgroups DP, DQ, and DR of the human class II HLA loci, oriented towards the centromere, are subdivided according to their composition of α- and β-chains (some genes between DM and DO are not shown). The DR locus contains two β chains, each of which can pair with the single α-chain. Thus, three sets of genes can produce four types of DR molecules. The class III loci, located between class I and class II, contain genes for complement factors C2 and C4 (C4A, C4B), steroid 21-hydroxylase (*CYP21 B*), and cytokines (tumor necrosis factor *TNFA* and lymphotoxin *LTA*, *LTB*). Several other class III genes are located in the class III region (not shown).

C. Structure of MHC molecules

The MHC class I and class II molecules have distinct structures. Class I molecules HLA-A, -B, and -C, consist of single membrane-bound polypeptides: an α chain with three domains, α1 at the extracellular N-terminal, α2, and α3 (**1**). The α chain is noncovalently bound to β$_2$-microglobulin, encoded by a gene on human chromosome 15. The three extracellular domains of the α chain have about 90 amino acids each. The α1 and α2 regions form the highly polymorphic peptide-binding region. The α3 domain and β$_2$-microglobulin structurally correspond to an immunoglobulin-like region. Class II MHC molecules (**2**) have two polypeptide chains, α and β, each with two domains, α1 and α2, and β1 and β2; each has 90 amino acids and a transmembrane region of 25 amino acids. The peptide-binding regions α1 and β1 are highly polymorphic. The crystalline structures of the MHC molecules have revealed details of the binding mechanisms. A small polypeptide-binding cleft of about 10–25 Å is present in both MHC class I and class II molecules.

(Three-dimensional figures in B and C from Janeway et al., 2005.)

References

Dausset J: The major histocompatibility complex in man: past, present, and future concepts. Science 213: 1469–1474, 1981.

Janeway CA, Travers P, Walport M, Shlomchik MJ: Immunobiology, 6th ed. The Immune System in Health and Disease. Garland Science, New York–London, 2005.

Klein J, Sato A: Advances in immunology. The HLA system. New Engl J Med 343: 702–709 (part I) and 782–786, 2000.

Trowsdale J: Genomic structure and function in the MHC. Trends Genet 9: 117–122, 1993.

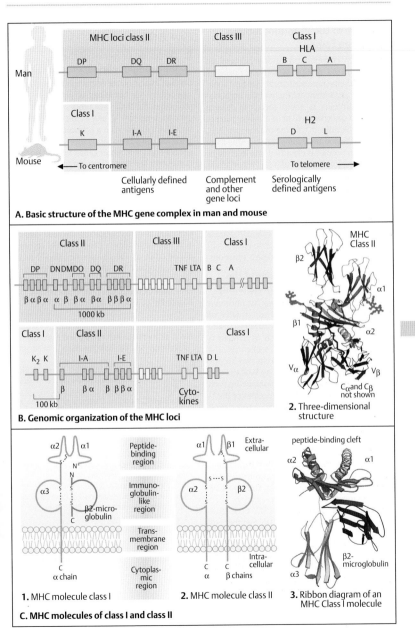

A. Basic structure of the MHC gene complex in man and mouse

Man

MHC loci class II — DP, DQ, DR
Class III
Class I — HLA — B, C, A

Mouse

Class I — K
I-A, I-E
D, L — H2

← To centromere   To telomere →

Cellularly defined antigens

Complement and other gene loci

Serologically defined antigens

B. Genomic organization of the MHC loci

Class II — DP DNDMDO DQ DR
β α β α   α β   β α   β α   β β β α
1000 kb

Class III — TNF LTA

Class I — B C A

Class I — K₂ K
Class II — I-A, I-E
β   β α   β β β α
100 kb

Class III — TNF LTA

Class I — D L
Cytokines

2. Three-dimensional structure

MHC Class II
β2, α1, β1, α2
Vα, Vβ
Cα and Cβ not shown

C. MHC molecules of class I and class II

1. MHC molecule class I
α2, α1, α3, β2-microglobulin
C, α chain

Peptide-binding region
Immunoglobulin-like region
Transmembrane region
Cytoplasmic region

2. MHC molecule class II
α1, β1, α2, β2
Extracellular
Intracellular
C α, β chains

3. Ribbon diagram of an MHC Class I molecule
peptide-binding cleft
α2, α1, α3, β2-microglobulin

Evolution of the Immunoglobulin Supergene Family

Owing to a common evolutionary origin, all molecules in the immune system and the genes encoding them share structural and functional features. A prominent attribute of the immunoglobulin proteins is their ability to adhere directly to another cell or a foreign protein through specialized cell membrane proteins, the cell adhesion molecules (CAMs). This large family of molecules is classified into four major families: the immunoglobulin (Ig) superfamily, cadherins, integrins, and selectins. CAMs are usually built up in repeats of domains with different properties. The Ig superfamiliy comprises a large group of molecules with globular domains. A gene superfamily is a collection of genes of common evolutionary origin that arose by gene duplications and subsequently diverged into genes with new, different functions.

A. Basic structure of proteins of the immunoglobulin supergene family

Characteristic structures shared by immunoglobulins are repeated domains, usually of about 70–110 amino acids with variable (V) and constant (C) domains (**1**). Each Ig domain is derived from conserved DNA sequences. The prototype, IgG, has three C domains in the heavy chain, one C domain in each of the two light chains, and a variable domain in each chain. The immunoglobulin molecules of the T-cell receptors (TCR) and the class I and class II MHC molecules are basically similar. Although their genes are located on different chromosomes, the gene products form functional complexes with each other. Others, such as the V, D, and J gene segments of all antigen receptors and their genes for the C domain form gene clusters. Genes of the MHC loci and for the two CD8 chains lie together.

Accessory molecules such as CD2, CD3, CD4, CD8, and thymosine 1 (Th-1) are members of this family with a relatively simple, but similar structure (**2**). Other members of the Ig superfamily are cell adhesion molecules, such as the Fc receptor II (FcRII); the polyimmunoglobulin receptor (pIgR), which transports antibodies through the membranes of epithelial cells; NCAM (neural cell adhesion molecules); and PDGFR (platelet-derived growth factor recep-

tor) (**3**). (Figure adapted from Hunkapiller & Hood, 1989.)

B. Evolution of genes of the immunoglobulin supergene family

Distinct evolutionary relationships can be recognized by the homology of genes for Ig-like molecules and their gene products. A precursor gene for a variant (V) and a constant (C) region must have arisen from a primordial cell by duplication and subsequently diverged into different cell surface receptor genes. Further duplication events resulted in structurally related gene segments encoding proteins with repeated domain structures. At an early stage, rearrangements between different gene segments occurred and became the standard for immunoglobulins, T-cell receptors, and CD8. Other members of this superfamily evolved into cell adhesion molecules without somatic recombination, such as the thymosine (Thy-1) or the poly-Ig receptor. Somatic recombination of the genes for antigen-binding molecules clearly has enormous evolutionary advantages. (Figure adapted from Hood et al., 1985.)

References

Abbas AK, Lichtman AH: Cellular and Molecular Immunology, 5th ed. WB Saunders, Philadelphia, 2005.

Hood L, Kronenberg M, Hunkapiller T: T-cell antigen receptors and the immunoglobin supergene family. Cell 40: 225–229, 1985.

Hunkapiller T, Hood L: Diversity of the immunoglobulin gene superfamily. Adv Immunol 44: 1–63, 1989.

Klein J, Sato A: The HLA system. Parts I and II. New Engl J Med 343: 702–709 and 782–786, 2000.

Klein J, Takahata N: Where do we come from? The Molecular Evidence for Human Descent. Springer, Berlin–Heidelberg–New York, 2002.

Shiina T et al: Molecular dynamics of MHC genes is unraveled by sequence analysis of the 1 796 938-bp HLA class I region. Proc Nat Acad Sci 96: 13282–13287, 1999.

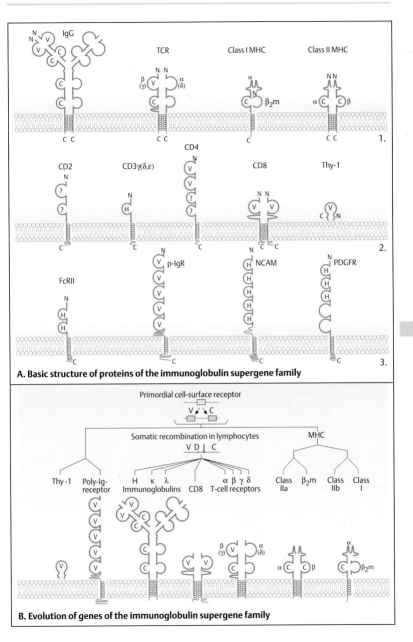

A. Basic structure of proteins of the immunoglobulin supergene family

B. Evolution of genes of the immunoglobulin supergene family

Hereditary Immunodeficiencies

Mutations in genes encoding the various proteins involved in the immune system cause severe, often life-threatening diseases. Genetic immunodeficiency diseases occur either isolated or as manifestations of multisystem diseases. Primary immunodeficiency disorders involve innate humoral immunity, the innate cell-mediated immunity, or disorders of humoral and cell-mediated adaptive immunity.

A. Hereditary immunodeficiency diseases: overview

This group of disease can be classified according to which branch of the immune system is primarily or solely involved. One group, Severe Combined Immune Deficiency (SCID), results from a genetic block of differentiation of precursor cells before they differentiate into B cells or T cells. Thus, both cell types are involved. It is a heterogeneous group of genetic disorders due to various defects in both B cell and T cell differentiation.

X-linked agammaglobulinemia type Bruton (MIM 300300) was the first hereditary immune deficiency described, in 1952 by Ogden Bruton (see part B). Here the first developmental step of B cell differentiation from pre-B to mature B cell is blocked by deficiency of Bruton tyrosine kinase (Btk) due to mutations in the *BTK* gene on the X chromosome (Xq22). Other forms involve later steps of differentiation (variable immune deficiency) or isolated Ig isotype (subclass) deficiencies. Several immune deficiency diseases involve T cells, such as TCR signal transduction, V(D)J recombination (mutations in *RAG1* and *RAG2* genes), cytokine signal transduction, and apoptosis regulation. Other disorders involve T cell activation and the function of one or both major subsets of T cells, CD4 or CD8. Predisposition to tumors of the lymphoid system and autoimmune dysfunction is relatively frequent in immunodeficiency diseases. Effective therapy by bone marrow transplantation is possible in some of the diseases.

B. Severe combined Immune Deficiency Disease

SCID is a heterogeneous group of disorders. The most common is X-linked SCID (MIM 308380, 300400). SCIDX1 (MIM 300400) results from mutations in the X-chromosomal gene *IL2 RG*

on Xq13.1, encoding the γ subunit of the interleukin 2 receptor (IL2Rγ). This subunit is shared with other interleukin receptors. e.g., IL4 and IL15. *IL2 RG* is an atypical member of the cytokine receptor family. It has 8 exons and spans 4.5 kb of genomic DNA. More than 200 different mutations have been recorded (Belmont & Puck, 2001).

C. DiGeorge syndrome and deletion 22q11

DiGeorge syndrome (MIM 188400) is a combination of T cell defects associated with a highly variable spectrum of congenital malformations involving absence of the thymus and other derivatives of the embryonic third and fourth branchial arches. If the parathyroid glands are absent, neonatal hypocalcemia is immediately life threatening. In addition, abnormalities of the face may be present. DiGeorge syndrome is now considered part of an overlapping spectrum of disorders involving various-sized deletions of 22q11 (microdeletion syndromes). This includes a previously recognized distinct disorder, the velocardiofacial syndrome of shprintzen (MIM 192430). Other examples of immune deficiency diseases are listed in a table in the appendix. (Figures in B and C adapted from Burmester & Pezzuto, 2003.)

References

Belmont JW, Puck JM: T cell and combined immunodeficiency disorders, pp 4751–4783. In: Scriver CR, Beaudet AL, Sly WS, Valle D (eds) The Metabolic and Molecular Bases of Inherited Disease, 8th ed. McGraw-Hill, New York, 2001.

Burmester G-R, Pezzutto: Color Atlas of Immunology. Thieme Medical Publ., Stuttgart–New York, 2003.

Cooper MD, Schroeder HW: Primary immune deficiency diseases, pp 1939–1947. In: Kasper DL et al (eds) Harrison's Principles of Internal Medicine, 16th ed. McGraw-Hill, New York, 2005.

Hong R: Inherited immune deficiency, pp 283–291. In: Jameson JL (ed) Principles of Molecular Medicine. Humana Press, Totowa, New Jersey, 1998.

Schwartz K et al: RAG mutations in human B cell-negative SCID. Science 274: 97–99, 1996.

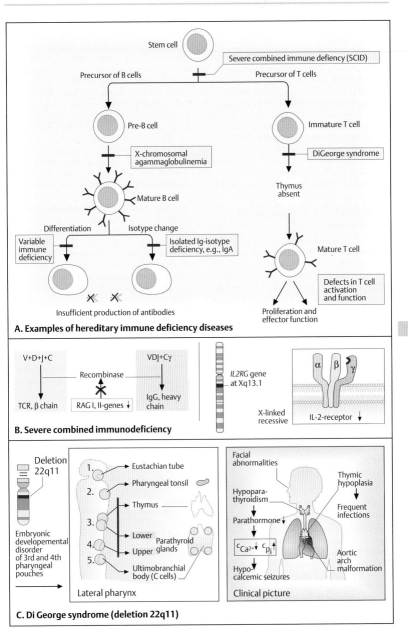

A. Examples of hereditary immune deficiency diseases

Stem cell

Severe combined immune deficiency (SCID)

Precursor of B cells

Precursor of T cells

Pre-B cell

Immature T cell

X-chromosomal agammaglobulinemia

DiGeorge syndrome

Mature B cell

Thymus absent

Differentiation

Isotype change

Variable immune deficiency

Isolated Ig-isotype deficiency, e.g., IgA

Mature T cell

Defects in T cell activation and function

Insufficient production of antibodies

Proliferation and effector function

B. Severe combined immunodeficiency

V+D+J+C

VDJ+Cγ

Recombinase

IL2RG gene at Xq13.1

α   β   γ

TCR, β chain

RAG I, II-genes ↓

IgG, heavy chain

X-linked recessive

IL-2-receptor ↓

C. Di George syndrome (deletion 22q11)

Deletion 22q11

1. → Eustachian tube

2. → Pharyngeal tonsil

→ Thymus

3. → Lower

4. → Upper   Parathyroid glands

5. → Ultimobranchial body (C cells)

Embryonic developmental disorder of 3rd and 4th pharyngeal pouches

Lateral pharynx

Facial abnormalities

Thymic hypoplasia

Hypoparathyroidism

Parathormone ↓

Frequent infections

$c_{Ca^{2+}}$↓ c_{P_i}↑

Hypocalcemic seizures

Aortic arch malformation

Clinical picture

Genetic Causes of Cancer: Background

Cancer cells (malignant cells) break two rules imposed on all cells in a multicellular organism: they and their progeny do not adhere to restrained cell division, and they invade and colonize tissues reserved for other cell types. A cancer is medically classified according to the cell type from which it originates: carcinoma from epithelial cells, sarcoma from connective tissue and muscle cells, leukemia from hematopoietic cells, and lymphoma from lymphoid cells. Genetically, cancer is either nonhereditary, due to somatic mutations, or hereditary, due to a predisposing mutation in the germline.

A. Multistep clonal expansion of malignancy

Nearly all cancers result from a single cell and subsequent clonal development of genetically unstable cells. The estimated rate of spontaneous mutations is about 10^{-6} per gene per cell division. With an estimated 10^{16} cell divisions in the human body, there are about 10^{10} occasions to undergo a mutation (Alberts et al., 2002, p. 1317). Not all mutations affect cell division control genes; most are recognized and repaired, or the cell is eliminated by apoptosis. Normal cells divide until they have matured into differentiated, specialized cells and then cease to divide. In tissues requiring continuous renewal, stem cells provide new cells. Various selective barriers detect and prevent unphysiological cell division. When a cell with one or more genetic changes (precursor tumor cell) breaks through a barrier, it will usually be eliminated. However, if it has accumulated sufficient genetic alterations, it will overcome the barrier and continue to develop into a tumor.

B. Four basic types of genetic alteration in tumor cells

The many genetic alterations affecting growth-controlling genes can be classified into four major categories: (i) change in the DNA sequence of a growth-controlling gene (somatic mutation), (ii) reciprocal chromosome translocation disrupting a gene expressed in a tissue that depends on controlled cell division (e.g., immune system, blood cell formation in bone marrow), (iii) gross alteration in chromosome number in somatic cells during tumor progression, and (iv) amplification of a growth-controlling gene. Here are four examples.

(1) *Change in DNA sequence.* A deletion of two adenines (A) in a series of ten in the gene *TGFBR2* for receptor type 2 of the transforming growth factor beta (TGFβ R2) in a colorectal cancer cell line changes the codon AAG (lysine) to GCC (alanine). This converts the subsequent codons into TGG (tryptophan) and TGA (stop codon), resulting in a truncated protein.

(2) *Chromosome translocation.* A reciprocal translocation between a chromosome 1 and a chromosome 17 in a neuroblastoma (MIM 256700) cell line disrupts genes involved in neuroblastoma located on chromosomes 1 and 17.

(3) *Gross chromosomal change.* Loss of a chromosome 3 and a chromosome 12 (yellow arrows) occurred in a clone of a cell line (SW837) of colorectal cancer cells (CRCs). Such gross changes are frequent during tumor progression.

(4) *Gene amplification.* In some tumor cells in culture, small chromosomal derivatives (double minutes) or homogeneously stained regions (HSRs) are visible. HSRs, first described by Biedler & Spengler (1976), are cytological manifestations of gene amplification. Specific DNA sequences are replicated to a disproportionately higher degree than normal. Here a metaphase from a clone of the CRC cell line SW837 expanded through 25 generations is shown. (Data and figures adapted from Lengauer et al., 1998.)

References

Alberts B et al: Molecular Biology of the Cell, 4th ed. Garland Publishing Co, New York, 2002.

Biedler JI, Spengler BA: Metaphase chromosome anomaly: association with drug resistance and cell-specific products. Science 191: 185–187, 1976.

Hahn WC, Weinberg RA: Rules for making human tumor cells. New Eng J Med 347: 1593–1603, 2002.

Hogarty MD, Brodeur GM: Gene amplification in human cancers: Biological and clinical significance, pp 115–128. In: Vogelstein B, Kinzler KW (eds): The Genetic Basis of Human Cancer, 2nd ed. McGraw-Hill, New York, 2002.

Lengauer C, Kinzler W, Vogelstein B: Genetic instabilities in human cancers. Nature 396: 643–649, 1998.

Weinberg RA: The Biology of Cancer. Garland Science, New York, 2006.

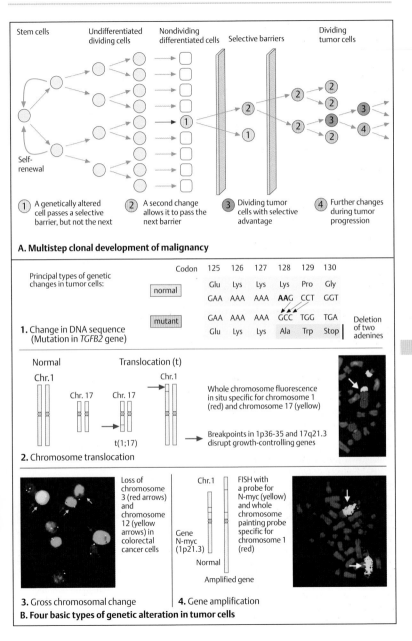

A. Multistep clonal development of malignancy

Stem cells — **Undifferentiated dividing cells** — **Nondividing differentiated cells** — **Selective barriers** — **Dividing tumor cells**

Self-renewal

① A genetically altered cell passes a selective barrier, but not the next

② A second change allows it to pass the next barrier

③ Dividing tumor cells with selective advantage

④ Further changes during tumor progression

Principal types of genetic changes in tumor cells:

Codon	125	126	127	128	129	130	
normal	Glu	Lys	Lys	Lys	Pro	Gly	
	GAA	AAA	AAA	**AA**G	CCT	GGT	
mutant	GAA	AAA	AAA	GCC	TGG	TGA	Deletion of two adenines
	Glu	Lys	Lys	Ala	Trp	Stop	

1. Change in DNA sequence (Mutation in *TGFB2* gene)

Normal   Translocation (t)

Chr. 1   Chr. 1

Chr. 17   Chr. 17

t(1;17)

Whole chromosome fluorescence in situ specific for chromosome 1 (red) and chromosome 17 (yellow)

Breakpoints in 1p36-35 and 17q21.3 disrupt growth-controlling genes

2. Chromosome translocation

Loss of chromosome 3 (red arrows) and chromosome 12 (yellow arrows) in colorectal cancer cells

Chr. 1

Gene N-myc (1p21.3)

Normal

Amplified gene

FISH with a probe for N-myc (yellow) and whole chromosome painting probe specific for chromosome 1 (red)

3. Gross chromosomal change

4. Gene amplification

B. Four basic types of genetic alteration in tumor cells

Categories of Cancer Gene

Cancer is a common genetic disease that affects 1 of every 4 individuals. More than 100 genes in the human genome contribute to cancer when altered by mutations. They are classified into three basic categories according to the effects of their mutations: too much activity of a gene product (oncogenes), insufficient activity (*tumor suppressor* genes), and disruption of *genome stability* genes.

A. Three categories of cancer genes

The first class is *proto-oncogenes*. Their mutant forms, called *oncogenes*, drive a cell to divide when it normally should not (gain-of-function mutations). A single activating mutation is the first step towards cancer (comparable to a stuck accelerator in a car). The second class is *tumor suppressor genes*. They require two mutational events to induce tumor development (comparable to a defective brake). The initial mutation predisposes the cell to become a cancer cell. The second mutation then inactivates the other allele (loss-of-function mutation) and results in loss of cell division control. Mutations in the third class of cancer genes, called *stability genes* or *caretakers*, affect the stability of the genome by disrupting one of the various repair processes.

B. Oncogene activation

Oncogenes serve in signal pathways controlling cell division. For example the *Ras* genes encode a family of related cell growth-controlling proteins. Ras proteins are GTPase-binding proteins functioning as switches, inactive when bound to GDP (guanosyldiphosphate) and active when bound to GTP (guanosyltriphosphate). Ras is activated by a receptor tyrosine kinase, which activates a guanine nucleotide exchange factor (GEF). GTPase-activating proteins (GAPs) increase the hydrolysis of GTP bound to Ras and inactivate Ras by removing GTP.

Mutant forms of Ras are hyperactive and do not respond to GAPs. Instead they remain bound to GTP, sending continuous cell division-promoting signals to the nucleus through several pathways and causing uncontrolled cell divisions. The genetic mechanisms that activate oncogenes include point mutations, chromosome rearrangement (chromosomal translocation), and gene amplification.

C. Tumor suppressor genes

Two successive mutational events are required within the same cell (**1**). The first event inactivates one allele and predisposes the cell to uncontrolled divisions. If the other allele is also inactivated by a mutation, cell division control is lost and a tumor develops. One of several mechanisms may be responsible: a further mutation, chromosome loss during cell division (mitotic nondisjunction), or mitotic recombination with gene conversion. Tumor suppressor genes can be assigned to two groups, *gatekeepers* and *caretakers*. *Gatekeeper* genes *directly* inhibit tumor growth. Inactivation of caretaker genes leads to genetic instability, indirectly promoting tumor growth.

Loss of one allele in a somatic cell carrying a mutation in the other allele can be visualized by Southern blot analysis (**2**). Whereas somatic cells heterozygous at a marker locus give two signals, tumor cells, which have lost both alleles of the gene, give one signal (loss of heterozygosity, LOH). LOH is a hallmark of a tumor suppressor gene. It occurs with variable frequency in different tumors and may be useful in detecting a mutation indirectly.

Mutations in tumor suppressor genes may be present in the zygote (by transmission or by new mutation) or occur in a somatic cell (**3**). A germline mutation predisposes all cells to develop into tumor cells. A somatic mutation predisposes a single cell. Germline mutations are the basis for hereditary forms of cancer; somatic mutations for the nonhereditary forms. A germline mutation occurring after the initial division of the fertilized egg may result in a mosaic of mutated and normal cells.

References

Alberts B et al: Molecular Biology of the Cell, 4th ed. Garland Publishing Co, New York, 2002.

Hahn WC, Weinberg RA: Modelling the molecular circuitry of cancer. Nature Rev Cancer 2: 331–341, 2002.

Hanahan D, Weinberg RA: The hallmarks of cancer. Cell 100: 57–70, 2000.

Vogelstein B, Kinzler KW: Cancer genes and the pathways they control. Nature Med 10: 789–799, 2004.

Vogelstein B, Kinzler KW (eds): The Genetic Basis of Human Cancer. McGraw-Hill, New York, 2002.

Weinberg RA: Tumor suppressor genes. Science 254: 1138–1146, 1991.

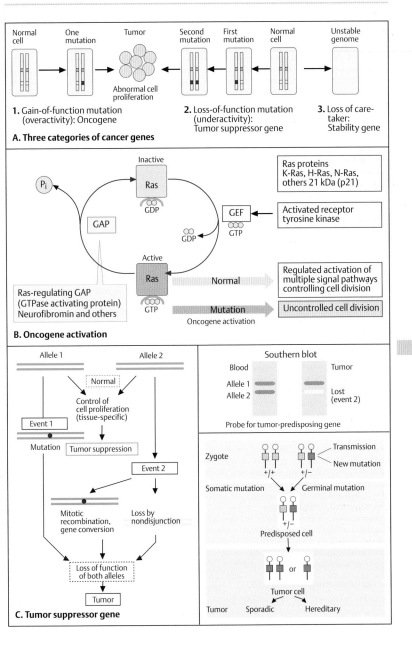

A. Three categories of cancer genes

1. Gain-of-function mutation (overactivity): Oncogene

2. Loss-of-function mutation (underactivity): Tumor suppressor gene

3. Loss of care-taker: Stability gene

Normal cell — One mutation — Tumor — Abnormal cell proliferation — Second mutation — First mutation — Normal cell — Unstable genome

B. Oncogene activation

Inactive Ras — GDP
GAP
P_i
GEF — GDP / GTP
Activated receptor tyrosine kinase
Ras proteins K-Ras, H-Ras, N-Ras, others 21 kDa (p21)
Active Ras — GTP
Ras-regulating GAP (GTPase activating protein) Neurofibromin and others
Normal → Regulated activation of multiple signal pathways controlling cell division
Mutation → Uncontrolled cell division
Oncogene activation

C. Tumor suppressor gene

Allele 1   Allele 2
Normal
Control of cell proliferation (tissue-specific)
Event 1
Mutation — Tumor suppression
Event 2
Mitotic recombination, gene conversion   Loss by nondisjunction
Loss of function of both alleles
Tumor

Southern blot
Blood   Tumor
Allele 1
Allele 2   Lost (event 2)
Probe for tumor-predisposing gene

Zygote   Transmission / New mutation
+/+   +/−
Somatic mutation   Germinal mutation
+/−
Predisposed cell
or
Tumor cell
Tumor   Sporadic   Hereditary

The *p53* Tumor Suppressor Gene

The *p53* tumor suppressor gene plays a central role in cell cycle control, apoptosis, and maintenance of genetic stability. It encodes a 53-kDa nuclear phosphoprotein translated from a 2.8-kb mRNA. The gene spans about 20 kb on the short arm of human chromosome 17 (17p13). The p53 protein binds to specific DNA sequences and controls the expression of different regulator genes involved in growth. It interacts with other proteins in response to DNA damage and mediates apoptosis (cell death) of the cell when the damage is beyond repair. Its basic function is to control entry of the cell into the S phase (see cell cycle control, p. 124). Somatic mutations in the *p53* gene occur in about half of all tumors.

A. The human p53 protein

The active form of the human p53 protein is a tetramer of four identical subunits. The p53 protein is a transcription factor. Normally it is extremely unstable and does not stimulate transcription because it is bound to a protein called Mdm2. Oncogenic mutations in *p53* have a dominant-negative effect. A mutation in one allele can abrogate all activity. Each subunit has 393 amino acids with five highly conserved functional domains, I–V. The carboxyl end beyond amino acid 300 has a nonspecific DNA interaction domain and the tetramerization domain. p53 protein function is inhibited by human papilloma virus protein E6, adenovirus protein E1 b, SV40, and others. Most mutations are clustered in the conserved domains II–V: in codons 129–146 (exon 4), 171–179 (exon 5), 234–260 (exon 7), and 270–287 (exon 8). Six mutations are strikingly frequent: a replacement of conserved amino acids arginine (R) in positions 175, 248, 249, 273, and 282, and glycine (G) in position 245. Mutations are missense, insertions, and deletions.

Knockout mice develop normally, but develop tumors at a high rate. Activated benzopyrene induces mutations at codons 175, 248, and 275 in cultured bronchial epithelial cells.

(Figure adapted from Lodish et al., 2005.)

B. Germline mutations of *p53*

Germline mutations lead to a familial form of multiple different cancers, the autosomal dominant Li–Fraumeni syndrome (MIM 114480) described in 1969 by Li and Fraumeni in families with individuals affected with diverse types of tumors, mainly soft-tissue sarcomas, early-onset breast cancer, brain cancers, cancer of the bone (osteosarcoma) and bone marrow (leukemias), and carcinoma of the lung, pancreas, and adrenal cortex. Lynch reported similar observations as "cancer family syndrome." The pedigree in panel 1 shows four individuals (II-2, II-3, III-1, III-2) affected by different types of tumor due to a mutation in codon 248 of the *p53* gene (CGG arginine, to TGG, tryptophan). The mutation is also present in individuals I-1 and III-5. They have not developed tumors, but are at increased risk. The absence of the mutation in individuals III-3 and III-4 indicates that they do not have an increased risk of cancer (data from D. Malkin). A subset of patients with Li–Fraumeni syndrome does not show p53 mutations.

C. Model of function of the *p53* gene

The normally inactive *p53* gene (**1**) is activated in response to DNA damage (**2**). Phosporylation at the Mdm2 binding site displaces Mdm2 and activates p53. Phosphorylation is induced by ATM (ataxia-telangiectasia, see p. 340) and probably also the ATR protein. Activated p53 binds to DNA and induces the transcription of p21, a cell-cycle controlling protein, which binds to Cdk complexes. As a consequence, the cell cannot enter the S phase. If DNA repair is successful, the cell can continue its cycle. If repair is not successful, the cell is sacrificed by apoptosis. Damaged cells with defective p53 protein are not arrested in G_1. (Figure adapted from D.P. Lane, 1992.)

References

Bell DW et al: Heterozygous germline *hCHK2* mutations in Li-Fraumeni syndrome. Science 286: 2828–2831, 1999.

Hanahan D, Weinberg RA: The hallmarks of cancer. Cell 100: 57–70, 2000.

Lane DP: p53, guardian of the genome. Nature 358: 15–16, 1992.

Lodish H et al: Molecular Cell Biology, 5th ed. WH Freeman & Co, New York, 2004.

Malkin D: The Li-Fraumeni syndrome, pp 387–401. In: Vogelstein B, Kinzler KW (eds) The Genetic Basis of Human Cancer, 2nd ed. McGraw-Hill, New York, 2002.

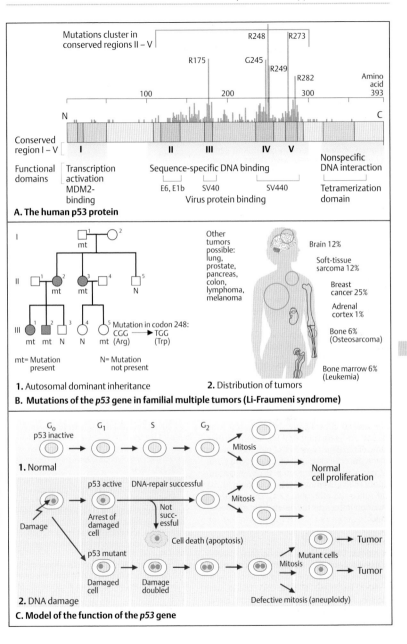

A. The human p53 protein

Mutations cluster in conserved regions II–V

R248   R273   R175   G245   R249   R282

Amino acid 393

Conserved region I–V

Functional domains

Transcription activation
MDM2-binding

Sequence-specific DNA binding

E6, E1b   SV40   SV440
Virus protein binding

Nonspecific DNA interaction
Tetramerization domain

B. Mutations of the *p53* gene in familial multiple tumors (Li-Fraumeni syndrome)

mt= Mutation present   N= Mutation not present

Mutation in codon 248:
CGG → TGG
(Arg)   (Trp)

1. Autosomal dominant inheritance

Other tumors possible: lung, prostate, pancreas, colon, lymphoma, melanoma

Brain 12%
Soft-tissue sarcoma 12%
Breast cancer 25%
Adrenal cortex 1%
Bone 6% (Osteosarcoma)
Bone marrow 6% (Leukemia)

2. Distribution of tumors

C. Model of the function of the *p53* gene

G₀ p53 inactive   G₁   S   G₂   Mitosis

1. Normal

Normal cell proliferation

p53 active   DNA-repair successful   Mitosis

Damage

Arrest of damaged cell

Not successful

Cell death (apoptosis)

Tumor

p53 mutant

Damaged cell   Damage doubled   Mitosis   Mutant cells   Tumor

Defective mitosis (aneuploidy)

2. DNA damage

The *APC* Gene and Polyposis coli

Cancer of the colon and rectum is the second leading cause of death from cancer. About 5% of the population is at risk to develop colorectal cancer. Most colorectal tumors arise from a series of somatic mutations in several genes. The *APC* gene (adenomatous polyposis coli) is a tumor suppressor gene in the Wnt/β-catenin signal pathway (see p. 274). Mutant APC protein does not bind to β-catenin. This induces transcription of several growth-controlling genes, including the oncogene *MYC* (MIM 190080). Germline mutations in the *APC* gene are the main cause of familial adenomatous polyposis of the colon (FAP).

A. Familial adenomatous polyposis

Familial polyposis (FAP, MIM 175100) is an autosomal dominant hereditary disease. In late childhood and early adulthood, up to 1000 and more polyps develop in the mucous membrane of the colon (**1**). Each polyp can develop into a carcinoma (**2**). In about 85% of affected persons, small hypertrophic areas not affecting vision are present in the retina (congenital hypertrophy of the retinal pigment, CHRPE, **3**). (Photographs 1 and 2 kindly provided by U. Pfeifer, Bonn; photo 3 by W. Friedl, Bonn.)

B. Structure and function of the *APC* gene

The *APC* gene, located on human chromsome 5 at q21–q22, has 8538 base pairs (bp) in 15 exons. It encodes a protein of 2843 amino acids with several alternatively spliced forms. Exon 15 has an exceptionally long open reading frame of 6579 bp. Over 95% of mutations result in a nonfunctional truncated protein with variable loss of the C-terminus due to nonsense mutations (40%), deletions (41%), insertions (12%), and splice site mutations (7%). The site of mutation influences the phenotypic manifestations.

C. Diagnosis in FAP

An indirect DNA analysis is the haplotype analysis using polymorphic DNA marker loci flanking the *APC* locus (**1**). Affected individuals I-1 and II-3 share haplotype 6–8 at loci D5S28 and D5S346. This haplotype must carry the mutation. Individual II-4 inherited this haplotype and is at risk of developing FAP. The protein

truncation test (**2**) detects the abnormal protein, which because of its smaller size migrates faster than the normal protein. (Data of Dr. W. Friedl, Bonn.)

D. Mutations in colorectal tumorigenesis

Tumor formation requires several stages, beginning with a mutation in one allele, followed by loss of the normal allele (LOH) or a mutation in the second allele. This leads to early stages of adenoma with less differentiated cells. Early polyps form at this stage. Additional mutations in other growth-controlling genes lead to malignant transformation and eventually to tumor development. The order of mutations seems to be important. About half of colorectal cancers have *RAS* mutations (MIM 603384). Genes involved include *DCC* (MIM 120470), *SMAD4* (MIM 600993), *SMAD2* (MIM 601366), *p53* (191170), and others. (Figure adapted from Fearon & Vogelstein, 1990.)

Hereditary nonpolyposis colorectal cancer (HNPCC, MIM 120435) affects about 1 in 200 to 1000 individuals (3% of all colorectal cancers). It results from germline mutation in one of the DNA-mismatch repair genes *hMSH1*, *hMLH2*, *hPMS1*, and *hPMS2* or related genes. Microsatellite instability is an important feature of HNPCC.

References
Boland CR, Meltzer SJ: Cancer of the colon and the gastrointestinal tract, pp 1824–1867. In: Rimoin DL, Connor JM, Pyeritz RE, Korf BR (eds) Emery and Rimoin's Principles and Practice of Medical Genetics, 4th ed. Churchill-Livingstone, Edinburgh, 2002.

Bronner CE et al: Mutation in the DNA mismatch repair gene homologue *hMLH1* is associated with hereditary nonpolyposis colon cancer. Nature 368: 258–261, 1994.

Chapelle A de la, Peltomäki P: The genetics of hereditary common cancers. Curr Opin Genet Develop 8: 298–303, 1998.

Fearon ER, Vogelstein B: A genetic model for colorectal tumorigenesis. Cell 61: 759–767, 1990.

Groden J et al: Identification and characterization of the familial adenomatous polyposis coli gene. Cell 66: 589–600, 1991.

Kinzler KW, Vogelstein B: Colorectal tumors, pp 583–612. In: Vogelstein B & Kinzler KW (eds) The Genetic Basis of Human Cancer, 2nd ed. McGraw-Hill, New York, 2002.

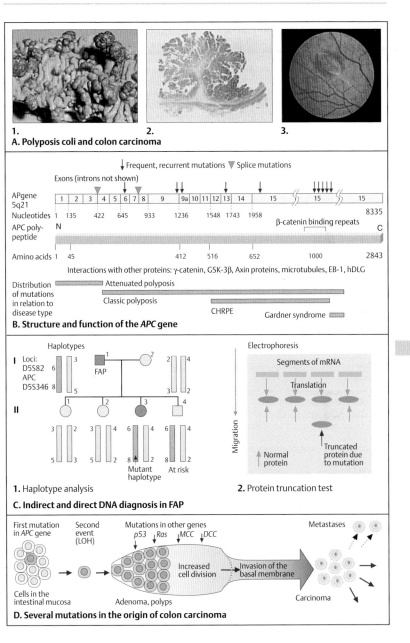

A. Polyposis coli and colon carcinoma

1.   2.   3.

↓ Frequent, recurrent mutations   ▽ Splice mutations

Exons (introns not shown)

APgene 5q21

| 1 | 2 | 3 | 4 | 5 | 6 | 7 | 8 | 9 | 9a | 10 | 11 | 12 | 13 | 14 | 15 | 15 | 15 |

Nucleotides 1   135   422   645   933   1236   1548 1743 1958   8335

APC poly-peptide   N   β-catenin binding repeats   C

Amino acids 1   45   412   516   652   1000   2843

Interactions with other proteins: γ-catenin, GSK-3β, Axin proteins, microtubules, EB-1, hDLG

Distribution of mutations in relation to disease type

Attenuated polyposis

Classic polyposis

CHRPE   Gardner syndrome

B. Structure and function of the *APC* gene

Haplotypes

I   Loci:
D5S82   6
APC
D5S346   8

3   FAP   1   2   2   4

3   2

5

II

1   2   3   4

3   2   3   4   6   4   6   4

5   3   5   2   8   2   8   2

Mutant haplotype   At risk

Electrophoresis

Segments of mRNA

Translation

Migration

Normal protein   Truncated protein due to mutation

1. Haplotype analysis   **2.** Protein truncation test

C. Indirect and direct DNA diagnosis in FAP

First mutation in *APC* gene

Second event (LOH)

Mutations in other genes
p53 ↓*Ras* ↓*MCC* ↓*DCC*

Metastases

Increased cell division

Invasion of the basal membrane

Cells in the intestinal mucosa

Adenoma, polyps

Carcinoma

D. Several mutations in the origin of colon carcinoma

Breast Cancer Susceptibility Genes

Breast cancer is one of the most common forms of cancer, accounting for 32% of all cancers in the Western world. Two genes confer susceptibility to breast and ovarian cancer when mutated, the breast cancer genes *BRCA1* and *BRCA2*. Both encode multifunctional proteins that play important roles in genomic stability, homologous recombination, and double-stranded and transcription-coupled DNA repair (see p.90). The BRCA1 and BRCA2 proteins interact and participate in cell cycle control (see p. 124). Germline mutations are the basis for familial occurrence. The mutations in different patients are distributed throughout the genes. The causative role of a particular sequence change may be difficult to assess in a given individual. In addition, polymorphic sequence variants are common. Mutations in other genes may be involved.

A. The breast cancer susceptibility gene *BRCA1*

The *BRCA1* gene on chromosome 17 at q21.1 accounts for 20–30% of inherited, autosomal dominant forms of breast cancer. This gene has 24 exons spanning 80 kb of genomic DNA and encodes a 7.8-kb mRNA transcript. Somatic mutations in breast tissue and germline mutations observed in unrelated patients are evenly distributed throughout the gene. About 55% of all mutations occur in the large (3.4-kb) exon 11. A deletion of an adenine (A) and a guanine (G) in nucleotide position 185 (185delAG) and an insertion of a cytosine in position 5382 (5382insC) are the most frequent, each accounting for about 10% of mutations. These mutations are particularly frequent in the Ashkenazi Jewish population.

The 1863-amino acid protein has distinct functional domains. Heterodimerization occurs at BARD1 (BRCA1-associated RING domain 1). Three protein-binding domains allow interaction with the p53 protein, the DNA recombination protein RAD51 (a human homolog of the bacterial RecA protein), and an RNA helicase. RAD50 and 51 are proteins involved in recombination during mitosis and meiosis, and in recombinational repair of double-stranded DNA breaks. The C-terminus contains a region involved in transcriptional activation and DNA repair. Two nuclear localization signals (NLS) are present at amino acid positions 500–508 and 609–615.

B. The breast cancer susceptibility gene *BRCA2*

Mutations in the *BRCA2* gene, at 13q12, occur throughout the gene. A deletion of thymine at nucleotide position 6174 (6174delT) is relatively frequent (1%) in the Ashkenazi Jewish population. The BRCA2 protein has distinct functional domains. A large central domain consists of eight copies of a 30–80-amino acid repeat, which are conserved in all mammalian BRCA2 proteins (BRC repeats). Four of these interact with the RAD51 protein.

(Figures based on Couch & Weber, 2002, and Welcsh et al., 2000.)

C. BRCA1-mediated effect on *p53*

BRCA1 functions as a coactivator of transcription of several genes, including RNA polymerase II holoenzyme through RNA helicase A and histone acetylase CREB-binding protein. Particularly important is the interaction with the p53 protein. Eukaryotic cells have two pathways to deal with double-stranded DNA breaks, nonhomologous end-joining and homologous recombination (HR) (the mechanism used for recombination during meiosis, see p. 118). BRCA1 is connected to HR. One model predicts that BRCA1 is involved in apoptosis induced by p53 in response to DNA damage. (Figure redrawn from Hohenstein & Giles, 2003.)

References

Couch FJ, Weber BL: Breast cancer, pp 549–581. In: Vogelstein B, Kinzler KW (eds) The Genetic Basis of Human Cancer, 2nd ed. McGraw-Hill, New York, 2002.

Hohenstein P, Giles RH: BRCA1: a scaffold for p53 response? Trends Genet 19: 489–494, 2003.

Miki Y et al: A strong candidate for the breast and ovarian cancer susceptibility gene *BRCA1*. Science 266: 66–71, 1994.

Welcsh PL, Schubert EL, King MC: Inherited breast cancer: an emerging picture. Clin Genet 54: 447–458, 1998.

Welcsh PL, Owens KN, King MC: Insights into the functions of BRCA1 and BRCA2. Trends Genet 16: 69–74, 2000.

Wooster R et al: Identification of the breast cancer susceptibility gene *BRCA2*. Nature 378: 789–792, 1995.

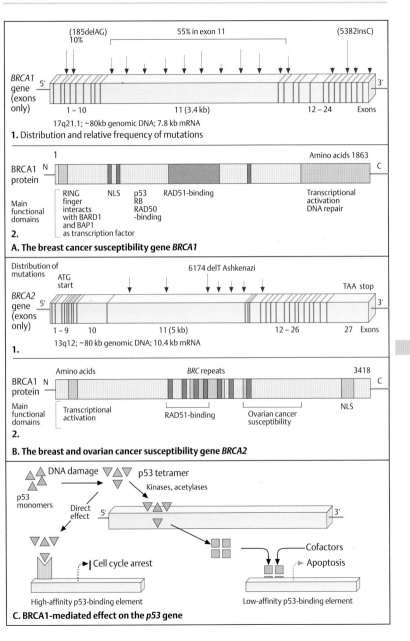

1. Distribution and relative frequency of mutations

BRCA1 gene (exons only) — 17q21.1; ~80kb genomic DNA; 7.8 kb mRNA

(185delAG) 10%
55% in exon 11
(5382insC)

1 – 10 | 11 (3.4 kb) | 12 – 24 | Exons

2.

BRCA1 protein — 1 / Amino acids 1863

Main functional domains:
- RING finger interacts with BARD1 and BAP1 as transcription factor
- NLS
- p53 RB RAD50 -binding
- RAD51-binding
- Transcriptional activation DNA repair

A. The breast cancer susceptibility gene *BRCA1*

1.

Distribution of mutations

ATG start
6174 delT Ashkenazi
TAA stop

BRCA2 gene (exons only) — 13q12; ~80 kb genomic DNA; 10.4 kb mRNA

1 – 9 | 10 | 11 (5 kb) | 12 – 26 | 27 Exons

2.

BRCA1 protein — Amino acids / BRC repeats / 3418

Main functional domains:
- Transcriptional activation
- RAD51-binding
- Ovarian cancer susceptibility
- NLS

B. The breast and ovarian cancer susceptibility gene *BRCA2*

DNA damage — p53 monomers — p53 tetramer — Kinases, acetylases

Direct effect

Cell cycle arrest — High-affinity p53-binding element

Cofactors — Apoptosis — Low-affinity p53-binding element

C. BRCA1-mediated effect on the *p53* gene

Retinoblastoma

Retinoblastoma (MIM 180200) is the most frequent malignant tumor of the eye in infancy and early childhood, with an incidence of about 1 in 15 000–25 000 live births. It results from loss of function of both alleles of the retinoblastoma gene, *RB1*. Two inactivating events are required for tumor initiation, as predicted by A. Knudson in 1971 (two-hit hypothesis). The first mutation predisposes the cell to develop a tumor; the second initiates tumor formation. The first mutation may be either a somatic mutation, in one retinal progenitor cell (retinoblast), or a germline mutation.

A. Phenotype

Retinoblastoma (Rb) may occur in one or both eyes (unilateral or bilateral). An early sign is a white shimmer, the "cat's eye" (**1**), and/or rapidly developing strabismus. One or several tumors (unifocal or multifocal, respectively) may be present in the retina of an affected eye (**2**) and progress rapidly (**3**). Early diagnosis and therapy are essential. About 60% of patients have somatic mutations (nonhereditary Rb) and these usually develop unilateral and unifocal Rb. About 40% of patients are heterozygous for an *RB1* mutation that is either transmitted from a parent (10–15%) or due to a new mutation, usually derived from the paternal allele (about 10:1). Heterozygous carriers for an oncogenic *RB1* mutation have a predisposition to Rb, which is transmitted as an autosomal dominant trait. In rare families carriers of an oncogenic mutation do not develop tumors (nonpenetrance). This low-penetrance phenotype is associated with specific *RB1* mutations. Milder phenotypic expression is also observed when the mutation is present in only a proportion of germ cells (mutational mosaicism).

B. Retinoblastoma locu

The *RB1* locus at 13q14.2 was first identified by microscopically visible interstitial deletions.

C. Retinoblastoma gene RB1 and its protein

The *RB1* gene is organized into 27 exons spanning 183 kb of genomic DNA (**1**). It is ubiquitously expressed and transcribed into a 4.7-kb mRNA (**2**). The main types of mutation in hereditary retinoblastoma are deletions (~26%), insertions (~9%), and point mutations (~65%), including splice-site mutations, distributed relatively evenly along the gene. Missense mutations have been associated with low penetrance retinoblastoma.

The gene product (pRB protein), a 100-kD phosphoprotein with 928 amino acids (**3**), has important functions in the regulation of the cell cycle (p. 124). It is activated by phosphorylation at about 12 distinct serine and threonine residues (P) during cell cycle progression from G_0 to G_1. Three functional domains, A, B, and C, bind cell cycle-dependently to transcription factors, including the oncoproteins Mdm-2 and c-abl. A nuclear localization signal (NLS) is located at the C terminus.

D. Diagnostic principles

Molecular analysis has greatly contributed to diagnosis and risk assessment. In about 3–5% of patients an interstitial deletion of 13q14 or a larger 13q area is visible by chromosomal analysis (**1**), usually associated with signs of developmental delay. In familial retinoblastoma indirect DNA diagnosis can be achieved by segregation analysis of DNA markers at the *RB1* locus. The affected girl (II-1) has inherited haplotype **a** from her unaffected father and haplotype **c** from her unaffected mother (**2**). Analysis of tumor cells, obtained from the affected eye, shows that only haplotype **a** is present (loss of heterozygosity, LOH). This represents the allele carrying the mutation. Direct DNA analysis is based on demonstrating a germline mutation in blood cells. Here, the affected individuals I-2 and II-2 (**3**) carry a C-to-T transversion in codon 575 (CAA glutamine to TAA stop codon). (Photographs courtesy of W. Höpping, and D. Lohmann, Essen.)

References

Knudson AG: Mutation and cancer: Statistical study of retinoblastoma. Proc Nat Acad Sci 68: 820–823, 1971.

Lohmann DR: RB1 gene mutations in retinoblastoma. Hum Mutat 14: 283–288, 1999.

Lohmann DR et al: Spectrum of RB1 germ-line mutations in hereditary retinoblastoma. Am J Hum Genet 58: 940–949, 1996.

Newsham IF, Hadjistilianou T, Cavenee WK: Retinoblastoma, pp 357–386. In: Vogelstein B, Kinzler KW (eds) The Genetic Basis of Human Cancer, 2nd ed. McGraw-Hill, New York, 2002.

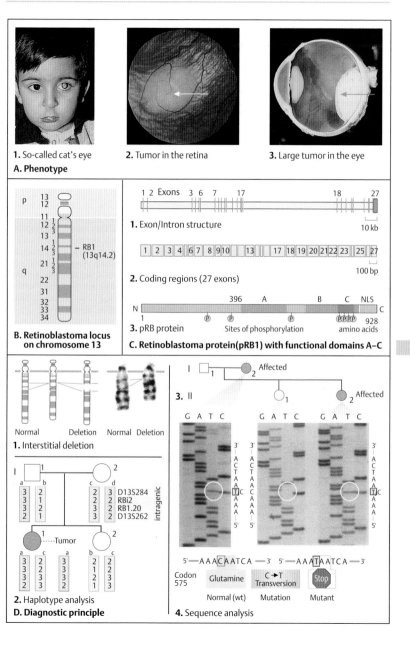

1. So-called cat's eye

2. Tumor in the retina

3. Large tumor in the eye

A. Phenotype

1. Exon/Intron structure

10 kb

2. Coding regions (27 exons)

100 bp

B. Retinoblastoma locus on chromosome 13

3. pRB protein   Sites of phosphorylation   amino acids

C. Retinoblastoma protein(pRB1) with functional domains A–C

Normal   Deletion   Normal Deletion

1. Interstitial deletion

3. II

Affected

Affected

		D13S284
		RBi2
		RB1.20
		D13S262

----Tumor

intragenic

2. Haplotype analysis

D. Diagnostic principle

5'—A A A C A A T C A — 3'   5'—A A A T A A T C A — 3'

Codon 575   Glutamine   C → T Transversion   Stop

Normal (wt)   Mutation   Mutant

4. Sequence analysis

The BCR/ABL Fusion Protein in CML

Chronic myelogenous leukemia (CML, MIM 151 410) is a malignant tumor in adults (1.5 per 100 000 per year) that originates by clonal expansion from a single myeloid cell of the bone marrow. The disease follows a chronic course. Acute crises develop intermittently and terminally. The hallmark of the disease is a reciprocal translocation between a chromosome 9 and a chromsome 22 in tumor cells. A small molecule can be used to interfere with an oncogenic protein produced by tumor cells. Other oncogenic translocations are listed in a table in the appendix.

A. Principal features

A strikingly increased number of myelocytes (white blood cells, stained blue) in the peripheral blood (**1**) and a translocation 9;22 (q34;q22) (**2**, arrows) characterize CML. The translocation results in a chromosome 22 of microscopically reduced size, called the Philadelphia chromosome (Ph[1]) after the city in which it was discovered in 1960, by P. Nowell and D. Hungerford. The Ph[1] chromosome is present in the bone marrow cells of most patients. If it is not present, the illness has a poorer prognosis and progresses more rapidly than usual. The Ph[1] chromosome is also present in 30–40 % of adults and 3–5 % of children with acute lymphocytic leukemia (ALL). Here, the presence of Ph[1] indicates a poor prognosis, and its absence is favorable. In ALL the fusion transcript, from which a fusion protein of 185–190 kDa is translated, is small (6.5–7.0 kb).

B. The Ph[1] translocation

About half of the long arm of a chromosome 22 is translocated to the long arm of a chromosome 9 (**1**) and a very small part of the distal long arm of a chromosome 9 is translocated to a chromosome 22 (**2**), but this is not visible by light microscopy. (Images by A. Schneider, Essen, and www.cmlsupport.com/cyto.jpg.)

C. Fusion of two genes, *BCR* and *ABL*

The breakpoints of the Ph[1] translocation are in the *BCR* gene at 22 q11 and the *ABL* gene at 9 q34. As a result these two genes are fused. The exact locations of the breakpoints differ, but in the *BCR* gene are limited to a small region of 5.8 kb (thus the designation *BCR*, or *b*reakpoint cluster *r*egion). In CML, the breakpoints lie in exons 10–12 of the *BCR* gene; in acute Ph[1]-positive leukemias (e.g., acute lymphocytic leukemia) they lie further in the 5′ direction in exon 1 or 2. The breakpoint region in the *ABL* gene extends over 180 kb between exons 1a and 1b.

D. The BCR/ABL fusion protein

The *ABL* gene is transcribed into two alternative mRNA transcripts of 7 kb (exon 1 b, 2–11) and 6 kb (exon 1 a, 2–11) length (**1**) and translated into a 145-kDa protein called p145^{abl} (**2**). This is a tyrosine protein kinase. The fused genes in CML are transcribed into an 8.5-kb mRNA transcript (**3**). The fusion protein (**4**) is a 210-kDa protein (p210$^{bcr/abl}$). Unlike the normal protein, the fusion protein has an inappropriately active ABL kinase domain. This causes hematopoietic cells in the bone marrow to proliferate excessivly, resulting in uncontrolled cell division and tumor growth. A small molecule, STI-571 (**5**), precisely fits into the fusion protein and abolishes its abnormal function (6). It used for therapy in CML under the name gleevec®. (Figure adapted from Schindler, 2000.)

References

Bartram CR et al: Translocation of c-abl oncogene correlates with the presence of a Philadelphia chromosome in chronic myelocytic leukaemia. Nature 306: 277–280, 1983.

Faderl S et al: The biology of chronic myeloid leukemia. New Engl J Med 341: 164–172, 1999.

Kurzrok R, Gutterman JU, Talpaz M: The molecular genetics of Philadelphia-positive leukemias. New Engl J Med 319: 990–998, 1988.

Sawyers CL: Chronic myeloid leukemia. New Engl J Med 340: 1330–1340, 1999.

Schindler T et al: Structural mechanism for STI-571 inhibition of Abelson tyrosine kinase. Science 289: 1938–1942, 2000.

Wetzler M, Byrd JC, Bloomfield CD: Acute and chronic myeloid leukemia, pp 631–641. In: Kasper DL et al (eds) Harrison's Principles of Internal Medicine, 16th ed. McGraw-Hill, New York, 2005.

Online information:
www.cmlsupport.com

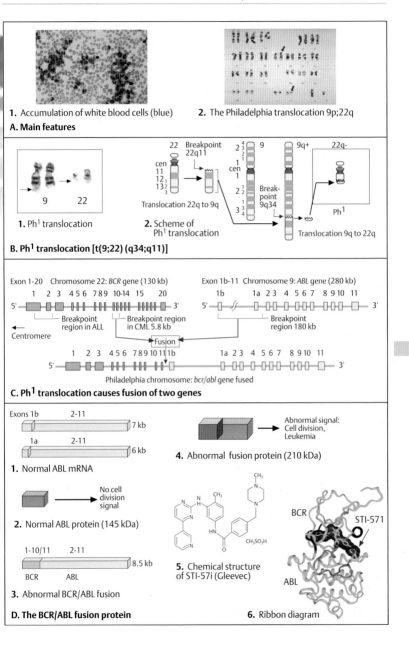

1. Accumulation of white blood cells (blue) **2.** The Philadelphia translocation 9p;22q

A. Main features

1. Ph^1 translocation

2. Scheme of Ph^1 translocation

22 Breakpoint 22q11

cen
11
12$_1$
13$\frac{2}{3}$

Translocation 22q to 9q

2$\frac{4}{3}$ 9
2$_1$
cen
1
2$\frac{1}{2}$ Break-point 9q34
3$\frac{1}{4}$

9q+ 22q-

Ph^1

Translocation 9q to 22q

B. Ph^1 translocation [t(9;22) (q34;q11)]

Exon 1-20 Chromosome 22: *BCR* gene (130 kb)   Exon 1b-11 Chromosome 9: *ABL* gene (280 kb)

1 2 3 4 5 6 7 8 9 10-14 15 20   1b   1a 2 3 4 5 6 7 8 9 10 11

5'————————————— 3'   5'——//——————————————— 3'

Breakpoint region in ALL   Breakpoint region in CML 5.8 kb   Breakpoint region 180 kb

←
Centromere

Fusion

1 2 3 4 5 6 7 8 9 10 11 1b   1a 2 3 4 5 6 7 8 9 10 11

5'—————————————————————————— 3'

Philadelphia chromosome: *bcr/abl* gene fused

C. Ph^1 translocation causes fusion of two genes

Exons 1b 2-11
7 kb
1a 2-11
6 kb

1. Normal ABL mRNA

No cell division signal

2. Normal ABL protein (145 kDa)

1-10/11 2-11
8.5 kb
BCR ABL

3. Abnormal BCR/ABL fusion

Abnormal signal: Cell division, Leukemia

4. Abnormal fusion protein (210 kDa)

CH_3
N
CH_3
N
N
HN
N
CH$_3$SO$_3$H
HN
N

5. Chemical structure of STI-57i (Gleevec)

BCR
STI-571
ABL

6. Ribbon diagram

D. The BCR/ABL fusion protein

Neurofibromatosis

The neurofibromatoses are a group of clinically and genetically different autosomal dominant hereditary diseases that predispose to benign and malignant tumors of the nervous system. The most important forms are neurofibromatosis-1 (NF1) and neurofibromatosis-2 (NF2, MIM 101000/607379), but others also exist.

A. Main manifestations of NF1

NF1 (von Recklinghausen disease, MIM 162200) affects 1 in 3000 individuals with very variable manifestations. Characteristic are Lisch nodules of the iris (**1**) in more than 90% of patients, café-au-lait spots (**2**) (more than five spots of more than 2 cm diameter are considered diagnostic) in more than 95%, and multiple neurofibromas (**3**) in more than 90% of patients, usually apparent between the ages of 4 and 15 years. About 2–3% of patients develop a neurofibrosarcoma or other malignancies.

B. The *NF1* gene

The *NF1* gene is a large gene spanning 350 kb of genomic DNA organized into at least 59 exons. It is localized on human chromosome 17 at q11.2. The gene was isolated in 1990 from a 600-kb *Nru*I restriction fragment by positional cloning. Two patients with a translocation involving the long arm of chromosome 17 with breakpoints at 17q11.2 and CpG islands provided anchor points. Three unrelated genes, *OMGP, EVI2B*, and *EVI2A*, are embedded within the *NF1* gene on the opposite DNA strand. About 50% of patients with NF1 have a new mutation. The mutations are deletions, insertions, base substitutions, and splice mutations. Currently mutations are found in about 60–70% of patients. (Figure adapted from Claudio & Rouleau, 1998.)

C. The NF1 gene product, neurofibromin-1

From multiple alternatively spliced transcripts 11–13 kb in size a 220–250 kDA protein, neurofibromin-1, with 2818 amino acids is translated and expressed in many tissues. It is a member of the GTPase-activating proteins (GAP) that down-regulate p21 in the Ras signal pathway (see p. 326). A region of GAP homology with a gene product in yeast *(S. cerevisiae)*, Ira1 (inhibitor of *ras* mutants), is between amino acids 840 and 1200. Mutations at the *NF1* locus, germline or somatic, interrupt the Ras signal pathway and result in loss of control of cell division (Figure adapted from Xu et al., 1990.)

D. The neurofibromatosis-2 gene *NF2*

This gene, localized on human chromosome 22 at q12, is organized into 16 constitutive exons and one alternatively spliced exon, spanning about 110 kb of genomic DNA. It encodes three mRNAs, of 2.6, 4.4, and 7.0 kb size. This widely expressed gene was identified in 1993 by Rouleau et al. and Trofatter et al. within a cosmid contig of YAC clones (yeast artificial chromosomes). Two deletions (Del 1 and Del 2) in unrelated patients aided in finding the gene. Mutations can be detected in more than 50% of patients (large deletions including the entire gene or several exons and small deletions are frequent). The gene product, neurofibromin-2 or schwannomin, is a member of the band 4.1 cytoskeleton-associated protein superfamily (see p. 382) called the ERM family (including ezrin, radixin, moesin and several protein tyrosine phosphatases). The basic function of these proteins is to maintain cellular integrity. (Figure adapted from Claudio & Rouleau, 1998.)

References

Carey JC, Viskochil DH: Neurofibromatosis Type 1: a model condition for the study of the molecular basis of variable expressivity in human disorders. Am J Med Genet (Semin Med Genet) 89: 7–13, 1999.

Claudio JO, Rouleau GA: Neurofibromatosis type 1 and type 2, pp 963–970. In: Jameson JL (ed) Principles of Molecular Medicine. Humana Press, Totowa, NJ, 1998.

Huson SM: What level of care for the neurofibromatoses? Lancet 353: 1114–1116, 1999.

Messiaen LM et al: Exhaustive mutation analysis of the *NF1* gene allows identification of 95% of mutations and reveals a high frequency of unusual splicing defects. Hum Mutat 15: 541–555, 2000.

Riccardi VM, Eichner JE: Neurofibromatosis. Phenotype, Natural History and Pathogenesis, 2nd ed. Johns Hopkins University Press, Baltimore, 1992.

Rouleau GA et al: Alteration in a new gene encoding a putative membrane-organizing protein causes neurofibromatosis type 2. Nature 363: 515–521, 1993.

Trofatter JA et al: A novel Moesin-, Ezrin-, Radixin-like gene is a candidate for the neurofibromatosis 2 tumor suppressor. Cell 72: 791–800, 1993.

Xu G et al: The neurofibromatosis type 1 gene encodes a protein related to GAP. Cell 62:599–608, 1990.

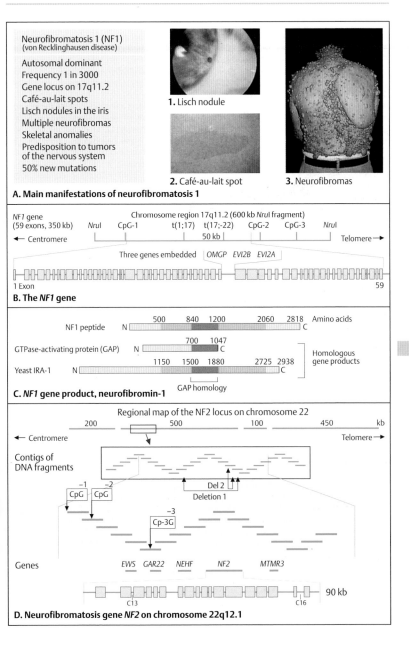

Neurofibromatosis 1 (NF1)
(von Recklinghausen disease)

Autosomal dominant
Frequency 1 in 3000
Gene locus on 17q11.2
Café-au-lait spots
Lisch nodules in the iris
Multiple neurofibromas
Skeletal anomalies
Predisposition to tumors
of the nervous system
50% new mutations

1. Lisch nodule

2. Café-au-lait spot

3. Neurofibromas

A. Main manifestations of neurofibromatosis 1

NF1 gene
(59 exons, 350 kb)   *Nru*I   CpG-1   t(1;17)   t(17;-22)   CpG-2   CpG-3   *Nru*I

Chromosome region 17q11.2 (600 kb *Nru*I fragment)

← Centromere | 50 kb | Telomere →

Three genes embedded   *OMGP*   *EVI2B*   *EVI2A*

1 Exon   59

B. The *NF1* gene

                              500   840   1200   2060   2818   Amino acids
NF1 peptide   N   C

                                        700   1047
GTPase-activating protein (GAP)   N   C   Homologous
                                                                           gene products
                              1150   1500   1880   2725   2938
Yeast IRA-1   N   C

                                         GAP homology

C. *NF1* gene product, neurofibromin-1

Regional map of the NF2 locus on chromosome 22

        200   500   100   450   kb

← Centromere   Telomere →

Contigs of
DNA fragments

                    −1   −2
                   CpG   CpG   Del 2
                                      Deletion 1

                            −3
                          Cp-3G

Genes   *EWS*   *GAR22*   *NEHF*   *NF2*   *MTMR3*

                                                                    90 kb
              C13   C16

D. Neurofibromatosis gene *NF2* on chromosome 22q12.1

Genomic Instability Diseases

Ataxia-telangiectasia, Fanconi anemia, and Bloom syndrome are important examples of hereditary diseases resulting from mutations in genes that contribute to genome stability. Different patterns of chromosomal breaks and rearrangements are visible by light microscopy of metaphase cells. The underlying genetic defects predispose affected individuals to different types of cancer.

A. Ataxia-telangiectasia (A-T)

A-T (MIM 208900) is a variable disease due to autosomal recessive mutations in the *ATM* gene at gene map locus 11q23. The main manifestations are immune defects, cerebellar ataxia, and characteristic telangiectasias of the conjunctivae (**1**), which develop in early childhood. Affected individuals are highly sensitive to irradiation and are prone to develop lymphomas and leukemias. The *ATM* gene has 66 exons spanning 150 kb of genomic DNA. A 3056-amino acid (350kD) protein kinase, ATM, is translated from its alternatively spliced 13-kb transcript. ATM is activated in response to double-strand DNA breaks. It has a central role in a network of proteins that regulate cellular responses to DNA damage and recombination. Mutations in a related gene result in the Nijmegen breakage syndrome (NBS1, MIM 251260).

B. Fanconi anemia (FA)

FA (MIM 227650) is a heterogeneous group of autosomal recessive and X-chromosomal diseases manifest in early childhood as pancytopenia, growth deficiency (**1**), hypoplastic radius often with hypoplastic or absent thumbs (**2**), and other malformations. About eight FA genes form a complementation group (see table in appendix). The proteins encoded by these genes form the FA complex. Together with other proteins, they detect DNA damage or errors in replication. The most prevalent mutation is of *FA-A* (also called *FANCA*), in about 65% of patients. FA cells are hypersensitive to DNA-crosslinking agents, such as diepoxybutane (DEB), which induces chromosomal breaks. (Diagram adapted from Rahman & Ashworth, 2004.)

C. Bloom syndrome (BLM)

BLM (210900) is a prenatal and postnatal growth deficiency disease (birth weight 2000g,

birth length 40 cm, adult height ca. 150 cm) with a distinct phenotype (**1**) including a narrow face, sunlight-induced facial erythema, variable immune deficiency, and a greatly increased risk of different malignancies (about 1 in 5 patients). Chemotherapy is very poorly tolerated.

The hallmark is a tenfold increase in the spontaneous rate of sister chromatid exchanges (SCE, see glossary) (**2**, **3**). Breaks in one or both chromatids and exchanges between homologous chromosomes occur in about 1–2% of metaphase cells.

BLM results from autosomal recessive mutations in the *BLM* gene at gene map locus 15q26.1, encoding a member of the RecQ family of DNA helicases. The 1417-amino acid BLM protein interacts with the FA complex and is involved in meiotic recombination. It is homologous to yeast Sgs1 (slow growth suppressor) and the human WRN protein (Werner syndrome, MIM 277700). Mainly protein-truncating nonsense mutations are distributed fairly evenly along the gene (**4**), but some missense mutations exist. Most distinct is a founder mutation in populations of Ashkenazi Jewish origin, consisting of a 6-bp deletion/7-bp insertion at nucleotide 2281. Homozygosity for mutations in the *BLM* gene results in an increased rate of somatic mutations.

References

Auerbach AD, Buchwald M, Joenje H: Fanconi anemia, pp. 289–306. In: Vogelstein B, Kinzler KW (eds) The Genetic Basis of Human Cancer, 2nd ed. McGraw-Hill, New York, 2002.

D'Andrea ADD, Grompe M: The Fanconi anaemia/BRCA pathway. Nature Rev Cancer 3: 23–34, 2003.

Gatti R: Ataxia-telangiectasia, pp 239–266. In: Vogelstein B, Kinzler KW (eds) The Genetic Basis of Human Cancer, 2nd ed. McGraw-Hill, New York, 2002.

German J, Ellis NA: Bloom syndrome, pp 301–315. In: Vogelstein B, Kinzler KW (eds) The Genetic Basis of Human Cancer, 2nd ed. McGraw-Hill, New York, 2002.

Rahman N, Ashworth A: A new gene on the X involved in Fanconi anemia. Nature Genet 36: 1142–1143, 2004.

Zhao S et al: Functional link between ataxia-telangiectasia and Nijmegen breakage syndrome gene products. Nature 405: 473–477, 2000.

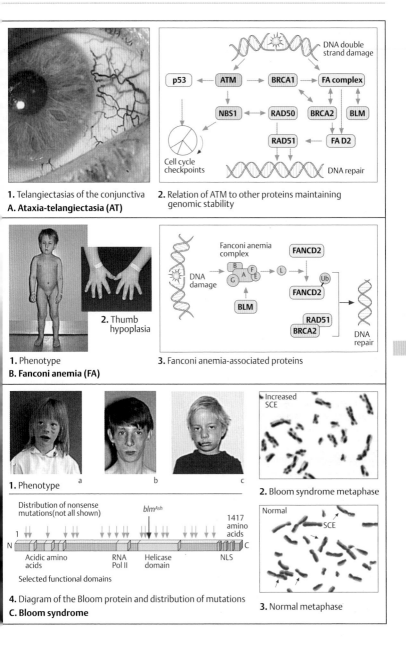

1. Telangiectasias of the conjunctiva
A. Ataxia-telangiectasia (AT)

2. Relation of ATM to other proteins maintaining genomic stability

2. Thumb hypoplasia

1. Phenotype
B. Fanconi anemia (FA)

3. Fanconi anemia-associated proteins

1. Phenotype

2. Bloom syndrome metaphase

Distribution of nonsense mutations(not all shown)

blm^Ash

1417 amino acids

Acidic amino acids

RNA Pol II

Helicase domain

NLS

Selected functional domains

4. Diagram of the Bloom protein and distribution of mutations
C. Bloom syndrome

3. Normal metaphase

Hemoglobin – Overview

Oxygen induces random alterations in biological molecules. Therefore it is toxic unless bound to biomolecules. Hemoglobin and myoglobin are proteins specialized in binding and transporting oxygen in vertebrates: myoglobin in muscle cells, hemoglobin in red blood cells. Myoglobin is a single globin polypeptide of about 150 amino acids with a single oxygen binding site. Hemoglobin, named by Hoppe-Seyler in 1862, is composed of four globin chains, each with an oxygen-binding site. Different types of hemoglobin molecules arose from myoglobin during the course of evolution. Hemoglobin was the first human protein to be defined by its amino acid sequence (Ingram 1956) and its three-dimensional structure at 5.5 Å resolution (Perutz 1960).

A. Types of hemoglobin

The subunits of hemoglobin are designated by the Greek letters α, β, γ, and δ. In addition specialized globin subunits exist in mammals during embryonic development, designated as ϵ (epsilon) and ζ (zeta). A hemoglobin molecule is stable only when the two pairs of globin differ. Adult hemoglobin (HbA) is composed of two alpha and two beta chains ($\alpha 2\beta 2$), fetal hemoglobin (HbF) of two alpha and two gamma chains ($\alpha 2\delta 2$). A minor adult hemoglobin, HbA$_2$, has a variant delta chain ($\alpha 2\gamma 2$). During early embryonic stages transient embryonic hemoglobins are present: Hb Gower 1 ($\zeta 2\epsilon 2$), Hb Gower 2 ($\alpha 2\epsilon 2$), and Hb Portland ($\zeta 2\gamma 2$). (Most hemoglobin names indicate where they were discovered.)

B. Hemoglobins in thalassemia

The thalassemias are a group of genetically determined disorders of hemoglobin synthesis. They are classified according to which chain is affected, the α chain (α-thalassemia, α-thal) or the β chain (β-thalassemia, β-thal). β-thalassemia affects HbA; α-thalassemia affects HbA and HbF. Hemoglobins with four identical globin chains are completely unstable and incompatible with life (HbH with four β chains [β4], Hb Bart's with four γ chains [γ4]).

C. Evolution of hemoglobin

The four oxygen binding sites in the $\alpha 2\beta 2$ HbA molecule interact, which permits an allosteric change in the molecule as it takes up and releases oxygen. The different types of hemoglobin existing in mammals today arose by duplication of genes encoding the globin chains. A primitive single-chain oxygen-binding protein, present early in evolution, adapted to the rising concentration of oxygen in the earth's atmosphere at an earlier time. It is assumed to be the ancestor of myoglobin and modern hemoglobins. About 500 million years ago, a hemoglobin gene underwent duplication, and from this the genes encoding the α type and the β type hemoglobins arose. In the evolution of mammals during the past 100 million years, hemoglobin evolved from an ancestral β type into different types by further duplication events.

D. Globin synthesis during embryonic development

The different types of globin chain are formed at different developmental stages at different anatomical sites: during the first six weeks in humans (embryonic hemoglobins), in the yolk sac; from the 6th week until prior to birth (fetal hemoglobin), in the liver and spleen; and in adult life, in red blood cell precursors in the bone marrow. They differ in oxygen-binding affinity. Thus, oxygen delivery is optimized for different phases of development, clearly an evolutionary selective advantage.

(Figure in A and B adapted from Lehmann & Huntsman, 1974; in D from Weatherall et al, 2001.)

References

Ingram VM: Specific chemical difference between the globin of normal and sickle-cell anaemia hemoglobin. Nature 178: 792–794, 1956.

Lehmann H, Huntsman RG: Man's Hemoglobins. North-Holland, Amsterdam, 1974.

Thein SL, Rochette J: Disorders of hemoglobin structure and synthesis, pp 179–190. In: Jameson JL (ed, Principles of Molecular Medicine. Humana Press, Totowa, New Jersey, 1998.

Weatherall DJ, Clegg JB, Higgs DR, Wood WG: The hemoglobinopathies, pp 4571–4636. In: Scriver CR et al (eds) The Metabolic and Molecular Bases of Inherited Disease, 8th ed. McGraw-Hill, New York 2001.

Weatherall DJ, Clegg JB: Genetic disorders of hemoglobin. Semin Hematol 36:2–37, 1999.

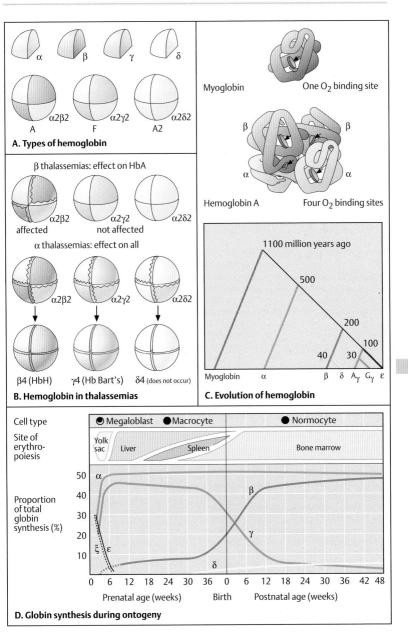

A. Types of hemoglobin

α β γ δ

α2β2 α2γ2 α2δ2
A F A2

B. Hemoglobin in thalassemias

β thalassemias: effect on HbA

α2β2 α2γ2 α2δ2
affected not affected

α thalassemias: effect on all

α2β2 α2γ2 α2δ2

β4 (HbH) γ4 (Hb Bart's) δ4 (does not occur)

Myoglobin One O₂ binding site

Hemoglobin A Four O₂ binding sites
β β
α α

C. Evolution of hemoglobin

1100 million years ago
500
200
100
40 30

Myoglobin α β δ Aγ Gγ ε

D. Globin synthesis during ontogeny

Cell type: Megaloblast Macrocyte Normocyte

Site of erythropoiesis: Yolk sac Liver Spleen Bone marrow

Proportion of total globin synthesis (%)

α β γ δ ξ ε

Prenatal age (weeks) Birth Postnatal age (weeks)
0 6 12 18 24 30 36 0 6 12 18 24 30 36 42 48

Hemoglobin Genes

Hemoglobins are encoded by clusters of α-type and β-type globin genes on human chromosomes 16 and 11, respectively. A specific gene is responsible for each type of the different globin polypeptide chains. Globin genes are under common control of an upstream (5′ direction) region (locus control region, LCR). They are arranged and expressed in a sequence, according to the time of activation during developmental stages (see previous page).

A. The β globin and α globin genes

The β globin-like genes (β, δ, γA, γG, and ε) of man are located as a gene cluster in the 3′ to 5′ direction on the short arm of chromosome 11 in region 1, band 5.5 (11p15.5). They span about 60 000 base pairs (bp), or 60 kb (kilobases), of DNA (1). There are two γ genes, γA and γG, which have arisen by a duplication event. They differ only in codon 136, which encodes an alanine in γA and a glycine in γG. A pseudogene (ψβ₁) is located between the γA gene and the δ gene. It is similar to the β gene, but contains deletions and internal stop codons and does not function. A long-range control region (LCR) located upstream (in the 5′ direction) jointly regulates these genes. As a result of their origin from a common ancestral gene, all globin genes have a similar structure.

Two α-globin genes are located on the short arm of human chromosome 16 (16p13.11 to 16p13.33) on a DNA segment of about 30 kb. A ζ gene, which is active only during the embryonic period, lies in the 5′ direction. Three pseudogenes, ψζ, ψα₂, and ψα₁, are located in between. A further gene, θ, with unknown function, has been identified in this region.

The β-globin gene (MIM 141900), the prototype, spans about 1.6 kb (1600 base pairs). It consists of three exons separated by a short and a long intron. Exon 1 encodes amino acids 1–30, exon 2 aa 31–104, and exon 3 aa 105–146 (2). Their coding sequences of the other beta-like genes also are arranged in three exons. Their exons are of similar size, but they differ in intron 2, which has 850–900 bp in the β genes. The alpha genes HBA1 and HBA2 (MIM 141800, 141850) span about 0.8 kb (800 bp), mainly due to a smaller intron 2 than in the beta group.

C. Tertiary structure of the β globin chain

The three-dimensional structures of myoglobin and of the hemoglobin α and β chains, shown schematically, are very similar, although their amino acid sequences correspond in only 24 of 141 positions. The β chain, with 146 amino acids, is somewhat longer than the α chain, with 141 aminoacids. The structural similarity is functionally significant: the oxygen-binding region resides inside the molecule, protected from the surrounding liquid. Oxygen uptake and release is reversible.

D. Functional domains of the β chain

Three functional and structural domains can be distinguished in all globin chains. They correspond to the three exons of the gene. Two domains, consisting of amino acids 1–30 and 105–146 (coded for by exons 1 and 3), are located on the outside. They are mainly formed of hydrophilic amino acids. A third domain, lying inside the molecule (coded for by exon 2), contains the oxygen-binding site and consists mainly of nonpolar hydrophobic amino acids.

The amino acid sequences of the hemoglobins of more than 60 investigated species are identical in nine positions. These invariant positions are especially important for the function of the molecule. Changes (mutations) in the invariant positions affect function so severely that they are not tolerated.

References

Antonarakis SE, Kazazian Jr HH, Orkin SH: DNA polymorphism and molecular pathology of the human globin gene clusters. Hum Genet 69: 1–14, 1985.

Stamatoyannopoulos G, Majerus PW, Perimutter RM, Varmus H, eds: The Molecular Basis of Blood Disease, 4th ed. Saunders, New York, 2001.

Weatherall DJ et al.: The hemoglobinopathies, p 4571–4636. In: CR Scriver et al., eds, The Metabolic and Molecular Bases of Inherited Disease. 8th ed. McGraw Hill, New York, 2001.

Weatherall DJ, Clegg JB: The Thalassaemia Syndromes. 4th ed. Blackwell Science, Oxford, 2001.

Weatherall DJ: Phenotype-genotype relationships in monogenic disease: Lessons from the thalassaemias. Nature Rev Genet 2: 245–255, 2001.

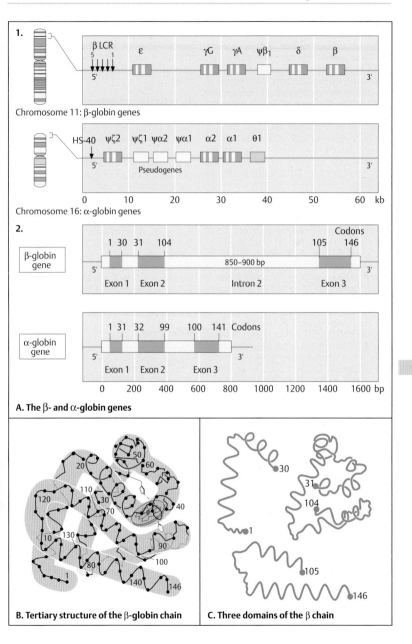

1.

Chromosome 11: β-globin genes

Chromosome 16: α-globin genes

2.

β-globin gene

α-globin gene

A. The β- and α-globin genes

B. Tertiary structure of the β-globin chain

C. Three domains of the β chain

Sickle Cell Anemia

Sickle cell anemia (Herrick, 1910) is a severe hemolytic anemia associated with many complications. It results from homozygosity for a mutation in the β globin gene. It is frequent in tropical regions where malaria is endemic (see p. 174). With a frequency of 1 in 500, it is an important cause of morbidity and mortality in these regions. Sickle cell anemia is the first human disease to be understood at the molecular level (Pauling et al., 1949). It is transmitted by autosomal recessive inheritance (Neel, 1949). Heterozygous carriers can readily be identified.

A. Sickle cells

In a blood smear under the light microscope, most erythrocytes (red blood cells) of affected persons are sickle-shaped (**1**). Small blood vessels are occluded by sickle cells (**2**). As a result, blood supply to many tissues is diminished. In a normal blood smear (**3**), erythrocytes appear as regular round disks of about 7 μm in diameter. In the course of the disease, acute crises called sickle crises occur, during which sickle-like cells are greatly increased and completely dominate the blood picture. Heterozygotes show occasional sickle cells but do not suffer from sickle crises, and at the most have only very mild signs and symptoms. (Images 1 and 3 by Dr. Daniel Nigro, Santa Ana College, California; image 2 by National Heart, Lung and Blood Institute, Bethesda, Maryland.)

B. Consequences of the sickle cell mutation

All manifestations of sickle cell anemia can be understood on the basis of the underlying mutation. In 1956 V.M. Ingram determined by amino acid sequence analysis that glutamic acid is replaced by valine in codon 6 of the β globin gene (mutation E6V). This alteration results from the transversion of an adenine (A) to a thymine (T), which changes codon GAG to GTG. Valine is a hydrophobic amino acid on the outside of the molcule. This makes sickle cell hemoglobin (HbS) less soluble than normal hemoglobin. HbS crystallizes in the deoxy state, forms small rods, and does not allow erythrocytes to deform when passing through small blood vessels. As a result, small arteries and capillaries become clogged and cause local oxygen deficiency in various organs. Chronic oxygen deficiency of the brain may lead to learning disability. Defective erythrocytes are destroyed (hemolysis). Chronic anemia results in numerous sequelae such as heart failure, liver damage, and infections.

C. Selective advantage for HbS heterozygotes in areas of malaria

About 1.5–2.5 million children die each year from malaria, mostly in Sub-Sahara Africa.

A.C. Allison in 1954 suggested that individuals heterozygous for the sickle cell mutation are protected from severe malaria infection; they are less frequently and less severely affected. Erythrocytes of heterozygotes for the sickle cell mutation are a less favorable environment for the malaria parasite than those of normal homozygotes. Heterozygotes have a higher probability of survival and of being able to reproduce. Sickle cell anemia is the best example in humans of a selective advantage in heterozygotes (see p. 174). The sickle cell mutation has arisen independently in at least four or five different malaria-infested regions and has been maintained in these populations.

References

Allison AC: Polymorphism and natural selection in human populations. Cold Spring Harb Symp Quant Biol 29: 137–149, 1954.

Ashley-Koch A, Yang Q, Olney RS: Sickle hemoglobin (HbS) allele and sickle cell disease: a HuGE review. Am J Epidemiol 15: 839–845, 2000.

Evans AG, Wellems TE: Coevolutionary genetics of *Plasmodium* malaria parasites and their human hosts. Integrative & Comparative Biology 42: 401–417, 2002.

Herrick JB: Peculiar elongated and sickle-shaped red blood cell corpuscles in a rare case of severe anemia. Arch Intern Med 6: 517–521, 1910.

Neel JV: The inheritance of sickle cell anemia. Science 110: 64–66. 1949.

Pauling L, Itano HA, Singer SJ, Wells IG: Sickle cell anemia, a molecular disease. Science 110: 543–548, 1949.

Pawlink R et al: Correction of sickle cell disease in transgenic mouse models by gene therapy. Science 294: 2368–2371, 2001.

Stuart MJ, Nagel RL: Sickle-cell anemia. Lancet 364: 1343–1360, 2004.

Vernick KD, Waters AP: Genomics and malaria control. New Eng J Med 351: 1901–1904, 2004.

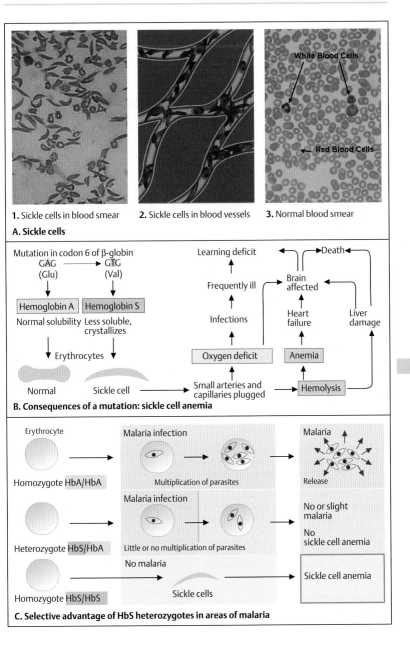

1. Sickle cells in blood smear   **2.** Sickle cells in blood vessels   **3.** Normal blood smear

A. Sickle cells

B. Consequences of a mutation: sickle cell anemia

Mutation in codon 6 of β-globin
GAG ⟶ GTG
(Glu)   (Val)

Hemoglobin A   Hemoglobin S
Normal solubility   Less soluble, crystallizes
⟶ Erythrocytes ⟶

Normal   Sickle cell ⟶ Small arteries and capillaries plugged ⟶ Hemolysis

Oxygen deficit   Anemia

Learning deficit
Frequently ill
Infections
Brain affected
Heart failure
Liver damage
Death

C. Selective advantage of HbS heterozygotes in areas of malaria

Erythrocyte

Homozygote HbA/HbA — Malaria infection — Multiplication of parasites — Malaria / Release

Heterozygote HbS/HbA — Malaria infection — Little or no multiplication of parasites — No or slight malaria / No sickle cell anemia

Homozygote HbS/HbS — No malaria — Sickle cells — Sickle cell anemia

Mutations in Globin Genes

More than 750 human hemoglobin variants with single amino acid substitutions in one of the globin chains are known. Many of them are associated with hemoglobin disorders of different types and severity. In addition, elongated or shortened globin chains and chains of fused parts of a β and a δ chain or of a β and a γ chain occur. The functional consequences vary, depending on the electrical charge and size of the substituted amino acid and its position in the polypeptide. Mutations may decrease the elasticity of the molecule, alter its oxygen affinity, or cause instability.

A. Structural changes in the β globin gene

More than 300 point mutations in the β globin gene and over 100 in one of the α globin genes have been documented. Two clinically important mutations affect codon 6: the sickle cell mutation, 6 Glu → Val (sickle cell hemoglobin, HbS, resulting in the incorporation of valine instead of glutamic acid) and 6 Glu → Lys (hemoglobin C, HbC, incorporating lysine instead of glutamic acid in codon 6). Compound heterozygotes with the HbS mutation on one chromosome and the HbC on the other (HbSC) are common. The marked methemoglobin formation in Hb$_{Zürich}$ and Hb$_{Saskatoon}$ results from substitutions for histidine (His) in codon 63, which alter the oxygen-binding region of the hemoglobin molecule. HbE ($\alpha 2\beta 2^{26\,Glu \to Lys}$) is very common in Thailand, Cambodia, and Vietnam.

B. Unequal crossing-over

The sequence homology between the globin genes may lead to nonhomologous pairing and unequal crossing-over during meiosis. A characteristic example is Hemoglobin$_{Gun\,Hill}$, described in 1968. This variant results from pairing of codon 90 with codon 95, 91 with 96, etc. As a result, codons 91–95 subsequently are deleted in one strand and duplicated in the other (not shown). This leads to an unstable hemoglobin. More than 90 unstable hemoglobins are known.

C. Fusion hemoglobins

Fused or hybrid hemoglobin variants probably also result from unequal crossing-over involving parts of adjacent genes. The first is Hb$_{Lepore}$,

described in 1962. Here the first 50–80 amino acids of the δ chain are fused with the last 60–90 residues of the normal C-terminal amino acids of the β chain. The complementary situation is Hb anti-Lepore, a δβ fusion gene together with the normal δ and β genes.

D. Hemoglobin with a chain elongation

More than ten globin chain variants with an elongated chain are known. Single base substitutions in a chain termination codon, frameshift mutations, or mutations affecting the translation initiator methionine are responsible for this type of variant. In Hb$_{Cranston}$ an insertion in codon 145 of the β chain changes UAU (tyrosine) to AGU (serine). This changes the stop codon after position 146 into ACU (tyrosine) and causes a read-through until codon 157, and consequently a chain that is ten amino acids too long (**1**).

In Hb $_{Constant\,Spring}$ (**2**), the α chain is elongated by a mutation in the stop codon UAA to CAA, which codes for glutamine (Gln). As a result the sequences that follow the stop codon are translated, resulting in a peptide that is 31 amino acids too long.

References

Baglioni C: The fusion of two peptide chains in hemoglobin Lepore and its interpretation as a genetic deletion. Proc Nat Acad Sci 48: 1880–1886, 1962.

Benz Jr EJ: Genotypes and phenotypes—another lesson from the hemoglobinopathies. New Engl J Med 351: 1490–1492, 2004.

Old J: Hemoglobinopathies and thalassemias, pp 1861–1898. In: Rimoin DL et al (eds) Emery and Rimoin's Principles and Practice of Medical Genetics, 4th ed. Churchill-Livingstone, London–Edinburgh, 2002.

Perutz MF, Lehmann H: Molecular pathology of human hemoglobin. Nature 219: 902–909, 1968.

Rieder RF, Bradley TB: Hemoglobin Gunn Hill: An unstable protein associated with chronic hemolysis. Blood 32: 355–369, 1968.

Stamatoyannopoulos G, Majerus PW, Perimutter RM, Varmus H (eds): The Molecular Basis of Blood Disease, 4th ed. Saunders, New York, 2001.

Thein SL, Rochette J: Disorders of hemoglobin structure and synthesis, pp 179–190. In: Jameson JL (ed) Principles of Molecular Medicine. Humana Press, Totowa, New Jersey, 1998.

Weatherall DJ et al: The hemoglobinopathies, pp 3417–3484. In: Scriver CR et al (eds) The Metabolic and Molecular Bases of Inherited Disease, 8th ed. McGraw-Hill, New York, 2001.

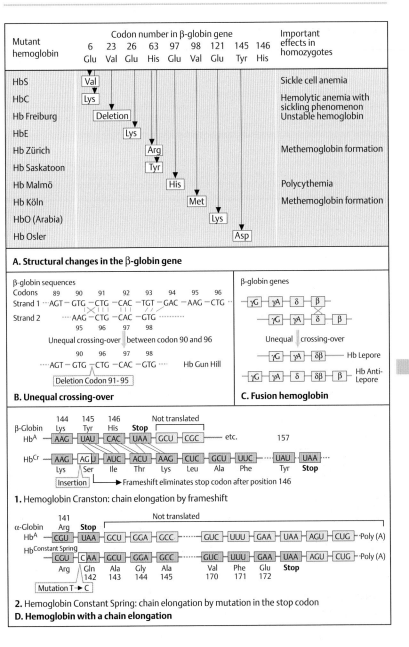

| Mutant hemoglobin | \multicolumn{9}{c}{Codon number in β-globin gene} | Important effects in homozygotes |
	6 Glu	23 Val	26 Glu	63 His	97 Glu	98 Val	121 Glu	145 Tyr	146 His	
HbS	Val									Sickle cell anemia
HbC	Lys									Hemolytic anemia with sickling phenomenon
Hb Freiburg		Deletion								Unstable hemoglobin
HbE			Lys							
Hb Zürich				Arg						Methemoglobin formation
Hb Saskatoon				Tyr						
Hb Malmö					His					Polycythemia
Hb Köln						Met				Methemoglobin formation
HbO (Arabia)							Lys			
Hb Osler								Asp		

A. Structural changes in the β-globin gene

β-globin sequences
Codons 89 90 91 92 93 94 95 96
Strand 1 ···AGT — GTG —CTG — CAC — TGT —GAC —AAG — CTG ···
Strand 2 ···· AAG —CTG — CAC — GTG ·········
95 96 97 98

Unequal crossing-over between codon 90 and 96

90 96 97 98
····AGT — GTG —CTG — CAC — GTG ···· Hb Gun Hill

Deletion Codon 91–95

B. Unequal crossing-over

β-globin genes

γG — γA — δ — β
γG — γA — δ — β

Unequal crossing-over

γG — γA — δβ — Hb Lepore
γG — γA — δ — δβ — β — Hb Anti-Lepore

C. Fusion hemoglobin

Not translated
144 145 146
β-Globin Lys Tyr His **Stop**
HbA AAG — UAU — CAC — UAA — GCU — CGC — etc. 157

HbCr AAG — AGU — AUC — ACU — AAG — CUC — GCU — UUC — UAU — UAA ···
Lys Ser Ile Thr Lys Leu Ala Phe Tyr **Stop**
Insertion ——→ Frameshift eliminates stop codon after position 146

1. Hemoglobin Cranston: chain elongation by frameshift

Not translated
141
α-Globin Arg **Stop**
HbA CGU — UAA — GCU — GGA — GCC —······— GUC — UUU — GAA — UAA — AGU — CUG — Poly (A)

HbConstant Spring CGU — C AA — GCU — GGA — GCC —······— GUC — UUU — GAA — UAA — AGU — CUG — Poly (A)
Arg Gln Ala Gly Ala Val Phe Glu **Stop**
142 143 144 145 170 171 172
Mutation T → C

2. Hemoglobin Constant Spring: chain elongation by mutation in the stop codon

D. Hemoglobin with a chain elongation

The Thalassemias

This is a heterogeneous group of hemoglobin diseases caused by decreased or absent formation of a globin chain. The thalassemias occur predominantly in regions where malaria is endemic (see p. 174). The term is derived from the Greek θαλασσα (sea).

A. Thalassemia, a chronic anemia

Thalassemia is a chronic anemia of various grades of severity, depending on the type. It is associated with extramedullary blood formation (outside the bone marrow) in the liver and spleen, causing both organs to be enlarged. Infections, malnourishment, and other signs add to severe illness. Thalassemias are classified into α-thalassemia and β-thalassemia. In addition, an unstable fusion δβ globin can be the cause. (Photographs: from Weatherall & Clegg, 2001.)

B. β-thalassemia and α-thalassemia

The thalassemias have a wide spectrum of different genotypes and phenotypes. In the β-thalassemias (**1**), complete absence of the β chain (β^0) is distinguished from decreased levels (β^+). Since there are four α globin loci, α-thalassemias (**2**) have a complex pattern of genotypes: αααα/ααα- (silent carrier), a mutation in two α loci on the same chromosome in *cis*: αα/α-α- (thal-1), or a mutation in two α loci on different chromosomes in *trans*: α-/α- (thal-2). Since the α gene loci are located within a 4-kb region of homology interrupted by small, nonhomologous regions, they are prone to mispairing and nonhomologous crossing-over during meiosis, resulting in one chromosome with one α locus (deletion) and one with three α loci (duplication). Thal-1 occurs mainly in Southeast Asia, thal-2, mainly in Africa.

C. Spectrum of mutations in β-thalassemia

The entire gene can be affected, including regulatory regions upstream (5′ direction), or nonsense, missense, or splice mutations may occur. A similarly wide spectrum exists for the α-globin genes.

D. Haplotypes in RFLP analysis

Many common mutations in the β globin gene are in linkage disequilibrium with polymorphic restriction sites in the β globin complex. Thus, certain mutations are linked to a particular haplotype defined by restriction fragment length polymorphisms (RFLPs, see p. 66). For example, seven polymorphic restriction sites define nine haplotypes, of which five are shown (A: + − − − − + +; B: − + + − + + +; etc.). This information can be used for simplified genetic diagnosis to identify a haplotype carrying a mutation in a given population in which this mutation is prevalent. The reason is that mutations have occurred independently in preexisting haplotypes and have remained linked to their neighboring DNA. (Data in C and D from Antonarakis et al., 1985.) α-Thalassemia may be associated with mental retardation. Two different syndromes exist: ATR-16 syndrome (MIM 141750) associated with a large (1–2 Mb) deletion at 19pter-p13.3 including the α globin gene cluster. The other is an X-linked disorder (MIM 301040, gene locus at Xq13) with a remarkably uniform phenotype and a mild form of HbH disease without α-globin deletion. A *trans*-acting regulatory factor appears to be encoded on the X chromosome (Gallego et al., 2005; Wada et al., 2005).

References

Antonarakis SE, Kazazian Jr HH, Orkin SH: DNA polymorphism and molecular pathology of the human globin gene clusters: Hum Genet 69: 1–14, 1985.

Cooley TB, Lee P: A series of splenomegaly in children with anemia and peculiar bone changes. Trans Am Pediat Soc 37: 29, 1925.

Gallego MS et al: ATR-16 due to a de novo complex rearrangement of chromosome 16. Hemoglobin 29: 141–150, 2005.

Olivieri NF: The thalassemias. New Engl J Med 341: 99–109, 1999.

Rund D, Rachmilewitz E: β-thalassemia. New Engl J Med 353: 1135–1146, 2005.

Wada T et al: Non-skewed X-inactivation may cause mental retardation in a female carrier of X-linked alpha-thalassemia/mental retardation syndrome (ATR-X): X-inactivation study of nine female carriers of ATR-X. Am J Med Genet 138 A: 18–20, 2005.

Weatherall DJ et al.: The hemoglobinopathies, pp 4571–4636. In: Scriver CR et al (eds) The Metabolic and Molecular Bases of Inherited Disease, 8th ed. McGraw–Hill, New York, 2001.

Weatherall DJ: Phenotype-genotype relationships in monogenic disease: Lessons from the thalassaemias. Nature Rev Genet 2: 245–255, 2001.

Weatherall DJ, Provan AB: Red cells I: inherited anaemias. Lancet 355: 1169–1175, 2000.

Weatherall DJ, Clegg JB: The Thalassemia Syndromes, 4th ed. Oxford, 2001.

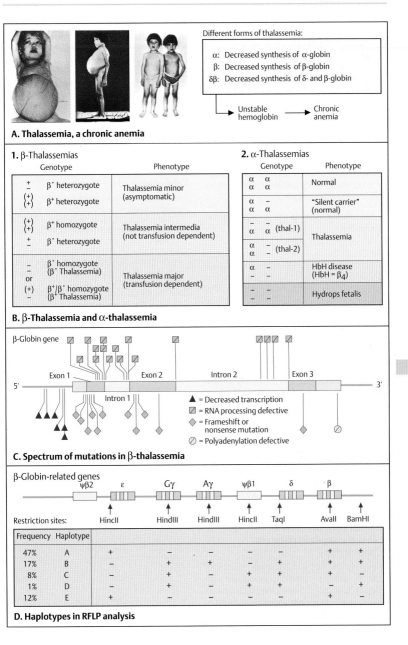

Different forms of thalassemia:

- α: Decreased synthesis of α-globin
- β: Decreased synthesis of β-globin
- δβ: Decreased synthesis of δ- and β-globin

Unstable hemoglobin → Chronic anemia

A. Thalassemia, a chronic anemia

1. β-Thalassemias

Genotype		Phenotype
+ / –	β° heterozygote	Thalassemia minor (asymptomatic)
(+) / (+)	β+ heterozygote	
(+) / (+)	β+ homozygote	Thalassemia intermedia (not transfusion dependent)
+ / –	β° heterozygote	
– / –	β° homozygote (β° Thalassemia)	Thalassemia major (transfusion dependent)
or (+)	β+/β° homozygote (β+ Thalassemia)	

2. α-Thalassemias

Genotype		Phenotype
α α / α α		Normal
α – / α α		"Silent carrier" (normal)
– α / – α (thal-1)		Thalassemia
α α / – – (thal-2)		
α – / – –		HbH disease (HbH = β4)
– – / – –		Hydrops fetalis

B. β-Thalassemia and α-thalassemia

β-Globin gene

5' — Exon 1 — Intron 1 — Exon 2 — Intron 2 — Exon 3 — 3'

▲ = Decreased transcription
▨ = RNA processing defective
◆ = Frameshift or nonsense mutation
⊘ = Polyadenylation defective

C. Spectrum of mutations in β-thalassemia

β-Globin-related genes

ψβ2   ε   Gγ   Aγ   ψβ1   δ   β

Restriction sites: HincII   HindIII   HindIII   HincII   TaqI   AvaII   BamHI

Frequency	Haplotype							
47%	A	+	–	–	–	–	+	+
17%	B	–	+	+	–	+	+	+
8%	C	–	+	–	+	+	+	–
1%	D	–	+	–	+	+	–	+
12%	E	+	–	–	–	–	+	–

D. Haplotypes in RFLP analysis

Hereditary Persistence of Fetal Hemoglobin (HPFH)

Hereditary persistence of fetal hemoglobin (HPFH) refers to a genetically heterogeneous group of diseases in which the temporal expression of the β globin genes during development is altered. Individuals with HPFH produce increased amounts of fetal hemoglobin (HbF). In some conditions, HbF may be the only β-globin-like gene product formed. Clinically, HPFH is relatively benign, although HbF is not optimally adapted to postnatal conditions. Analysis of HPFH has yielded insight into the control of globin gene transcription and the effects of mutations in noncoding sequences.

A. Large deletions in the β-globin gene cluster

A number of very large deletions in the β-globin gene cluster region are known, especially in the 3′ direction. The deletions are distributed differently in different ethnic populations, reflecting that they originated at different points in time. δβ-Thalassemia and failure of β-globin production have resulted in some cases.

B. Mutations in noncoding sequences of the promoter region

Hereditary persistence of fetal hemoglobin can result from mutations in noncoding sequences of the promoter region upstream (5′ direction) of the β-globin gene cluster (on the 5′ side of the γ globin genes). Even though the highly conserved sequences CACCC, CCAAT, or ATAAA are not affected, the number of observed mutations substantiates the significance of the remaining noncoding sequences (long-range transcription control). They are probably required for the changes in transcription control in the different gene loci that occur during embryonic and fetal development. (Figure adapted from Gelehrter & Collins, 1990.)

C. Frequent mutations of β-thalassemia in different populations

Heterozygotes for β-thalassemia mutations occur in different ethnic populations with different frequencies. Since a few mutations are quite frequent in certain populations, preventive diagnostic programs for assessing the risk of disease are possible.

According to estimates of the WHO (Bull World Health Org, 1983) about 275 million persons are heterozygotes for hemoglobin diseases worldwide. Substantial numbers are due to the β-thalassemias in Asia (over 60 million), α^0-thalassemia in Asia (30 million), HbE/β-thalassemia in Asia (84 million), and sickle cell heterozygosity in Africa (50 million), India, the Caribbean, and the USA (about 50 million). At least 200000 severely affected homozygotes are born annually, about 50% with sickle cell anemia and 50% with thalassemia (Weatherall, 1991).

References

Antonarakis SE, Kazazian Jr HH, Orkin SH: DNA polymorphism and molecular pathology of the human globin gene clusters. Hum Genet 69: 1–14, 1985.

Gelehrter TD, Collins F: Principles of Medical Genetics. Williams & Wilkins, Baltimore, 1990.

Kan YW, Holland JP, Dozy AM, Charache S, Kazazian Jr H: Deletion of the beta globin structural gene in hereditary persistence of fetal hemoglobin. Nature 258: 162–163, 1975.

Orkin SH, Kazazian HH: The mutation and polymorphism of the human β-globin gene and its surrounding DNA. Ann Rev Genet 8: 131–171, 1984.

Stamatoyannopoulos G et al (eds): The Molecular Basis of Blood Diseases, 4th ed. WB Saunders, Philadelphia, 2001.

Weatherall DJ et al: The hemoglobinopathies, pp 4571–4636. In: Scriver CR et al (eds) The Metabolic and Molecular Bases of Inherited Disease, 8th ed. McGraw-Hill, New York, 2001.

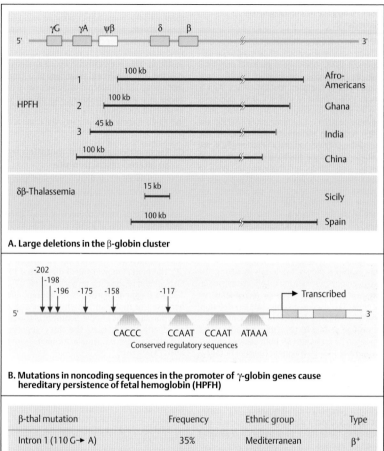

A. Large deletions in the β-globin cluster

B. Mutations in noncoding sequences in the promoter of γ-globin genes cause hereditary persistence of fetal hemoglobin (HPFH)

β-thal mutation	Frequency	Ethnic group	Type
Intron 1 (110 G → A)	35%	Mediterranean	β⁺
Codon 39 (C → T)	27%	Mediterranean	β°
TATA-Box (-29 A → G)	39%	Afro-Americans	β⁺
Poly A (T → C)	26%	Afro-Americans	β⁺
Intron 1 (5 G → C)	36%	India	β⁺
Partial deletion (619 nt)	36%	India	β°
Codon 71-72 frameshift	49%	China	β°
Codon 71-72 frameshift	49%	China	β°
Intron 2 (654 C → T)	38%	China	β°
Codon 41-42 -CTTT	frequent	Southeast Asia	β°

C. Frequent mutations of β-thalassemia in different populations

DNA Analysis in Hemoglobin Disorders

The techniques of recombinant DNA were first applied to the diagnosis of hemoglobin disorders in 1978, including in prenatal diagnosis and in screening populations with high frequencies of common mutant alleles (Kan & Dozy, 1978). Initially indirect DNA diagnosis was based on haplotype analysis using polymorphic restriction enzyme sites (RFLPs, see p. 66) linked to mutations. Subsequently the use of the polymerase chain reaction (PCR, see p. 60) facilitated the diagnosis of both structural hemoglobin defects and thalassemias. With so many mutations known, a direct approach is possible in many cases, but the indirect approach is still valuable to identify rare mutations. Here the principle of indirect diagnosis based on RFLP analysis is illustrated.

A. Detection of a deletion by RFLP analysis

This example shows a deletion involving the two α-globin genes within a 14.5-kb restriction fragment (red arrows). A probe recognizing the α2 gene (**1**) is used to demonstrate the deletion, which results in a restriction fragment of 10.5 kb instead of 14.5 kb (**2**). Three possible genotypes can be expected (**3**): both alleles carrying the normal 14.5-kb fragment (normal homozygote), one allele carrying the normal 14.5-kb fragment and the other the 10.5-kb fragment indicating the deletion (heterozygous), or both alleles carrying the 10.5-kb fragment, i.e., the deletion (homozygous abnormal). The corresponding Southern blot pattern (**4**) clearly distinguishes these three possibilities and allows a precise diagnosis.

B. Haplotype analysis of a mutation

In this situation a polymorphic restriction enzyme site is utilized for diagnosis. One variant consists of two restriction fragments of 7 kb and 6 kb, the other of one 13-kb fragment (**1**). A probe recognizing the 7-kb fragment is used for Southern blot analysis (**2**). The analysis rests on prior knowledge of which fragment carries the mutation (see below). Here the mutation is represented by the 13-kb fragment. The three possible genotypes (**3**) can be readily recognized by Southern blot analysis. The normal homozygotes carry two 7-kb fragments, the heterozygotes one 7-kb and one 13-kb fragment, and the homozygous affected two 13-kb fragments (**4**).

The prerequisite for this type of indirect analysis is prior knowledge of which allele carries the mutation. This is established by an analysis of affected and unaffected members of a family (not shown). If all affected members carry two alleles of the 13-kb fragment, but none of the unaffected members, this would indicate that the 13-kb fragment must carry the mutant allele.

C. Recognition of a point mutation by an altered restriction site

A mutation might alter a restriction site by either abolishing an existing one or creating a new one in the mutant allele. For example, the sickle cell mutation in codon 6 of the β globin gene results in loss of a restriction site (**1**). The mutation changes the recognition sequence CCTNTGG of the enzyme *Mst*II to CCTNAGG because an adenine (A) has replaced a thymine (T) (**2**). Therefore, following *Mst*II digestion a 1.15-kb fragment represents the normal allele (β^A). A 1.35-kb fragment represents the mutant allele because the mutation has eliminated the restriction site in the middle. Southern blot analysis clearly distinguishes the three genotypes and allows a precise, simple, inexpensive diagnosis (**3**).

References

Housman D: Human DNA polymorphism. New Engl J Med 332: 318–320, 1995.

Kan YW, Dozy AM: Polymorphism of DNA sequence adjacent to human beta-globin structural gene: relationship to sickle mutation. Proc Nat Acad Sci 75: 5631–5635, 1978.

Kan YW, Dozy AM: Antenatal diagnosis of sickle-cell anaemia by DNA analysis of amniotic-fluid cells. Lancet II: 910–912, 1978.

Kan YW et al: Polymorphism of DNA sequence in the beta-globin gene region: application to prenatal diagnosis of beta-zero-thalassemia in Sardinia. New Engl J Med 302: 185–188, 1980.

Old J: Hemoglobinopathies and thalassemias, pp 1861–1898. In: Rimoin DL et al (eds) Emery and Rimoin's Principles of Medical Genetics, 4th ed. Churchill-Livingstone, London–Edinburgh, 2002.

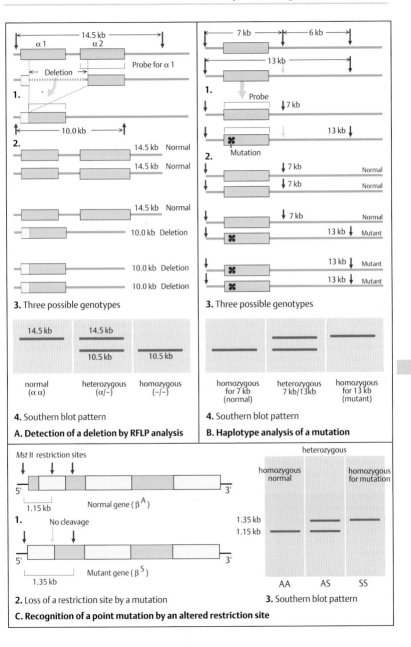

1.

2.

3. Three possible genotypes

14.5 kb | 14.5 kb | |
| 10.5 kb | 10.5 kb |

normal (α α) | heterozygous (α/–) | homozygous (–/–)

4. Southern blot pattern

A. Detection of a deletion by RFLP analysis

1.

2.

3. Three possible genotypes

homozygous for 7 kb (normal) | heterozygous 7 kb/13kb | homozygous for 13 kb (mutant)

4. Southern blot pattern

B. Haplotype analysis of a mutation

Mst II restriction sites

5' ————— 3'

1.15 kb   Normal gene (β^A)

1.

No cleavage

5' ————— 3'

1.35 kb   Mutant gene (β^S)

2. Loss of a restriction site by a mutation

heterozygous

homozygous normal | | homozygous for mutation

1.35 kb
1.15 kb

AA | AS | SS

3. Southern blot pattern

C. Recognition of a point mutation by an altered restriction site

Lysosomes

Lysosomes are membrane-enclosed intracellular vesicles with a diameter of 0.05–0.5 μm, which are required for the intracellular digestion of large molecules. They contain more than 50 active hydrolytic enzymes (acid hydrolases), such as glycosidases, sulfatases, phosphatases, lipases, phospholipases, proteases, and nucleases (collectively called lysosomal enzymes) in an acid milieu (pH about 5). Lysosomal enzymes enter a lysosome by means of a recognition signal (mannose-6-phosphate) and a corresponding receptor.

A. Receptor-mediated endocytosis and lysosome formation

Extracellular macromolecules to be degraded are taken into the cell by endocytosis. First, the molecules are bound to specific cell surface receptors (receptor-mediated endocytosis). The loaded receptors are concentrated in an invagination of the plasma membrane (coated pit). This separates from the plasma membrane and forms a membrane-enclosed cytoplasmic compartment (coated vesicle). The cytoplasmic lining of the vesicle consists of a network of a trimeric protein, clathrin. The clathrin coat is removed within the cell, forming an endosome. The receptor and the molecule to be degraded (the ligand) are separated and the receptor is recycled to the cell surface.

A multivesicular body (endolysosome) forms and takes up acid hydrolases arriving in clathrin-enclosed vesicles. Hydrolytic degradation takes place in the lysosome. Parts of the membrane also are also recycled.

A mannose-6-phosphate receptor serves as a recognition signal for uptake into the endolysosome, which will also be recyled back into the Golgi apparatus. The acid milieu in the lysosomes is maintained by a hydrogen pump in the membrane, which hydrolyzes ATP and uses the energy produced to move H^+ ions into the lysosome. Some of the mannose-6-phosphate receptors are transported back to the Golgi apparatus.

B. Mannose-6-phosphate receptors

There are two types of mannose-6-phosphate receptor molecules, which differ in their binding properties and their cation dependence. They consist of either 2 or 16 extracellular domains with different numbers of amino acids. The cDNA of Ci-MPR (cation-independent mannose-6-phosphate receptor) is identical with insulinlike growth factor II (IGF-2). Thus, Ci-MPR is a multifunctional binding protein.

C. Biosynthesis

Two enzymes are essential for the biosynthesis of mannose-6-phosphate recognition signals: a phosphate transferase and a phosphoglycosidase. The phosphate is delivered by uridine-diphosphate-N-acetylglucosamine (UDP-GlcNAc, uridine-5′-diphosphate-N-acetylglucosamine-glycoprotein-N-acetylglucosaminylphosphotransferase). A second enzyme, N-ace-tylglucosamine-1-phosphodiester-N-acetyl-glucosaminidase) removes the N-acetylglucosamine, leaving the phosphate residue at position 6 of the mannose. (Figures adapted from Sabatini & Adesnik, 2001, and C. de Duve, 1984.)

Medical relevance

Lysosomal storage disorders are a large group of genetic diseases involving glycogen, mucopolysaccharides, acid lipases, acid ceramidases, acid sphingomyelonidases, sphingolipid activator proteins, multiple sulfatases, β-galactosidase, gangliosides, and others (Hopkin & Grabowski, 2005; Scriver et al., 2001; Gilbert-Barness & Barness, 2000).

References

de Duve C: A Guided Tour of the Living Cell, 2 vols., Scientific American Books, Inc, New York, 1984.

Gilbert-Barness E, Barness L: Metabolic Diseases. Foundations of Clinical Management, Genetics, and Pathology, 2 vols., Eaton Publishing, Natick, Massachussetts, 2000.

Hopkin RJ, Grabowsi GA: Lysosomal storage diseases, pp 2315–2319. In: Kasper DL et al (eds) Harrison's Principles of Internal Medicine, 16th ed. McGraw-Hill, New York, 2005.

Sabatini DD, Adesnik MB: The biogenesis of membranes and organelles, pp 433–517. In: Scriver CR et al (eds) The Metabolic and Molecular Bases of Inherited Disease, 8th ed. McGraw-Hill, New York, 2001.

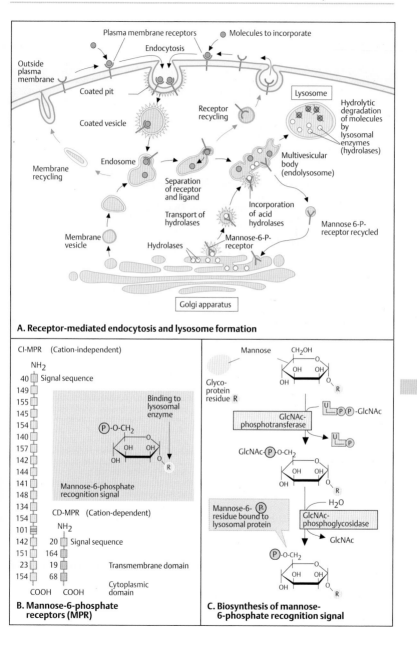

A. Receptor-mediated endocytosis and lysosome formation

B. Mannose-6-phosphate receptors (MPR)

C. Biosynthesis of mannose-6-phosphate recognition signal

Diseases Due to Lysosomal Enzyme Defects

Mutations in genes encoding enzymes that degrade complex macromolecules in lysosomes (lysosomal enzymes) cause a large group of different diseases. Their clinical signs and biochemical and cellular manifestations depend on the function normally carried out by the enzyme involved. Macromolecules that normally are degraded remain in the lysosome, resulting in *lysosomal storage diseases*. This occurs at different rates, so that each disease has its own characteristic course. Twelve groups of genetically determined disorders of specific lysosomal functions are known, each with about three to ten individually defined diseases.

A. Defective uptake of enzymes into lysosomes: I-cell disease

Mucolipidosis type II, also called *I-cell disease* because of conspicuous cytoplasmic inclusions first described by Leroy & DeMars (1967), is a disorder of abnormal lysosomal transport and protein sorting, manifest in mesenchymal cells (MIM 252500). The first of a two-step reaction in the Golgi apparatus is defective due to mutations in the *GNPTA* gene (MIM 607840), located at 12q23.3, encoding the lysosomal enzyme N-acetylglucosamine-1-phosphotransferase (GlcNAc-phosphotransferase, see previous page). As a result, the recognition marker that binds mannose-6-phosphate is lacking and mucolipids accumulate in mesenchymal cells (**1**), but not in normal fibroblasts (**2**). The vesicular inclusions consist of hydrolases that cannot enter the lysosomes because the mannose-6-phosphate recognition signal is absent. Lysosomes lack several enzymes, whereas their concentration outside the cells is increased. A severe, progressive disease usually becomes apparent in the first 6 months of life (**3**). The *GNPTA* gene has 21 exons spanning 85 kb of DNA (Tiede et al., 2005). Mutations result in premature translational termination. Two complementation groups have been delineated. Other types of mucolipidosis exist (MIM 252600; 252650).

B. Degradation of heparan sulfate

Lysosomal enzymes are bond-specific, not substrate-specific. Thus, they also degrade other glycosaminoglycans, such as dermatan sulfate, keratan sulfate, and chondroitin sulfate (mucopolysaccharides). Ten specific enzyme defects cause the mucopolysaccharide storage diseases (see next page). Heparan sulfate is an example of a macromolecule that is degraded stepwise by eight different lysosomal enzymes. The first step in heparan sulfate degradation is the removal of sulfate from the terminal iduronate group by an iduronate sulfatase. A defect in the gene encoding this enzyme leads to the X-chromosomal mucopolysaccharide storage disease type II (MPS II, type Hunter). All other mucopolysaccharidoses are autosomal recessive. The second enzymatic step removes the terminal iduronate by an α-L-iduronidase. A homozygous mutation in the responsible gene leads to mucopolysaccharidosis type I (MPS I, Hurler/Scheie). Defective function in the next three enzymatic steps causes three of the four subtypes of mucopolysaccharidosis type III (Sanfilippo, MPS IIIA, IIIC, and IIIB). MPSIIID results from a defect in the last (8th) step. MPS type VII (Sly) is due to a defect of β-glucuronidase (step 7) with a different phenotype than MPS types I, II, and III.

References

Kornfeld S, Sly WS: I-cell disease and Pseudo-Hurler polydystrophy: Disorders of lysosomal enzyme phosphorylation and localization, pp 3469–3482. In: CR Scriver et al (eds) The Metabolic and Molecular Bases of Inherited Disease, 8th ed. McGraw-Hill, New York, 2001.

Kudo M et al: Mucolipidosis II (I-cell disease) and mucolipidosis IIIA (classical pseudo-Hurler polydsystrophy) are caused by mutations in the GlcNAc-phosphotransferase α/β-subunits precursor gene. Am J Hum Genet 78: 451–463, 2006.

Leroy JG, DeMars RI: Mutant enzymatic and cytological phenotypes in cultured human fibroblasts. Science 157: 804–806, 1967.

Neufeld EF, Muenzer J: The mucopolysaccharidoses, pp 3421–3452. In: Scriver CR et al (eds) The Metabolic and Molecular Bases of Inherited Disease, 8th ed. McGraw-Hill, New York, 2001.

Olkkonen VM, Ikonen E: Genetic defects of intracellular-membrane transport. New Engl J Med 343: 1095–1104, 2000.

Tiede S et al: Mucolipidosis II is caused by mutations in GNPTA encoding the alpha/beta GlcNAc-1-phosphotransferase. Nature Med 11: 1109–1112, 2005.

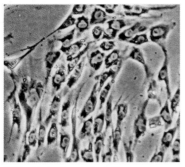

1. Fibroblast culture in I-cell disease

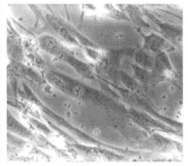

2. Normal fibroblast culture

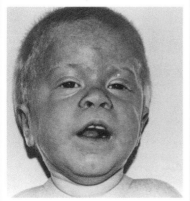

3. Patient with I-cell disease

A. Defective uptake of enzymes in lysosomes: I-cell disease

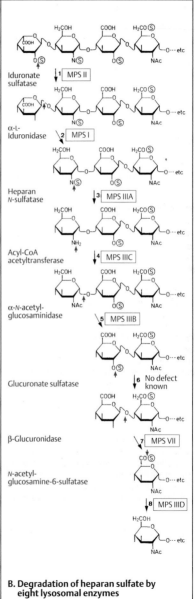

B. Degradation of heparan sulfate by eight lysosomal enzymes

Iduronate sulfatase   ↓ 1 MPS II

α-L-Iduronidase   ↓ 2 MPS I

Heparan N-sulfatase   ↓ 3 MPS IIIA

Acyl-CoA acetyltransferase   ↓ 4 MPS IIIC

α-N-acetyl-glucosaminidase   ↓ 5 MPS IIIB

Glucuronate sulfatase   ↓ 6 No defect known

β-Glucuronidase   ↓ 7 MPS VII

N-acetyl-glucosamine-6-sulfatase   ↓ 8 MPS IIID

Mucopolysaccharide Storage Diseases

The mucopolysaccharide storage diseases (the mucopolysaccharidoses) are a clinically and genetically heterogeneous group of lysosomal storage diseases caused by defects in different enzymes for mucopolysaccharide degradation (glycosaminoglycans). All are transmitted by autosomal recessive inheritance, except for mucopolysaccharide storage disease type II (Hunter).

A. Mucopolysaccharide storage disease type I (Hurler)

Infants with MPS IH (MIM 252800) at first seem normal. Early signs of the disease occur at about 1–2 years of age, with increasing coarsening of the facial features, retarded mental development, limited joint mobility, enlarged liver, umbilical hernia, and other signs. Radiographs show coarsening of skeletal structures (dysostosis multiplex). The photographs show the same patient at different ages (own data). MPS IS (Scheie) is a clinically different, less severe allelic disease.

B. Mucopolysaccharide storage disease type II (Hunter)

This type of mucopolysaccharidosis is transmitted by X-chromosomal inheritance (MIM 309900). Four cousins from one pedigree are shown in the diagram. Clinically, the disease is similar to, but less rapidly progressive than, MPS type I. Molecular diagnosis is possible in most cases. (Photos: Passarge et al., 1974.)

Diagnosis

The diagnosis of MPS is based on the patient's history, careful clinical and radiological evaluation, and biochemical analysis of the urine. In MPS the urinary concentration of one of several types of glycosaminoglycan is increased, depending on the type of MPS. DNA diagnosis must take the extreme genetic heterogeneity into account.

References

Hopkin RJ, Grabowsi GA: Lysosomal storage diseases, pp 2315–2319. In: Kasper DL et al (eds) Harrison's Principles of Internal Medicine, 16th ed. McGraw-Hill, New York, 2005.

McKusick VA: Mendelian Inheritance in Man, 12th ed. 1998 (OMIM at www.ncbi.nlm.nih/Omim).

Neufeld EF, Muenzer J: The mucopolysaccharidoses, pp 3421–3452. In: Scriver CR et al (eds) The Metabolic and Molecular Bases of Inherited Disease, 8th ed. McGraw-Hill, New York, 2001.

Passarge E et al: Krankheiten infolge genetischer Defekte im lysosomalen Mucopolysaccarid-Abbau. Dtsch Med Wschr 99: 144–158, 1974.

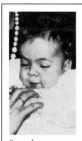
8 weeks

7 months

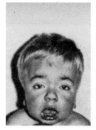
2 1/4 years

3 3/4 years

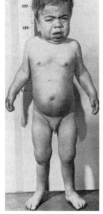
5 years

8 years

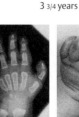
Dysostosis
multiplex

Joint
contractures

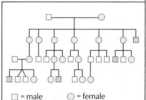

□ = male   ○ = female
▩ = Hunter syndrome

X-Chromosomal inheritance

A. Mucopolysaccharide storage disease type I (Hurler)

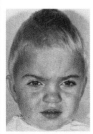
4 1/2 years

10 years

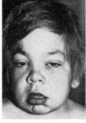
13 years

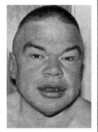
21 years

B. Mucopolysaccharide storage disease type II (Hunter)

Peroxisomal Biogenesis Diseases

Peroxisomes are small membrane-bound intracellular organelles of about $0.5-1.0\,\mu m$ diameter, somewhat smaller than mitochondria. They are involved in anabolic and catabolic metabolic functions. Their name is derived from hydrogen peroxide, which is formed as an intermediary product of oxidative metabolism. Most cells, especially in the liver and kidney, contain about 100–1000 peroxisomes. A number of genetic defects in peroxisome biogenesis or peroxisome enzymes lead to severe diseases in humans (peroxisomal diseases).

Peroxisomes are surrounded by a single-layer granular matrix, which contains about 50 matrix enzymes. They are involved in β-oxidation of fatty acids, biosynthesis of phospholipids and bile acids, and other functions. Peroxisome biogenesis involves the synthesis of matrix proteins (peroxines) and their receptor-mediated transfer into the organelle under the control of at least 23 identified *PEX* genes and peroxisomal targeting signals (PTS). About 50 matrix enzymes are involved in β-oxidation of fatty acids, biosynthesis of phospholipids and bile acids, and other functions.

A. Biochemical reactions

The electron micrograph (**1**) shows three peroxisomes in a rat liver cell. The dark striated structures within the organelles are urates, a result of an enzyme that oxidizes uric acid. Peroxisomes have both catabolic (degrading) and anabolic (synthesizing) functions (**2**). Two biochemical reactions are especially important: a peroxisomal respiratory chain and the β-oxidation of very long-chain fatty acids. In the peroxisomal respiratory chain (**3**), certain oxidases and catalases act together. Specific substrates of the oxidases are organic metabolites of intermediary metabolism. Very long-chain fatty acids are broken down by β-oxidation (**4**) in a cycle with four enzymatic reactions. Energy production in peroxisomes is relatively inefficient compared with that of mitochondria. While free energy in mitochondria is mainly preserved in the form of ATP (adenosine triphosphate), in peroxisomes it is mostly converted into heat. Peroxisomes are probably a very early adaption of living organisms to oxygen. (Photograph from de Duve, 1984.)

B. Peroxisomal diseases

Six important examples of peroxisomal diseases are listed, all autosomal recessive. Patients with neonatal adrenoleukodystrophy do not form sufficient amounts of plasmalogens and cannot adequately degrade phytanic acid and pipecolic acid. More than 12 complementation groups have been delineated by fused cultured fibroblasts from different affected individuals. Hybrid cells from different subtypes correct each other (see p. 128).

C. Zellweger cerebrohepatorenal syndrome

This is a characteristic autosomal recessive disease resulting from mutations of genes located at 1p36, 1q22, 2p15, 6q23–q24, 7q21–q22, 12p13.3 and 22q11.21 (MIM 214100). It is recognizable by a characteristic facial appearance (**1–4**), extreme muscle weakness (**5**), and a number of accompanying manifestations such as calcified stippling of the joints on radiographs (**6**), renal cysts (**7, 8**), and clouding of the lens and cornea. The severe form of the disease usually leads to death before the age of one year. (Photographs 1–5 from Passarge & McAdams, 1967.)

References

de Duve C: A Guided Tour through the Living Cell. Scientific American Books, New York, 1984.

Gould SJ, Raymond BV, Valle D: The peroxisome biogenesis disorders, pp 3181–3217. In: Scriver CR et al (eds) The Metabolic Bases of Inherited Disease, 8th ed. McGraw-Hill, New York, 2001.

Muntau AC et al: Defective peroxisome membrane synthesis due to mutations in human PEX3 causes Zellweger syndrome, complementation group G. Am J Hum Genet 67: 967–975, 2000.

Passarge E, McAdams AJ: Cerebro-hepato-renal syndrome. A newly recognized hereditary disorder of multiple congenital defects, including sudanophilic leukodystrophy, cirrhosis of the liver, and polycystic kidneys. J Pediat 71: 691–702, 1967.

Shimozawa N et al: A human gene responsible for Zellweger syndrome that affects peroxisome assembly. Science 255: 1132–1134, 1992.

Warren DS et al: Phenotype-genotype relationships in *PEX10*-deficient peroxisome biogenesis disorder patients. Hum Mutat 15: 509–521, 2000.

Online information:
www.peroxisome.org at Johns Hopkins University School of Medicine, Stephen J. Gould

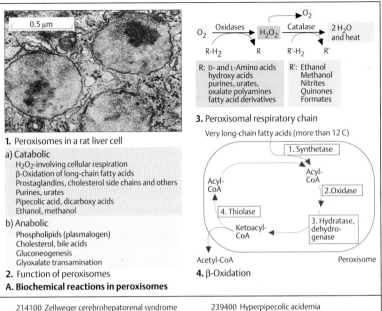

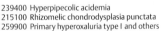

1. Peroxisomes in a rat liver cell

a) Catabolic
H₂O₂-involving cellular respiration
β-Oxidation of long-chain fatty acids
Prostaglandins, cholesterol side chains and others
Purines, urates
Pipecolic acid, dicarboxy acids
Ethanol, methanol

b) Anabolic
Phospholipids (plasmalogen)
Cholesterol, bile acids
Gluconeogenesis
Glyoxalate transamination

2. Function of peroxisomes

3. Peroxisomal respiratory chain

4. β-Oxidation

A. Biochemical reactions in peroxisomes

214100 Zellweger cerebrohepatorenal syndrome	239400 Hyperpipecolic acidemia
202370 Neonatal adrenoleukodystrophy	215100 Rhizomelic chondrodysplasia punctata
266510 Infantile Refsum disease	259900 Primary hyperoxaluria type I and others

B. Examples of peroxisomal diseases

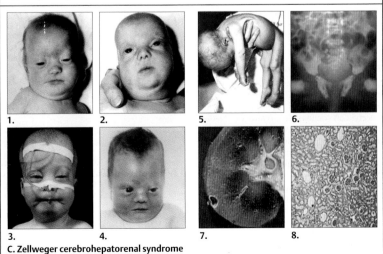

C. Zellweger cerebrohepatorenal syndrome

Cholesterol Biosynthesis

Cholesterol is a precursor of many steroid hormones and a major constituent modulating the fluidity of cell membranes in eukaryotes. In 1932 Wieland and Dane elucidated its structure as a monosaturated 27-carbon sterol. The biosynthetic pathway of cholesterol requires about 30 enzymatic reactions regulated by 22 genes in a series including oxidation with molecular oxygen, reductions, demethylations, and alterations in double bonds. Konrad Bloch was awarded the Nobel Prize in 1954 for elucidating this pathway. Cholesterol evolved after the atmosphere of the earth became enriched in oxygen. During the past 10 years a number of hereditary diseases resulting from mutations in genes encoding enzymes of the cholesterol biosynthesis pathway have been discovered.

A. Malformation syndromes due to defects in cholesterol metabolism

About six different genetic diseases are known to result from a block of the cholesterol biosynthesis pathway (see next page). Three examples are shown here: (1) the autosomal recessive Smith–Lemli–Opitz syndrome (MIM 270400), (2) X-linked chondrodysplasia punctata type 2 (CDPX2, Conradi-Hünermann syndrome, MIM 302960), and (3) autosomal dominant Greenberg skeletal dysplasia (MIM 215140). (Photo 1 courtesy of the parents; photo 2 courtesy of Dr. Richard I. Kelley, Baltimore; photo 3 courtesy of Dr. David L. Rimoin, Los Angeles.)

B. Cholesterol biosynthesis overview

Cholesterol biosynthesis begins with acetyl coenzyme A (acetyl CoA), from which all 27 carbon atoms are derived. Acetyl CoA and acetoacetyl CoA condense to 3-hydroxy-3-methylglutaryl-CoA. This is converted by 3-hydroxy-3-methylglutaryl-CoA reductase to mevalonate. This is the precursor of isoprene, which is synthesized in three steps (not shown). Squalene, a 30-carbon linear isoprenoid, is synthesized from six isoprene units. Isopentyl pyrophosphate is the starting point of a reaction C5 → C10 → C15 → C30. With squalene, the distal (post-squalene) part of the cholesterol biosynthesis pathway begins.

Mevalonic aciduria (MIM 251170) results from a block in mevalonate kinase. This variable auto-somal recessive disease is characterized by increased urinary excretion of mevalonic acid associated with failure to thrive, psychomotor retardation, vomiting, diarrhea, episodes of fever and dysmorphic facial features.

C. Squalene to lanosterol

Initially, squalene is circularized through a reactive intermediate, squalene epoxide (not shown), to lanosterol, the first post-squalene sterol intermediate. Squalene epoxide is closed by a cyclase to lanosterol, a 30-carbon sterol. This requires movements of electrons through four double bonds and the migration of two methyl groups. Removal of the 24–25 double bond results in dihydrolanosterol, the other precursor of cholesterol.

References

Farese jr RV, Herz J: Cholesterol metabolism and embryogenesis. Trends Genet 14: 115–120, 1998.

Fitzky BU et al.: Mutations in the delta-7-sterol reductase gene in patients with the Smith–Lemli–Opitz syndrome. Proc Nat Acad Sci 95: 8181–8186, 1998.

Goldstein JL, Brown MS: Regulation of the mevalonate pathway. Nature 343: 425–430, 1990.

Greenberg CR et al.: A new autosomal recessive lethal chondrodystrophy with congenital hydrops. Am J Med Genet 29: 623–632, 1988.

Herman GE et al: Characterization of mutations in 22 females with X-linked dominant chondrodysplasia punctata (Happle syndrome). Genet Med 4: 434–438, 2002.

Herman GE: Disorders of cholesterol biosynthesis: prototypic malformation syndromes. Hum Mol Genet 12(R1): R75–R88, 2003.

Kelley RI et al: Abnormal sterol metabolism in patients with Conradi–Hünerman–Happle syndrome and sporadic chondrodysplasia punctata. Am J Med 83: 231–219, 1999.

Porter FB: Malformations due to inborn errors of cholesterol synthesis. J Clin Invest 110: 715–724, 2002.

Smith DW, Lemli L, Opitz JH: A newly recognized syndrome of multiple congenital anomalies. J Pediat 64: 210–217, 1964.

Waterham HR et al.: Autosomal recessive HEM/Greenberg skeletal dysplasia is caused by 3-beta-hydroxysterol delta (14)-reductase deficiency due to mutations in the lamin B receptor gene. Am J Hum Genet 72: 1013–1017, 2003.

Waterman HR: Inherited disorders of cholesterol biosynthesis. Clin Genet 61: 393–403, 2002.

Witsch-Baumgartner M et al.: Maternal apo E genotype is a modifier of the Smith–Lemli–Opitz syndrome. J Med Genet 41: 577–584, 2004.

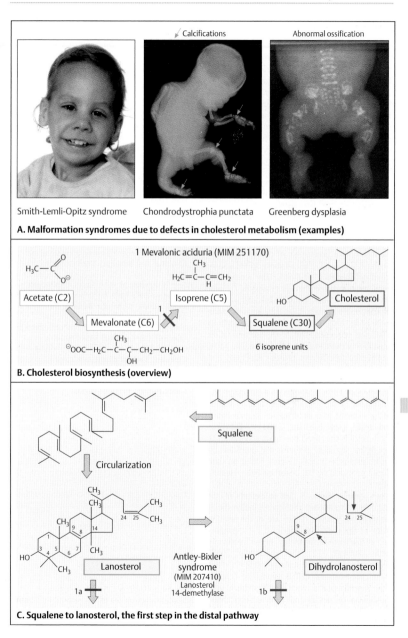

Smith-Lemli-Opitz syndrome   Chondrodystrophia punctata   Greenberg dysplasia

A. Malformation syndromes due to defects in cholesterol metabolism (examples)

1 Mevalonic aciduria (MIM 251170)

Acetate (C2)

Isoprene (C5)

Cholesterol

Mevalonate (C6)

Squalene (C30)

6 isoprene units

B. Cholesterol biosynthesis (overview)

Squalene

Circularization

Lanosterol

Antley-Bixler syndrome (MIM 207410) Lanosterol 14-demethylase

Dihydrolanosterol

1a

1b

C. Squalene to lanosterol, the first step in the distal pathway

Distal Cholesterol Biosynthesis Pathway

In the distal part of the cholesterol biosynthesis pathway (post-squalene) lanosterol and dihydrolanosterol (see previous page) are converted to desmosterol and 7-dihydrocholesterol (DHCR7), the immediate precursors of cholesterol. Mutations in genes encoding the enzymes required for cholesterol biosynthesis have been found at all steps. They cause rare genetic diseases characterized by developmental delay and abnormalities of the skeletal and other systems.

A. Distal cholesterol biosynthesis pathway and diseases

Lanosterol and dihydrolanosterol (see previous page) are converted by four enzymatic reactions through four intermediate metabolites to the immediate precursors of cholesterol, desmosterol (cholesta-5(6), 24-dien-3β-ol) and 7-dihydrocholesterol (7-DHC). The enzymatic steps remove three methyl groups at C4 and C14, open one double bond at C24, and shift the C8–C9 double bond to C7–C8 by an isomerase. Some of the enzymatic reactions must occur in a defined sequence: the Δ^8–Δ^7 isomerization must follow the C14α demethylation. The pathway is tied to different cellular functions and signaling pathways. Lanosterol and its two subsequent intermediates, 4,4-dimethyl-5α-cholesta-8,14,24-trien-3β-ol and 4,4-dimethyl-5α-cholesta-8,24-dien-3β-ol, have meiosis-stimulating activity and accumulate in the ovary and testis (see Herman, 2003). 7-Dihydrocholesterol is the direct precursor of vitamin D. Hedgehog signaling proteins are modified by cholesterol (see p. 364). (Figure based on data kindly provided by Dr. Dorothea Haas, Heidelberg.)

Seven genetic diseases due to defects in the distal (post-squalene) cholesterol biosynthesis pathway are known. These are in descending order of the pathway reactions: (1) a proportion of patients with Antley–Bixler syndrome (MIM 207410), (2) the prenatally lethal Greenberg skeletal dysplasia (MIM 215140), (3) X-chromosomal CHILD syndrome (congenital hemidysplasia with ichthyosiform erythroderma or nevus and limb defects; MIM 308050), (4) X-chromosomal dominant chondrodysplasia punctata type 2 (Conradi–Hüner-mann syndrome, MIM 302960), (5) lathosterolosis (MIM 607330), (6) Smith–Lemli–Opitz syndrome (MIM 270400), and (7) desmosterolosis (602398). The main features of these diseases are summarized in the table (see appendix).

References

Farese jr RV, Herz J: Cholesterol metabolism and embryogenesis. Trends Genet 14: 115–120, 1998.

Greenberg CR, Rimoin DL, Gruber HE et al: A new autosomal recessive lethal chondrodystrophy with congenital hydrops. Am J Med Genet 29: 623–632, 1988.

Herman GE: Disorders of cholesterol biosynthesis: prototypic metabolic malformation syndromes. Hum Mol Genet 12(R1): R75–R88, 2003.

Herman GE, Kelley RI, Pureza V et al: Characterization of mutations in 22 females with X-linked dominant chondrodysplasia punctata (Happle syndrome). Genet Med 4: 434–438, 2002.

Kelley RI, Herman GE: Inborn errors of sterol metabolism. Ann Rev Genomics Hum Genet 2: 299–341, 2001.

Kelley RI, Hennekam RCM: Smith–Lemli–Opitz syndrome, pp 6183–6201. In: Scriver CR et al (eds) The Metabolic and Molecular Bases of Inherited Disease, 8th ed. McGraw-Hill, New York, 2001.

Waterman HR: Inherited disorders of cholesterol biosynthesis. Clin Genet 61: 393–403, 2002.

Waterham HR et al: Autosomal recessive HEM/Greenberg skeletal dysplasia is caused by 3-beta-hydroxysterol delta (14)-reductase deficiency due to mutations in the lamin B receptor gene. Am J Hum Genet 72: 1013–1017, 2003.

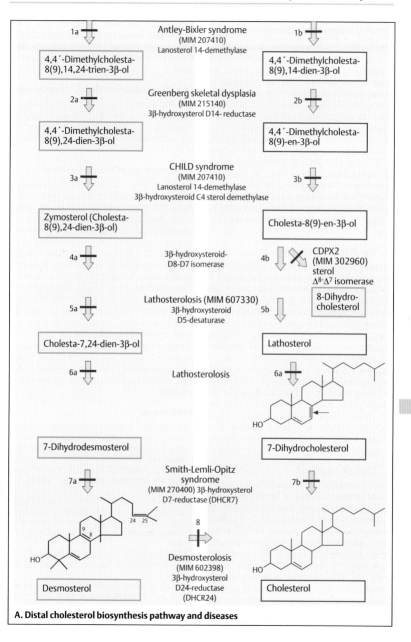

A. Distal cholesterol biosynthesis pathway and diseases

Familial Hypercholesterolemia

Familial hypercholesterolemia (FH) is a hereditary disorder due to increased plasma concentration of low density lipoprotein (LDL), the main cholesterol-transporting lipoprotein. Lipoproteins transport cholesterol and triglycerides in bodily fluids. The protein components make the hydrophobic lipids accessible to cells. Lipoproteins are classified according to increasing density: chylomicrons, very low density lipoproteins (VLDL), low density lipoproteins (LDL), and high density lipoproteins (HDL). Different genetic forms of FH are distinguished according to the step of the metabolic pathway involved and the type of mutation. Many other forms of hyperlipidemia exist, often caused by multigenic, monogenic, and environmental factors.

A. The disease phenotype

Familial hypercholesterolemia (MIM 143890) occurs, in the heterozygous state, in about 1 in 500 individuals (**1**). Plasma cholesterol levels are greatly increased (**2**). Important clinical signs are lipid deposits around the iris in the eye (**3**) and deposits of cholesterol esters in the tendons, especially the Achilles tendon, and the skin (xanthomas, **4**). The activity of functional LDL receptors per cell is decreased by about 50% and is nearly absent in homozygotes (**5**). In the rare homozygous state (1 in 10^6) FH leads to death during the first or second decade of life. Several other, related forms at other gene loci are known (MIM 144010, 107730, 603776, 607786). (Figures in 2 and 5 adapted from Goldstein et al., 2001; 3 and 4 author's own observations.)

B. The LDL receptor

The LDL receptor is a cell surface receptor encoded by a gene located at 19p13.2, consisting of 18 exons spanning 45 kb of genomic DNA and transcribed into a 5.3-kb mRNA. It is a membrane-bound 160-kD protein of 839 amino acids with six distinct functional domains. Seven cysteine-rich units of 40 amino acids each form the ligand-binding region (exons 2–6). The EGF precursor homology domain (exons 7–14) allows the lipoprotein to dissociate from the receptor in the endosome (see next page). The intracellular domain encoded by part of exon 17 and the 5′ end of exon 18 contains signals for localizing the receptor in the coated pits during endocytosis and targeting within hepatocytes. (Figure adapted from Goldstein et al., 2001, according to Hobbs et al., 1990.)

C. LDL receptor-mediated endocytosis

The LDL receptor mediates the endocytosis of LDL. Receptors loaded with LDL accumulate in a coated pit (**1**), which forms an endocytotic vesicle inside the cell (**2**). (Photograph: Anderson et al., 1977.)

D. Homology with other proteins

The mammalian LDL receptor is highly conserved in evolution (90% identical within mammals, 79% identity between man and shark), dating back at least 500 million years. It is a member of a family of genes related by common evolutionary origin. The proximal halves of the extracellular domains of the LDL receptor family are structurally related to the epidermal growth factor family (EGF). Both are related to protease of the blood coagulation system, factors IX and X, protein C, and complement C9. Other related genes (not shown) are the VLDL (very low density lipoprotein) receptor, the ApoE receptor 2 (ApoER2), the LDL receptor-related protein (LRP), and megalin. LRP and megalin are multifunctional and bind diverse ligands such as lipoproteins, proteases and their inhibitors, peptide hormones, and carrier proteins of vitamins.

References

Anderson RGW, Brown MS, Goldstein JL: Role of the coated endocytic vesicle in the uptake of receptor-bound low density lipoprotein in human fibroblasts. Cell 10: 351–364, 1977.

Brown MS, Goldstein JL: A receptor-mediated pathway for cholesterol homeostasis. Science 232: 34–47, 1986.

Goldstein JL, Brown MS, Hobbs HH: Familial hypercholesterolemia, pp 2863–2913. In: Scriver CR et al (eds) The Metabolic and Molecular Bases of Inherited Disease, 8th ed. McGraw-Hill, New York, 2001.

Hobbs HH et al: The LDL receptor locus in familial hypercholesterolemia. Ann Rev Genet 24: 133–170, 1990.

Rader DJ, Hobbs HH: Disorders of lipoprotein metabolism, pp 2286–2298. In: Kasper DL et al (eds) Harrison's Principles of Internal Medicine, 16th ed. McGraw-Hill, New York, 2005.

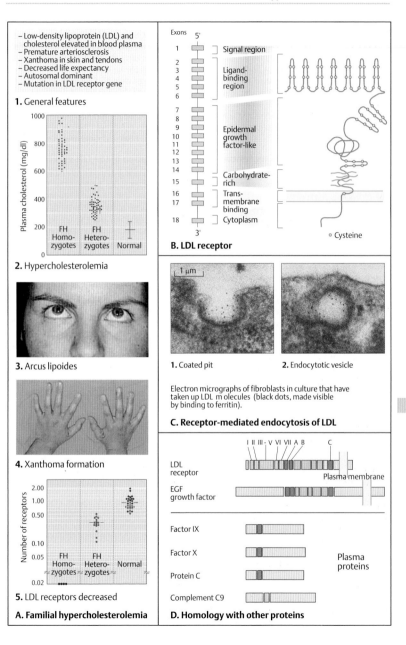

- Low-density lipoprotein (LDL) and cholesterol elevated in blood plasma
- Premature arteriosclerosis
- Xanthoma in skin and tendons
- Decreased life expectancy
- Autosomal dominant
- Mutation in LDL receptor gene

1. General features

2. Hypercholesterolemia

3. Arcus lipoides

4. Xanthoma formation

5. LDL receptors decreased

A. Familial hypercholesterolemia

Exons 5'

Signal region

Ligand-binding region

Epidermal growth factor-like

Carbohydrate-rich

Trans-membrane binding

Cytoplasm

3'

° Cysteine

B. LDL receptor

1 µm

1. Coated pit   **2.** Endocytotic vesicle

Electron micrographs of fibroblasts in culture that have taken up LDL molecules (black dots, made visible by binding to ferritin).

C. Receptor-mediated endocytosis of LDL

I II III - V VI VII A B   C

LDL receptor

Plasma membrane

EGF growth factor

Factor IX

Factor X

Protein C

Plasma proteins

Complement C9

D. Homology with other proteins

LDL Receptor Mutations

Low-density lipoprotein (LDL) is the main transport agent of cholesterol in the blood. Its hydrophobic core contains about 1500 esterified cholesterol molecules surrounded by an outer layer of phospholipids and unesterified cholesterols containing a single apoB-100 lipoprotein molecule. LDL delivers cholesterol to peripheral tissues and regulates de novo cholesterol synthesis there.

A. LDL receptor classes of mutation

The receptor–LDL complex enters the cell by endocytosis (see p. 354) within a coated vesicle. The receptor is separated from the LDL including the apoB-100 in the endosome. The LDL receptor is recycled to the cell surface. In the lysosome the LDL is broken down into amino acids and cholesterol. Free cholesterol activates the enzyme acetyl-CoA cholesterol transferase (ACAT), which catalyzes the esterification. The key enzyme for endogenous cholesterol synthesis is 3-hydroxy-3-methylglutaryl-CoA reductase (HMG-CoA reductase). This enzyme is downregulated by exogenous LDL uptake. LDL receptor mutations interrupt this feedback mechanism and result in increased endogenous cholesterol synthesis.

Five principal classes of LDL receptor mutations can be distinguished: (**1**) receptor null mutations (R^0) due to lack of receptor protein synthesis in the endoplasmic reticulum (ER), (**2**) defective intracellular transport to the Golgi apparatus, (**3**) defective extracellular ligand binding, (**4**) defective endocytosis (R^+ mutations), and (**5**) failure to release the LDL molecules inside the endosome (recycling-defective mutations). (Figure adapted from Goldstein et al., 2001.)

B. Mutational spectrum

More than 350 mutations have been recorded in the LDL receptor gene (Varret et al., 1998). Of these, 63% are missense mutations. Mutations occur in all parts of the gene, but there is a relative excess of mutations in exons 4 and 9. Exons 13 and 15 are involved less often than expected. A high proportion of mutations (74%) located in the ligand-binding domain (exons 2–6) involve amino acids conserved in evolution (Varret et al., 1998; data also available online at http://www.umd.necker.fr). In addition to point mutations, several deletions of various

sizes and locations, and insertions have been described. Depending on the intragenic location of a mutation, different effects can be observed, including absent mRNA synthesis, defective intracellular transport due to abolished binding (**1**) or defective receptor recycling (**2**) resulting in transport failure (**3**), reduced membrane anchorage (**4**), and defective internalization (**5**). Alu repeats may be involved as a cause of intragenic deletions. (Figure adapted from Hobbs et al., 1990, and Goldstein et al., 2001.)

C. A mutation in the LDL receptor gene

Direct sequencing demonstrates a mutation in exon 9. First, exon 9 is amplified by PCR (P1 and P2 = primers 1 and 2). The mutation in codon 408, GTG (valine) to GTA (methionine), produces a recognition site (N) for *Nla*III (GATC) that is not normally present. This results in two fragments of 126 and 96 base pairs (bp) instead of the usual 222-bp fragment (**1**). Thus, affected individuals (1 and 3 in the pedigree) have two smaller fragments of 126 and 96 kb in addition to the 222-kb fragment (**2**). Sequence analysis of the patient (individual 1 in the pedigree) demonstrates the mutation by the presence of an additional adenine (A, yellow circle) next to the normal guanine (**3**). Once the type of mutation has been established in an affected individual, all other members of a family can easily be screened for presence or absence of the mutation after appropriate genetic counseling. (Photographs kindly provided by Dr. H. Schuster, Berlin.)

References

Goldstein JL, Brown MS, Hobbs HH: Familial hypercholesterolemia, pp 2863–2913. In: Scriver CR et al (eds) The Metabolic and Molecular Bases of Inherited Disease, 8th ed. McGraw-Hill, New York, 2001.

Rader DJ, Hobbs HH: Disorders of lipoprotein metabolism, pp 2286–2298. In: Kasper DL et al (eds) Harrison's Principles of Internal Medicine, 16th ed. McGraw-Hill, New York, 2005.

Varret M et al.: LDLR database (second edition): new additions to the database and the software, and results of the first molecular analysis. Nucleic-Acids Res 26: 248–252, 1998.

Online information:
Universal Mutation Database (UMD) at
www.umd.necker.fr/

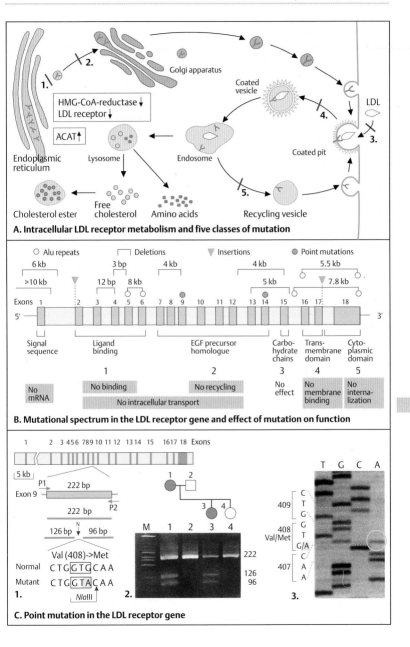

A. Intracellular LDL receptor metabolism and five classes of mutation

B. Mutational spectrum in the LDL receptor gene and effect of mutation on function

C. Point mutation in the LDL receptor gene

Diabetes Mellitus

Diabetes mellitus (DM) is an etiologically heterogeneous group of more than 60 individually defined disorders characterized by high fasting levels of glucose in the blood (> 125 mg/dl, hyperglycemia). The two principal types according to pathogenesis are type 1 (insulin-dependent diabetes mellitus, IDDM, MIM 222100) and type 2 (non-insulin-dependent, NIDDM, MIM 125853). In type 1 the β-cells in the pancreas have been destroyed by an autoimmune process, resulting in lack of insulin. In type 2, insulin resistance in the peripheral tissues and dysfunction of β-cells with reduced insulin secretion cause hyperglycemia. Diabetes mellitus is a common health problem in many parts of the world, affecting up to 1–2% of the population. Hyperglycemia causes numerous complications such as vascular damage, myocardial infarction, stroke, renal failure, ulcerative lesions of the legs often requiring amputation, and blindness.

A. Insulin biosynthesis

Human insulin is encoded by a gene with two exons and a signal sequence in the 5′ direction of the gene, located on the short arm of chromosome 11 (11p15.5). It has a β-cell-specific enhancer and a variable number of tandem repeats (VNTR) upstream (5′) of the gene. The primary transcript is spliced into mRNA and translated into preproinsulin (1430 amino acids). Then the signal sequence (24 amino acids) and the C-peptide are removed. The A chain and the B chain are joined by two disulfide bonds. This gene is expressed exclusively in the β-cells of the pancreas.

B. Insulin receptor

The receptor consists of two extracellular α-chains and two transmembrane β-chains, which are connected by disulfide bridges at specific sites. Distinct functional domains in both extracellular and intracellular parts of the receptor reflect the various functions. Insulin receptor substrate IRS and Shc proteins mediate the principal functions of insulin: activating growth factors, protein synthesis, glycogen synthesis, and glucose transport.

C. Diabetes mellitus (simplified model)

DM type 1 is caused by external factors, such as certain viral infections, which directly or by autoimmune reactions cause destruction of β-cells on a background of genetic susceptibility. DM type 2 results from genetic factors but is associated with lifestyle as a strong environmental component. Monozygotic twins are concordant for DM type 1 in about 25%, for type 2 in about 40–50%. First-degree relatives of individuals with DM type 1 are affected in about 2–7%, depending on the relationship and age at onset of disease. Several types of autosomal dominant forms of DM type 2 in adults exist (Maturity Onset Diabetes of the Young, MODY 1–6, MIM 125850). DM is a secondary manifestation of a number of genetically determined diseases involving insulin receptor defects (insulin resistance syndromes).

D. Genetic susceptibility

A major influence on the genetic susceptibility to DM type 1 is exerted by certain alleles of class I MHC genes (see p. 318). Several alleles of the DR3 and DR4 loci are associated with susceptibility to diabetes type I, especially DR3/DR4 heterozygotes. Some alleles of DR2 confer relative resistance to DM. Numerous genes or loci conferring susceptibility to DM type 2 have been identified at several sites in the human genome.

References

Bell GI, Polonsky KS: Diabetes mellitus and genetically programmed defects in β-cell function. Nature 414: 788–791, 2002.

Daneman D: Type 1 diabetes. Lancet 367: 847–858, 2006.

Lowe Jr WL: Diabetes mellitus, pp 433–442. In: Jameson JL (ed) Principles of Molecular Medicine. Humana Press, Totowa, New Jersey, 1998.

Maclaren NK, Kukreja A: Type 1 diabetes, pp 1471–1488. In: Scriver CR et al (eds) The Metabolic and Molecular Bases of Inherited Diseases, 8th ed. McGraw-Hill, New York, 2001.

O'Rahilly S, Barroio I, Wareham NJ: Genetic factors in type 2 diabetes: The end of the beginning. Science 307: 370–372, 2005.

Schwartz MW et al: Leptin- and insulin-receptor signalling. Nature 404: 663, 2000.

Stumvoll M, Goldstein BJ, van Haeften TW: Type 2 diabetes: principles of pathogenesis and therapy. Lancet 365: 1333–1346, 2005.

Taylor SI: Insulin action, insulin resistance, and type 2 diabetes mellitus, pp 1433–1469. In: Scriver CR et al (eds) The Metabolic and Molecular Bases of Inherited Diseases, 8th ed. McGraw-Hill, New York, 2001.

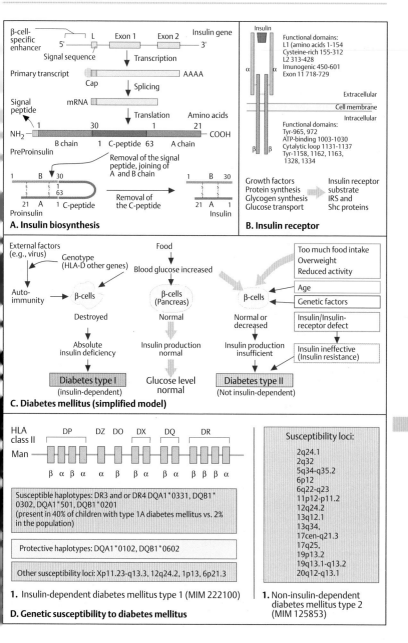

A. Insulin biosynthesis

B-cell-specific enhancer — L — Exon 1 — Exon 2 — Insulin gene
5' — Signal sequence — 3'
Transcription
Primary transcript — AAAA
Cap — Splicing
mRNA
Signal peptide — Translation — Amino acids
NH_2 — 1 — 30 — 1 — 21 — COOH
B chain — 1 — C-peptide — 63 — A chain
PreProinsulin

Removal of the signal peptide, joining of A and B chain
1 — B — 30
63
21 — A — 1 — C-peptide
Proinsulin

Removal of the C-peptide
1 — B — 30
21 — A — 1
Insulin

B. Insulin receptor

Insulin
Functional domains:
L1 (amino acids 1-154
Cysteine-rich 155-312
L2 313-428
Imunogenic 450-601
Exon 11 718-729
Extracellular
Cell membrane
Intracellular
Functional domains:
Tyr-965, 972
ATP-binding 1003-1030
Cytalytic loop 1131-1137
Tyr-1158, 1162, 1163, 1328, 1334

Growth factors — Insulin receptor
Protein synthesis — substrate
Glycogen synthesis — IRS and
Glucose transport — Shc proteins

C. Diabetes mellitus (simplified model)

External factors (e.g., virus) — Genotype (HLA-D other genes)
Food — Blood glucose increased
Too much food intake / Overweight / Reduced activity
Age
Genetic factors
Insulin/Insulin-receptor defect

Auto-immunity → β-cells
β-cells (Pancreas)
β-cells
Insulin ineffective (Insulin resistance)

Destroyed — Normal — Normal or decreased
Absolute insulin deficiency — Insulin production normal — Insulin production insufficient

Diabetes type I (insulin-dependent) — Glucose level normal — Diabetes type II (Not insulin-dependent)

D. Genetic susceptibility to diabetes mellitus

HLA class II — DP — DZ — DO — DX — DQ — DR
Man
β α β α — α β — β α — β α — β β β α

Susceptible haplotypes: DR3 and or DR4 DQA1*0331, DQB1*0302, DQA1*501, DQB1*0201 (present in 40% of children with type 1A diabetes mellitus vs. 2% in the population)

Protective haplotypes: DQA1*0102, DQB1*0602

Other susceptibility loci: Xp11.23-q13.3, 12q24.2, 1p13, 6p21.3

1. Insulin-dependent diabetes mellitus type 1 (MIM 222100)

Susceptibility loci:
2q24.1
2q32
5q34-q35.2
6p12
6q22-q23
11p12-p11.2
12q24.2
13q12.1
13q34,
17cen-q21.3
17q25,
19p13.2
19q13.1-q13.2
20q12-q13.1

1. Non-insulin-dependent diabetes mellitus type 2 (MIM 125853)

Protease Inhibitor α₁-Antitrypsin

α₁-Antitrypsin (α₁-AT) is an essential protease inhibitor in blood plasma, first described by Laurell and Ericksson in 1963. As a member of the serine protease inhibitor superfamily (SERPIN) it has a major role in activating neutrophil elastase and other proteases to maintain protease/antiprotease balance. It binds to a wide range of proteases, such as elastase, trypsin, chemotrypsin, thrombin, and bacterial proteases. Its most important physiological effect is the inhibition of leukocyte elastase in the bronchial system.

A. α₁-Antitrypsin

Human α₁-antitrypsin is a 52-kDa glycoprotein composed of 394 amino acids and 12% carbohydrate content. It is encoded by a 12.2-kb gene with five exons on chromosome 14 (14q32.1).

B. α₁-Antitrypsin deficiency

α₁-Antitrypsin deficiency (MIM 107400) results in a variable disease with chronic obstructive pulmonary emphysema as the main sign (visible by increased darkness of the chest X-ray, kindly provided by N. Konietzko, Essen). The cause is uninhibited activity of leukocyte elastase on the elastin of the pulmonary alveoli. The concentration of α₁-antitrypsin in alveolar fluid is greatly reduced in heterozygotes (red squares) and more so in homozygotes (yellow triangles). This can be corrected by intravenous administration of α₁-antitrypsin. A related, allelic disorder is α₁-antichemotrypsin deficiency (MIM 107280).

Oxidizing substances have an inhibitory effect and inactivate the molecule. Smokers have a much more rapid course of α₁-AT deficiency disease (onset of dyspnea at 35 years of age instead of 45–50).

C. α₁-Antitrypsin mutations

More than 100 alleles are known. The α₁-antitrypsin protein has three oligosaccharide side chains at positions 46, 83, and 247 and is highly polymorphic due to differences in the amino acid sequence and variations in the carbohydrate side chains. The reactive site is located at position 358/359 (methionine/serine). The gene has four coding exons, 2, 3, 4, and 5, and three noncoding exons, 1 a, 1 b, and 1 c. The α₁-AT alleles are grouped into four classes: (i) nor-

mal, (ii) deficiency, (iii) null alleles, and (iv) dysfunctional alleles. The typical normal allele is Pi*M; the most important deficiency alleles are Pi*Z, Pi*P, and Pi*S.

The most frequent deficiency allele, Pi*Z, causes plasma concentrations of α₁-AT of about 12–15% of normal in the homozygous genotype Pi*ZZ and 64% in the heterozygote (Pi*MZ). MS heterozygotes have 86% of the MM homozygote activity. The most frequently affected sites of mutations are codons 213 (PI*Z), 256 (PI*P), 264 (Pi*S), 342 (Pi*Z), and 357 (Pi[Pittsburgh]). The molecular genetic diagnosis is facilitated by the presence of variant restriction enzyme sites.

D. Synthesis of α₁-antitrypsin

The α₁-AT gene is expressed in liver cells (hepatocytes). The gene product is channeled through the Golgi apparatus and released from the cell (secreted). The Z mutation leads to aggregation of the enzyme in the liver cells, with too little of it being secreted. The S mutation leads to premature degradation. About 2–4% of the population in Central and Northern Europe are MZ heterozygotes.

E. Reactive center of protease inhibitors

α₁-Antitrypsin is a member of a family of protease inhibitors that show marked homology, especially at their reactive centers. Oxidizing substances have an inhibitory effect and inactivate the molecule. (Figures in C–E adapted from Cox, 2001, and Owen et al., 1983.)

References

Cox DW: α₁-Antitrypsin deficiency, pp 5559–5584. In: Scriver CR et al (eds) The Metabolic and Molecular Bases of Inherited Disease, 8th ed. McGraw-Hill, New York, 2001.

Laurell C-B, Eriksson S: The electrophoretic alpha-1-globulin pattern of serum in alpha-1-antitrypsin deficiency. Scand J Clin Lab Invest 15: 132–140, 1963.

Lomas DA et al.: The mechanism of Z α₁-antitrypsin accumulation in the liver. Nature 357: 605–607, 1992.

Owen MC et al: Mutation of antitrypsin to antithrombin: α₁-antitrypsin Pittsburgh (358→Arg), a fatal bleeding disorder. N Engl J Med 309: 694–698, 1983.

Stoller JK, Aboussovan LS: α₁-Antitrypsin deficiency. Lancet 365: 2225–2236, 2005.

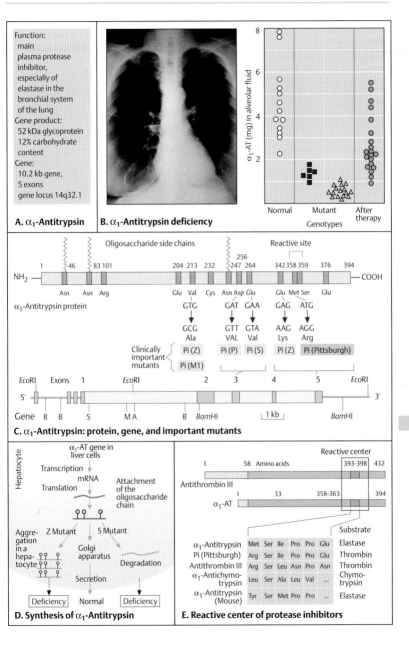

A. α₁-Antitrypsin

Function:
main
  plasma protease
  inhibitor,
  especially of
  elastase in the
  bronchial system
  of the lung
Gene product:
  52 kDa glycoprotein
  12% carbohydrate
  content
Gene:
  10.2 kb gene,
  5 exons
  gene locus 14q32.1

B. α₁-Antitrypsin deficiency

C. α₁-Antitrypsin: protein, gene, and important mutants

D. Synthesis of α₁-Antitrypsin

E. Reactive center of protease inhibitors

Blood Coagulation Factor VIII (Hemophilia A)

Hemophilia is a severe X-chromosomal bleeding disorder resulting from inactivity of blood coagulation factor VIII (hemophilia A, MIM 306700) or factor IX (hemophilia B, MIM 306900). Factors VIII and IX function in the blood coagulation cascade. Factor VIII functions as a cofactor in the activation of factor X to factor Xa. Hemophilia was the first major disease recognized to be genetically determined. The Talmud refers to its increased occurrence in males in certain families. The term hemophilia was introduced by F. Hopff, in a medical thesis in 1828 in Würzburg, Germany.

A. Inheritance of hemophilia A

Hemophilia A is an X-chromosomal disease with a frequency of about 1 in 10000 males. Its X-chromosomal inheritance is readily apparent in some royal families in Europe.

B. Blood coagulation factor VIII

When activated by thrombin, factor VIII protein consists of five subunits (A1, A2, A3, C1, C2) held together by calcium ions (Ca^{++}, **1**). The inactive factor VIII protein (**2**) contains three domains (A, B, C). Domain A occurs in three homologous copies (A1, A2, A3), domain C in two (C1, C2), and domain B in one copy. In humans, the *F8* gene (**3**) is located at Xq28, near the gene for factor IX. It consists of 26 exons and spans 186 kb of DNA. Noteworthy are the large exon 14 (3106 base pairs), which encodes the B domain, and the large intron 22 (32000 base pairs), between exons 22 and 23. Most mutations involve the dinucleotide CG in DNA sequences TCGA. This mutates easily to TTGA because the cytosine of this dinucleotide is frequently methylated. Subsequent deamination of methyl cytosine leads to a C-to-T transition. This creates a stop codon (TGA), resulting in a truncated factor VIII protein. Even a stop codon at position 2307 leads to severe hemophilia, although only the last 26 amino acids are missing. Since TCGA is the recognition sequence for the restriction enzyme *Taq*I, RFLP (restriction fragment length polymorphisms) can be utilized for molecular genetic diagnosis (**4**). A polymorphic variant in the recognition sequence for the restriction enzyme *Bcl*II in the region of exons 17 and 18 produces a fragment of 879 bp and a fragment

of 286 bp; when it is absent, a single fragment of 1165 base pairs results. In the family pedigree (**5**) two affected males (II-1 and III-2) carry the 879-bp fragment. Thus, this fragment indicates presence of the mutation.

Common types of mutation in hemophilia A are nonsense in 14%, small deletions and insertions in 15%, splice mutations in 4%, and inversion flip tip mutations in 42% (see D).

C. Clinical manifestations

Frequent acute bleeding episodes (**1**) resulting from minor trauma will result in severe functional consequences such as stiff knee joints (**2**) or elbows, or extensive soft tissue hematoma. Factor VIII activity below 2% results in very severe disease with spontaneous bleeding into joints, muscle, and internal organs (in 48%), 2–10% of activity in moderate severity (31%), and 10–30% of activity in relatively mild hemophilia (in 21%). (Figures from www.pathguy.com/lectures/ (1) and Gulnara Huseinova, Baku, Azerbaijan (2) obtained at Google.)

D. Factor VIII inversion

A frequent cause is an inversion between an unrelated gene A in intron 22 and one of two homolgous distal genes A, followed by nonhomologous crossingover. This flip tip inversion results in disruption of the *F8* gene between exon 22 and exon 23.

References

Dahlbäck B: Blood coagulation. Lancet 355: 1627–1632, 2000.

Gitschier J et al: Detection and sequence of mutations in the factor VIII gene of haemophiliacs. Nature 315: 427–430, 1985.

Graw J et al.: Haemophilia A: From mutation analysis to new therapies. Nature Rev Genet 6: 488–501, 2005.

Hopff F: Über die Haemophilie oder die erbliche Anlage zu tötlichen Blutungen. Inaugural Dissertation, Universität Würzburg, 1828.

Kazazian Jr HH, Tuddenham EGD, Antonarakis SE: Hemophilia A: Deficiencies of coagulation factors VIII, pp 4367–4392. In: Scriver CR et al (eds) The Metabolic and Molecular Bases of Inherited Disease, 8th ed. McGraw-Hill, New York, 2001.

Online information:
National Hemophilia Foundation
  http://www.hemophilia.org/NHFWeb/

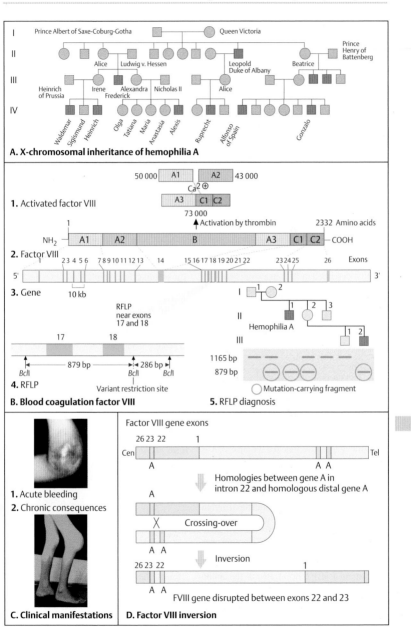

A. X-chromosomal inheritance of hemophilia A

1. Activated factor VIII

50 000 — A1 — A2 — 43 000

Ca²⊕

A3 — C1 — C2

73 000

2. Factor VIII

1 — NH₂ — A1 — A2 — B — A3 — C1 — C2 — COOH — 2332 Amino acids

▲ Activation by thrombin

3. Gene

1 2 3 4 5 6 7 8 9 10 11 12 13 14 15 16 17 18 19 20 21 22 23 24 25 26 Exons

5'   3'

10 kb

RFLP near exons 17 and 18

17 18

879 bp 286 bp

BclI BclI BclI

4. RFLP

Variant restriction site

B. Blood coagulation factor VIII

I 1 2
II 1 2 3
Hemophilia A
III 1 2

1165 bp
879 bp

◯ Mutation-carrying fragment

5. RFLP diagnosis

1. Acute bleeding
2. Chronic consequences

C. Clinical manifestations

Factor VIII gene exons

26 23 22 1

Cen A A A Tel

Homologies between gene A in intron 22 and homologous distal gene A

A

Crossing-over

A A

Inversion

26 23 22 1

A A FVIII gene disrupted between exons 22 and 23

D. Factor VIII inversion

Von Willebrand Disease

Hereditary dysfunction of a complex, multimeric glycoprotein (named *von Willebrand* factor, vWF) in blood plasma, thrombocytes (platelets), and subendothelial mesenchymal tissue of blood vessels result in a heterogeneous group of common bleeding disorders, grouped together under the name of von Willebrand disease or von Willebrand–Jürgens syndrome (MIM 193400). Von Willebrand factor has two basic biological functions: it binds to specific receptors on the surface of platelets and subendothelial connective tissue, and it forms bridges between platelets and damaged regions of a vessel. It binds to clotting factor VIII and stabilizes it. Deficiency of vWF leads to reduced or absent platelet adhesion and to secondary deficiency of factor VIII. Hereditary deficiency of vWF is the most common bleeding disorder in man, with a frequency of about 1 : 250 for all forms, and about 1 : 8000 for severe forms. It was first described in 1926 in a large family on the Åland islands in Finland by Erik von Willebrand.

A. Von Willebrand factor glycoprotein

Von Willebrand factor is formed in endothelial cells, in megakaryocytes, and possibly in some other tissues. It is encoded by a large (178 kb) gene with 52 exons of various sizes on chromosome 12 (12p12pter). Along its 8.7-kb cDNA several polymorphic restriction sites (red arrows) exist (**1**). From the mRNA a primary peptide (prepro-vWF) of 2813 amino acids is translated (**2**). Prepro-vWF is a polypeptide of 2813 amino acids with a signal peptide of 22 amino acids and 5 repetitive functional domains, A-D and CK, which are distributed from the amino terminal in the sequence D1, D2, D3, A1, A2, A3, D4, B1, B2, B3, C1, C2, and CK. A segment of 741 amino acids in D1 and D2 corresponds to the vW antigen II. The various domains contain binding sites for factor VIII, heparin, collagen, thrombocytes, and thrombin. A tetrapeptide sequence Arg-Gly-Asp-Ser (RGDS) located near the C-terminus serves as a binding site for vWF. The vW factor contains 8.3% cysteine (234 of 2813 amino acids), concentrated at the amino and carboxy ends, whereas the three A domains are cysteine-poor (**3**). After posttranslational modification, the mature plasma vWF contains 12 oligosaccharide side chains (**4**), yielding a carbohydrate weight proportion of 19% of the vWF molecule.

B. Maturation of the von Willebrand factors (vWF)

A prepropeptide is translated from the vWF mRNA. After the signal peptide is removed in the endoplasmic reticulum, the two pro-vWF units attach to each other at their carboxy ends by means of numerous disulfide bridges to form a dimer. The dimers represent the repetitive units, or protomers, of mature vWF. The pro-vWF dimers are transported to the Golgi apparatus, where the pro-vWF antigen (vW antigen II or vWagII) is removed. Mature vWF and vWagII are stored in Weibel–Palade bodies in epithelial cells. The mature subunits and vWagII contain binding sites for factor VIII, heparin, collagen, ristocetin + platelets, and thrombin-activated platelets.

C. Classification

The bleeding time is prolonged, but coagulation time is normal. Bleeding occurs mainly into mucocutaneous tissues rather than joints. The disease is divided into several subtypes. In types I and III, the defect is quantitative; in type II, qualitative. Dominant and recessive (MIM 277480) phenotypes with vWF deficiency are often difficult to distinguish, because heterozygosity may not be manifest and can only be determined by laboratory tests. Type I with subtypes A and B is the most frequent group (70% of all patients). (Figures adapted from Sadler, 2001.)

References

James AH: Von Willebrand disease. Obstet Gynecol Survey 61: 136–145 2006.

Manusco DJ et al: Structure of the gene for human von Willebrand factor. J Biol Chem 264: 19514–19527, 1989.

Sadler JE: Von Willebrand disease, pp 4415–4431. In: Scriver CR et al (eds) The Metabolic and Molecular Bases of Inherited Disease, 8th ed. McGraw-Hill, New York, 2001.

von Willebrand EA: Hereditär pseudohemofili. Fin Laekaresaellsk Hand 68: 87–112, 1926.

von Willebrand EA, Jürgens R: Über ein neues vererbbares Blutungsübel. Dtsch Arch Klin Med 175: 453–483, 1933.

Wise RJ et al: Autosomal recessive transmission of hemophilia A due to a von Willebrand factor mutation. Hum Genet 91: 367–372, 1993.

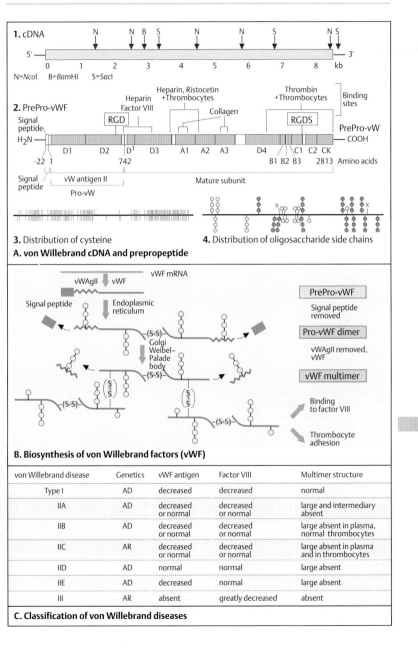

1. cDNA

N=*Nco*I   B=*Bam*HI   S=*Sac*I

2. PrePro-vWF

Heparin, Ristocetin +Thrombocytes

Heparin Factor VIII

Collagen

Thrombin +Thrombocytes

Binding sites

Signal peptide

RGD

RGDS

PrePro-vW

H₂N —

— COOH

D1   D2   D¹   D3   A1   A2   A3   D4   C1   C2   CK

-22   1   742   B1 B2 B3   2813 Amino acids

Signal peptide

vW antigen II

Mature subunit

Pro-vW

3. Distribution of cysteine

4. Distribution of oligosaccharide side chains

A. von Willebrand cDNA and prepropeptide

vWF mRNA

vWAgII   vWF

Signal peptide

Endoplasmic reticulum

Golgi Weibel–Palade body

(S-S)

PrePro-vWF

Signal peptide removed

Pro-vWF dimer

vWAgII removed, vWF

vWF multimer

Binding to factor VIII

Thrombocyte adhesion

B. Biosynthesis of von Willebrand factors (vWF)

von Willebrand disease	Genetics	vWF antigen	Factor VIII	Multimer structure
Type I	AD	decreased	decreased	normal
IIA	AD	decreased or normal	decreased or normal	large and intermediary absent
IIB	AD	decreased or normal	decreased or normal	large absent in plasma, normal thrombocytes
IIC	AR	decreased or normal	decreased or normal	large absent in plasma and in thrombocytes
IID	AD	normal	normal	large absent
IIE	AD	decreased	normal	large absent
III	AR	absent	greatly decreased	absent

C. Classification of von Willebrand diseases

Pharmacogenetics

Pharmacogenetics, a term introduced by Motulsky (1957) and Vogel (1959), refers to a blend of genetics and pharmacology concerned with genetically determined modifications of individual pharmacological responses. The genetic principle underlying pharmcogenetics is the common presence of polymorphic variants at virtually each gene locus. As a result, individuals differ in their response to chemical substances used for therapy. Enzymes encoded by genes with different alleles may have different metabolic rates. Some individuals metabolize a certain chemical compound slower or faster than others. Unforeseen, undesirable side effects may jeopardize a patient. *Pharmacogenomics* refers to the scientific study of variable drug response based on the entire genome.

A. Malignant hyperthermia (MH)

Malignant hyperthermia (MIM 145600, 154275, 180901) is a severe, life-threatening complication of anesthesia that occurs in persons who are hypersensitive to halothane and similar agents used in general anesthesia. Normally a nerve impulse depolarizes the plasma membrane of a nerve ending at the nerve–muscle endplate (**1**). The influx of calcium into the cell triggers the release of acetylcholine. This temporarily opens the receptor-controlled cation (Na^+) channels (see p. 280). This opens calcium channels located in the sarcoplasmic reticulum of the muscle cell and causes the myofibrils to contract. The calcium channels in the sarcoplasmic reticulum are regulated by the ryanodine receptor (**2**). Mutations in the ryanodine receptor, a protein with four transmembrane domains, cause greatly increased sensitivity to halothane and other anesthetic agents (**3**). They induce muscle rigidity, drastic elevation of temperature (hyperthermia), acidosis, and cardiac arrest (**4**). Malignant hyperthermia is inherited as an autosomal dominant trait (**5**). Gene loci relevant for MH are located at 19q13.1 (MIM 145600, MH1), 17q11.2-q24 (154275, MH2), 7q21-q22 (154276, MH3), 3q13.1 (600467, MH4), 1q32 (601887, MH5), and 5p (161888, MH6). In a given family the mutant haplotype can be determined by segregation analysis.

B. Pharmacogenetic variants of butyrylcholinesterase

About 1 in 200 individuals reacts to muscle relaxants, such as suxamethonium (succinylcholine), with prolonged muscle relaxation and respiratory arrest. Such persons do not degrade suxamethonium because of insufficient activity of serum butyrylcholinesterase (MIM 177400, formerly acetylcholinesterase). This enzyme hydrolyzes butyrylcholine more readily than acetylcholine (MIM 100740). Individuals at risk cannot be identified by enzyme activity alone (**1**). However, all three genotypes can be determined after dibucaine, an inhibitory substance, is added (**2**). Individuals at risk (red squares) have only 20% activity, heterozygotes have 50–70%, and homozygous normal persons about 80%. Butyrylcholinesterase is encoded by genes located at 3q26.1-q26.2 and 7q22. (Figure adapted from Harris, 1975.)

C. Adverse reactions to drugs

Numerous human pharmacogenetic disorders are known (see opposite table and Nebert, 2003).

References

Arranz MJ et al: Pharmacogenetic and pharmacogenomic research in psychiatry: current advances and clinical applications. Current Pharmacogenomics 1: 151–158, 2003.

Denborough M: Malignant hyperthermia. Lancet 352: 1131–1136, 1998.

Kalow W, Grant DM: Pharmacogenetics, pp 225–255. In: Scriver CR et al (eds) The Metabolic and Molecular Bases of Inherited Disease, 8th ed. McGraw-Hill, New York, 2001.

McLennan DH, Britt BA: Malignant hyperthermia and central core disease, pp 949–954. In: Jameson JL (ed) Principles of Molecular Medicine. Humana Press, Totowa, New Jersey, 1998.

Meyer VA: Pharmacogenetics – five decades of therapeutic lessons from genetic diversity. Nature Rev Genet 5: 669–676, 2004.

Motulsky AG: Drug reactions, enzymes and biochemical genetics. JAMA 165: 835–837, 1957

Nebert DW: Pharmacogenetics and pharmacogenomics. Nature Encyclopedia Human Genome 4: 558–567, 2003.

Roses AD: Pharmacogenetics and the practice of medicine. Nature 405: 857–865, 2000.

Vogel F: Moderne Probleme der Humangenetik. Ergeb Inn Med Kinderheilk 12: 52–125, 1959.

Weinhilsboum R: Inheritance of drug response. New Engl J Med 348: 529–573, 2003.

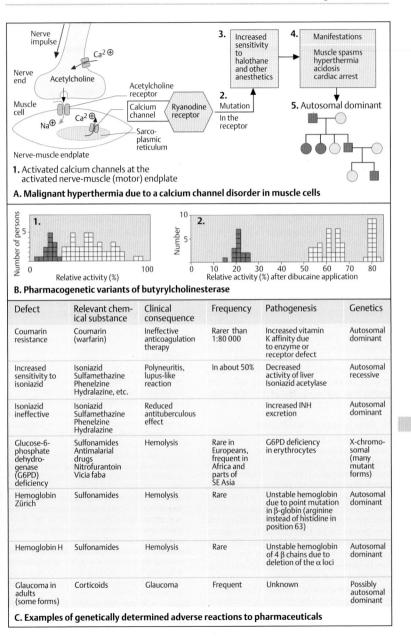

1. Activated calcium channels at the activated nerve-muscle (motor) endplate

A. Malignant hyperthermia due to a calcium channel disorder in muscle cells

B. Pharmacogenetic variants of butyrylcholinesterase

Defect	Relevant chemical substance	Clinical consequence	Frequency	Pathogenesis	Genetics
Coumarin resistance	Coumarin (warfarin)	Ineffective anticoagulation therapy	Rarer than 1:80 000	Increased vitamin K affinity due to enzyme or receptor defect	Autosomal dominant
Increased sensitivity to isoniazid	Isoniazid Sulfamethazine Phenelzine Hydralazine, etc.	Polyneuritis, lupus-like reaction	In about 50%	Decreased activity of liver Isoniazid acetylase	Autosomal recessive
Isoniazid ineffective	Isoniazid Sulfamethazine Phenelzine Hydralazine	Reduced antituberculous effect		Increased INH excretion	Autosomal dominant
Glucose-6-phosphate dehydrogenase (G6PD) deficiency	Sulfonamides Antimalarial drugs Nitrofurantoin Vicia faba	Hemolysis	Rare in Europeans, frequent in Africa and parts of SE Asia	G6PD deficiency in erythrocytes	X-chromosomal (many mutant forms)
Hemoglobin Zürich	Sulfonamides	Hemolysis	Rare	Unstable hemoglobin due to point mutation in β-globin (arginine instead of histidine in position 63)	Autosomal dominant
Hemoglobin H	Sulfonamides	Hemolysis	Rare	Unstable hemoglobin of 4 β chains due to deletion of the α loci	Autosomal dominant
Glaucoma in adults (some forms)	Corticoids	Glaucoma	Frequent	Unknown	Possibly autosomal dominant

C. Examples of genetically determined adverse reactions to pharmaceuticals

Cytochrome P450 (CYP) Genes

The cytochrome P450 system refers to a group of 57 genes (*CYP* genes) known in the human genome and the enzymes with different functions they encode. These enzymes derive their name from their maximal light absorption at 450 nm after binding to CO. Cytochrome P450 enzymes degrade complex chemical substances, such as drugs or plant toxins, by an oxidation system (monooxygenases) in the microsomes of the liver and mitochondria of the adrenal cortex. *CYP* genes are members of a large system of evolutionarily related genes encoding P450 proteins with different enzymatic specificity in mammals.

A. Cytochrome P450 system

The cytochrome P450 enzymes carry out phase I of a detoxification pathway (**1**): a substrate (RH) is oxidized to ROH utilizing atmospheric oxygen (O_2), with water (H_2O) being formed as a byproduct. A reductase delivers hydrogen ions (H^+) either from NADPH or NADH. In phase II, ROH is further degraded and thereafter eliminated. The P450 enzymes have a wide spectrum of activity (**2**). Characteristically, a single P450 protein can oxidize a number of structurally different chemical substances or several P450 enzymes can degrade a single chemical substrate. The enzyme activities of phases I and II have to be well coordinated, since toxic intermediates occasionally arise in the initial stages of phase II.

B. Debrisoquine metabolism

Debrisoquine is an isoquinoline-carboxamidine used to treat high blood pressure until it was found to cause severe side effects in 5–10% of the population. These individuals have reduced activity of debrisoquine-4-hydroxylase (CYP2D). This enzyme degrades several pharmacological substances such as β-adrenergic blockers, antiarrhythmics, and antidepressives. Two groups can be distinguished in the population: those with normal and those with slow degradation (**1**). Individuals with low activity are at increased risk for untoward toxic reactions. Individuals with a slow rate of degradation show an increased ratio of debrisoquin/4-hydrodebrisoquin. This enzyme is encoded by the *CYP2D6* gene (MIM 124030/608902) located at 22q13.1. Certain mutations of the primary transcript of this gene, with 9 exons, produce aberrant splicing (**2**). As a result, variant mRNAs still contain an intron and produce proteins with reduced enzymatic activity. (Figure adapted from Gonzales et al., 1988.)

C. *CYP* gene superfamily

The *cytochrome P450* genes in mammals make up a superfamily of genes that resemble each other in exon/intron structure and that encode related enzymes. An evolutionary pedigree based on the similarity of their cDNA sequences reveals that the *CYP* gene family arose during the last 1500–2000 million years. The largest *P450* family in mammals is *CYP2*, with 16 genes in humans. It is assumed that the *CYP2* family developed in response to toxic substances in plants that had to be detoxified by animal organisms. At least 30 gene duplications and gene conversions have led to an unusually diverse repertoire of *CYP* genes. Most important for drug metabolism are the enzymes CYP2C8, -2C9, -2C18, and -2C19, which together metabolize more than 50 compounds. *CYP3A4*, *CYP2D6*, and *CYP2C9* are responsible for 50%, 25%, and 5% of drug metabolism, respectively. (Figures adapted from Gonzales et al., 1988, and Gonzales & Nebert, 1990.)

References

Gonzalez FJ et al: Characterization of the common genetic defect in humans deficient in debrisoquine metabolism. Nature 331: 442–446, 1988.

Gonzalez FJ, Nebert DW: Evolution of the P450 gene superfamily: animal-plant "warfare," molecular drive, and human genetic differences in drug oxidation. Trends Genet 6: 182–186, 1990.

Nebert DW, Russell DW: Clinical importance of the cytochromes P450. Lancet 360: 1155–1162, 2002.

Nebert DW, Nelson DR: Cytochrome P450 (*CYP*) gene superfamily. Nature Encyclopedia Hum Genome 1: 1028–1037, 2003.

Nelson DR: Comparison of cytochrome P450 (CYP) genes from the mouse and human genomes, including nomenclature recommendations for genes, pseudogenes and alternative-splice variants. Pharmacogenetics 14: 1–18, 2004.

Panserat S et al: DNA haplotype-dependent differences in the amino acid sequence of debrisoquine 4-hydroxylase (CYP2D6): evidence for two major allozymes in extensive metabolisers. Hum Genet 94: 401–406, 1994.

Sachse C et al: Cytochrome P450 2 D6 variants in a Caucasian population: allele frequencies and phenotypic consequences. Am J Hum Genet 60: 284–295, 1997.

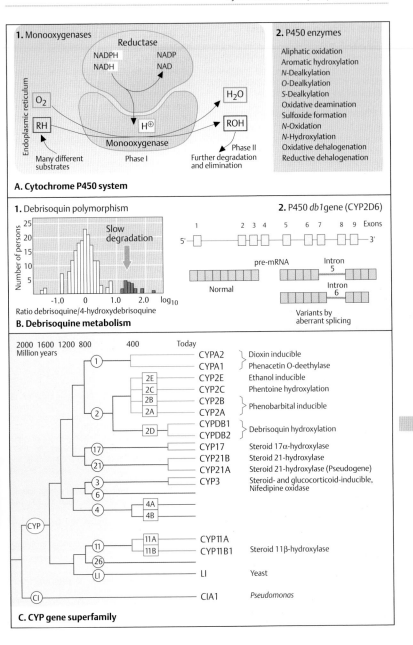

1. Monooxygenases

Reductase

NADPH   NADP
NADH   NAD

O₂

Endoplasmic reticulum

RH

H⊕

H₂O

ROH

Monooxygenase

Many different substrates   Phase I

Phase II

Further degradation and elimination

2. P450 enzymes

Aliphatic oxidation
Aromatic hydroxylation
N-Dealkylation
O-Dealkylation
S-Dealkylation
Oxidative deamination
Sulfoxide formation
N-Oxidation
N-Hydroxylation
Oxidative dehalogenation
Reductive dehalogenation

A. Cytochrome P450 system

1. Debrisoquin polymorphism

Number of persons

Slow degradation

Ratio debrisoquine/4-hydroxydebrisoquine $\log_{10}$

B. Debrisoquine metabolism

2. P450 *db1* gene (CYP2D6)

1   2 3 4   5   6 7   8 9   Exons

5'   3'

pre-mRNA

Normal

Intron 5

Intron 6

Variants by aberrant splicing

2000 1600 1200 800   400   Today
Million years

CYPA2   Dioxin inducible
CYPA1   Phenacetin O-deethylase
2E   CYP2E   Ethanol inducible
2C   CYP2C   Phentoine hydroxylation
2B   CYP2B   Phenobarbital inducible
2A   CYP2A
2D   CYPDB1   Debrisoquin hydroxylation
CYPDB2
17   CYP17   Steroid 17α-hydroxylase
21   CYP21B   Steroid 21-hydroxylase
CYP21A   Steroid 21-hydroxylase (Pseudogene)
3   CYP3   Steroid- and glucocorticoid-inducible,
6   Nifedipine oxidase
4   4A
4B
CYP   11   11A   CYP11A
11B   CYP11B1   Steroid 11β-hydroxylase
26
LI   LI   Yeast
CI   CIA1   *Pseudomonas*

C. CYP gene superfamily

Amino Acid Degradation and Urea Cycle Disorders

Mutations in genes encoding the various enzymes involved in amino acid metabolism and urea cycle cause a large group of disorders. Two examples are presented here.

A. Phenylalanine degrading system

Phenylketonuria (PKU, MIM 261600) is a deficiency of the enzyme phenylalanine hydroxylase (PAH). This enzyme converts phenylalanine (Phe) to tyrosine. PKU results from mutations in the *PAH* gene and is inherited as an autosomal recessive trait. The complex Phe hydroxylating system includes tetra-hydrobiopterin (BH_4) cofactor, which requires several enzymes for recycling, including dihydropteridine reductase (MIM 261630) and pterin-4α-carbinolamine dehydratase (MIM 264070). The *PAH* gene, located on chromosome 12 at q24.1, has 13 exons spanning 90 kb of DNA, with complex $5'$ untranslated *cis*-acting, *trans*-activated regulatory elements. It has developmental and tissue-specific transcription and translation. The hepatic and renal gene product is a 452-amino acid polypeptide. Several polymorphic sites (RFLPs and SNPs, and a tetranucleotide short tandem repeat) are in linkage disequilibrium and define haplotypes.

Hyperphenylalaninemia is defined as a plasma Phe concentration of $> 120\,\mu M$ (2 mg/dl). Long-term exposure to Phe concentrations $> 600\,\mu M$ results in severe mental retardation, as in classical PKU. This condition, first described in 1934 by Asbjørn Følling in Norway, is the prototype of a treatable metabolic disease. Here, dietary phenylalanine restriction must be instituted in the neonatal period. PKU is detected by neonatal screening. Maternal PKU causes various developmental problems if the maternal Phe level is not well controlled throughout pregnancy.

B. Distribution of *PAH* mutations

Almost 500 disease-causing mutations have been recorded at the *PAH* locus, many unevenly distributed in different populations. Other mutations are observed in Asian populations (China, Korea). PKU is rare in Finnish, Ashkenazi, American aboriginal, and Japanese populations. (Data derived from Scriver et al., 2003, and Zschocke, 2003.)

C. Urea cycle defects

Terrestrial vertebrates synthesize urea by means of the urea cycle, the first metabolic pathway to be described (by Krebs and Henseleit, a medical student, in 1932). Arginine, the immediate precursor of urea, is hydrolyzed by arginase (MIM 207830) to urea and ornithine. Ornithine transcarbamoylase (OTC, MIM 310461) transfers carbamoyl phosphate to ornithine, which results in citrulline. Carbamoyl phosphate is synthesized from NH_4^+, CO_2, H_2O, and ATP by carbamoyl phosphate synthetase (MIM 237300). Argininosuccinate synthetase (MIM 215700) catalyzes the condensation of citrulline and aspartate to argininosuccinate. This is cleaved by argininosuccinase (MIM 202900) into arginine and fumarate. Five metabolic disorders result from mutations of genes encoding the enzymes of the urea cycle. They are characterized by high plasma levels of ammonium, which is highly neurotoxic. Affected children may show progressive lethargy and coma leading to death in the neonatal period. A late-onset form may present with signs of encephalopathy. The most common form is X-linked OTC resulting from mutations in the *OTC* gene located at Xp21.1. It has 10 exons spanning 73 kb of DNA. (Figure adapted from Stryer, 1995.)

References

Brusilow SW, Horwich AL: Urea cycle enzymes, pp 1909–1963. In: Scriver CR et al (eds) The Metabolic and Molecular Bases of Inherited Disease, 8th ed. McGraw-Hill, New York, 2001.

Scriver CR, Kaufman S: Hyperphenylalaninemia: Phenylalanine hydroxylase deficiency, pp 1667–1724. In: Scriver CR et al (eds) The Metabolic and Molecular Bases of Inherited Disease, 8th ed. McGraw-Hill, New York, 2001.

Scriver CR et al: *PAHdb* 2003: What a locus-specific knowledgebase can do. Hum Mutat 21: 333–344, 2003.

Stryer I: Biochemistry, 4th ed. WH Freeman, New York, 1995.

Zschocke J: Phenylketonuria mutations in Europe. Hum Mutat 21: 345–356, 2003.

Online information:

PHD database (http://www.pahdb.mcgill.ca/).

National Society for Phenylketonuria http://www.nspku.org/

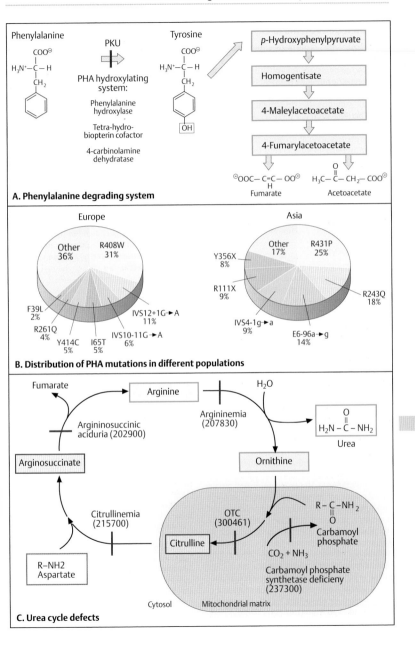

A. Phenylalanine degrading system

B. Distribution of PHA mutations in different populations

C. Urea cycle defects

Cytoskeletal Proteins in Erythrocytes

The cytoskeleton is an intracellular system of proteins with a fibrillar structure. The three main components are microfilaments (7.9 nm diameter), intermediate filaments (10 nm), and microtubuli (24 nm). They consist of polymers of small subunits. Microfilaments and membrane-binding proteins serve as a skeleton of the cell under the plasma membrane. A middle-sized protein is actin. It accounts for 1–5% of all cellular proteins, in muscle cells for 10%. Actins form an evolutionarily conserved gene family. Erythrocytes have to meet extreme requirements: they traverse small capillaries with diameters less than that of the erythrocytes themselves about a half million times during a 4-month lifespan. Membrane flexibility is also essential for muscle cell function.

A. Erythrocytes

A normal erythrocyte is maintained in a characteristic biconcave discoid form by the cytoskeletal proteins. Genetic defects in different cytoskeletal proteins lead to characteristic erythrocyte deformations: as ellipses (elliptocytes), as spheres (spherocytes), or as cells with a mouthlike area (stomatocytes) or thornlike projections (acanthocytes). The various forms are the result of defects of different cytoskeleton proteins (see D). (Figures based on scanning electron micrographs of Davies & Lux, 1989.)

B. Proteins in the erythrocyte membrane

Spectrin is a 200-nm long rod-shaped protein, which runs parallel to the erythrocyte plasma membrane. The characteristic shape of an erythrocyte is maintained by attachment of the spectrin-actin cytoskeleton to the erythrocyte plasma membrane. This is accomplished by two specific integral membrane proteins: ankyrin and band 4.1 protein. Ankyrin connects to band 3 protein, an anion transport protein in the plasma membrane. Band 4.1 protein binds to glycophorin. Glycophorins (A, B, C) are transmembrane proteins with several carbohydrate units. Glycophorin A, the major protein marker in the erythrocyte, is a single-pass transmembrane sialoglycoprotein. The anion channels in erythrocytes are important for CO_2 transport. (Figures adapted from Luna & Hitt, 1992.)

C. α- and β-Spectrin

Spectrin is the main component of erythrocyte cytoskeletal proteins (1). It is a long protein composed of a 260-kDa α-chain and a 225-kDa β-chain. The chains consist of 20 (α-chain) and 18 (β-chain) subunits, respectively (2), each with 106 amino acids. Each subunit is composed of three α-helical protein strands running counter to one another. Subunit 10 and subunit 20 of the α chain consist of five, instead of three, parallel chains. The individual subunits are assigned to different domains (I–V in the α chain and I–IV in the β chain).

D. Erythrocyte skeletal proteins

SDS polyacrylamide gel electrophoresis differentiates numerous membrane-associated erythrocyte proteins. Each band of the gel is numbered, and the individual proteins are assigned to them. The main proteins include α- and β-spectrin, ankyrin, an anion-channel protein (band-3 protein), proteins 4.1 and 4.2, actin, and others.

Medical relevance

Inherited red cell membrane disorders are spherocytosis (MIM 182900, 270970), elliptocytosis, including the rare pyropoikilocytosis (MIM 130500), acanthosis (MIM 109270), and stomatocytosis type 1 and type 2 (MIM 185000, 185010). With the possible exception of an autosomal recessive spherocytosis (MIM 270970), all have autosomal dominant inheritance at loci shown in the figure.

References

Davies KA, Lux SE: Hereditary disorders of the red cell membrane skeleton. Trends Genet 5: 222–227, 1989.

Delaunay J: Disorders of the red cell membrane, pp 191–196. In: Jameson JL (ed) Principles of Molecular Medicine. Humana Press, Totowa, New Jersey, 1998.

Luna EJ, Hitt AL: Cytoskeleton plasma membrane interactions. Science 258: 955–964, 1992.

Tse T, Lux SE: Hereditary spherocytosis and hereditary elliptocytosis, pp 4665–4727. In: Scriver CR et al (eds) The Metabolic and Molecular Bases of Inherited Disease, 8th ed. McGraw-Hill, New York, 2001.

Tse WT, Lux SE: Red blood cell membrane disorders. Br J Haematol 104: 2–13, 1999.

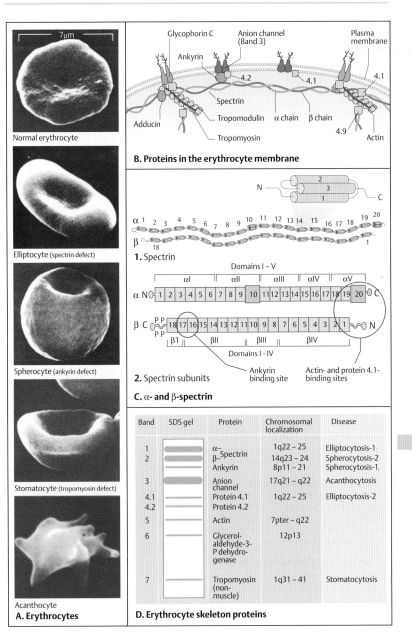

A. Erythrocytes

- Normal erythrocyte
- Elliptocyte (spectrin defect)
- Spherocyte (ankyrin defect)
- Stomatocyte (tropomyosin defect)
- Acanthocyte

B. Proteins in the erythrocyte membrane

Glycophorin C — Anion channel (Band 3) — Plasma membrane — Ankyrin — 4.2 — 4.1 — 4.1 — Spectrin — Adducin — Tropomodulin — α chain — β chain — 4.9 — Actin — Tropomyosin

1. Spectrin

N — 2 — 3 — 1 — C

α 1 2 3 4 5 6 7 8 9 10 11 12 13 14 15 16 17 18 19 20

β 18 1

2. Spectrin subunits

Domains I – V

α N 1 2 3 4 5 6 7 8 9 10 11 12 13 14 15 16 17 18 19 20 C

αI αII αIII αIV αV

β C P P 18 17 16 15 14 13 12 11 10 9 8 7 6 5 4 3 2 1 N

β1 βII βIII βIV

Domains I - IV

Ankyrin binding site

Actin- and protein 4.1-binding sites

C. α- and β-spectrin

D. Erythrocyte skeleton proteins

Band	SDS gel	Protein	Chromosomal localization	Disease
1		α– Spectrin	1q22 – 25	Elliptocytosis-1
2		β– Spectrin	14q23 – 24	Spherocytosis-2
		Ankyrin	8p11 – 21	Spherocytosis-1.
3		Anion channel	17q21 – q22	Acanthocytosis
4.1		Protein 4.1	1q22 – 25	Elliptocytosis-2
4.2		Protein 4.2		
5		Actin	7pter – q22	
6		Glycerol-aldehyde-3-P dehydro-genase	12p13	
7		Tropomyosin (non-muscle)	1q31 – 41	Stomatocytosis

Hereditary Muscle Diseases

Hereditary neuromuscular diseases are classified into muscular dystrophies, congenital and other myopathies, spinal muscular atrophies, motor neuron diseases, and others. They are genetically heterogeneous and clinically variable, with more than 50 distinct forms listed in McKusicks's catalogue of Mendelian Inheritance of Man Online (OMIM at www3.ncbi.nih.gov/Omim/).

A. The dystrophin–glycan complex

The dystrophin–glycan complex is a multifarious system of six proteins connected with each other and bound to the muscle cell plasma membrane. They belong to the group of dystroglycans and sarcoglycans. Laminins connect with the extracellular matrix. The central protein is dystrophin, a large elongated protein of 175 nm with a specific structure of two subunits, which are connected to the thin myofilament F-actin (filamentous actin) at the N-terminus and to dystrobrevin and syntrophin at the C-terminus. Dystrophin provides a bridge between the intracellular cytoskeleton involved in the contractile myofilaments and the extracellular matrix. The largest of the interconnecting proteins, α-dystroglycan (156 kDa), is located outside the cell. It is connected to the extracellular matrix by a heterotrimeric protein, laminin-2. Its partner, β-dystroglycan (43 kDa), is embedded in the sarcolemma and connected to a series of other cytoskeletal proteins, which are divided into the sarcoglycan and syntrophin subcomplexes. Two dystrophin molecules connect neighboring dystrophin-glycan complexes. All members of the sarcoglycan complex are involved specific types of muscular dystrophies.

Several types of congenital muscular dystrophies are known. The complex group of limb girdle muscular dystrophies (MIM 159000, 159001, 253600, 253601) is classified by molecular analyses according to the type of sarcoglycan involved, as indicated in the figure.

B. Model of the dystrophin molecule

Dystrophin, the largest member of the spectrin superfamily, is composed of 3685 amino acids (molecular mass 427 kDa), which form four functional domains: (i) the N-terminal actin-binding domain of 336 amino acids; (ii) 24 long repeating units, each consisting of 88- to 126-amino-acid triple-helix segments as in spectrin; (iii) a 135-amino-acid cysteine-rich domain, which binds to the sarcolemma proteins; and (iv) the C-terminal domain of 320 amino acids with binding sites to syntrophin and dystrobevin. The triple helix segments form the central rod domain, which is 100–125 nm long. (Figure adapted from Koenig et al., 1988.)

C. The dystrophin gene

The human *dystrophin* gene (*DMD*, MIM 310200) is located on the short arm of the X chromosome in region 2, band 1.1 (Xp21.1) (**1**). *Dystrophin* is the by far largest known gene in man, spanning 2.4 million base pairs (2.3 Mb or 2300 kb) in 79 exons (**2**). The large DMD transcript has 14 kb. The dystrophin gene contains at least seven intragenic promoters. The primary transcript is alternatively spliced into a variety of different mRNAs that encode smaller proteins expressed in other tissues than muscle cells, especially in the central nervous system.

D. Distribution of deletions

Deletions in the *DMD* gene (60% of patients) are unevenly distributed. Most frequently involved are exons 43–55 and exons 1–15, roughly corresponding to the F-actin binding site and the dystroglycan-binding site. Duplications of one or more exons (in 6% of patients) and point mutations also occur. (Data kindly provided by Professor C. R. Müller-Reible, Würzburg.)

References

Ahu AW, Kunkel LM: The structural and functional diversity of dystrophin. Nature Genet 3: 283–291, 1993.

Duggan DJ et al: Mutations in the sarcoglycan genes in patients with myopathy. N Engl J Med 336: 618–624, 1997.

Flanigan KM et al: Rapid direct sequence analysis of the dystrophin gene. Am J Hum Genet 72: 931–939, 2003.

Koenig M, Monaco AP, Kunkel LM: The complete sequence of dystrophin predicts a rod-shaped cytoskeletal protein. Cell 53: 219–228, 1988.

Worton R: Muscular dystrophies: diseases of the dystrophin-glycoprotein complex. Science 270: 755–756, 1995.

Online information:

Diseases of the Musculoskeletal System. The Stanford Health Library (http://healthlibrary.stanford.edu/resources/internet/bodysystems/musc_muscle.html).

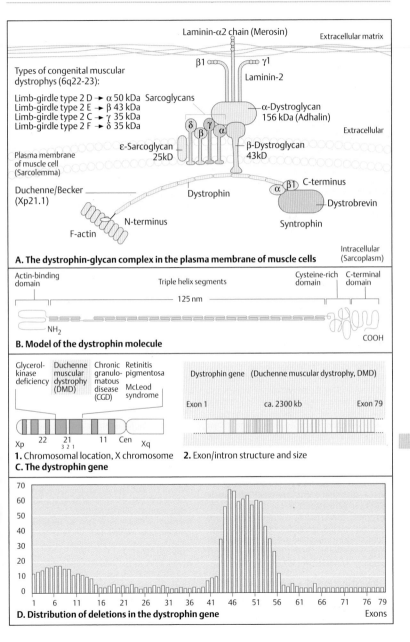

Types of congenital muscular
dystrophys (6q22-23):

Limb-girdle type 2 D → α 50 kDa Sarcoglycans
Limb-girdle type 2 E → β 43 kDa
Limb-girdle type 2 C → γ 35 kDa
Limb-girdle type 2 F → δ 35 kDa

Laminin-α2 chain (Merosin)   Extracellular matrix

β1   γ1
Laminin-2

α-Dystroglycan
156 kDa (Adhalin)   Extracellular

ε-Sarcoglycan   β-Dystroglycan
25kD   43kD

Plasma membrane
of muscle cell
(Sarcolemma)

Duchenne/Becker   C-terminus
(Xp21.1)

Dystrophin   Dystrobrevin

N-terminus   Syntrophin

F-actin

Intracellular
(Sarcoplasm)

A. The dystrophin-glycan complex in the plasma membrane of muscle cells

Actin-binding
domain   Triple helix segments   Cysteine-rich   C-terminal
                                                        domain   domain

                          125 nm

NH₂

COOH

B. Model of the dystrophin molecule

Glycerol- Duchenne Chronic   Retinitis
kinase   muscular granulo- pigmentosa
deficiency dystrophy matous
           (DMD)   disease   McLeod
                     (CGD)   syndrome

Dystrophin gene   (Duchenne muscular dystrophy, DMD)

Exon 1   ca. 2300 kb   Exon 79

        22   21   11   Cen
Xp   3 2 1   Xq

1. Chromosomal location, X chromosome   **2.** Exon/intron structure and size
C. The dystrophin gene

D. Distribution of deletions in the dystrophin gene

Exons

Duchenne Muscular Dystrophy

Duchenne muscular dystrophy (DMD, MIM 310200) is the most common of the muscular dystrophies, with a frequency of 1 in 3500 live-born males. It is named after the French neurologist Guillaume Duchenne (1806–1875), who described this disease in 1861. It is caused by mutations in the *DMD* gene, either by a new mutation or by transmission of the mutation from a heterozygous mother. Germline mosaicism has been observed (i.e., a female carrying a *DMD* mutation in a variable proportion of her germ cells). The mutation rate is high.

A. Clinical signs

The age of onset is usually less than 3 years and signs are evident at 4–5 years; the patient requires a wheelchair by 12 years and usually succumbs to the disorder by age 20. Progressive muscular weakness of the hips, thighs, and back causes difficulties in walking and in using steps. Lumbar lordosis and enlarged but weak calves (pseudohypertrophy) are visible (**1**). The affected child performs a characteristic series of maneuvers to rise from a kneeling position (Gower's sign, **2**).

A clinically milder variant, Becker muscular dystrophy (BMD), is an allelic disorder with a milder course and a later age of onset of the disease. The clinical difference results from the type of rearrangement of the *DMD* gene. In DMD the reading frame is changed, whereas in BMD it is maintained. Thus, even a relatively large deletion is compatible with residual muscle function if the reading frame is not changed. (Drawings by Duchenne, 1861, and Gowers, 1879, from Emery, 1993.)

B. Dystrophin analysis in muscle cells

Dystrophin is normally located along the plasma membrane (sarcolemma) of muscle cells (**1**). In patients it is absent (**2**). Female heterozygotes show a patchy distribution of groups of normal and defective muscle cells (**3**) as a result of X inactivation (see p. 234). (Photographs kindly provided by Dr. R. Gold, Department of Neurology, University of Würzburg.)

C. Investigation of a family with DMD

In about one-third of patients a mutation may not be detectable. Several alternative methods are available. These include single-strand con-formational polymorphism (see p. 76), heteroduplex analysis, reverse transcriptase PCR (see p. 60), and the protein truncation test (see p. 418). In a family with more than one affected male, indirect DNA analysis can be performed. The panel shows a simplified example of a two-allele system (marker DXS7) based on data kindly provided by Dr. C. R. Müller-Reible, Institute of Human Genetics, University of Würzburg. Two patients (III-1 and III-2) carry the allele 1. Their mothers, II-1 and II-2, can be considered heterozygotes for the mutation, as is the case for their mother, I-2. An unaffected male (II-4) has allele 2. This indicates that allele 2 does not carry the mutation.

Two males, III-3 and III-4, are not affected. This can be explained by recombination in their mother, II-5. In current practice one uses a set of several linked markers flanking the disease locus to avoid an erroneous diagnosis due to recombination. Female heterozygotes show mild clinical signs in 8%. About 23% of mothers of isolated patients are not carriers of the mutation, in agreement with the Haldane rule that about 1/3 of patients with a severe X-chromosomal disease have new mutations.

D. Other forms of muscular dystrophy

Several other forms of genetically determined muscular dystrophy are known in man. Course, diagnosis, and molecular genetic analysis depend on the basic disorder. Selected examples are listed.

References

Emery AEH: Duchenne Muscular Dystrophy, 2nd ed. Oxford University Press, Oxford, 1993.

Emery AEH: The muscular dystrophies. Fortnightly review. Brit J Med 317: 991–995, 1998 (Online at http://bmj.bmjjournals.com/cgi/content/full/317/7164/991).

Hoffman EP: Muscular dystrophies, pp 859–868. In: Jameson JG (ed) Principles of Molecular Medicine. Humana Press, Totowa, NJ, 1998.

Tennyson CN, Klamut HJ, Worton RG: The human dystrophin gene requires 16 hours to be transcribed and is cotranscriptionally spliced. Nature Genet 9: 184–190, 1995.

Worton RG et al: The X-linked muscular dystrophies, pp 5493–5523. In: Scriver CR et al (eds) The Metabolic and Molecular Bases of Inherited Disease, 8th ed. McGraw-Hill, New York, 2001.

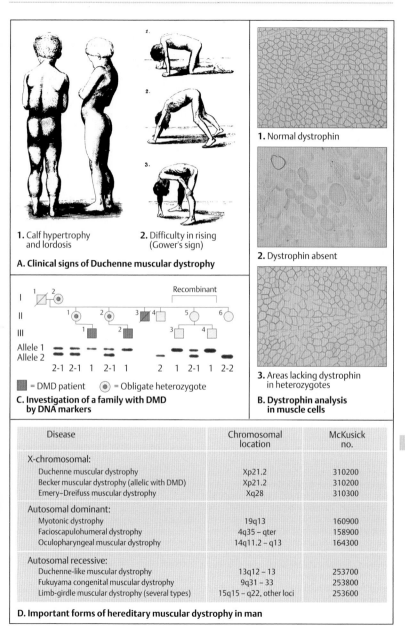

1. Calf hypertrophy and lordosis   **2.** Difficulty in rising (Gower's sign)

A. Clinical signs of Duchenne muscular dystrophy

1. Normal dystrophin

2. Dystrophin absent

3. Areas lacking dystrophin in heterozygotes

B. Dystrophin analysis in muscle cells

Recombinant

Allele 1
Allele 2
    2-1   2-1   1   2-1   1   2   1   2-1   1   2-2

■ = DMD patient   ◉ = Obligate heterozygote

C. Investigation of a family with DMD by DNA markers

Disease	Chromosomal location	McKusick no.
X-chromosomal:		
Duchenne muscular dystrophy	Xp21.2	310200
Becker muscular dystrophy (allelic with DMD)	Xp21.2	310200
Emery–Dreifuss muscular dystrophy	Xq28	310300
Autosomal dominant:		
Myotonic dystrophy	19q13	160900
Facioscapulohumeral dystrophy	4q35 – qter	158900
Oculopharyngeal muscular dystrophy	14q11.2 – q13	164300
Autosomal recessive:		
Duchenne-like muscular dystrophy	13q12 – 13	253700
Fukuyama congenital muscular dystrophy	9q31 – 33	253800
Limb-girdle muscular dystrophy (several types)	15q15 – q22, other loci	253600

D. Important forms of hereditary muscular dystrophy in man

Collagen Molecules

Collagen is a large group of insoluble, extracellular glycoproteins with specialized functions in the maintenance of tissue shape and structure. Collagen is the most abundant protein in mammals, constituting about one-quarter of the total body protein. Collagens occur in skin, bones, tendons, cartilage, ligaments, blood vessels, teeth, basement membranes, and supporting tissues of the internal organs. Most collagens have a remarkable mechanical resilience, with interlinked, insoluble threads (fibrils) of unusual strength. The 27 known types of collagens, numbered I-XXVII, are classified according to their main structural and functional features. The most important in humans are fibrillar collagen types I, II, III, V, and XI, and basal membrane collagen (type IV). Type I is the main component of tendons, ligaments, and bones; type II occurs in cartilage and the notochord of vertebrate embryos; type III occurs in arteries, the intestine, and the uterus; and type IV, in the basal lamina of epithelia, in particular around the glomeruli of the kidney, and in blood capillaries.

A. Collagen structure

Collagen is made up of three chains forming a triple helix, synthesized in seven principal steps. The amino acid sequence of collagen is simple and periodic (**1**). Every third amino acid is glycine (Gly). Other amino acids alternate between the glycines. The general structural motif is $(Gly–X–Y)_n$. X is either proline or hydroxyproline; Y is either lysine or hydroxylysine (**2**). Three chains of procollagen form a triple helix (**3**). In collagen type I, the helix is composed of two identical $\alpha1$ and $\alpha2$ chains. First a precursor molecule, *procollagen* is formed (**4**). Procollagen peptidases remove peptides at the N-terminal and C-terminal ends to form *tropocollagen* (**5**). Tropocollagen molecules are connected by numerous hydroxylated proline and lysine residues to form a *collagen fibril* (**6**). Each fibril consists of staggered, parallel rows of end-to-end tropocollagen molecules, separated by gaps (**7**). Collagen fibrils are visible as transverse stripes under the electron microscope (**8**). A fiber of 1-mm diameter can hold a weight of almost 10 kg. (Photograph: Stryer, 1995.)

B. Prototype of a gene for procollagen

Procollagen type II is made up of a triple helix of three procollagen type II chains, designated as $\alpha1[II]$. The corresponding gene, *COL2A1*, consists of 52 exons of different sizes, each encoding 5, 6, 11, 12, or 18 Gly-X-Y units. The translated part of exon 1 (85 base pairs) encodes a signal peptide necessary for secretion. The genes for procollagen types I, II, and III differ in that some exons are fused, but otherwise they are similar, especially for the three main fibrillar collagen types (I, II, III).

C. Gene structure of procollagen type $\alpha1(I)$

Procollagen type I consists of two $\alpha1$ chains and one $\alpha2$ chain, designated $[\alpha1(I)_2\alpha2(II)]$. The corresponding genes are *COL1A1* and *COL1A2*. The 52 exons of the gene encoding procollagen type I, *COL1A1*, correspond to the different domains (A to G) of procollagen $\alpha1(I)$. The *COL1A2* gene for procollagen $\alpha2(I)$ is about twice as large (~ 40 kb) as the *COL1A1* gene because the introns between the exons are on average twice as long as in *COL1A1*.

Medical relevance

More than ten distinct human diseases are caused by mutations in one of the genes encoding collagen (see table in the appendix).

References

Byers PH: Disorders of collagen synthesis and structure, pp 5241–5285. In: Scriver CR et al (eds) The Metabolic and Molecular Bases of Inherited Disease, 8th ed. McGraw-Hill, New York, 2001.

Chu M-L, Prockop DJ: Collagen gene structure, pp 149–165. In: Broyce PM, Steinmann B (eds) Connective Tissue and Its Heritable Disorders. Wiley-Liss, New York, 1993.

De Paepe A: Heritable collagen disorders: from phenotype to genotype. Verh K Acad Geneeskd Belg 65: 463–482, 1998.

Myllyharju J, Kiviriko KI: Collagens, modifying enzymes and their mutations in humans, flies, and worms. Trends Genet 20: 33–43, 2004.

Rauch F, Glorieux FH: Osteogenesis imperfecta. Lancet 363: 1377–1385, 2004

Online information:
Kimball's Biology Pages at www.biology-pages.info

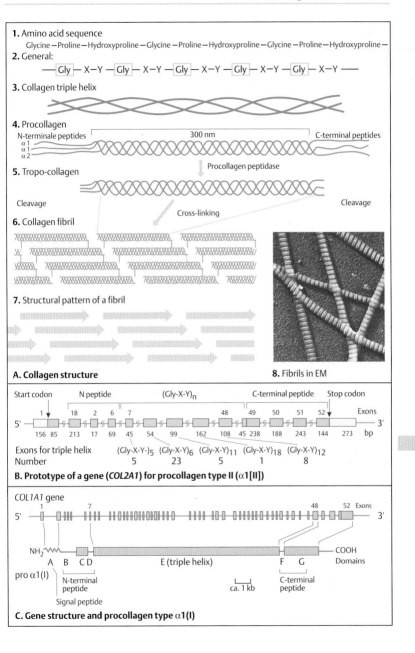

1. Amino acid sequence

Glycine — Proline — Hydroxyproline — Glycine — Proline — Hydroxyproline — Glycine — Proline — Hydroxyproline —

2. General:

— Gly — X—Y — Gly — X—Y — Gly — X—Y — Gly — X—Y — Gly — X—Y —

3. Collagen triple helix

4. Procollagen

N-terminale peptides   300 nm   C-terminal peptides
α 1
α 1
α 2

Procollagen peptidase

5. Tropo-collagen

Cleavage   Cleavage

Cross-linking

6. Collagen fibril

7. Structural pattern of a fibril

A. Collagen structure   **8.** Fibrils in EM

Start codon   N peptide   (Gly-X-Y)n   C-terminal peptide   Stop codon

1   1B   2   6   7   48   49   50   51   52   Exons

5'   3'

156   85   213   17   69   45   54   99   162   108   45   238   188   243   144   273   bp

Exons for triple helix   (Gly-X-Y-)5   (Gly-X-Y)6   (Gly-X-Y)11   (Gly-X-Y)18   (Gly-X-Y)12
Number   5   23   5   1   8

B. Prototype of a gene (*COL2A1*) for procollagen type II (α1[II])

COL1A1 gene

1   7   48   52   Exons

5'   3'

NH_2 COOH

A   B   C D   E (triple helix)   F   G   Domains

pro α1(I)   N-terminal   ca. 1 kb   C-terminal
peptide   peptide

Signal peptide

C. Gene structure and procollagen type α1(I)

Osteogenesis Imperfecta

Osteogenesis imperfecta (OI, MIM 120150), or "brittle bone disease", is a heterogeneous group of clinically and genetically different types of diseases with a total frequency of about 1 in 10 000 individuals. Common manifestations include spontaneously occurring bone fractures, bone deformity, small stature, defective dentition (dentinogenesis imperfecta), hearing impairment due to faulty formation of the auditory ossicles, and blue sclerae. The latter results from a shift of refracted light to the blue from the abnormally thin conjunctivae. The manifestations vary depending on the type of OI. OI has been demonstrated in a seventh-century Egyptian mummy (Byers, 1993). In most, but not all patients, collagen type I is defective owing to a mutation in one of the two genes encoding the two chains, *COL1A1* or *COL1A2*.

A. Molecular mechanisms

In normal procollagen type I the two chains, pro α1(I) and α2(I), are produced in equal amounts to form a normal procollagen (**1**). Some types of mutation lead to reduced production of pro α1(I) (**2**). In this case the imbalance causes degradation of α2(I) chains. Thus, less procollagen than normal is formed, but it is not defective. The types of mutation in the genes for pro α1(I) and α2(I) (*COL1A1* and *COL1A2*) causing defective procollagen (**3**) include a deletion in a *COL1A1* allele, a splicing defect, and others. Mutations in the *COL1A1* gene are more severe than mutations in the *COL1A2* gene because more defective collagen is formed with the former. (Figure adapted from Wenstrup et al., 1990.)

B. Mutations and phenotype

The location of a mutation in the gene influences the phenotype. Generally, mutations in the 3′ region are more serious than mutations in the 5′ region (*position effect*). Mutations of the pro α1(I) chain are more severe than those in the pro α2(I) chain (*chain effect*). The substitution of a larger amino acid for glycine, which is indispensable for the formation of the triple helix, leads to severe disorders (*size effect*). Different types of mutations may occur, such as deletions, mutations in the promoter or enhancer, and splicing mutations. The codons (AAG, AAA) for the amino acid lysine, which occurs frequently in collagen, are readily transformed into a stop codon by substitution of the first adenine by a thymine (TAG or TAA), so that a short, unstable procollagen is formed. Splicing mutations may lead to the loss of exons (exon skipping). (Figure adapted from Byers, 2001.)

C. Different forms

Osteogenesis imperfecta is classified into four phenotypes, I-IV, the Sillence classification. Although the classification does not correspond to the types of mutation, in general it has proved clinically useful. OI types I and IV are less severe than type II (fatal in infancy) and type III. Three radiographs show a relatively mild (but for the patient nevertheless very disabling) deformity of the tibia and fibula in OI type IV (**1**); severe deformities in the tibia and fibula in OI type III (**2**); and the distinctly thickened and shortened long bones in the lethal OI type II (**3**). Mutations in OI are autosomal dominant, the severe forms being due to *de novo* mutations. Germline mosaicism has been shown to account for rare instances of affected siblings being born to unaffected parents.

Two new types of OI were described in 2002: type V (MIM 120215, gene locus at 9q34.2-q34.3), with signs of osteomalacia, and type VI, with hyperplastic callus formation (Glorieux et al., 2002).

References

Byers PH: Osteogenesis imperfecta, pp 137–350. In: Broyce PM, Steinmann B (eds) Connective Tissue and Its Heritable Disorders. Wiley-Liss, New York, 1993.

Byers PH: Disorders of collagen synthesis and structure, pp 5241–5285. In: Scriver CR et al (eds) The Metabolic and Molecular Bases of Inherited Disease, 8th ed. McGraw-Hill, New York, 2001.

Chu M-L, Prockop DJ: Collagen gene structure, pp 149–165. In: Broyce PM, Steinmann B (eds) Connective Tissue and Its Heritable Disorders. Wiley-Liss, New York, 1993.

Glorieux FH et al: Osteogenesis imperfecta type VI: A form of brittle bone disease with a mineralization defect. J Bone Min Res 17: 30–38, 2002.

Kocher MS, Shapiro F: Osteogenesis imperfecta. J Am Acad Orthop Surg 6: 225–236, 1998.

Sillence DO, Senn A, Danks DM: Genetic heterogeneity in osteogenesis imperfecta. J Med Genet 16: 101–116, 1979.

Wenstrup J et al: Distinct biochemical phenotypes predict clinical severity in nonlethal variants of osteogenesis imperfecta. Am J Hum Genet 46: 975–982, 1990.

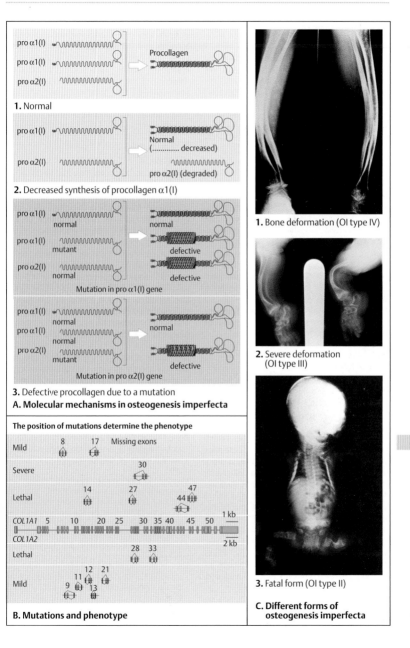

1. Normal

2. Decreased synthesis of procollagen α1(I)

Mutation in pro α1(I) gene

Mutation in pro α2(I) gene

3. Defective procollagen due to a mutation

A. Molecular mechanisms in osteogenesis imperfecta

1. Bone deformation (OI type IV)

2. Severe deformation (OI type III)

The position of mutations determine the phenotype

Mild	8	17	Missing exons			
Severe			30			
Lethal		14	27		47 44	

COL1A1 5   10   20   25   30 35 40   45   50   1 kb

COL1A2   2 kb

Lethal		28	33	
Mild	9 11 12 21 13			

B. Mutations and phenotype

3. Fatal form (OI type II)

C. Different forms of osteogenesis imperfecta

Molecular Basis of Bone Development

Bone develops from three mesodermal cell lineages committed to differentiate into three specialized cell types: *chondrocytes* (cartilage-forming cells), *osteoblasts* (bone-forming cells), and *osteoclasts* (bone-degrading cells). Two major processes form bone (osteogenesis): (i) direct conversion of mesenchymal tissue into bone tissue (*intramembranous* or *dermal ossification*) and (ii) *enchondral ossification*, with cartilage intermediates produced by chondrocytes, which later are replaced by bone cells (osteoblasts). Osteoblasts produce most of the proteins for the extracellular bone matrix and control its mineralization. The osteoblast cell lineage involves osteoblast-specific transcription factors (OSFs). One such transcription factor is a major regulator of osteoblast differentiation in direct intramembranous bone formation: the core-binding factor Cbfa1. This has been renamed Runx2 after the *Drosophila* pair-rule gene *runt*. The mouse Runx2 transcription factor (the human counterpart is RUNX2) is a member of the *runt* domain family. The *runt* domain is a DNA-binding domain homologous to *Drosophila runt*.

A. Effects of homozygous *Runx2* mutations in mice

Targeted disruption of the *Runx2* gene, located on mouse chromosome 17, in the homozygous state (–/–) results in a severe phenotype affecting the entire skeleton (**1**). In contrast to normal mice (+/+, a, c), the mutant mice (–/–) completely lack bone development, as shown by lack of alizarin red staining (b, e). These mice are small and die at birth from respiratory failure. The humerus and humeral tuberosity (circle) in heterozygous mice (+/–, d) show reduced ossification of the long bones and severe hypoplasia, respectively. The skull and thorax are also severely affected (**2**). Heterozygous (+/–) mice lack ossification of the skull (b). Normal calcified bone is stained red by alizarin red, here at embryonic day 17.5, three and a half days before birth. Cartilage is stained blue by alcian blue. Heterozygous mice lack clavicles (d, arrows) in contrast to normal mice (c).

B. Cleidocranial dysplasia in humans

Cleidocranial dysplasia (CCD, MIM 118980) is an autosomal dominant skeletal disease caused by mutation in the human *RUNX2* gene, localized at 6p21. It is characterized by absence of the clavicles and deficient bone formation of the skull. Radiological findings show generalized underossification. Patients can oppose their shoulders (**1**) due to absence of the clavicles (**2**, photograph by Dr. J. Warkany, Cincinnati). The calvarium (skull case) is enlarged, with a poorly ossified midfrontal area (**3**). (Figure in 3 from Mundlos et al., 1997.)

C. The human *RUNX2* gene

The *RUNX2* gene at 6p21 encodes a transcription factor of the core-binding factor (CBF) family (MIM 600211). It has nine exons, not seven as previously determined. It contains two alternative transcription initiation sites with two promoters, P1 and P2. Part of exons 1, 2, and 3 encode the DNA-binding *runt* domain; exons 4, 5, 6, and 7 encode the transcriptional activation and repression domains. The nuclear localization signal (NLS) is located at the 5′ end of exon 3. Exon 6 is alternatively spliced and unique to *RUNX2*. The role of the *RUNX2* gene also includes a major regulatory function in chondrocyte differentiation during endochondral bone formation (Mundlos, 1999). As such, it functions as a "master gene" in bone development.

All mutations result in loss of function, i.e., haploinsufficiency causes the CCD phenotype. (Figures in A, B, and D kindly provided by Dr. Stefan Mundlos, Institute for Medical Genetics, Humboldt University, Berlin.)

References

Komori T et al: Targeted disruption of Cbfa1 results in a complete lack of bone formation owing to maturational arrest of osteoblasts. Cell 89: 755–764, 1997.

Mundlos S: Cleidocranial dysplasia: clinical and molecular genetics. J Med Genet 36:177–182, 1999.

Mundlos S et al: Mutations involving the transcription factor CBFA1 cause cleidocranial dysplasia. Cell 89: 773–779, 1997.

Zheng Q et al: Dysregulation of chondrogenesis in human cleidocranial dysplasia. Am J Hum Genet 77: 305–312, 2005.

Zou G et al: *CBFA1* mutations analysis and functional correlation with phenotypic variability in cleidocranial dysplasia. Hum Mol Genet 8: 2311– 2316, 1999.

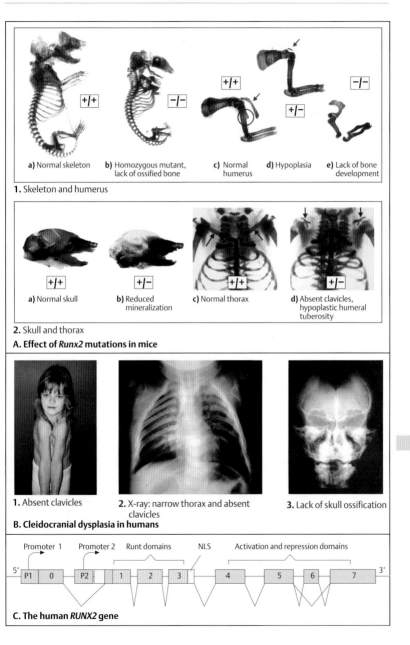

a) Normal skeleton   **b)** Homozygous mutant, lack of ossified bone   **c)** Normal humerus   **d)** Hypoplasia   **e)** Lack of bone development

1. Skeleton and humerus

a) Normal skull   **b)** Reduced mineralization   **c)** Normal thorax   **d)** Absent clavicles, hypoplastic humeral tuberosity

2. Skull and thorax

A. Effect of *Runx2* mutations in mice

1. Absent clavicles   **2.** X-ray: narrow thorax and absent clavicles   **3.** Lack of skull ossification

B. Cleidocranial dysplasia in humans

Promoter 1   Promoter 2   Runt domains   NLS   Activation and repression domains

5'   P1   0   P2   1   2   3   4   5   6   7   3'

C. The human *RUNX2* gene

Mammalian Sex Determination

Genes of the sex determination pathway regulate the developmental differentiation into a male or a female. In this sequential process, genes are expressed in relevant tissues at the right time. At each step a binary decision is made towards one or the other gender. First, the future gender is determined by the presence or absence of a Y chromosome. Thereafter, in different stages of the gonads and of the internal and external genital organs differentiate into those of the female or male gender.

A. Role of the mammalian Y chromosome

The crucial role of the Y chromosome in man became evident in 1959 from chromosomal analyses of two disorders in man, Turner syndrome and Klinefelter syndrome. Individuals with Klinefelter syndrome have two X chromosomes and a Y chromosome, and show a male phenotype, although incompletely developed (p. 414). Even the presence of several X chromosomes does not result in a female phenotype as long as one functional Y chromosome is present. Individuals with Turner syndrome have only one X chromosome and no Y chromosome, and show a female phenotype, although incompletely developed and usually accompanied by malformations (p. 414). In the 1940s, the French embryologist Alfred Jost observed that removal of the testes of a fetal rabbit prior to external male differentiation resulted in the development of a female. Thus, the Y chromosome and the fetal testes are required to induce male differentiation.

B. Sex-determining region SRY

Experiments in animals and clinical observations of human males with various sized deletions of the Y chromosome indicate that only a small region of the distal short arm of the Y chromosome is required to induce male development. This region is named SRY (sex-related Y).

SRY is a small region on the short arm of the human Y chromosome. Within this region, the gene SRY (sex-determining region Y) was identified. It is located just proximal to the pseudoautosomal region 1 (PAR1) within interval 1A1. The PAR1 is homologous to the distal segment of the short arm of the X chromosome. Homologous pairing occurs here with crossing-over during male meiosis.

C. SRY gene

The SRY gene (MIM 480000), located at Yp11.32, consists of a single exon. It has a TATAAA motif for binding transcription factor TFIID. It is transcribed into a 1.1-kb RNA. From a coding region of 612 bp, a 204-amino-acid protein with a molecular weight of 23.9 kDa is translated (1). SRY is a member of the SOX family of transcription factors. It contains a conserved high-mobility group (HMG) motif, which binds to DNA and causes reversible bending (2). The bending opens the double helix and permits access of transcription factors. HMG proteins are nonhistone DNA-binding proteins. (Figure in 2 kindly provided by Dr. Michael A. Weiss, Cleveland, from Li et al., 2006.)

D. Sry-transgenic male XX mouse

Experimental evidence in mice confirms the crucial role of Sry. When a 14-kb DNA fragment containing the mouse Sry gene is inserted into the blastocyst of a chromosomally female (XX) transgenic mouse, a male mouse develops. (Figure from Koopman et al., 1991.)

E. Time pattern of Sry expression

In the mouse embryo with XY chromosomes, Sry is expressed only between days 10.5 and 12.5 of embryonic development. (Figure from Koopman & Gubbay, 1991.)

References

Brennan J, Capel B: One tissue, two fates: molecular genetic events that underlie testis versus ovary development. Nature Rev Genet 5: 509–521, 2004.

Erickson RP: The sex determination pathway, pp 482–501. In: Epstein CJ, Erickson RP, Wynshaw-Boris A (eds) Inborn Errors of Development. The Molecular Basis of Clinical Disorders of Morphogenesis. Oxford University Press, Oxford, 2004.

Koopman P et al: Male development of chromosomally female mice transgenic for Sry. Nature 351: 117–121, 1991.

Li B et al: SRY-directed DNA bending and human sex reversal: Reassessment of a clinical mutation uncovers a global coupling between the HMB box and its tail. J Mol Biol, in press.

MacLaughlin DT, Donahoe PK: Sex determination and differentiation. New Engl J Med 350: 367–378, 2004.

Scherer G & Schmid M, eds: Genes and Mechanisms in Vertebrate Sex Determination. Birkhäuser, Basel, 2001.

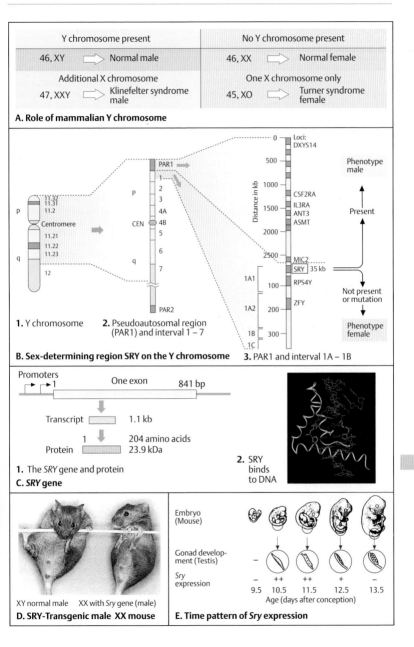

Y chromosome present		No Y chromosome present	
46, XY ⇒	Normal male	46, XX ⇒	Normal female
Additional X chromosome		One X chromosome only	
47, XXY ⇒	Klinefelter syndrome male	45, XO ⇒	Turner syndrome female

A. Role of mammalian Y chromosome

1. Y chromosome **2.** Pseudoautosomal region (PAR1) and interval 1 – 7

B. Sex-determining region SRY on the Y chromosome **3.** PAR1 and interval 1A – 1B

Loci:
DXYS14
CSF2RA
IL3RA
ANT3
ASMT
MIC2
SRY 35 kb
RPS4Y
ZFY

Phenotype male

Present

Not present or mutation

Phenotype female

Promoters
1 One exon 841 bp
Transcript 1.1 kb
1 204 amino acids
Protein 23.9 kDa

1. The *SRY* gene and protein

C. *SRY* gene

2. SRY binds to DNA

Embryo (Mouse)

Gonad development (Testis)

Sry expression − ++ ++ + −
9.5 10.5 11.5 12.5 13.5
Age (days after conception)

XY normal male   XX with *Sry* gene (male)

D. SRY-Transgenic male XX mouse   **E. Time pattern of *Sry* expression**

Sex Differentiation

Sex differentiation is a series of consecutive developmental processes during early embryogenesis resulting in either the female or the male gender. Initially all anatomical structures involved are undifferentiated. Under the influence of various genes they develop into either sex.

A. Gonads and external genitalia

The gonads (**1**), the efferent (mesonephric and paramesonephric) ducts (**2**), and the external genitalia (**3**) all develop from an indifferent anlage. At about the end of the sixth week of pregnancy in humans, after the primordial germ cells of the embryo have migrated into the initially undifferentiated gonads, an inner portion (medulla) and an outer portion (cortex) of the gonads can be distinguished (**1**). In XY embryos, early embryonic testes develop at about the 10th week of pregnancy under the influence of a testis-determining factor (TDF), the *SRY* gene. If this is not present, ovaries develop. The early embryonic testis produces two hormones, testosterone, with a male differentiating effect, and the Müllerian inhibition factor MIF (anti-Müllerian hormone). MIF inhibits the development of female anatomical structures. The excretory ducts differentiate under the influence of the hormones produced by the early gonads (**2**). The Müllerian ducts, precursors of the Fallopian tubes, the uterus, and the upper vagina, develop when a male differentiating influence is absent. The Wolffian ducts, precursors of the male efferent ducts (vas deferens, seminal vesicles, and prostate), develop under the influence of testosterone, a male steroid hormone formed in the fetal testis. If testosterone is absent or ineffective, the Wolffian ducts degenerate. The external genitalia develop after the gonads have differentiated into testes or ovaries. In humans this occurs relatively late, in the 15th to 16th week (**3**). Full development of male external genitalia depends on a derivative of male-inducing testosterone, 5-dihydrotestosterone, a metabolite of testosterone produced by the enzymatic action of 5α-reductase. The differentiation of the gonadal ridge into the bipotential gonad, and this into ovary or testis, requires several genes as shown in this simplified scheme (**4**, adapted from Gilbert, 2003).

B. Sequence of events in sex differentiation

Four levels can be schematically defined: (i) genetic, (ii) gonadal, and (iii) anatomical, as prenatal stages, and (iv) from early childhood on, psychological. A fifth, the legal gender, recorded as "female" or "male" in all legal documents, can be added. Each level is reached in a series of temporally regulated successive steps. First the primordial germ cells differentiate into early embryonic testes under the influence of testis-determining factors (TDF), mainly the *SRY* gene (in humans) or the *Sry* gene in other mammals and other genes. Male differentiation includes suppression of the Müllerian ducts by the Müllerian Inhibitor Factor. In the absence of *SRY* no testes develop and no subsequent male differentiation stages occur. In the absence of testes, ovaries develop, the Wolffian ducts degenerate, and the Müllerian ducts differentiate into Fallopian tubes, uterus, and the upper vagina. The male differentiating effect of testosterone depends on the function of an intracellular androgen receptor (see next page). Testosterone also has an effect on the central nervous system by influencing the psychosexual orientation apparent later in life ("brain imprinting"). When testosterone is absent or ineffective due to a receptor defect, gender orientation is female.

Other genes involved in a dosage-dependent manner include *DAX1* (MIM 300473) on Xp21.3-p21.2, *SOX9* (MIM 608160) on 17q24.3-q25.1, and others. An ovary-determining gene may be *WNT4* (MIM 603490) acting together with *DAX1* to control female development.

References

Acherman JC, Jameson JL: Disorders of sexual differentiation, pp 2214–2220. In: Kasper DL et al (eds) Harrison's Principles and Practice of Internal Medicine, 16th ed. McGraw-Hill, New York, 2005.

Goodfellow PN et al: SRY and primary sex-reversal syndromes, pp 1213–1221. In: Scriver CR et al (eds) The Metabolic and Molecular Bases of Inherited Disease, 8th ed. McGraw-Hill, New York, 2001.

Su H, Lau Y-FC: Identification of the transcriptional unit, structural organization, and promoter sequence of the human sex-determining region Y (SRY) gene, using a reverse genetic approach. Am J Hum Genet 52: 24–38, 1993.

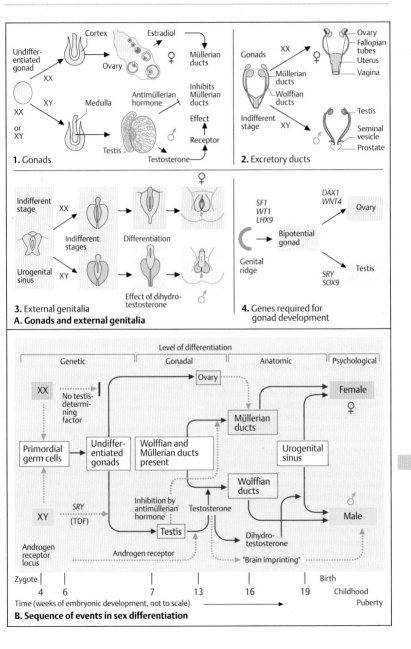

1. Gonads

2. Excretory ducts

3. External genitalia

A. Gonads and external genitalia

4. Genes required for gonad development

B. Sequence of events in sex differentiation

Disorders of Sexual Development

Disorders of sexual development occur at any level of sex determination or differentiation (see previous page) due to mutations or rearrangements of the Y and the X chromosome. Occasionally the gonads do not correspond to the genital duct system or external genitalia, and the external genitalia may be ambiguous, i.e., partly female and partly male (pseudohermaphroditism). In true hermaphroditism gonads contain both testicular and ovarian tissues. Therapy requires that the underlying developmental error is correctly diagnosed.

A. XX males and XY females

Normally, the male-determining Y-specific DNA sequences (*SRY* gene) remain on the Y chromosome during the homologous pairing and crossing-over during meiosis. Since *SRY* is located very close to the pseudoautosomal region 1 (PAR1), crossing-over outside PAR1 transfers the *SRY* region to the X chromosome and results in a male individual with an XX karyotype (XX male syndrome, MIM 278850). The complementary Y chromosome does not receive *SRY*, which results in a female phenotype with XY chromosomes (XY female gonadal dysgenesis, MIM 306100).

B. Point mutations in the *SRY* gene

The human *SRY* gene encodes a 204-amino-acid protein. This contains a DNA-binding domain of 79 highly conserved amino acids, the HMG box (high mobility group protein), in the middle section between amino acids 58 and 137. Point mutations and deletions in this region result in complete or partial gonadal dysgenesis. Sex reversal associated with campomelic dysplasia results from mutations in the *SOX9* gene (SRY-related HMG-box gene, MIM 608160, located at 17q24.3-q25.1. (Figure adapted from McElreavey & Fellous, 1999.)

C. Androgen receptor

Testosterone produced by the fetal testes can only exert its effect if it binds to an intracellular receptor, the androgen receptor (**1**). Likewise, dihydrotestosterone (DHT), converted from testosterone in the urogenital sinus by 5α-reductase, requires this receptor. The activated hormone–receptor complex (TR and DR) acts as a transcription factor for genes that regulate the differentiation of the Wolffian ducts and the urogenital sinus. Mutations in the androgen receptor result in androgen insensitivity. Affected XY individuals have the *SRY* gene and testes, and they produce testosterone. But since testosterone cannot exert any effect, the resulting phenotype is female (**2**). This variable condition is the androgen insensitivity syndrome, also called testicular feminization, TFM (MIM 300068). The androgen receptor gene (MIM 313700) is located on Xq11-q12.

References

Acherman JC, Jameson JL: Disorders of sexual differentiation, pp 2214–2220. In: Kasper DL et al (eds) Harrison's Principles and Practice of Internal Medicine, 16th ed. McGraw-Hill, New York, 2005.

Erickson RP: Introduction to the sex determining pathway: Mutations in many genes lead to sexual ambiguity and reversal, pp 482–491. In: Epstein CJ, Erickson RP, Wynshaw-Boris A (eds) Inborn Errors of Development. The Molecular Basis of Clinical Disorders of Morphogenesis. Oxford University Press, Oxford, 2004.

Foster JW et al: Campomelic dysplasia and autosomal sex reversal caused by mutations in an SRY-related gene. Nature 372: 525–530, 1994.

Goodfellow PN, Camerino G: DAX-1, an "antitestis" gene. Cell Mol Life Sci 55: 857–863, 1999.

Gottlieb B et al: Androgen insensitivity. Am J Med Genet (Semin Med Genet) 89: 210–217, 1999.

Griffin JE et al: The androgen resistance syndromes: Steroid 5α-reductase deficiency, testicular feminization, and related disorders, pp 4117–4146. In: Scriver CR et al (eds) The Metabolic and Molecular Bases of Inherited Disease, 8th ed. McGraw-Hill, New York, 2001.

McElreavey K, Fellous, M: Sex determination and the Y chromosome. Am J Med Genet 89: 176–185, 1999.

Sharp A et al: Variability of sexual phenotype in 46,XX (SRY+) patients: the influence of spreading X inactivation versus position effects. J Med Genet 42: 420–427, 2005.

Wagner T et al: Autosomal sex reversal and campomelic dysplasia are caused by mutations in and around the SRY-related SOX9 gene. Cell 79: 1111–1120, 1994.

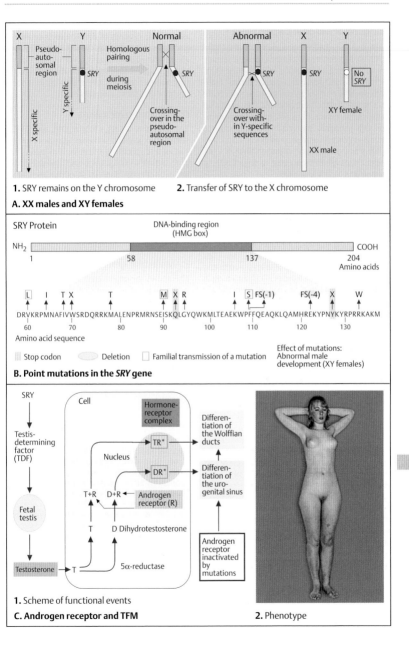

1. SRY remains on the Y chromosome   **2.** Transfer of SRY to the X chromosome

A. XX males and XY females

SRY Protein

DNA-binding region
(HMG box)

NH₂ — COOH

1   58   137   204
Amino acids

L   I   T X   T   M X R   I   S FS(-1)   FS(-4)   X   W

DRVKRPMNAFIVWSRDQRRKMALENPRMRNSEISKQLGYQWKMLTEAEKWPFFQEAQKLQAMHREKYPNYKYRPRRKAKM

60   70   80   90   100   110   120   130

Amino acid sequence

▦ Stop codon   ⬭ Deletion   ☐ Familial transmission of a mutation

Effect of mutations:
Abnormal male
development (XY females)

B. Point mutations in the *SRY* gene

SRY

↓

Testis-
determining
factor
(TDF)

↓

Fetal
testis

↓

Testosterone → T

Cell

Hormone-
receptor
complex

T+R → TR* → Differen-
tiation of
the Wolffian
ducts

D+R → DR* → Differen-
tiation of
the uro-
genital sinus

Nucleus

Androgen
receptor (R)

T   D Dihydrotestosterone

5α-reductase

Androgen
receptor
inactivated
by
mutations

1. Scheme of functional events

C. Androgen receptor and TFM

2. Phenotype

Congenital Adrenal Hyperplasia

CAH, also called adrenogenital syndrome (MIM 201910), is a genetically determined deficiency of cortisol due to insufficient biosynthesis of steroid 21-hydroxylase, a microsomal cytochrome P450 enzyme. A compensatory increase in adrenocortical hormone (ACTH) excretion causes hyperplasia of the adrenal cortex with increased prenatal production of androgenic steroids. With a frequency of about 1 in 5000 newborns it is the most common form of CAH, the heterozygote frequency being 1:50 in Europe and North America.

A. Overview

The main characteristics of this disease are listed (**1**). CAH occurs in three manifestations: (i) a severe salt-losing form, in about 60–75% of patients, (ii) a virilizing form without salt loss, and (iii) an attenuated, late-onset form. They depend in part on the type of mutation present. The severe salt-losing form is life threatening in newborns. Ambiguous or virilized genitalia are present in newborn girls (**2**). The biochemical defect results in hyperplasia of the adrenals (**3**). Inheritance is autosomal recessive (**4**). Untreated or poorly treated children show advanced growth and early puberty, but are short as adults due to early closure of the epiphyseal plates. Such girls develop a male physical appearance (**5**). Prenatal therapy by maternal dexamethasone before the 6th week of gestation may reduce virilization of a female fetus.

B. Biochemical defect

The enzymatic block is at the conversion of progesterone to deoxycortisol (DOC) by hydroxylation at position 21 (steroid 21-hydroxylase). As a result, the plasma concentration of 17-hydroxy-progesterone is increased.

C. The *CYP21* gene structure

The *CYP21* gene is located at 6p21.3 in tandem with three closely linked genes, the complement 4A (*C4A*) gene, a *CYP* pseudogene (*CYP21P*) without function, and the complement 4B (*C4B*) gene. These lie within the class III genes of the major histocompatibilty complex MHC (see p. 318). The *CYP21* gene spans nearly 6 kb. Deletions and nonsense or frameshift mutations result in the severe salt-losing form. Missense mutations occur in both forms. Most patients are compound heterozygotes. In general the severity of the disease is related to the type of mutation. Seven mutations are shown. Deletions account for about 20% of the classical salt-losing form, the majority occurring between exons 3 and 8. Duplications occur without clinical consequences.

D. Crossing-over events

The four genes, *C4A*, *CYP21P*, *C4B*, and *CYP21*, share structural and sequence similarity as a result of a duplication event in evolution. This predisposes to mispairing and unequal crossing-over at meiosis. A variety of crossover products may result in partial or a whole gene deletion of *CYP21* or its duplication. Gene conversion is an important mechanism. Mismatch between the functional *CYP21* gene and the nonfunctional pseudogene *CYP21P* converts part of the *CYP21* gene to *CYP21P*.

E. Molecular genetic analysis

This shows the detection of an 8-bp deletion by a semiquantitative PCR method. In normal individuals (lanes 2, 4, 6–8) the intensity of the 952-bp fragment, derived from exons 1–3, and of the 200-bp fragment is about the same. In heterozygotes (lanes 5 and 9), the 952-bp fragment is less intense than in controls, and in the patient it is absent (lane 3). The fragments at the bottom of each lane represent parts of the β-actin gene for control. (Figure courtesy of Dr. Alireza Baradaran, Mashhad, Iran, from Vakili et al., 2005.)

References

Donohoue PA et al: Congenital adrenal hyperplasia, pp 4077–4115. In: Scriver CR et al (eds) The Metabolic Basis of Inherited Disease, 8th ed. McGraw-Hill, New York, 2001.

Höppner W: 21-Hydroxylase-Mangel und andere Ursachen des kongenitalen adrenogenitalen Syndroms. Medgenet 16: 292–298, 2004.

Merke DP, Bornstein SR: Congenital adrenal hyperplasia. Lancet 365: 2125–2136, 2005.

New I, Wilson RC: Genetic disorders of the adrenal gland, pp 2277–2314. In: Rimoin DL et al (eds) Emery and Rimoin's Principles and Practice of Medical Genetics, 4th ed. Churchill-Livingstone, Edinburgh, 2002.

Vakili R et al: Molecular analysis of the *CYP21* gene and prenatal diagnosis in families with 21-hydroxylase deficiency in Northeastern Iran. Hormone Res 63: 119–124, 2005.

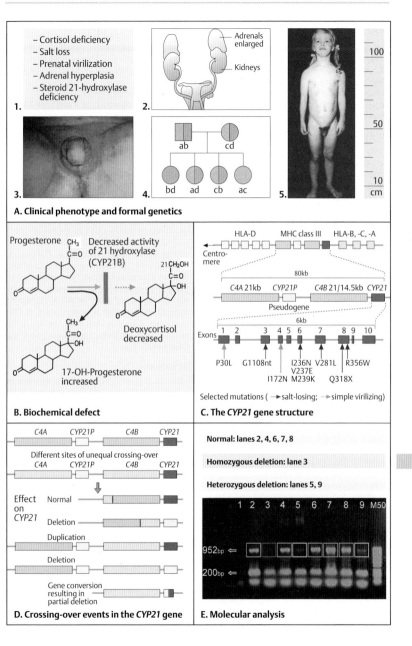

1.
- Cortisol deficiency
- Salt loss
- Prenatal virilization
- Adrenal hyperplasia
- Steroid 21-hydroxylase deficiency

2. Adrenals enlarged / Kidneys

3.

4. ab cd / bd ad cb ac

5. 100 50 10 cm

A. Clinical phenotype and formal genetics

Progesterone CH₃ C=O — Decreased activity of 21 hydroxylase (CYP21B)

21 CH₂OH C=O OH — Deoxycortisol decreased

CH₃ C=O OH — 17-OH-Progesterone increased

B. Biochemical defect

HLA-D   MHC class III   HLA-B, -C, -A
Centro-mere

80kb
C4A 21kb   CYP21P   C4B 21/14.5kb   CYP21
Pseudogene

6kb
Exons 1 2 3 4 5 6 7 8 9 10
P30L G1108nt I172N I236N V237E M239K V281L Q318X R356W

Selected mutations (→ salt-losing; → simple virilizing)

C. The CYP21 gene structure

C4A   CYP21P   C4B   CYP21

Different sites of unequal crossing-over
C4A   CYP21P   C4B   CYP21

Effect on CYP21
Normal
Deletion
Duplication
Deletion
Gene conversion resulting in partial deletion

D. Crossing-over events in the CYP21 gene

Normal: lanes 2, 4, 6, 7, 8

Homozygous deletion: lane 3

Heterozygous deletion: lanes 5, 9

1 2 3 4 5 6 7 8 9 M50
952bp ⇐
200bp ⇐

E. Molecular analysis

Unstable Repeat Expansion

Pathogenic mutations involving nucleotide repeats cause a distinct class of diseases affecting the central nervous system. In most, a repeat of three nucleotides at a specific location of a gene is expanded, in others, a tetranucleotide or pentanucleotide repeat. Nearly 20 unstable repeat diseases are known.

A. Types of disease

Unstable repeat diseases can result from expanded repeats in the untranslated regions of a gene (5′ UTR; 3′ UTR), in an exon, or in an intron. Functional consequences are loss of protein function or gain of an abnormal RNA level or protein, including altered translational control (FRAXA), signaling (FRAXAE), or mitochondrial function (Friedreich ataxia, FRDA). A tract within a coding region may code for expanded glutamine repeats in the protein (polyglutamine diseases, e.g., Huntington disease).

B. Huntington disease

Huntington disease (HD, MIM 143100), described 1872 by George Huntington in a family from Long Island, occurs in about 4–7 individuals per 100000. It is an autosomal dominant late-onset progressive neurodegenerative disease resulting from neuronal cell death and leading to complete loss of motor control and intellectual abilities within 5–10 years (1). It usually begins around the age of 40–50 years with uncoordinated movements (chorea, St. Vitus' dance), excitation, hallucinations, and psychological changes. The mutation is completely penetrant without differences in individuals heterozygous or homozygous for the mutation. The 210-kb gene, located on the distal short arm of human chromosome 4 at 4p16.3 between markers D4S127 and D4S125 (2), has 67 exons. Two transcripts of 10.3 and 13.6 kb encode a 3144-amino-acid protein, huntingtin, involved in neuron function and survival. The 5′ coding region of the gene contains 4–35 copies of a trinucleotide repeat of cytosine, adenine, and guanine (CAG), a codon for glutamine. In patients this is expanded to 40–250 CAG repeats. The CAG repeat length is inversely correlated with the age of onset. A diagnostic test distinguishes expanded and normal (CAG)$_n$ repeats (3). As this can be used for a predictive diagnosis many years prior to the onset of the first clinical signs, comprehensive genetic counseling according to established guidelines has to ensure informed consent before such a test is carried out. (Figure in 3 from Zühlke et al., 1993, kindly provided by Prof. W. Engel, Göttingen.)

C. Myotonic dystrophy

Dystrophia myotonica type 1 and type 2 (DM1, MIM 160900/605377, and DM2, MIM 602668) are multisystem autosomal dominant neurological diseases caused by different unstable repeat expansions. An early form occurs in young children and infants. Diagnostic signs are muscular weakness, cataracts, abnormal cardiac conduction, testicular atrophy, and other signs (1). A masklike facies is characteristic (2). The DM1 mutation affects the dystrophia myotonica protein kinase (*DMPK*) gene. It causes an increased number of CTG repeats in the 3′ UTR (3). This causes an altered RNA function resulting from RNAs with long CUG tracts. Patients have 50–2000 copies compared with 5–35 in normal individuals. The number of repeats influences the clinical severity. Southern blot analysis using probe pBBO7 at the marker locus D19S95 reveals increased sizes of DNA fragments after digestion with the restriction endonuclease *Eco*RI (4). (Figure based on data of Harley et al., 1992.)

References

Gatchel JR, Zoghbi HY: Diseases of unstable repeat expansion: Mechanisms and common principles. Nature Rev Genet 6: 743–755, 2005.

Harley HG et al: Unstable DNA sequence in myotonic dystrophy. Lancet 339: 1125–1130, 1992.

Harper P, Johnson K: Myotonic dystrophy, pp 5525–5550. In: Scriver CR et al. (ers) The Metabolic and Molecular Bases of Inherited Disease, 8th ed. McGraw-Hill, New York, 2001.

Hayden MR, Kremer B: Huntington disease, pp 5677–5701. In: Scriver CR et al (eds) The Metabolic and Molecular Bases of Inherited Disease, 8th ed. McGraw-Hill, New York, 2001.

Zühlke C et al: Mitotic stability and meiotic variability of the (CAG)n repeat in the Huntington disease gene. Hum Mol Genet 2: 2063–2067, 1993.

Online information:
Huntington disease:
  www.huntington-study-group.org
Hereditary Disease Foundation:
  www.hdfoundation.org
Muscular Dystrophy Association:
  www.mdausa.org

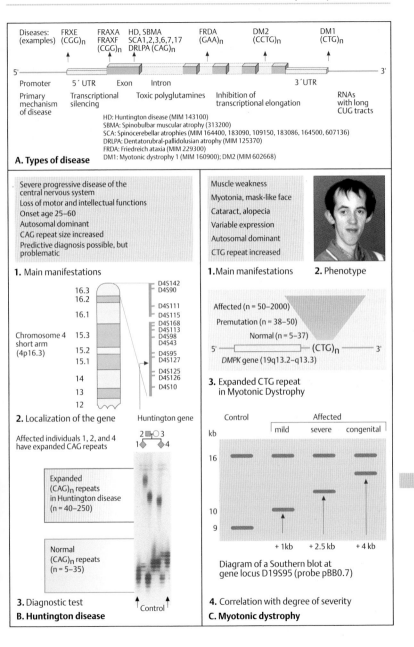

A. Types of disease

Diseases: (examples)
- FRXE (CGG)n
- FRAXA FRAXF (CGG)n
- HD, SBMA SCA1,2,3,6,7,17 DRLPA (CAG)n
- FRDA (GAA)n
- DM2 (CCTG)n
- DM1 (CTG)n

5′ — Promoter — 5′UTR — Exon — Intron — 3′UTR — 3′

Primary mechanism of disease:
- Transcriptional silencing
- Toxic polyglutamines
- Inhibition of transcriptional elongation
- RNAs with long CUG tracts

HD: Huntington disease (MIM 143100)
SBMA: Spinobulbar muscular atrophy (313200)
SCA: Spinocerebellar atrophies (MIM 164400, 183090, 109150, 183086, 164500, 607136)
DRLPA: Dentatorubral-pallidolusian atrophy (MIM 125370)
FRDA: Friedreich ataxia (MIM 229300)
DM1: Myotonic dystrophy 1 (MIM 160900); DM2 (MIM 602668)

B. Huntington disease

1. Main manifestations

- Severe progressive disease of the central nervous system
- Loss of motor and intellectual functions
- Onset age 25–60
- Autosomal dominant
- CAG repeat size increased
- Predictive diagnosis possible, but problematic

2. Localization of the gene

Chromosome 4 short arm (4p16.3)

16.3 — D4S142 / D4S90
16.2
16.1 — D4S111 / D4S115
15.3 — D4S168 / D4S113 / D4S98 / D4S43
15.2 — D4S95 / D4S127
15.1
14 — D4S125 / D4S126
13 — D4S10
12

Huntington gene

3. Diagnostic test

Affected individuals 1, 2, and 4 have expanded CAG repeats

Expanded (CAG)n repeats in Huntington disease (n = 40–250)

Normal (CAG)n repeats (n = 5–35)

Control

C. Myotonic dystrophy

1. Main manifestations

- Muscle weakness
- Myotonia, mask-like face
- Cataract, alopecia
- Variable expression
- Autosomal dominant
- CTG repeat increased

2. Phenotype

3. Expanded CTG repeat in Myotonic Dystrophy

Affected (n = 50–2000)
Premutation (n = 38–50)
Normal (n = 5–37)

5′ — (CTG)n — 3′
DMPK gene (19q13.2–q13.3)

4. Correlation with degree of severity

Control / Affected (mild, severe, congenital)

kb
16
10
9

+ 1kb   + 2.5 kb   + 4 kb

Diagram of a Southern blot at gene locus D19S95 (probe pBB0.7)

Fragile X Syndrome

Fragile X syndrome (MIM 309550; synonyms: fragile X mental retardation syndrome, FMR1, or Martin-Bell syndrome) is a frequent form of mental retardation with a prevalence of about 1 : 3000–6000 males. It is caused by expanded CGG trinucleotide repeats in the 5′ untranslated region of the *FMR1* gene on the distal long arm of the X chromosome at Xq27.3. The increased number of CGG repeats results in transcriptional silencing, loss of the FMR1 protein, and defective translational control of neuronal synaptic proteins. It was identified in 1991 as the first human disease to be caused by expansion of an unstable trinucleotide repeat sequence (Oberle et al., 1991; Verkerk et al., 1991).

A. Phenotype

Individuals with fragile X syndrome have varying degrees of intellectual developmental delay associated with behavioral and physical features. Some patients have connective tissue weakness. The testes are usually enlarged.

B. Fragile site FRAXA

The syndrome's name is derived from a cytogenetically visible fragile site at band Xq27.3 of the X chromosome in lymphocytes cultured in folic acid-deficient medium. Other fragile sites exist, e.g., FRAXE at Xq28 (MIM 309548).

C. *FMR1* gene and protein

The *FMR1* gene has 17 exons spanning 38 kb. Its transcript is alternatively spliced and translated into at least 20 protein isoforms (FMRPs). The protein (FMRP) selectively binds RNA. At least two functional domains are RNA-binding sites, KH2 and RGG (KH domains consist of 40–60 amino acids with invariant hydrophobic leucine, isoleucine, or methionine; RGG, of a 20–30 amino acid motif with arginine-glycine-glycine residues). In Drosophila *Fmr1* is a component of the RNA-induced silencing complex (RISC, see p. 224).

D. Inheritance and genetic testing

Southern blot analysis (**1**) can distinguish individuals carrying a full mutation allele (<200 CGG trinucleotides), a premutation (59–200), or a normal allele (6–50). The visible bands (arrows) are DNA fragments harboring the CGG repeat region, derived from genomic DNA cleaved by restriction endonuclease *Pst*I, and hybridized to the radiolabelled *FMR1* probe Ox0.55. PCR protocols measure the lengths of normal and premutation alleles. Expansion leads to hypermethylation and transcriptional silencing of the *FMR1* gene. (Figure kindly provided by Prof. P. Steinbach, Ulm, Germany.)

In the pedigree (**2**) the number of CGG repeats at the *FMR1* locus is shown for each individual. A premutation can be transmitted by either a female (I-2, II-3, III-2) or a male (II-2). A premutation allele may expand into a full mutation when passed from a mother to her children. All daughters of a normal male transmitter will be heterozygous. Full mutation males transmit a premutation to all their daughters. Carriers of a premutation allele do not usually have signs of fragile X syndrome, but 50–60% of girls with a full mutation have significant cognitive deficits. (Figure based on information kindly provided by Prof. P. Steinbach, Ulm.)

References

Gatchel JR, Zoghbi HY: Diseases of unstable repeat expansion: Mechanisms and common principles. Nature Rev Genet 6: 743–755, 2005.

Ishizuka A, Siomi MC, Siomi H: A *Drosophila* fragile X protein interacts with components of RNAi and ribosomal proteins. Genes Dev 16: 2497–2508, 2002.

Nolin SL et al. Expansion of the fragile X CGG repeat in females with premutation or intermediate alleles. Am J Hum Genet 72: 454–464, 2003.

Oberle I et al: Instability of a 550-base pair DNA segment and abnormal methylation in fragile X syndrome. Science 252: 1097–1102, 1991.

Verkerk A et al: Identification of a gene (FMR-1) containing a CGG repeat coincident with a breakpoint cluster region exhibiting length variation in fragile X syndrome. Cell 65: 905–914, 1991.

Warren ST, Sherman SL: The fragile X syndrome, pp. 1257–1289. In: Scriver CR et al, eds: The Metabolic and Molecular Bases of Inherited Disease. 8th ed. McGraw-Hill, New York, 2001.

Wöhrle D et al: Demethylation, reactivation, and destabilization of human fragile X full-mutation alleles in mouse embryocarcinoma cells. Am J Hum Genet 69: 504–515, 2001.

Zalfa F et al: The fragile X syndrome protein FMRP associates with *BC1* RNA and regulates the translation of specific mRNAs at synapses. Cell 112: 317–327, 2003.

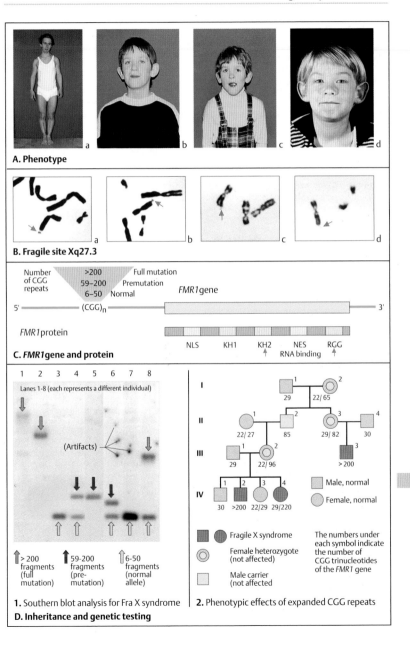

A. Phenotype

B. Fragile site Xq27.3

C. FMR1 gene and protein

Number of CGG repeats
>200 Full mutation
59–200 Premutation
6–50 Normal

FMR1 gene

5'———(CGG)ₙ———————————————3'

FMR1 protein

NLS   KH1   KH2   NES   RGG
RNA binding

Lanes 1-8 (each represents a different individual)

(Artifacts)

⇧ > 200 fragments (full mutation)
⬆ 59-200 fragments (pre-mutation)
⇧ 6-50 fragments (normal allele)

1. Southern blot analysis for Fra X syndrome

Male, normal

Female, normal

Fragile X syndrome

Female heterozygote (not affected)

Male carrier (not affected

The numbers under each symbol indicate the number of CGG trinucleotides of the FMR1 gene

2. Phenotypic effects of expanded CGG repeats

D. Inheritance and genetic testing

Imprinting Diseases

Imprinting diseases are caused by different mechanisms affecting one or more active genes normally expressed in only one parental allele in an imprinted region. Best known are Prader–Willi syndrome (MIM 176270), Angelman syndrome (MIM 105830), and Beckwith–Wiedemann syndrome (MIM 130650), at 11p15.5.

A. Prader–Willi and Angelman syndromes

Prader–Willi syndrome (PWS) and Angelman syndrome (AS) are neurogenetic developmental disorders resulting from different genetic lesions in an imprinted region of human chromosome 15 (15q11–13) extending over 2 Mb. The effect of an interstitial deletion of this region depends on whether it involves the chromosome 15 of paternal origin (resulting in PWS) or the chromosome 15 of maternal origin (resulting in AS). Prader–Willi syndrome is characterized by neonatal muscular weakness and feeding difficulties, followed in early childhood by reduced or lack of satiation control, which leads to massive obesity in many patients. In Angelman syndrome the developmental retardation is usually severe, with nearly complete lack of speech development, an abnormal electroencephalogram and tendency to seizures, and hyperactivity.

B. Deletion and uniparental disomy

The functional result of a deletion and of uniparental disomy (UPD) of an imprinted region is the same. If a deletion of 15q11-q13 involves the chromosome 15 of paternal origin (loss of one paternal allele 2 in the diagram of a Southern blot on the left), PWS results. If it involves the chromosome 15 of maternal origin, AS results (**1**). In uniparental disomy (UPD) both chromosomes are of the same parental origin (**2**). In *isodisomy* they are identical (1–1 in lane 1 on the left); in *heterodisomy* they are of the same parental origin but differ (1–2 in lane 3 on the right). (See table in the appendix.)

C. Parent-of-origin effect

In the imprinted region 15q11-q13 some genes are expressed depending on their parental origin (**1**). Prader–Willi syndrome results from loss of function of paternally expressed genes in the PWS region (**2**). Angelman syndrome results from loss of function of the maternally expressed *UBE3A* gene (**3**). In addition, point mutations in this region may cause PWS in about 5–10% of patients. This can result in familial occurrence.

D. Imprinted chromosomal region

This simplified figure shows the genetic map of the chromosomal region 15q11–13. Loss of expression of paternally expressed genes (blue) results in PWS. Angelman syndrome results from loss of function of the *UBE3A* gene (ubiquitin-protein ligase E3, MIM 600012), which is expressed from the maternal copy only (red), although mono-allelic expression occurs in brain cells only.

The imprinting center (IC), controlling the entire imprinted region, appears to consist of two elements. One is required for the maintenance of the paternal imprint during early embryogenesis, the other for maternal imprinting in the female germline.

(Figure kindly provided by K. Buiting and B. Horsthemke, Essen.)

References

Constância M, Kelsey G, Reik W: Resourceful imprinting. Nature 432: 53–57, 2004.

Horsthemke B, Dittrich B, Buiting K: Imprinting mutations on human chromosome 15. Hum Mutat 10: 329–337, 1997.

Horsthemke B, Buiting K: Imprinting in Prader–Willi and Angelman syndromes, pp 245–258. In: Jorde LB, Little PFR, Dunn MJ, Subramaniam S (eds) Encyclopedia of Genetics, Genomics, Proteomics, and Bioinformatics, vol 1. Wiley & Sons, Chichester, 2005.

Horsthemke B, Buiting K: Imprinting defects on human chromosome 15. Cytogenet Genome Res 113: 292–299, 2006.

Lossie AC et al: Distinct phenotypes distinguish the molecular classes of Angelman syndrome. J Med Genet 38: 834–845, 2001.

Nicholls RD, Knepper JL: Genome organization: Function and imprinting in Prader–Willi and Angelman syndromes. Ann Rev Genom Hum Genet 2: 53–175, 2001.

Varela MC et al: Phenotypic variability in Angelman syndrome: comparison among different deletion classes and between deletion and UPD subjects. Eur J Hum Genet 12: 987–992, 2004.

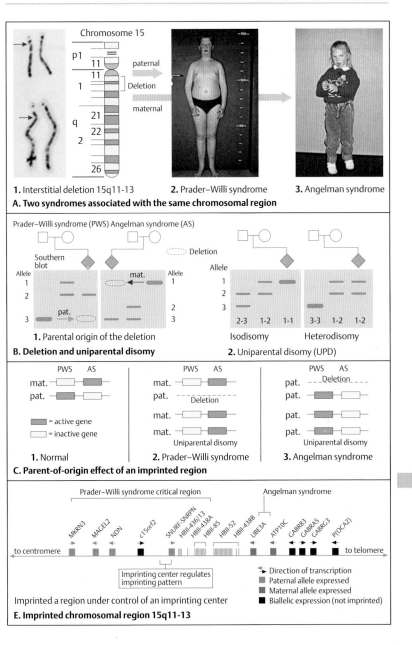

Chromosome 15

p1
11
11
1
Deletion
paternal
maternal
q
21
22
2
26

1. Interstitial deletion 15q11-13   **2.** Prader–Willi syndrome   **3.** Angelman syndrome

A. Two syndromes associated with the same chromosomal region

Prader–Willi syndrome (PWS) Angelman syndrome (AS)

Southern blot
  ····· = Deletion

Allele
1
2
3
mat.
pat.

Allele
1
2
3

Allele
1
2
3

2-3 1-2 1-1   3-3 1-2 1-2

1. Parental origin of the deletion   Isodisomy   Heterodisomy

B. Deletion and uniparental disomy   **2.** Uniparental disomy (UPD)

PWS   AS
mat.
pat.

■ = active gene
□ = inactive gene

1. Normal

PWS   AS
mat.
pat.
Deletion
mat.
mat.
Uniparental disomy

2. Prader–Willi syndrome

PWS   AS
pat. ····· Deletion ·····
pat.
pat.
pat.
Uniparental disomy

3. Angelman syndrome

C. Parent-of-origin effect of an imprinted region

Prader–Willi syndrome critical region   Angelman syndrome

MKRN3 MAGEL2 NDN c15orf2 SNURF-SNRPN HBII-436/13 HBII-438A HBII-85 HBII-52 HBII-438B UBE3A ATP10C GABRB3 GABRA5 GABRG3 P(OCA2)

to centromere   to telomere

Imprinting center regulates imprinting pattern

→ Direction of transcription
■ Paternal allele expressed
■ Maternal allele expressed
■ Biallelic expression (not imprinted)

Imprinted a region under control of an imprinting center

E. Imprinted chromosomal region 15q11-13

Autosomal Trisomies

Trisomy arises during meiosis I or II (see p. 118) by nondisjunction (faulty distribution of one member of a chromosome pair) as a *prezygotic* event. It may, less commonly, occur after fertilization as a *postzygotic* event in the early embryo during a somatic cell division (mitosis). In this case the additional chromosome is present only in a certain proportion of cells. This condition is called *chromosomal mosaicism*. Of the 22 human autosomes, only trisomy for chromosomes 13, 18, and 21 regularly occurs in liveborn infants.

A. Trisomy in *Datura stramonium*

The phenotypic effects of trisomy were first discovered in a plant by A.F. Blakeslee in 1922. He studied jimsonweed plants, also called common thorn apple (*Datura stramonium*), and observed that plants with three copies of one of the 12 chromosomes had a characteristic appearance that was specific for each of the 12 chromosomes of this plant. (Figure adapted from Blakeslee, 1922.)

B. Trisomies in the mouse

Specific phenotypic effects of autosomal trisomies and monosomies are also observed in mice. In the 1970s, A. Gropp and co-workers found that each trisomy and monosomy in the mouse results in a developmental profile and is associated with morphological changes and malformations that are characteristic for each trisomy (**1**). Embryos with a monosomy died very early, within the first 8 days of the 21-day gestation. The intrauterine survival time depends on the chromosome that is trisomic. The examples show a mouse embryo with trisomy 12 (**2**) and the brain of a mouse with trisomy 19 at birth (**3**), each compared with a normal control. The mouse embryo with trisomy 12 shows an open skullcap and other malformations on the 14th day of fetal development. Only trisomy 19 is compatible with survival until birth, but the brain is too small. Growth retardation is a common feature in all trisomies. The trisomic mice were generated by breeding mice with translocations among various chromosomes. The resulting trisomies were observed at different stages of fetal development. (Figure in 1 from Gropp, 1982; in 2 and 3 courtesy of Dr. H. Winking, Lübeck, Germany.)

C. Autosomal trisomies in man

Human trisomies occur as trisomy 21 (about 1 : 600), trisomy 18 (about 1 : 5000), and trisomy 13 (about 1 : 8000). Each has a distinct pattern of congenital malformations associated with variable degrees of mental impairment in trisomy 21 (Down syndrome) and complete lack of mental development in trisomies 18 and 13. Only trisomy 21 is compatible with survival into adulthood, although the overall life expectancy is about half of that of the normal population.

D. Nondisjunction as a cause

All three human trisomies occur more frequently with advanced maternal age (**1**). The age of the father has no or very little influence. If nondisjunction occurs in meiosis I, the three chromosomes will be different $(1 + 1 + 1)$, whereas if nondisjunction occurs during meiosis II, two of the three chromosomes will be identical $(2 + 1)$. In humans, about 70% of nondisjunctions occur in meiosis I, and 30% in meiosis II.

References

Antonarakis SE: Down syndrome, pp 1069–1078. In: Jameson JL (ed) Principles of Molecular Medicine. Humana Press, Totowa, New Jersey, 1998.

Blakeslee AF: Variation in Datura due to changes in chromosome number. Am Naturalist 56: 16–31, 1922.

Boué A, Gropp A, Boué J: Cytogenetics of pregnancy wastage. Adv Hum Genet 14: 1–57, 1985.

Epstein CJ: Down syndrome (trisomy 21), pp 1223–1256. In: Scriver CR et al (eds) The Metabolic and Molecular Bases of Inherited Disease, 8th ed. McGraw-Hill, New York, 2001.

Gropp A: Value of an animal model for trisomy. Virchows Arch Pathol Anat 395: 117–131, 1982.

Miller OJ, Therman E: Human Chromosomes, 4th ed. Springer, New York–Heidelberg, 2001.

Roizen NJ, Patterson D: Down's syndrome. Lancet 361: 1281–1289, 2003.

Tolmie JJ: Down syndrome and other autosomal trisomies, pp 1129–1183. In: Emery and Rimoins's Principles and Practice of Medical Genetics, 4th ed. Churchill-Livingstone, London–New York, 2002.

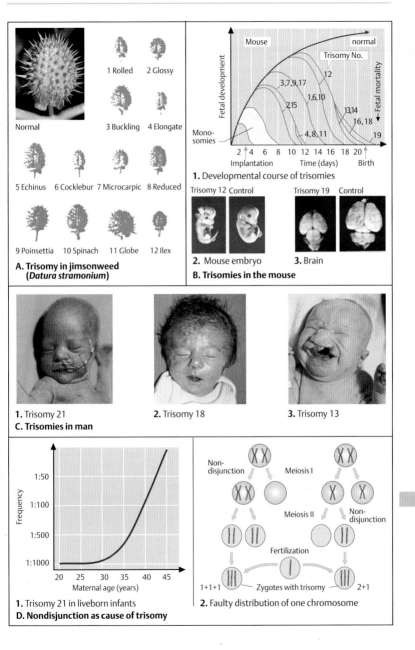

1 Rolled **2 Glossy**

Normal

3 Buckling **4 Elongate**

5 Echinus **6 Cocklebur** **7 Microcarpic** **8 Reduced**

9 Poinsettia **10 Spinach** **11 Globe** **12 Ilex**

A. Trisomy in jimsonweed
(Datura stramonium)

1. Developmental course of trisomies

Trisomy 12 Control   Trisomy 19 Control

2. Mouse embryo   **3. Brain**

B. Trisomies in the mouse

1. Trisomy 21   **2. Trisomy 18**   **3. Trisomy 13**

C. Trisomies in man

1. Trisomy 21 in liveborn infants
D. Nondisjunction as cause of trisomy

Non-disjunction   Meiosis I

Meiosis II   Non-disjunction

Fertilization

1+1+1 — Zygotes with trisomy — 2+1

2. Faulty distribution of one chromosome

Other Numerical Chromosomal Deviations

Other conditions associated with an abnormal number of chromosomes involve either an entire additional set of chromosomes (triploidy or tetraploidy) or the X chromosome or Y chromosome. Deviations from the normal number of X or Y chromosomes comprise about half of all chromosomal aberrations in man (total frequency about 1 : 400).

A. Triploidy

Triploidy may be either one paternal and two maternal sets (karyotype 69,XXY or 69,XXX) or one maternal and two paternal sets (69,XXX, 69,XYY or 69,XXY). Triploidy is associated with severe developmental failure and congenital malformations (**1**). The fetus shows numerous severe malformations (**2**), such as cardiac defects, cleft lip and palate, skeletal defects, and others. Triploidy accounts for about 17% of spontaneous abortions (**3**). The causes include a diploid spermatocyte, a diploid oocyte, or fertilization of an egg cell by two spermatozoa (dispermy).

B. Monosomy X (Turner syndrome)

Monosomy X (karyotype 45,X) is frequent, about 5% at conception. However, of 40 zygotes with monosomy X, only one will develop to birth. Monosomy X causes Turner syndrome with a very wide phenotypic spectrum, ranging from severe to very mild. The phenotype during the fetal stage is usually massive lymphedema of the head and neck, large multilocular thin-walled lymphatic cysts (**1**). Congenital cardiovascular defects, especially involving the aorta, and kidney malformations are frequent. Important is fetal degeneration of the ovaries into connective tissue, as streak gonads. Small stature is always present, with an average adult height of about 150 cm. In many patients the manifestations are mild (**2**), but in others, webbing of the neck (pterygium colli) may be present as a residue of the fetal lymphedema (**3**). Most patients have chromosomal mosaicism 45,X/46,XX, i.e., some of their cells have normal chromosomal complements or a deletion Xp or an isochromosome for the long arm [i(Xq)]. The loss of genes on the short arm of the X chromosome (Xp) is responsible for the phenotype (see *SHOX* genes, MIM 312865).

C. Additional X or Y chromosome

An additional X chromosome in males (47,XXY) causes Klinefelter syndrome after puberty if untreated (**1**). This includes tall stature, absent or decreased development of male secondary sex characteristics, and infertility due to absent spermatogenesis. Testosterone substitution beginning at puberty is necessary. In contrast, an additional Y chromosome (47,XYY) does not result in a recognizable phenotype (**2**). Girls with three X chromosomes (47,XXX) are physically unremarkable (**3**). However, learning disorders and delayed speech development have been observed in some of these children.

D. Chromosomal aberrations in human fetuses

A wide spectrum of trisomies (and monosomies) occur at conception and lead to spontaneous abortion during the second and third months of pregnancy. The relative proportions of the various trisomies observed in fetuses after spontaneous abortion differ. The most frequent is trisomy 16, which accounts for about 5% of all autosomal trisomies. (Data from Lauritsen, 1982.)

References

DeGrouchy J, Turleau C: Clinical Atlas of Human Chromosomes, 2nd ed. John Wiley & Sons, New York, 1984.

Lauritsen JG: The cytogenetics of spontaneous abortion. Res Reproduct 14: 3–4, 1982.

Menasha J: Incidence and spectrum of chromosome abnormalities in spontaneous abortions: new insights from a 12-year study. Genet Med 7: 251–263, 2005.

Miller OJ, Therman E: Human Chromosomes, 4th ed. Springer, New York–Heidelberg, 2001.

Ranke MB, Saenger P: Turner's syndrome. Lancet 358: 309–314, 2001.

Schinzel A: Catalogue of Unbalanced Chromosome Aberrations in Man, 2nd ed. W de Gruyter, Berlin, 2001.

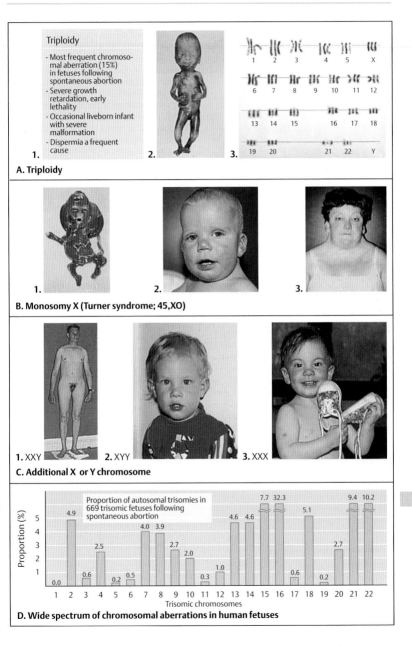

Triploidy

- Most frequent chromosomal aberration (15%) in fetuses following spontaneous abortion
- Severe growth retardation, early lethality
- Occasional liveborn infant with severe malformation
- Dispermia a frequent cause

A. Triploidy

B. Monosomy X (Turner syndrome; 45,XO)

C. Additional X or Y chromosome

1. XXY   2. XYY   3. XXX

Proportion of autosomal trisomies in 669 trisomic fetuses following spontaneous abortion

D. Wide spectrum of chromosomal aberrations in human fetuses

Autosomal Deletion Syndromes

Cytologically visible deletions or duplications cause developmental disturbances and congenital malformations, usually in a recognizable pattern (Brewer et al., 1998). Important large terminal deletions occur at 4p–, 5p–, 9p–, 11p–, 11q–, 13q–, 18p–, and 18q–. An important category is subtelomeric deletions detectable only by fluorescence *in-situ* hybridization (FISH) and other molecular methods (see references). Most deletions and duplications occur de novo.

A. Deletion 5p–: cri-du-chat syndrome

In 1963, Lejeune and his co-workers in Paris described children with a partial deletion of the short arm of a chromosome 5 (5p–) and a characteristic pattern of dysmorphic facial features associated with impaired mental development, the cri-du-chat syndrome (MIM 123450). The critical deletion region is 5p15.2 with variable sized deletions. Affected infants have a prolonged high-pitched cry resembling that of a kitten.

In about 12% one of the parents has a translocation involving a chromosome 5.

B. Deletion 4p–: Wolf–Hirschhorn syndrome

This characteristic phenotype (MIM 194190) results from a partial deletion of variable size of the short arm of a chromosome 4. It was described in 1964 independently by U. Wolf and K. Hirschhorn and their co-workers. Variable but considerable mental and statomotoric retardation is associated with characteristic facial features (**1**, **2**), midline defects (cleft palate, hypospadias), coloboma of the iris, congenital heart defects, and other malformations. In some patients the deletion is too small to be detectable in a conventional karyotype and requires FISH analysis to confirm the deletion. The critical chromosomal region (Wolf–Hirschhorn critical region, WHSCR) is 4p16 (**3**). (Figure in 3 adapted from Wright et al., 1999.)

C. Microdeletion syndromes

Microdeletion syndromes involve very small deletions of adjacent gene loci in a defined chromosomal region (also called contiguous gene syndromes). They can usually only be detected by molecular cytogenetic methods.

Of the more than 20 different microdeletion syndromes known (see table in the appendix;

Budarf & Emanuel, 1997) three are shown here. The Williams–Beuren syndrome (**1**, MIM 194050, 130160) usually presents with characteristic facial features ("elfinlike"), infantile hypercalcemia, supravalvular aortic stenosis, growth retardation, and impaired mental development. Deletion of 22q11 (**2**) leads to a group of clinically different but overlapping disorders (DiGeorge syndrome, MIM 188400), characterized by absence or hypoplasia of the thymus and the parathyroid glands and malformations of the aortic arch; velocardiofacial syndrome (MIM 192430); conotruncal cardiac defects (MIM 217095); and others. The Rubinstein–Taybi syndrome (MIM 180849) is characterized by typical facial features (**3**), broad thumbs and toes with associated radiological changes, mental retardation, and other features. Mutations in the *CREBBP* gene encoding the CREB-binding protein (MIM 600140) cause this disorder. A deletion of 16p13.3 is detectable in about 12% of patients.

D. Phenotype of duplication 5q at different ages

A unique duplication illustrates the similar facial phenotypes at different ages: in a fetus at 22 weeks' gestation (1), in a 5-month-old infant (2), and in an 8-year-old child (3). The affected individuals are siblings with mental retardation. The partial duplication 5q33-qter resulted from a paternal reciprocal translation (Passarge et al., 1982).

References

Brewer C et al: A chromosomal deletion map of human malformations. Am J Hum Genet 63: 1153–1159, 1998.

De Vries BBA et al: Telomeres: a diagnosis at the end of chromosomes. J Med Genet 40: 385–398, 2003.

Linardopoulou EV et al: Human subtelomeres are hot spots of interchromosomal recombination and segmental duplication. Nature 437: 94–100, 2005.

Miller OJ, Therman E: Human Chromosomes, 4th ed. Springer, New York–Heidelberg, 2001.

Passarge E et al: Fetal manifestation of a chromosomal disorder: partial duplication of the long arm of chromosome 5 (5q33-qter). Teratology 25: 221–225, 1982.

Schinzel A: Catalogue of Unbalanced Chromosome Aberrations in Man, 2nd ed. W de Gruyter, Berlin, 2001.

Wright TJ et al: Comparative analysis of a novel gene from the Wolf-Hirschhorn/Pitt-Rogers-Danks syndrome critical region. Genomics 59: 203–212, 1999.

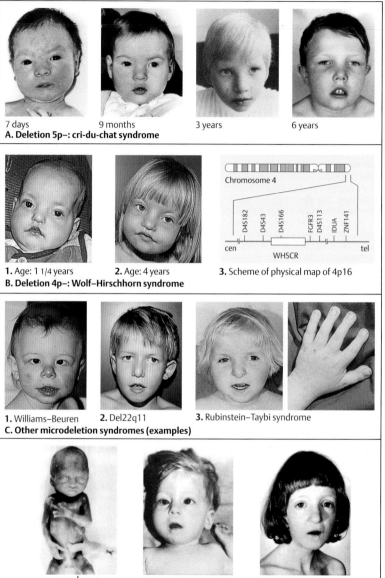

7 days   9 months   3 years   6 years
A. Deletion 5p−: cri-du-chat syndrome

1. Age: 1 1/4 years   **2.** Age: 4 years   **3.** Scheme of physical map of 4p16

Chromosome 4

D4S182 D4S43 D4S166 FGFR3 D4S113 IDUA ZNF141
cen   tel
                    WHSCR

B. Deletion 4p−: Wolf–Hirschhorn syndrome

1. Williams–Beuren   **2.** Del22q11   **3.** Rubinstein–Taybi syndrome
C. Other microdeletion syndromes (examples)

1. Fetus: 22nd week   **2.** 5 months   **3.** 8 years
D. Phenotype of duplication 5q at different ages

Principles of Genetic Diagnostics

The diagnosis of a genetic disease requires a systematic approach that takes many clinical and genetic considerations into account. It begins with an analysis of the phenotype, i.e., the disorder in question. The McKusick system, MIM, with the Morbid Anatomy of the Human Genome as well as Mutation Databases have to be consulted for each case.

A. Genetic diagnosis, a multistep procedure

Genetic diagnosis requires a sequence of steps with binary decisions. The first decision to be made is whether a pattern can be recognized in the manifestations. If a disease pattern can be recognized, the next decision concerns the category of disease. Although difficult to establish in practice, this decision is the basis for subsequent steps. A particular phenotype may be caused by mutations at different loci (*locus heterogeneity*) or by different mutant alleles at the same locus (*allele heterogeneity*). All genetic diagnostic procedures should be preceded by genetic counseling, which properly includes obtaining (informed) consent from the persons involved.

B. Genotype analysis by PCR typing

Genotype analysis by PCR typing of a polymorphic restriction site is preferred to the more laborious Southern blot hybridization (see p. 74). (Figure adapted from Strachan and Read, 2004.)

C. Protein truncation test (PTT)

This is a test for frameshift, splice, or nonsense mutations that leads to a truncated protein due to an early stop codon created downstream of the mutation. The truncated protein is detected by an in vitro translation system. The translation will be interrupted at a premature stop codon resulting from the mutation. The size of the newly translated protein is determined by gel electrophoresis. PTT is useful in studying genes with frequent nonsense mutations, such as the *APC*, *BRCA1*, and *BRCA2* genes. (Figure adapted from Strachan & Read, 2004.)

References

Aase JM: Diagnostic Dysmorphology. Plenum Medical Book Company, New York, 1990.

Hochedlinger K, Jaenisch R: Nuclear reprogramming and pluripotency. Nature 441: 1061–1067, 2006

Horaitis R, Scriver CR, Cotton RGH: Mutation databases: Overview and catalogues, pp 113–125. In: Scriver CR et al (eds) The Metabolic and Molecular Bases of Inherited Disease, 8th ed. McGraw–Hill, New York, 2001.

Jones KL: Smith's Recognizable Patterns of Human Malformation, 6th ed. WB Saunders, Philadelphia, 2006.

McKusick VA: Mendelian Inheritance in Man. A Catalog of Human Genes and Genetic Disorders, 12th ed. Johns Hopkins University Press, Baltimore, 1998 (Online at OMIM www.ncbi.nlm.nih.gov/Omim with links to diagnostic laboratories at www.genetests.com).

Misfeldt S, Jameson JL: The practice of genetics in clinical medicine, pp 386–391. In: Kapser DS et al (eds) Harrison's Principles of Internal Medicine, 16th ed. McGraw–Hill, New York, 2005.

Passarge E, Kohlhase J: Genetik, pp 4–66. In: Siegenthaler W & Blum HE (eds) Klinische Pathophysiologie, 9 Aufl. Thieme Verlag, Stuttgart-New York, 2006.

Pelz J, Arendt V, Kunze J: Computer assisted diagnosis of malformation syndromes: an evaluation of three databases (LDDB, POSSUM, and SYNDROC). Am J Med Genet 63: 257–267, 1996.

Rimoin DL, Connor JM, Pyeritz RE, Korf BR: Emery and Rimoins's Principles and Practice of Medical Genetics, 5th ed. Elsevier Churchill-Livingstone, London–New York, 2006.

Stevenson RE, Hall JG (eds): Human Malformations and Related Anomalies, 2nd ed. Oxford Univ. Press, Oxford, 2006.

Strachan T, Read AP: Human Molecular Genetics, 3rd ed. Garland Science, London–New York, 2004.

van der Luijt R et al: Rapid detection of translation terminating mutations at the adenomatous polyposis (APC) gene locus by direct protein truncation test. Genomics 20: 1–4, 1994.

Online information about gene tests:
http://www.geneclinics.org

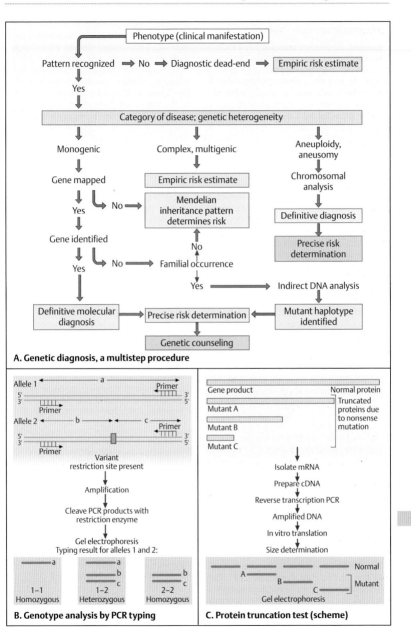

A. Genetic diagnosis, a multistep procedure

B. Genotype analysis by PCR typing

C. Protein truncation test (scheme)

Gene and Stem Cell Therapy

Treatment of genetic diseases is still limited to selected disorders in clinical trials. Gene therapy aims at replacing a defective gene by a normal allele to treat a disease or to delay its course. Stem cell therapy would apply pluripotent, renewable cells to organs that are irreversibly damaged by a disease. Considerable technical difficulties and side effects have to be overcome, and ethical considerations have to be taken into account.

A. Principle of gene therapy

Somatic gene therapy involves delivering a correcting gene to somatic cells in the tissues affected. Two basic forms of somatic gene therapy are distinguished, (i) *ex vivo* and (ii) *in vivo*. In the first, genes are transferred into cells outside the body, and subsequently introduced into tissues that need to be corrected. In the second, a gene is introduced directly. The gene may be transferred by a viral vector or by a nonviral method. The advantage of viral vectors is the relative ease with which they could enter the recipient's cells. However, controlling viral production, carrying capacity, dependence on cell proliferation, and other aspects make application difficult.

One strategy for introducing a correcting gene into the hematopoietic system is illustrated here. Blood is taken from the patient (**1**); red blood cells and white blood cells are separated from the blood, and the red blood cells reinfused (**2**). From the white blood cells (**3**) immunologically competent CD34 cells are separated (**4**), placed together with a virus vector carrying the desired normal gene (**5**), and propagated together in a cell culture (**6**). Once the defective cells have incorporated the viral vector and are thereby corrected, these cells are reintroduced into the recipient (**7**). (Figure adapted from J.A. Barranger, Pittsburgh, with permission, at www.gaucher.mgh.harvard.edu/ genetherapy2 big.gif.)

B. Stem cells

Stem cells are undifferentiated progenitor cells that can develop into specialized cells. They differ according to their tissue destiny, replicative capacity, and differentiation potential. *Totipotent* stem cells can develop into a complete embryo and form a placenta. This ability is limited to cells derived from the first few divisions of the zygote. *Pluripotent* stem cells can form tissues derived from the endoderm, mesoderm, and ectoderm germinal layers. *Embryonic stem cells* (ESC) fall into this category. Stem cells divide symmetrically into two identical stem cells (self-renewal). They form a pool of cells from which progenitor cells of specialized cells develop by asymmetric division. Such cells lose the ability to undergo cell division and self-renewal.

C. Stem cell therapy

Stem cell therapy would have the advantage of providing the recipient with a permanent supply of genetically corrected cells. This is especially important in organs where cells are constantly lost and replaced, such as the bone marrow (hematopoietic system) and epithelial cell systems (e.g., in the gastrointestinal tract). Future stem cell therapy will be applicable to a wide spectrum of tissues and types of diseases. Whether adult stem cells will be sufficient or embryonic stem cells will be required has not been established. (Figure adapted from G.J. Nabel, 2004.)

References

Bodine D, Jameson JL, McKay R: Stem cell and gene therapy in clinical medicine, pp 392–397. In: Kasper DL et al: Harrison's Principles of Internal Medicine. 16th ed. McGraw-Hill, New York, 2005.

Gilbert-Barness E, Barness L: Metabolic Diseases. Foundations of Clinical Management, Genetics, and Pathology. Eaton Publishing, Natick, MA 01760, USA, 2000.

Hochedlinger K, Jaenisch R: Nuclear transplantation, embryonic stem cells, and the potential for cell therapy. New Eng J Med 349:275–286, 2003.

Jiang Y et al: Pluripotency of mesenchymal stem cells derived from adult marrow. Nature 418: 41–49, 2002.

Nabel GJ: Genetic, cellular and immune approaches to disease therapy: past and future. Nature Med 10: 135–141, 2004.

Strachan T, Read AP: Human Molecular Genetics. 3rd ed. Garland Science, London–New York, 2004.

Online information:

Database of clinical trials, J Gene Med
www.wiley.co.uk/genetherapy/clinical

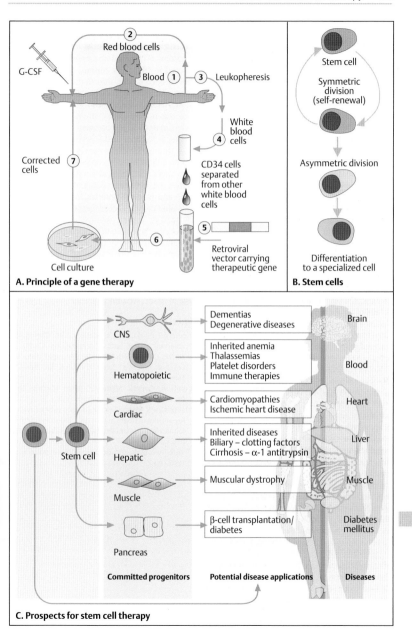

A. Principle of a gene therapy

G-CSF

Blood (1)

(2) Red blood cells

(3) Leukopheresis

White blood cells (4)

CD34 cells separated from other white blood cells

(5)

(6) Retroviral vector carrying therapeutic gene

(7) Corrected cells

Cell culture

B. Stem cells

Stem cell

Symmetric division (self-renewal)

Asymmetric division

Differentiation to a specialized cell

C. Prospects for stem cell therapy

Stem cell

Committed progenitors

CNS → Dementias / Degenerative diseases → Brain

Hematopoietic → Inherited anemia / Thalassemias / Platelet disorders / Immune therapies → Blood

Cardiac → Cardiomyopathies / Ischemic heart disease → Heart

Hepatic → Inherited diseases / Biliary – clotting factors / Cirrhosis – α-1 antitrypsin → Liver

Muscle → Muscular dystrophy → Muscle

Pancreas → β-cell transplantation/diabetes → Diabetes mellitus

Potential disease applications   Diseases

Chromosomal Location of Human Genetic Diseases

Nowhere is the growth of knowledge about disease-causing mutations in human genes more apparent than in the map of gene loci on all chromosomes relating to the about 3000 individually defined disease phenotypes. Of these, more than 1000 are understood at the molecular level. This gene map is referred to as *The Morbid Anatomy of the Human Genome* (McKusick, 1998; Amerberger et al, 2001). For an understanding of the genetic causes of human diseases this can be compared to the impact on medicine made by the seven volumes of *De humani coporis fabrica libri septa* (die "Fabrica") 1543 by Andreas Vesalius (1514–1564) and the causal analysis of diseases *De Sedibus et Causis Morborum per Anatomen Indagatis* 1761 by Giovanni Morgagni (1682–1771). The progress in the knowledge of genetic diseases is documented in twelve published editions of *Mendelian Inheritance in Man. A Catalog of Human Genes and Disorders* (MIM) by Victor A. McKusick, M.D. of Johns Hopkins University School of Medicine. Its first edition in 1966 contained a total of 1487 entries. The second edition (1545 entries) in 1968 included the first autosomal gene mapped. The subsequent editions reveal an entry-doubling time of about 15 years (3368 in the 6th edition (1983), 5710 entries in the 10th edition (1992), 8587 entries in the 12th edition (1998) and 16774 on 13 May, 2001. Since 1987 the McKusick catalog has been internationally available online from the National Library of Medicine (Online Mendelian Inheritance in Man, OMIM, see references). Regularly updated, OMIM is a major source of information on human genes and genetic diseases. Each entry has a unique 6-digit identifying number and is assigned to one of five catalogs according to genetic category: (1) autosomal dominant, (2) autosomal recessive, (3) X-chromosomal, (4) Y-chromosomal, and (5) mitochondrial. Autosomal entries initiated since 1994 begin with the digit 6. The McKusick catalog provides a systematic basis for the genetics of man comparable to the first periodic table of chemical elements by Dimitrij I. Mendelyev in 1869 or to the *Chronologisch-thematisches Verzeichnis sämtlicher Tonwerke Wolfgang Amade Mozarts* by Ludwig Alois Ferdinand Köchel in 1862.

A special feature of the McKusick catalog is a map of disease-related gene loci assigned to specific chromosomal sites, called The Morbid Anatomy of the Human Genome. This first appeared in 1971 (3rd edition) on a single page, but the complete information can now no longer be presented in a readable printed version. The map of disease loci presented here on the next five pages, therefore, represents selected entries. For complete information, the reader is referred to the network of data available through OMIM. However, the maps shown on the following pages do provide an overview. The McKusick catalog also reflects an important difference between customary clinical medicine and medical genetics. Whereas medicine classifies diseases according to their main manifestations, organ systems, age, gender, and other criteria related to the phenotype, medical genetics focuses on the genotype. The gene locus involved, the type of mutation, and genetic heterogeneity provide the basis for disease classification. This expands the concept of disease beyond the clinical manifestation and age of onset (see Childs, 1999).

References

Amberger JS, Hamosh A, McKusick VA: The morbid anatomy of the human genome, pp 47–111. In: Scriver CR et al (eds) The Metabolic and Molecular Bases of Inherited Disease, 8th ed. McGraw-Hill, New York, 2001.

Childs B: Genetic Medicine. Johns Hopkins Univ. Press, Baltimore, 1999.

McKusick VA: Mendelian Inheritance in Man. A Catalog of Human Genes and Genetic Disorders, 12th ed. Johns Hopkins University Press, Baltimore, 1998 (Online at OMIM www.ncbi.nlm.nih.gov/Omim with links to diagnostic laboratories at www.genetests.com).

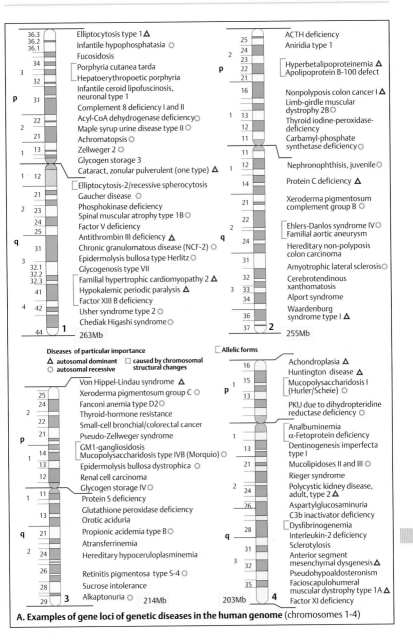

Chromosome 1 (263Mb)

Elliptocytosis type 1 △
Infantile hypophosphatasia ○
Fucosidosis
Porphyria cutanea tarda
Hepatoerythropoetic porphyria
Infantile ceroid lipofuscinosis, neuronal type 1
Complement 8 deficiency I and II
Acyl-CoA dehydrogenase deficiency ○
Maple syrup urine disease type II ○
Achromatopsis ○
Zellweger 2 ○
Glycogen storage 3
Cataract, zonular pulverulent (one type) △
Elliptocytosis-2/recessive spherocytosis
Gaucher disease ○
Phosphokinase deficiency
Spinal muscular atrophy type 1B ○
Factor V deficiency
Antithrombin III deficiency △
Chronic granulomatous disease (NCF-2) ○
Epidermolysis bullosa type Herlitz ○
Glycogenosis type VII
Familial hypertrophic cardiomyopathy 2 △
Hypokalemic periodic paralysis △
Factor XIII B deficiency
Usher syndrome type 2 ○
Chediak Higashi syndrome ○

Chromosome 2 (255Mb)

ACTH deficiency
Aniridia type 1
Hyperbetalipoproteinemia △
Apolipoprotein B-100 defect
Nonpolyposis colon cancer I △
Limb-girdle muscular dystrophy 2B ○
Thyroid iodine-peroxidase-deficiency
Carbamyl-phosphate synthetase deficiency ○
Nephronophthisis, juvenile ○
Protein C deficiency △
Xeroderma pigmentosum complement group B ○
Ehlers-Danlos syndrome IV ○
Familial aortic aneurysm
Hereditary non-polyposis colon carcinoma
Amyotrophic lateral sclerosis ○
Cerebrotendinous xanthomatosis
Alport syndrome
Waardenburg syndrome type I △

Diseases of particular importance
△ autosomal dominant   □ caused by chromosomal
○ autosomal recessive   structural changes

□ Allelic forms

Chromosome 3 (214Mb)

Von Hippel-Lindau syndrome △
Xeroderma pigmentosum group C ○
Fanconi anemia type D2 ○
Thyroid-hormone resistance
Small-cell bronchial/colorectal cancer
Pseudo-Zellweger syndrome
GM1-gangliosidosis
Mucopolysaccharidosis type IVB (Morquio) ○
Epidermolysis bullosa dystrophica ○
Renal cell carcinoma
Glycogen storage IV ○
Protein S deficiency
Glutathione peroxidase deficiency
Orotic aciduria
Propionic acidemia type B ○
Atransferrinemia
Hereditary hypoceruloplasminemia
Retinitis pigmentosa type S-4 ○
Sucrose intolerance
Alkaptonuria ○

Chromosome 4 (203Mb)

Achondroplasia △
Huntington disease △
Mucopolysaccharidosis I (Hurler/Scheie) ○
PKU due to dihydropteridine reductase deficiency ○
Analbuminemia
α-Fetoprotein deficiency
Dentinogenesis imperfecta type I
Mucolipidoses II and III ○
Rieger syndrome
Polycystic kidney disease, adult, type 2 △
Aspartylglucosaminuria
C3b inactivator deficiency
Dysfibrinogenemia
Interleukin-2 deficiency
Sclerotylosis
Anterior segment mesenchymal dysgenesis △
Pseudohypoaldosteronism
Facioscapulohumeral muscular dystrophy type 1A △
Factor XI deficiency

A. Examples of gene loci of genetic diseases in the human genome (chromosomes 1-4)

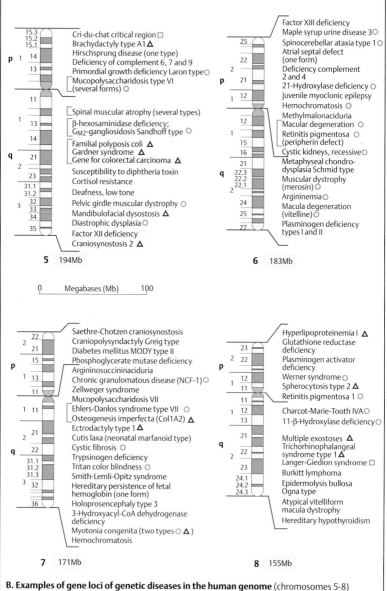

B. Examples of gene loci of genetic diseases in the human genome (chromosomes 5–8)

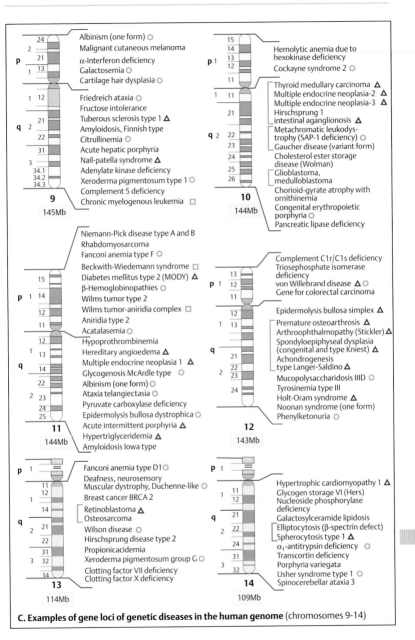

Chromosome 9 (145Mb)

- Albinism (one form) ○
- Malignant cutaneous melanoma
- α-Interferon deficiency
- Galactosemia ○
- Cartilage hair dysplasia ○
- Friedreich ataxia ○
- Fructose intolerance
- Tuberous sclerosis type 1 △
- Amyloidosis, Finnish type
- Citrullinemia ○
- Acute hepatic porphyria
- Nail-patella syndrome △
- Adenylate kinase deficiency
- Xeroderma pigmentosum type 1 ○
- Complement 5 deficiency
- Chronic myelogenous leukemia □

Chromosome 10 (144Mb)

- Hemolytic anemia due to hexokinase deficiency
- Cockayne syndrome 2 ○
- Thyroid medullary carcinoma △
- Multiple endocrine neoplasia-2 △
- Multiple endocrine neoplasia-3 △
- Hirschsprung 1 intestinal aganglionosis △
- Metachromatic leukodystrophy (SAP-1 deficiency) ○
- Gaucher disease (variant form)
- Cholesterol ester storage disease (Wolman) ○
- Glioblastoma, medulloblastoma
- Chorioid-gyrate atrophy with ornithinemia
- Congenital erythropoietic porphyria ○
- Pancreatic lipase deficiency

Chromosome 11 (144Mb)

- Niemann-Pick disease type A and B
- Rhabdomyosarcoma
- Fanconi anemia type F ○
- Beckwith-Wiedemann syndrome □
- Diabetes mellitus type 2 (MODY) △
- β-Hemoglobinopathies ○
- Wilms tumor type 2
- Wilms tumor-aniridia complex □
- Aniridia type 2
- Acatalasemia ○
- Hypoprothrombinemia
- Hereditary angioedema △
- Multiple endocrine neoplasia 1 △
- Glycogenosis McArdle type ○
- Albinism (one form) ○
- Ataxia telangiectasia ○
- Pyruvate carboxylase deficiency
- Epidermolysis bullosa dystrophica ○
- Acute intermittent porphyria △
- Hypertriglyceridemia △
- Amyloidosis Iowa type

Chromosome 12 (143Mb)

- Complement C1r/C1s deficiency
- Triosephosphate isomerase deficiency
- von Willebrand disease △ ○
- Gene for colorectal carcinoma
- Epidermolysis bullosa simplex △
- Premature osteoarthrosis △
- Arthroophthalmopathy (Stickler) △
- Spondyloepiphyseal dysplasia (congenital and type Kniest) △
- Achondrogenesis type Langer-Saldino △
- Mucopolysaccharidosis IIID ○
- Tyrosinemia type III
- Holt-Oram syndrome △
- Noonan syndrome (one form)
- Phenylketonuria ○

Chromosome 13 (114Mb)

- Fanconi anemia type D1 ○
- Deafness, neurosensory
- Muscular dystrophy, Duchenne-like ○
- Breast cancer BRCA 2
- Retinoblastoma △
- Osteosarcoma
- Wilson disease ○
- Hirschsprung disease type 2
- Propionicacidemia
- Xeroderma pigmentosum group G ○
- Clotting factor VII deficiency
- Clotting factor X deficiency

Chromosome 14 (109Mb)

- Hypertrophic cardiomyopathy 1 △
- Glycogen storage VI (Hers)
- Nucleoside phosphorylase deficiency
- Galactosylceramide lipidosis
- Elliptocytosis (β-spectrin defect)
- Spherocytosis type 1 △
- α₁-antitrypsin deficiency ○
- Transcortin deficiency
- Porphyria variegata
- Usher syndrome type 1 ○
- Spinocerebellar ataxia 3

C. Examples of gene loci of genetic diseases in the human genome (chromosomes 9–14)

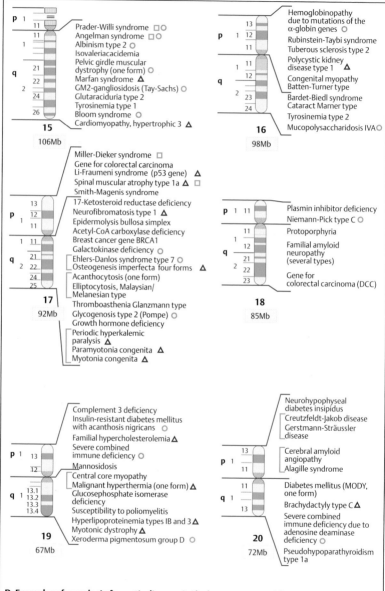

15
106Mb

Prader-Willi syndrome □○
Angelman syndrome □○
Albinism type 2 ○
Isovaleriacacidemia
Pelvic girdle muscular
dystrophy (one form) ○
Marfan syndrome △
GM2-gangliosidosis (Tay-Sachs) ○
Glutaraciduria type 2
Tyrosinemia type 1
Bloom syndrome ○
Cardiomyopathy, hypertrophic 3 △

16
98Mb

Hemoglobinopathy
due to mutations of the
α-globin genes ○
Rubinstein-Taybi syndrome
Tuberous sclerosis type 2
Polycystic kidney
disease type 1 △
Congenital myopathy
Batten-Turner type
Bardet-Biedl syndrome
Cataract Marner type
Tyrosinemia type 2
Mucopolysaccharidosis IVA○

17
92Mb

Miller-Dieker syndrome □
Gene for colorectal carcinoma
Li-Fraumeni syndrome (p53 gene) △
Spinal muscular atrophy type 1a △ □
Smith-Magenis syndrome
17-Ketosteroid reductase deficiency
Neurofibromatosis type 1 △
Epidermolysis bullosa simplex
Acetyl-CoA carboxylase deficiency
Breast cancer gene BRCA1
Galactokinase deficiency ○
Ehlers-Danlos syndrome type 7 ○
Osteogenesis imperfecta four forms △
Acanthocytosis (one form)
Elliptocytosis, Malaysian/
Melanesian type
Thromboasthenia Glanzmann type
Glycogenosis type 2 (Pompe) ○
Growth hormone deficiency
Periodic hyperkalemic
paralysis △
Paramyotonia congenita △
Myotonia congenita △

18
85Mb

Plasmin inhibitor deficiency
Niemann-Pick type C ○
Protoporphyria
Familial amyloid
neuropathy
(several types)
Gene for
colorectal carcinoma (DCC)

19
67Mb

Complement 3 deficiency
Insulin-resistant diabetes mellitus
with acanthosis nigricans ○
Familial hypercholesterolemia △
Severe combined
immune deficiency ○
Mannosidosis
Central core myopathy
Malignant hyperthermia (one form) △
Glucosephosphate isomerase
deficiency
Susceptibility to poliomyelitis
Hyperlipoproteinemia types IB and 3 △
Myotonic dystrophy △
Xeroderma pigmentosum group D ○

20
72Mb

Neurohypophyseal
diabetes insipidus
Creutzfeldt-Jakob disease
Gerstmann-Sträussler
disease
Cerebral amyloid
angiopathy
Alagille syndrome
Diabetes mellitus (MODY,
one form)
Brachydactyly type C △
Severe combined
immune deficiency due to
adenosine deaminase
deficiency ○
Pseudohypoparathyroidism
type 1a

D. Examples of gene loci of genetic diseases in the human genome (chromosomes 15-20)

21
50Mb

Cerebral arterial amyloidosis (Dutch type)
Alzheimer disease (an autosomal dominant form) △
Amyotrophic lateral sclerosis (one form) ○
Homocystinuria (vitamin B6-dependent and nondependent types) ○
Hemolytic anemia due to phospho-fructokinase deficiency
Progressive myoclonic epilepsy

22
56Mb

Cat eye syndrome □
DiGeorge syndrome □
Velocardiofacial syndrome
α-N-acetyl-galactosaminidase deficiency
Glutathionuria
BCR chromosomal region for chronic myelogenous leukemia □
Neuroepithelioma
Ewing sarcoma
Debrisoquin hypersensitivity
Susceptibility to Parkinsonism
Neurofibromatosis type 2 (acoustic neurinoma) △
Meningioma
Glucose/galactose malabsorption
Transcobalamin-II deficiency
Metachromatic leukodystrophy ○

Y
51Mb

XY gonadal dysgenesis (mutations in the *SRY* gene)
Spermatogenesis factor AZ Fa-c

2p:
22.3
22.2
22.1
21.3
21.2
21.1
11.4
11.3
11.23
11.22
11.21

Amelogenesis imperfecta
Steroid sulfatase deficiency (ichthyosis)
Kallmann syndrome
Chondrodysplasia punctata
Hypophosphatemia
Ocular albinism type 1 *
Retinoschisis
Adrenal cortical hypoplasia (glycerol kinase deficiency)
Chronic granulomatous disease
Retinitis pigmentosa-3 *
Duchenne muscular dystrophy *
Becker muscular dystrophy *
Ornithine transcarbamylase deficiency
Norrie syndrome
Retinitis pigmentosa-2 *
Incontinentia pigmenti
Wiskott-Aldrich syndrome
Menkes syndrome

Xq:
11
12
13
21.1
21.2
21.3
22.1
22.2
22.3
23
24
25
26
27
28

TFM androgen receptor defect *
Aarskog syndrome
Phosphoglucokinase deficiency
Hypohidrotic ectodermal dysplasia *
Agammaglobulinemia type Bruton
Spinal and bulbar muscular atrophy (type Kennedy)
Spinal muscular atrophy
Choroideremia *
Spastic paraplegia, X-chromosomal form
Impaired hearing due to stapes fixation
Pelizaeus-Merzbacher disease
Hereditary nephritis (Alport syndrome) *
Fabry disease
Lowe syndrome
Hyper-IgM immunodeficiency
Lymphoproliferative syndrome
Lesch-Nyhan syndrome
Hemophilia B *
Albinism-deafness syndrome
Fragile X syndrome *
Mucopolysaccharidosis type II (Hunter) *
Hemophilia A *
G6PD deficiency *
Nephrogenic diabetes insipidus
Adrenoleukodystrophy
Red-green blindness *
Dyskeratosis congenita
Adrenoleukodystrophy
Muscular dystrophy (Emery-Dreifuss)
Otopalatodigital syndrome type I
Rett syndrome

(X Chromosome 25% enlarged)

X
163Mb

* relatively frequent

E. Examples of gene loci of genetic diseases in the human genome (chromosomes 21, 22, X, Y)

(Alphabetical list to the maps on pp. 423–427;
"ch." = chromosome)

Aarskog syndrome (X ch.)
Acanthocytosis (one form) (ch. 17)
Acatalasemia (ch. 11)
Acetyl-CoA carboxylase deficiency (ch. 17)
Achromatopsia (ch. 1)
Achondroplasia (ch. 4)
ACTH deficiency (ch. 2)
Acute hepatic porphyria (ch. 9)
Acute intermittent porphyria (ch. 11)
Acyl-CoA dehydrogenase deficiency (ch. 1)
Adenylate kinase deficiency (ch. 9)
Adrenal cortical hypoplasia with glycerol
   kinase deficiency (X ch.)
Adrenoleukodystrophy (X ch.)
α-Interferon deficiency (ch. 9)
Agammaglobulinemia (X ch.)
Alagille syndrome (ch. 20)
Albinism (one form) (ch. 9)
Albinism (one form) (ch. 11)
Albinism type 2 (ch. 15)
Albinism – deafness syndrome (X ch.)
Alkaptonuria (ch. 3)
Alpha-N-acetylgalactosaminidase deficiency
   (ch. 22)
Alpha-1-antitrypsin deficiency (ch. 14)
Alpha-fetoprotein deficiency (ch. 4)
Alport syndrome (ch. 2)
Alzheimer disease (one form) (ch. 21)
Amelogenesis imperfecta (X ch.)
Amyloidosis, Finnish type (ch. 9)
Amyloidosis type Iowa (ch. 11)
Amyotrophic lateral sclerosis (one form)
   (ch. 21)
Amyotrophic lateral sclerosis, juvenile (ch. 2)
α-N-acetylgalactosaminidase deficiency
   (ch. 22)
Analbuminemia (ch. 4)
Androgen receptor defect (X ch.)
Angelman syndrome (ch. 15)
Aniridia type 1 (ch. 2)
Aniridia type 2 (ch. 11)
Anterior segmental mesenchymal dysgenesis
   (ch. 4)
Antithrombin III deficiency (ch. 1)
Apert syndrome (ch.4)
Apolipoprotein B-100 defect (ch. 2)
Argininemia (ch. 6)
Argininosuccinicaciduria (ch. 7)
Arthroophthalmopathy (Stickler syndrome)
   (ch. 12)

Aspartylglucosaminuria (ch. 4)
Ataxia-telangiectasia (ch. 11)
Atransferrinemia (ch. 3)
Atrial septal defect (one form) (ch. 6)
Atypical vitelliform macular dystrophy (ch. 8)
Bardet–Biedl syndrome (ch. 16)
BCR chromosomal region for chronic myelo-
   genous leukemia (ch. 22)
Becker muscular dystrophy (X ch.)
Beckwith–Wiedemann syndrome (ch. 11)
Beta-hemoglobinopathies (ch. 11)
Beta-hexosaminidase deficiency; GM^2–gan-
   gliosidosis type Sandhoff (ch. 5)
11-Beta-hydroxylase deficiency (ch. 8)
Bloom syndrome (ch. 15)
Brachydactyly type A1 (ch. 5), type C (ch. 20)
Breast cancer gene BRCA1 (ch. 17)
Breast cancer gene BRCA 2 (ch. 13)
Burkitt lymphoma (ch. 8)
Carbamylphosphate synthetase–I deficiency
   (ch. 4)
Cardiomyopathy, familial hypertrophic type 3
   (ch. 15)
Cartilage – hair dysplasia (ch. 9)
Cat eye syndrome (ch. 22)
C3b-inactivator deficiency (ch. 4)
Central core myopathy (ch. 19)
Cerebral amyloid angiopathy (ch. 20)
Cerebral arterial amyloidosis (Dutch type)
   (ch. 21)
Cerebrotendinosis xanthomatosis (ch. 2)
Charcot–Marie–Tooth neuropathy type1 b
   and type 2 (ch. 17) (Spinal muscular atrophy)
Charcot–Marie–Tooth neuropathy type IVa
   (ch. 8)
Chediak-Higashi syndrome (ch. 1)
Cholesteryl ester storage disease (Wolman)
   (ch. 10)
Chondrodysplasia punctata (X ch.)
Choroid gyrate atrophy with ornithinemia
   (ch. 10)
Choroideremia (X ch.)
Chronic granulomatous disease (NCF-1) (ch. 7)
Chronic granulomatous disease (NCF-2
   deficiency) (ch. 1)
Chronic granulomatous disease (X ch.)
Chronic myelogenous leukemia (ch. 9)
Citrullinemia (ch. 9)
Clotting factor VII deficiency (ch. 13)
Clotting factor X deficiency (ch. 13)
Cockayne syndrome 2 (ch. 10)
Colon cancer, familial nonpolyposis type 1
   (ch. 2)

Colorectal adenocarcinoma (ch. 12)
Colorectal carcinoma (ch. 5 and 18)
Colorectal carcinoma/Li–Fraumeni syndrome
  (ch. 17)
Complement 2 and 4 deficiency (ch. 6)
Complement 3 deficiency (ch. 19)
Complement 5 deficiency (ch. 9)
Complement 6, 7, and 9 deficiency (ch. 5)
Complement 8 deficiency 1 and 2 (ch. 1)
Complement C1 r/C1 s deficiency (ch. 12)
Congenital erythropoietic porphyria (ch. 10)
Congenital myopathy, Batten–Turner type
  (ch. 16)
Cortisol resistance (ch. 5)
Craniopolysyndactyly, Greig type (ch. 7)
Craniosynostosis type 2 (ch. 5)
Creutzfeldt–Jakob disease (ch. 20)
Cri du chat critical region (ch. 5)
Crigler–Najjar syndrome (ch. 1)
Crouzon craniofacial dysostosis (ch. 4)
Cutis laxia (neonatal marfanoid type) (ch. 7)
Cystic fibrosis (ch. 7)
Deafness, low-tone (ch. 5)
Deafness, neurosensory (ch. 13)
Debrisoquine hypersensitivity (ch. 22)
Dentinogenesis imperfecta type I (ch. 4)
Diabetes mellitus type MODY (ch. 11, ch 20)
Diabetes mellitus, MODY type II (ch. 7)
Diastrophic dysplasia (ch. 5)
DiGeorge syndrome (ch. 22)
Duchenne muscular dystrophy (X ch.)
Dysfibrinogenemia (ch. 4)
Dyskeratosis congenita (X ch.)
Ectrodactyly type 1 (ch. 7)
Ehlers–Danlos syndrome type 4 (ch. 2)
Ehlers–Danlos syndrome type 7 (ch. 7)
Ehlers–Danlos syndrome type 7 A1 (ch. 17)
Elliptocytosis (β-spectrin defect) (ch. 14)
Elliptocytosis, Malaysian/Melanesian type
  (ch. 17)
Elliptocytosis-2/recessive spherocytosis (ch. 1)
Epidermolysis bullosa dystrophica (ch. 3)
Epidermolysis bullosa simplex (ch. 12, ch. 17)
Epidermolysis bullosa type Herlitz (ch. 1)
Epidermolysis bullosa type Ogna (ch. 8)
Ewing sarcoma (ch. 22)
Fabry disease (X ch.)
Facioscapulohumeral muscular dystrophy
  (ch. 4)
Factor V deficiency (ch. 1)
Factor XI deficiency (ch. 4)
Factor XII deficiency (ch. 5)
Factor XIIIa deficiency (ch. 6)

Factor XIII B deficiency (ch. 1)
Familial amyloid neuropathy (several types)
  (ch. 18)
Familial aortic aneurysm (ch. 2)
Familial hypercholesterolemia (ch. 19)
Familial hypertrophic cardiomyopathy (ch. 1)
Familial polyposis coli (ch. 5)
Fanconi anemia type D2 (ch. 6), type F
  (ch. 15), D1 (ch. 17)
Fragile X syndrome (X ch.)
Friedreich's ataxia (ch. 9)
Fructose intolerance (ch. 9)
Fucosidosis (ch. 1)
G6PD deficiency (X ch.)
Galactokinase deficiency (ch. 17)
Galactose epimerase deficiency (ch. 1)
Galactosemia (ch. 9)
Galactosylceramide lipidosis (ch. 14)
Gardner syndrome (ch. 5)
Gaucher disease (ch. 1)
Gaucher disease (variant form) (ch. 10)
Gerstmann–Sträussler disease (ch. 20)
Glioblastoma, medulloblastoma (ch. 10)
Glucose/galactose malabsorption (ch. 22)
Glucosephosphate isomerase deficiency
  (ch. 19)
Glutaricaciduria type 2 (ch. 15)
Glutathione peroxidase deficiency (ch. 3)
Glutathione reductase deficiency (ch. 8)
Glutathionuria (ch. 22)
Glycogenosis type VII (ch. 1)
Glycogenosis, McArdle type (ch. 11)
Glycogenosis type 2 (Pompe) (ch. 17)
Glycogen storage type 3 (ch. 1)
Glycogen storage type 4 (ch. 3)
Glycogen storage VI (Hers) (ch. 14)
GM1-gangliosidosis (ch. 3)
GM2-gangliosidosis (Tay–Sachs) (ch. 15)
GM2-gangliosidosis, Sandhoff type (ch. 5)
Growth hormone deficiency (ch. 17)
Hemochromatosis (ch. 7)
Hemoglobinopathies due to mutations of
  the α-globin genes (ch. 16)
Hemolytic anemia due to hexokinase defi-
  ciency (ch. 10)
Hemolytic anemia due to phosphofructokinase
  deficiency (ch. 21)
Hemophilia A (X ch.)
Hemophilia B (X ch.)
Hepatoerythropoietic porphyria (ch. 1)
Hereditary angioedema (ch. 11)
Hereditary congenital hypothyroidism (ch. 8)
Hereditary hypoceruloplasminemia (ch. 3)

Hereditary nephritis (Alport syndrome) (X ch.)
Hereditary non-polyposis colon carcinoma (ch. 2)
Hereditary persistence of fetal hemoglobin (one form) (ch. 7)
Hirschsprung disease (chs. 10 and 13)
Holoprosencephaly type 3 (ch. 7)
Holt–Oram syndrome (ch. 12)
Homocystinuria (B6-responsive and B6-nonresponsive forms) (ch. 21)
Huntington disease (ch. 4)
3-Hydroxyacyl-CoA dehydrogenase deficiency (ch. 7)
21–Hydroxylase deficiency (ch. 6)
Hyperbetalipoproteinemia (ch. 2)
Hyper-IgM immune deficiency (X ch.)
Hyperlipoproteinemia type 1 (ch. 8)
Hyperlipoproteinemia type 1 b (ch. 19)
Hyperlipoproteinemia type 3 (ch. 19)
Hypertriglyceridemia (ch. 11)
Hypertrophic cardiomyopathy (ch. 14)
Hypochondroplasia (ch. 4)
Hypohidrotic ectodermal dysplasia (X ch.)
Hypophosphatemia (X ch.)
Hypoprothrombinemia (ch. 11)
Immune deficiency, severe combined (ch. 19)
Immunodeficiency due to ADA deficiency (ch. 20)
Impaired hearing (lower frequencies) (ch. 5)
Impaired hearing due to stapes fixation (X ch.)
Infantile ceroid lipofuscinosis, neuronal type (ch. 1)
Infantile hypophosphatasia (ch. 1)
Insulin-resistant diabetes mellitus with acanthosis nigricans (ch. 19)
Interleukin 2 deficiency (ch. 4)
Intestinal aganglionosis (Hirschsprung) (ch. 10 and 13)
Isovalericacidemia (ch. 15)
Juvenile myoclonic epilepsy (ch. 6)
Kallmann syndrome (X ch.)
17-Ketosteroid reductase deficiency (ch. 17)
Lamellar cataract (one type) (ch. 1)
Langer–Giedion syndrome (ch. 8)
Lesch–Nyhan syndrome (X ch.)
Li–Fraumeni syndrome (ch. 17)
Limb – girdle muscular dystrophy 2 b (ch. 2)
Lowe syndrome (X ch.)
Lymphoproliferative syndrome (X ch.)
Macular degeneration, vitelline (ch. 6)
Malignant cutaneous melanoma (ch 9)
Malignant hyperthermia (ch. 19, others)
Mandibulofacial dysostosis

(Franceschetti–Klein syndrome) (ch. 5)
Mannosidosis (ch. 19)
Maple syrup urine disease type 2 (ch. 1)
Maple syrup urine disease type 3 (ch. 6)
Marfan syndrome (ch. 15)
Meningioma (ch. 22)
Menkes syndrome (X ch.)
Metachromatic leukodystrophy (ch. 22)
Metachromatic leukodystrophy (SAP–1 deficiency) (ch. 10)
Metaphyseal chondrodysplasia type Schmid (ch. 6)
Methylmalonicaciduria (ch. 6)
Miller–Dieker syndrome (ch. 17)
Morquio syndrome B (ch. 3)
Mucolipidosis types II and III (ch. 4)
Mucopolysaccharidosis type I (Hurler/Scheie) (ch. 4)
Mucopolysaccharidosis type II (X ch.)
Mucopolysaccharidosis type IVa (ch. 16)
Mucopolysaccharidosis type IVb (ch. 3)
Mucopolysaccharidosis type VI (Maroteaux–Lamy) (ch. 5)
Mucopolysaccharidosis type VII (ch. 7)
Multiple endocrine neoplasia type 1 (ch. 11)
Multiple endocrine neoplasia type 2 (ch. 10)
Multiple endocrine neoplasia type 3 (ch. 10)
Multiple exostoses (ch. 8)
Muscular dystrophy, Becker type (X ch.)
Muscular dystrophy, Duchenne type (X ch.)
Muscular dystrophy, Emery–Dreifuss type (X ch.)
Muscular dystrophy, Duchenne-like (ch. 13)
Muscular dystrophy, merosin (ch. 6)
Myotonia congenita (ch. 17)
Myotonia congenita (two types) (ch. 7)
Myotonic dystrophy (ch. 19)
Myotubular myopathy (X ch.)
Nail–patella syndrome (ch. 9)
Nephrogenic diabetes insipidus (X ch.)
Nephronophthisis, juvenile (ch. 2)
Neuroepithelioma (ch. 22)
Neurofibromatosis type 1 (ch. 17)
Neurofibromatosis type 2 (acusticus neurinoma) (ch. 22)
Neurohypophyseal diabetes insipidus (ch. 20)
Niemann–Pick disease type A and B (ch. 11)
Niemann–Pick type C (ch. 18)
Noonan syndrome (one locus) (ch. 12)
Norrie syndrome (X ch.)
Nucleoside phosphorylase deficiency (ch. 14)
Ocular albinism (X ch.)
Ornithine transcarbamylase deficiency (X ch.)

Oroticacidemia (ch. 3)
Osteogenesis imperfecta (ch. 17)
Osteogenesis imperfecta (COL1 A2) (ch. 7)
Osteosarcoma (ch. 13)
Otopalatodigital syndrome type 1 (X ch.)
Pancreatic lipase deficiency (ch. 10)
Paramyotonia congenita (ch. 17)
Pelizaeus–Merzbacher disease (X ch.)
Pelvic girdle muscular dystrophy
    (ch. 5 and ch. 15)
Periodic hyperkalemic paralysis (ch. 17)
Phenylketonuria (PKU) (ch. 12)
Phosphoglucokinase deficiency (X ch.)
Phosphoglycerate mutase deficiency (ch. 7)
Phosphokinase deficiency (ch. 1)
PKU due to dihydropteridine reductase
    deficiency (ch. 4)
Plasmin inhibitor deficiency (ch. 18)
Plasminogen activator deficiency (ch. 8)
Plasminogen deficiency, types I and II (ch. 6)
Polycystic kidney disease (ch. 16 and 4)
Porphyria cutanea tarda (ch. 1)
Porphyria variegata (ch. 14)
Prader–Willi syndrome (ch. 15)
Primordial growth deficiency, Laron type
    (ch. 5)
Progressive myoclonic epilepsy (ch. 21)
Propionicacidemia type A (ch. 13)
Propionicacidemia type B (ch. 3)
Protein C deficiency (ch. 2)
Protein S deficiency (ch. 3)
Protoporphyria (ch. 18)
Pseudoaldosteronism (ch. 4)
Pseudohypoparathyroidism type 1 a (ch. 20)
Pseudo-Zellweger syndrome (ch. 3)
Red – green blindness (X ch.)
Renal cell carcinoma (ch. 3)
Retinitis pigmentosa (perpherin defect) (ch. 6)
Retinitis pigmentosa type 1 (ch. 8)
Retinitis pigmentosa type 2 (X ch.)
Retinitis pigmentosa type 3 (X ch.)
Retinitis pigmentosa type 4 (ch. 3)
Retinoblastoma (ch. 13)
Retinoschisis (X ch.)
Rett syndrome (X ch.)
Rhabdomyosarcoma (ch. 11)
Rieger syndrome (ch. 4)
Rubinstein–Taybi syndrome (ch. 16)
Saethre–Chotzen craniosynostosis (ch. 7)
Sclerotylosis (ch. 4)
Sex reversal (XY females due to mutation in
    the SRY gene) (Y ch.)

Small cell bronchial carcinoma/colorectal
    cancer (ch. 3)
Smith–Lemli–Opitz syndrome (ch. 7)
Smith–Magenis syndrome (ch. 17)
Spastic paraplegia (X chromosomal form)
    (X ch.)
Spermatogenesis factors AZFa - c (Y ch.)
Spherocytosis type 1 (ch. 14)
Spherocytosis type 2 (ch. 8)
Spinal muscular atrophy (X ch.)
Spinal muscular atrophy Ia (ch. 17)
Spinal muscular arophy type 1 B (ch. 1)
Spinal muscular atrophy IVa (ch. 8)
Spinal muscular atrophy Werdnig–Hoffmann
    and other types (ch. 5)
Spinocerebellar ataxia type 1 (ch. 6)
Spinocerebellar ataxia type 3 (ch. 14)
Spondyloepiphyseal dysplasia
    (congenital type) (ch. 12)
Spondyloepiphyseal dysplasia (Kniest type)
    (ch. 12)
Steroid sulfatase deficiency (ichthyosis) (X ch.)
Sucrose intolerance (ch. 3)
Susceptibility to diphtheria toxin (ch. 5)
Susceptibility to Parkinsonism (ch. 22)
Susceptibility to poliomyelitis (ch. 19)
T cell leukemia/lymphoma (ch. 14)
TFM androgen receptor defect (X ch.)
Thrombasthenia, Glanzmann type (ch. 17)
Thyroid hormone resistance (ch. 3)
Thyroid iodine peroxidase deficiency (ch. 2)
Thyroid medullary carcinoma (ch. 10)
Transcobalamin II deficiency (ch. 22)
Transcortin deficiency (ch. 14)
Trichorhinophalangeal syndrome type 1 (ch. 8)
Triosephosphate isomerase deficiency (ch. 12)
Tritan color blindness (ch. 7)
Trypsinogen deficiency (ch. 7)
Tuberous sclerosis type 1 (ch. 9)
Tuberous sclerosis type 2 (ch. 19)
Tyrosinemia type 1 (ch. 14), type 2 (ch. 16),
    type 3 (ch. 12)
Usher syndrome type 1 (ch. 14), type 2 (ch. 1)
Velocardiofacial syndrome (ch. 22)
Vitelline macular dystrophy (ch. 6)
Von Hippel–Lindau syndrome (ch. 3)
Von Willebrand disease (ch. 12)
Waardenburg syndrome type 1 (ch. 2)
Werner syndrome (ch. 8)
Williams–Beuren syndrome (ch. 4)
Wilms tumor–aniridia complex (ch. 11)
Wilms tumor type 2 (ch. 11)
Wilson's disease (ch. 13)

Xeroderma pigmentosum complementation
   group B (ch. 2)
Xeroderma pigmentosum group C (ch. 3)
Xeroderma pigmentosum group D (ch. 19)
Xeroderma pigmentosum group G (ch. 13)
Xeroderma pigmentosum type 1 (ch. 9)
XY gonadal dysgenesis (Y ch.)
Zellweger syndrome (ch. 7)
Zellweger syndrome type 2 (ch. 1)
Zonular cataract (ch. 1)

Caveat: There are numerous similar disorders
caused by mutations of genes at other loci,
sometimes with other modes of inheritance.
This list and the corresponding maps are not
complete, but only examples.
For a complete map and list see OMIM
(www.ncbi.nlm.nih.gov/Omim).

Appendix—Supplementary Data

The following tables provide additional information on selected plates and their accompanying texts for the following topics:

1. Replication of Viruses (p. 104)

Examples of representative viruses important in human diseases

Virus class	Virus type	Genome	Lipid envelope
Parvovididae	Parvovirus B19	ss DNA	No
Papovaviridae	Human papilloma	ds DNA	No
Adenoviridae	Human adenoviruses	ds DNA	No
Herpesviridae	Herpes simplex Varicella zoster Cytomegalovirus Epstein-Barr virus	ds DNA	Yes
Hepadnaviridae	Hepatitis B virus	de DNA, ss part	Yes
Poxviridae	Smallpox (variola)	ds DNA	Yes
Picornaviridae	Poliovirus Coxsackievirus Echovirus Hepatitis A virus	RNA (+)	No
Flaviviridae	Yellow fever virus Hepatitis C and G Dengue fever virus	RNA (+)	Yes
Togaviridae	Rubella virus	RNA (+)	Yes
Coronaviridae	Coronaviruses, SARS	RNA (+)	Yes
Rhabdoviridae	Rabies virus	RNA (−)	Yes
Filoviridae	Marburg, Ebola virus	RNA (−)	Yes

Continued ▶

Examples of representative viruses important in human diseases (continued)

Virus class	Virus type	Genome	Lipid envelope
Paramyxoviridae	Mumps, measles	RNA (-)	Yes
Orthomyxoviridae	Influenza A, B, C	RNA (-) 8 segments	Yes
Reoviridae	Rotavirus	ds RNA 10–12 segm.	No
Retroviridae	HIV-1, -2; HTLV-1, -2	RNA (+) 2 segments	Yes

ss – single strand; ds –double strand. (Data from Wang & Kieff, 2005)

2. Apoptosis (p. 126)

Examples of the main players in apoptosis[1]

Protein	Synonym	Effect on apoptosis	Gene locus
Fas	APT1, CD95, Apo-1, Fas1	+	10q24
FADD	MORT-1	+	11q13
Caspase 2	ICH1, NEDD2	+	7q35
Caspase 3	CPP32B; NEDD2	+	4q33
Caspase 4	TX, ICH-2, ICE-rel-II	+	11q22
Caspase 6	MCH2	+	4q25
Caspase 7	MCH3, ICE-LAP3, CMH3	+	10q25
Caspase 8	MACH, MCH5, FLICE	+	2q33
Caspase 9	APAF3, MCH6, ICE-LAP6	+	?
Caspase 10	MCH4	+	2q33
Apaf-1	CED4	+	?
Bcl-2		–	18q21
Bak1	Bcl-2L7	+	6p21
Bax		+	19q13
Bid		+/-	22q11
Bik	NBK	+ ?	

[1] Data from S. Nagata, 2005

3. Mitochondrial Diseases (p. 136)

Examples of diseases due to mutations or deletions in human mitochondrial DNA[1]

MIM	Disease name	Abbreviation
530000	Kearns-Sayre syndrome (ophthalmoplegia, pigmentary retinal degeneration, cardiomyopathy)	KSS
535000	Leber hereditary optic atrophy	LHON
540000	Mitochondrial myopathy, encephalopathy, lactic acidosis	MELAS
545000	Myoclonius epilepsy with ragged red fibers in muscle	MERRF
551500	Neuropathy, ataxia, retinitis pigmentosa	NARP
603041	Mitochondrial neurogastrointestinal encephalopathy[2]	MNGIE
557000	Pearson marrow-pancreas syndrome	PEAR
515000	Chloramphenicol-induced toxicity	
580000	Deafness, aminoglycoside-induced (mutation A1555 G)	
520000	Diabetes-deafness syndrome, maternally transmitted	

1) Data from OMIM (www.ncbi.nlm.nih.gov/Omim).
2) This may be autosomal recessive.

4. The Banding Pattern of Human Chromosomes (p. 190)

Examples of types of commonly used chromosome bands

Type	Goal of application	Principal method
G	Shows euchromatic light bands and heterochromatic dark bands	Pretreatment with trypsin
Q	Darkfield fluorescence bright fluorescence of heterochromatin	Quinacrine-induced, Hoechst 33258
R	Reverse of G-bands	Pretreatment with alkali at 80–90° C
C	Preferential stain of centromeres and constitutive heterochromatin	Pretreatment with acid and barium hydroxide
T	Preferential stain of telomeres	Telomere-specific DNA probes
DAPI	AT-specific fluorescence (4′,6-diamidino-2-phenylindole)	Enhanced chromosome fluorescence

5. Karyotypes of Man and Mouse (p. 192)

Human chromosome nomenclature

Abbreviation	Explanation in detail
46,XX	46 chromosomes including two X chromosomes (normal female karyotype)
46,XY	46 chromosomes including one X and one Y chromosome (normal male karyotype)
47,XXY	47 chromosomes including two Xs and one Y chromosome
47,XXX	47 chromosomes including three X chromosomes
47, XY,+21	47 chromosomes including an X and a Y chromosome and an additional chromosome 21 (trisomy 21)
13p	The *short* arm of a chromosome 13
13q	The *long* arm of a chromosome 13
13q14	Region 1, band 4 of a chromosome 13
13q14.2	Region 1, band 4.2 of a chromosome 13
2q−	Loss of chromosome material from the long arm of a chromosome 2
t(2;5)(q21;q31)	Reciprocal translocation between a chromosome 2 and a chromosome 5 with breakpoints in 2q21 and 5q31
t(13q14q)	Centric fusion between a chromosome 13 and a chromosome 14
der	A derivative chromosome as a result of a rearrangement
dic(Y)	Dicentric Y chromosome (two centromeres)
del(2)(q21-qter)	Deletion in chromosome 2 from q21 until the telomere
fra(X)(q27.3)	Fragile site at q27.3 of an X chromosome
dup(1)	Duplication in a chromosome 1
h	Heterochromatin, constitutive (at centromeres)
i	Isochromosome, e.g. i(Xq) for the long arm of an X chromosome
ins(5;2)(p14;q22q32)	Insertion of the region q22–q32 of a chromosome 2 into region p14 of a chromosome 5
inv (9)(p11q21)	Inversion of a chromosome 9 between p11 and q21 (indicating the breakpoints)
invdup(15)	Inverted duplication in a chromosome 15
mat	Maternal origin
pat	Paternal origin
r(13)	Ring-shaped chromosome 13 (due to a partial deletion)

(from ICN 2005)

6. Inhibitors of Transcription and Translation (p. 208)

Selected examples

Inhibitor	Effect
Prokaryotes	
Actinomycin	Intercalation between adjacent G-C base pairs
Chloramphinicol	Inhibition of peptidyltransferase in the 70S ribosome subunit
Daunomycin	Intercalation between adjacent G-C base pairs
Erthromycin	Arrest of 70S ribosome, binding to 50S ribosome subunit
Neomycin	Inhibition of tRNA, binding to 30S ribosomal subunit
Puromycin	Early termination of translation by imitating an aminoacyl-tRNA
Rifamycin	Premature peptide chain termination
Streptomycin	Same as erythromycin
Tetracycline	Inhibition of binding of tRNA to 30S ribosomal subunit
Eukaryotes	
α-Amanitin	Inhibition of RNA polymerase II
Aminoglycosides	Inhibition of all stages of translation
Chloramphincol	Inhibition of peptidyltransferase of mitochondrial ribosomes
Cycloheximide	Inhibition of peptidyltransferase
Diphtheria toxin	Inhibition of elongation factor 2 (eEF2)

(Data from Singer M & Berg P. Genes and Genomes. Blackwell Scientific, Oxford University Press, 1991)

7. Genetic Diseases and FGF Receptors (p. 270)

Examples of genetic disorders due to mutations in FGF receptors[1]

Gene	Location	Disease	Main manifestation	MIM[2]
FGFR1	8p11.2	Pfeiffer syndrome	Craniosynostenosis, broad thumbs	101600
FGFR2	10q25.3	Apert syndrome	Craniosynostosis, fused digits	101200
		Crouzon syndrome	Craniosynostosis, ocular proptosis	123500
		Pfeiffer syndrome	Craniosynostenosis, broad thumbs	101600
FGFR3	4p16.3	Achondroplasia	Short stature, bone dysplasias	100800
		Hypochondroplasia	mild form of achondroplasia	146000
		Muenke syndrome	Asymmetric coronal stenosis	602849

1) Many more exist; some with unique mutations
2) Mendelian Inheritance in Man, online at: www.ncbi.nlm.nih.gov/Omim/

8. Diseases and G Proteins (p. 272)

Diseases caused by mutations in G protein-coupled receptor or G proteins[*]

Disease	OMIM	Mutated protein	Type of inheritance
McCune-Albright syndrome	174800	$G\alpha_s$ (gain-of-function)	somatic
Hypoparathyroidism	145980	Ca^{++}-sensing receptor	AD
Diabetes insipidus	304800	$G\alpha_s$ (loss-of-function)	X-chromosomal
Congenital hypothyroidism	275200	TSH receptor	AR
	188545	Thyrotropin releasing hormone	AD

[*] Examples only. See Online Mendelian Inheritance in Man (www.ncbi.nlm.nih.gov/Omim)
   AD: autosomal dominant; AR: autosomal recessive

9. Diseases Related to the Hedgehog Signaling Network (p. 276)

Hedgehog and related proteins

Name	Abbreviation	Gene locus	MIM
Sonic Hedgehog	SHH	7q36	600725
Indian Hedgehog	IHH	2q33-q35	600726
Desert Hedgehog	DHH	12q13.1	605423
Hedgehog Acetyltransferase	HHAT	1q32	605743
Hedgehog Interacting Protein	HHIP	4q28-q32	606178
Smoothened	SMOH	7q31-q33	601500
Patched	PTCH	9q22.3	601309

(Data from OMIM at www.ncbi.nlm.nih.gov/Omim)

Mutations and deletions in more than 10 human genes in the hedgehog gene network result in a group of malformation syndromes, holoprosencephaly (MIM 236100, 142945, and others, see table below) and accompanying brain malformations (reviewed by Muenke & Beachy, 2001, and Cohen, 2003). Mutations in PTCH (9q22) cause basal cell nevus syndrome, Gorlin-Goltz type (MIM 109400). Mutations in SMO (7q31) are found in some basal cell carcinomas and medulloblastomas. The segment polarity gene *Ci* is an ortholog of the *Gli* gene family in vertebrates, consisting of three distinct Gli proteins, Gl1, Gl2, and Gl3. Human *GLI* is expressed in nearly all isolated basal cell carcinoma (MIM 139150). *GLI3* mutations cause cephalopolydactyly, Greig type (MIM 175700, 145400) and Pallister-Hall syndrome (hypothalamic hamartoblastoma, MIM 146510). Loss-of-function mutations of Ptch and Smo in *Drosophila* have very similar phenotypes.

Gene loci involved in the causes of holoprosencephaly

Type of holoprosencephaly	Gene locus	MIM
HPE1	21q22.3, 2q37.1-q37.3	236100
HPE2	2p21	157170
HPE3	7q36	142945
HPE4	18p11.3	142946
HPE5	13q32	609637
HPE6	2q37.1-q37.3	605934
HPE7	9q22.3	601309
HPE8	14q13	609408
Pseudotrisomy syndrome	?13q22	264480

10. Inherited Cardiac Arrhythmias (p. 282)

Inherited cardiac arrhythmias (long QT syndromes)[1]

Type	Gene	Locus	Common age of onset	OMIM[2]
LQT 1	KCNQ1	11p15.5	Childhood (90% by age of 20)	192500
LQT 2	KCNH2	7q35-q36	Young adults (gene formerly HERG)	152427
LQT 3	SCN5A	3p21–24	Young adults	603830
LQT 4	ANK2 [3]	4q25	Adults	600919
LQT 5	KCNE1	21q22	Children	176261
LQT 6	KCNE2	21q22	Adults	603796
LQT 7	KCNJ2	17q23	Adults (Andersen syndrome)	600681
LQT 8	CACNA1C	12p13.3	Syndactyly, immune defect	601005

Syndromic forms associated with deafness: Romano-Ward syndrome (OMIM 192500) and Jervell and Lange-Nielsen syndrome (OMIM 220400)

[1] Other forms of inherited arrhythmias exist. [2] Online Mendelian Inheriatnce in Man at (www.ncbi.nl,.nih.gov/Omim). [3] Ankyrin-2 (Ank-B in mice).

11. Genes Involved in Inherited Deafness (p. 292)

Examples of genes and protein involved in hereditary deafness

Type of protein	Main function	Gene	DFN	MIM	Mouse mutant
Cytoskeletal proteins					
Myosin 6	Motor protein	MYO6	DFNB37/A22	600970	Snell's waltzer
Myosin 7A	Motor protein	MYO7A	DFNB2/A11	276903	Shaker-1
Myosin 15	Motor protein	MYO15	DFNB3	600316	Shaker-2
Ion transporters					
Connexin 26	Gap junction	GJB2/CX26	DFNB1/A3	220290	
Connexin 30	Gap junction	BJB6/CX30	DFNB1/A3	604418	
KCNQ4	K⁺ channel	KCNQ4	DFNA2	600101	
Pendrin	Iodide-chloride	SLC26A4	DFNB4	605646	
Structural proteins					
α-Tectorin	Tectorial membrane	TECTA	DFNB21/A8/A12	602574	
Collagen XI	Extracellular matrix	COL11 A2	DFNB53	609706	
Cochlin	Extracellular matrix	COCH	DFNA9	603196	
POU3F4		POU3F4	DFN3	300039	
Mitochondrial					
12S RNA			DFNA5	600994	
Unknown					
Diaphanous	Actin polymerization in hair cells = DIAPH1		DFNA1	602121	

(Data according to Petit et al, 2001; and Petersen & Willems, 2006)

12. Genetic Immune Deficiency Diseases (p. 322)

Hereditary immune deficiency diseases (examples)

Disease	MIM	Gene locus	Gene	Inheritance
Innate immunity				
Disorders of the complement system	106100	11q11-q13.1	C1NH	AD
	217000	6p21.3	C2	AR
	120700	19p13.3-p13.2	C3	AD
Chronic granulomatous disease	306400	Xp21.1	CYBB	X
	300481			

Continued ▶

Hereditary immune deficiency diseases (examples) (continued)

Disease	MIM	Gene locus	Gene	Inheritance
Adaptive immunity				
Agammglobulinemia, Bruton type	300300	Xq21.3-q22	*BTK*	X
Severe combined immunodeficiency disease (T- and B-cells)	308300 300400	Xq13.1	*IL2RG*	X
Adenosine deaminase deficiency	608950	20q13.1	*ADA*	AR
μ-chain deficiency	147020	14q32	Ig H-chain	AR
DiGeorge syndrome	188400	del22q11	several	sporadic
Wiskott-Aldrich syndrome	301000	Xp11.23	*WAS*	X
Ataxia-telangiectasia	208900	11q23	*ATM*	AR

AD: autosomal dominant; AR: autosomal recessive; X: X-chromosomal

13. Oncogenic Chromosome Translocations (p. 336)

Examples of oncogenic chromosome translocations

Translocation	Type of tumor	Genes involved
(9;22)(q34;q11)	Chronic myelogenous leukemia	*ABL/BCR*
(14;18)(q32;q21)	Follicular lymphoma	*BCL2, IgH*
(14;19)(q32;q13)	B-cell lymphocytic leukemia	*BCL3, IgH*
(8;14)(q24;q32)	Burkitt lymphoma, B-cell ALL	*MYC , IgH*
(11;14)(q13;q32)	Mantle cell lymphoma	*BCL1, IgH*
(1;7)(p34;q35)	T-cell acute lymphocytic leukemia	*LCK, TCRB*
(4;11)(q21;q23)	Acute lymphocytic leukemia	*MLL, ALL1, HRX*
(3;21)(q26;q22)	Acute myeloid leukemia	*AML1, EAP, EV11*
(1;14)(p22;q32)	Mucosa-associated lymphoma (MALT)	*BCL10*
(21;22)(q22;q12)	Ewing sarcoma	*EWS, ERG*
(11;22)(q24;q12)	Ewing sarcoma	*EWS, FL11*

(Data from P. J. Morin: Cancer Genetics, p. 519, in Harrison's Principles of Internal Medicine, 16th edition, 2005)

14. Genes Involved in Fanconi Anemia (p. 340)

Gene	Locus	MIM
FA-A	16q24.3	227650
FA-B	Xp22.31	300514
FA-C	9q22.3	227445
FA-D1	13q12.3	605724
FA-D2	3p25.3	227646
FA-E	6p22-p21	600901
FA-F	11p15	603467
FA-G	9p13	602956
FA-I		609053
FA-J	17q22	609054
FA-L	2p16.1	608111
FA-M	14.q21.3	609644
FA-ZF	19q13.1	605859

(Data from OMIM at www.ncbi.nlm.nih.gov/Omim/)

15. Mucopolysaccharide Storage Diseases (p. 360)

Classification of mucopolysaccharidoses

Type	MIM	Enzyme defect	Gene locus	Main manifestations
IH Hurler	252800	α-L-Iduronidase	4p16.3	Dysostosis multiplex, developmental delay, corneal clouding
IS Scheie	252800	α-L-Iduronidase	4p16.3	Stiff joints, normal development
II Hunter	309900	Iduronate sulfatase	Xq28	Similar to MPS IH, no corneal clouding
MPS III (Sanfilippo)				
IIIA	252900	Heparan-N-sulfatase	17q25.3	Progressive mental retardation
IIIB	252920	α-N-Acetyl-glucosaminidase		Same
IIIC	252930	Acetyl-CoA: α-glucosaminidase		Same

Continued ▶

Classification of Mucopolysaccharidoses (continued)

Type	MIM	Enzyme defect	Gene locus	Main manifestations
IIID	252940	*N*-Acetylglu-cosamine-6-sul-fatase		Same
IV Morquio				
IVA	253000	*N*-actetylgalac-tosamine-6-sul-fatase	16q24.3	Skeletal abnormalities,
IVB	253010	β-galactosidase	3p21.33	Short stature
VI (Maroteaux-Lamy)	253200	*N*-Acetlygalac-tosamine-4-sulfate	5q13-q14	Dysostosis multiplex, normal mental development
VII (Sly)	253220	β-glucuronidase	7q21.11	Dysostosis multiplex, corneal clouding
IX	601492	Hyaluronidase	3p21.2-p21.3	Soft tissue masses, short stature

(Data adapted from Neufeld & Muenzer, 2001)

16. Distal Cholesterol Biosynthesis Pathway (p. 366)

Diseases of the distal (post-squalene) cholesterol biosynthesis pathway

Disease	OMIM	Locus	Gene	Main manifestations
Antley-Bixler syndrome (some)	207410, 601637	7q21.2	*CYP51*	Skeletal dysplasias, choanal atresia, radioulnar synostosis
Greenberg dysplasia	215140 600024	1q42.1	*LBR*	Hydrops fetalis, ectopic calcifi-cations, prenatal lethal
CHILD syndrome	308050 300275	Xq28	*NSDHL*	Hemidysplasia, ichthyosiform erythrodermia, limb defects
Chondrodysplasia punctata type 2	302960	Xp11.2	*EBP*	Skeletal dysplasias, calcifica-tions over joints, short ex-tremities
Lathosterolosis	607330	11q13	*SC5DL*	Facial dysmorphism, simlar to SLOS, mental retardation
Smith—Lemli—Opitz syndrome (SLOS)	270400 602858	11q13	*DHCR7*	Skeletal anomalies, facial dys-morphia, malformations
Desmosterolosis	603398	1p31 – 33	*DHCR24*	Facial dysmorphism, short limbs, perinatal lethal

For enzymes involved, see opposite plate.
Abbreviations: CYP – cytochrome p51; LBR – laminin receptor B; NSDHL – NADH steroid dehydrogenese-like; EBP – emopamil binding protein; SC5DL – sterol C5-desaturaselike; DHCR7 – 7-dehydrocholesterol; DHCR24 – desmosterol reductase. (For details, see OMIM entry or GeneReviews at www.geneclinics.org)

17. Collagen Molecules and Diseases (p. 392)

Important types of human collagen and diseases causes by mutations in their genes

Type	Molecular structure	Gene	Gene locus	Disease	MIM
I	$[\alpha1(I)_2\alpha2(II)]$	COL1A1	17q21-22	Osteogenesis imperfecta	120150
		COL1A2	7q22	Ehlers-Danlos syndr.	130000
II	$[\alpha1(II)_3]$	COL2A1	12q13.1	Stickler syndrome	108300
				Spondylo-epiphys. dyspl.,	183900
				achondrogenesis, others	200600
III	$[\alpha1(III)_3]$	COL3A1	2q31	Ehlers-Danlos syndr. IV	225350
IV	$[\alpha1(IV)\alpha2(IV)]$ and others	COL4A1,A2	13q34		
		A3,A4	2q36	Alport syndr. autsomal	203780
		A5,A6	Xq22		
				Alport syndr. X-chromosomal	301050
V	$[\alpha1(V)_2\alpha2(V)]$	COL5A1	9q34.2	Ehlers-Danlos syndr. I + II	130000
		COL5A2	2q31		

(Data from Byers, 2001; and McKusick, Mendelian Inheritance in Man, MIM, Online at OMIM: www.ncbi.nlm.nih.gov/Omim/)

18. Disorders of Sexual Development (p. 402)

Overview of disorders of sex differentiation

1. Defects of sex determination due to mutation or structural aberration of SRY (XX males, XY gonadal dysgenesis, and others)
2. Disorders of testis development (SF1, DAX1, WNT4, SOX9, and other genes)
3. Defects of androgen biosynthesis (e.g., 21-hydroxylase deficiency)
4. Defect in steroid 5a-reductase (dihydrotestosterone deficiency)
5. Defects of androgen receptor (testicular feminization)
6. Defects of the Müllerian inhibition factor (hernia uteri syndrome)
7. XO/XY gonadal dysgenesis
8. Turner syndrome (45,X), Klinefelter syndrome (47,XXY)
9. True hermaphroditism XX/XY

19. Imprinting Diseases (p. 410)

Relative frequency of genetic lesions in Prader–Willi and Angelman syndromes

	Deletion 15q11-q13	UPD	Imprinting defect	UBE3A gene	Unknown defect
PWS	70%	29%	1%	Does not apply	
AS	70%	1–3%	2–4%	10–15%	10–15%

20. Microdeletion Syndromes[1] (p. 414)

Chromosomal region	Name	Comments	MIM
4p16.3	4p- or Wolf-Hirsch-horn syndrome	Not always visible by G band analysis	194190
5p15.2–15.3	5p- or cri-du-chat syndrome	Familial occurrence by translocation in 12–15%	123450
5q35	Sotos syndrome*	Overgrowth, retardation, seizures; *NSD1* gene deletionens; frequent in Japan	117550
7q11.23	Williams–Beuren syndrome	Elastin gene und other gene involved; deletion shown in 70%	194050
11p13	Wilms tumor - aniridia - genitourinary anomalies (WAGR)	*WT1* and *PAX6* genes involved	194072
15q11–13	Prader-Willi syndrome	Paternal chromosome 15 involved, deletion in70%, 25% UPD, 1–2% imprinting mutations	176270
15q11–13	Angelman syndrome*	Maternal chromosome 15 involved, deletion in 70%, 2–4% imprinting-mutations, 1% UPD, 25% other changes incl. *UBE3A* gen mutations	105830
16p13.3	Rubinstein-Taybi-syndrome*	Gene für *CREB* binding protein involved	180849
17p13.3	Miller-Dieker syndrome	Lissencephaly, *L1S1* gene deletion in about 90%	247200
17p11.2	Smith-Magenis syndrome	Complex malformation syndrome	182290
20p12.1	Alagille syndrome*	Arteriohepatic dysplasia and other systemic manifestationen, *JAG1* gene mutations	118450
22q11	DiGeorge/Shprintzen syndrome*	Immune defects, neonatal hypocalcemia, congenital heart defects, wide clinical spectrum, *TBX1* gene deletion in 70–90%,	192430

[1] Clinical spectrum and size of deletions variable, often requiring molecular cytogenetic analysis.

21. Chromosomal Location of Disease Genes (p. 422)

Chromosomal location of gene loci of genetic diseases

	Autosomal	X-chrom.	Y-chrom.	mtDNA	Total
Gene with known sequence	10212	472	48	37	10769
Gene with known sequence and phenotype	349	31	0	0	380
Phenotype and molecular basis known	1709	153	2	26	1891
Mendelian phenotype/locus, molecular basis unknown	1384	134	4	0	1522
Other phenotypes with suspected Mendelian basis	2066	145	2	0	2212
Total	15720	935	56	63	16774

(Data from OMIM, 13 May, 2006)

Number of gene loci on each human chromosome

1: 922 / 2: 601 / 3: 525 / 4: 364 / 5: 464 / 6: 598 / 7: 441 / 8: 336 / 9: 351
10: 333 / 11: 605 / 12: 514 / 13: 183 / 14: 293 / 15: 283 / 16: 369 / 17: 559
18: 141 / 19: 635 / 20: 231 / 21: 129 / 22: 237 / X: 563 / Y:44
Total number of loci: 972

(Data from OMIM, 13 May, 2006)

Definitions of Genetic Terms

Acentric—refers to a chromosome or chromatid without a centromere.

Acrocentric (White, 1945)—refers to a chromosome with a centromere that lies very close to one of the ends, dividing the chromosome into a long and a very short arm.

Actin—a structural protein interacting wih many other proteins. In muscle cells as F actin it interacts with myosin during contraction.

Activator—a protein that as transcription factor stimulates gene expression.

Active site—the region of a protein that is mainly responsible for its functional activity.

Alkyl group—covalently joined carbon and hydrogen atoms as in a methyl or ethyl group.

Allele (Johannsen, 1909) or **allelomorph** (Bateson and Saunders, 1902)—one of several alternative forms of a gene at a given gene locus.

Allelic exclusion—expression from only one allele.

Alternative splicing—production of different mRNAs from one transcript.

Alu sequences—a family of related DNA sequences, each about 300 base pairs long and containing the recognition site for the Alu restriction enzyme; about 1.2 million copies of Alu sequences are dispersed throughout the human genome.

Amber codon—the stop codon UAG (a word play on the discoverer Bernstein – amber)

Ames test—a mutagenicity test carried out with a mixture of rat liver and mutant bacteria.

Amino acid—an organic compound with an amino ($-NH_2$) and a carboxyl (-COOH) group.

Aminoacyl tRNA—a transfer RNA carrying an amino acid.

Amplification—production of additional copies of DNA sequences.

Anaphase (Strasburger, 1884)—a stage of mitosis and of meiosis I and II. Characterized by the movement of homologous chromosomes (or sister chromatids) toward opposite poles of the cell division spindle.

Aneuploidy (Täckholm, 1922)—deviation from the normal number of chromosomes by gain or loss (see trisomy and monosomy).

Aneusomy—deviation from the normal presence of homologous chromosomal segments. Aneusomy by recombination refers to the duplication/deficiency resulting from crossing-over within an inversion (inverted region).

Annealing—hybridizing complementary single strands of nucleic acid to form double-stranded molecules (DNA with DNA, RNA with RNA, or DNA with RNA).

Antibody—a protein (immunoglobin) that recognizes and binds to an antigen as part of the immune response.

Anticodon—a trinucleotide sequence in tRNA that is complementary to a codon for a specific amino acid in mRNA.

Antigen—a substance with a molecular surface structure that triggers an immune response, i.e., the production of antibodies, and/or that reacts with (its) specific antibodies (antigen-antibody reaction).

Antisense RNA—an RNA strand that is complementary to mRNA, which it can prevent from being used as a template for normal translation. The term antisense is generally used to refer to a sequence of DNA or RNA that is complementary to mRNA.

Apoptosis—programmed cell death, characterized by a series of regulated cellular events resulting in cell death, to eliminate a damaged cell or a normal cell no longer needed during development.

Archea – Archaebacteria, one of the three evolutionary lineages of organisms living today, a domain distinct from prokaryotes and eukaryotes, able to live in harsh environmental conditions.

Attenuator—terminator sequence regulating the termination of transcription, involved in controlling the expression of some operons in bacteria.

Australopithecus—the genus of fossil Hominidae from Eurasia. Walked erect; brain size between that of modern man and other modern primates; large and massive jaw. Lived about 4–5 million years ago.

Autonomously replicating sequence (ARS)—a DNA sequence that is required to induce replication.

Autoradiography (Lacassagne and Lattes, 1924)—photographic detection of a radioactive substance incorporated into cells or tissue. The distribution of the radioactively labeled substance can be demonstrated, e.g., in tissue, cells, or metaphase chromosomes by placing a photographic film or photographic emulsion in close contact with the preparation.

Autosome (Montgomery, 1906)—any chromosome except a sex chromosome (the latter usually designated X or Y). Autosomal refers to genes and chromosomal segments that are located on autosomes.

Auxotrophic (Ryan and Lederberg, 1946)—refers to cells or cell lines that cannot grow on minimal medium unless a certain nutritive substance is added (cf. prototrophic).

Backcross—cross of a heterozygous animal with one of its homozygous parents. In a double backcross, two heterozygous gene loci are involved.

BAC—bacterial artificial chromosome; a synthetic DNA molecule that contains bacterial DNA sequences for replication and segregation (see YAC).

Bcl-2 family—a family of proteins localized to mitochondria involved in regulation of apoptosis.

Bacteriophage—a virus that infects bacteria. Usually abbreviated as phage.

Banding pattern (Painter, 1939)—staining pattern of a chromosome consisting of alternating light and dark transverse bands. Each chromosomal segment of homologous chromosomes shows the same specific banding pattern, characterized by the distribution and size of the bands, which can be used to identify that segment. The term was introduced by Painter in 1939 for the linear pattern of strongly and weakly staining bands in polytene chromosomes of certain diptera (mosquitoes, flies). Each band is defined relative to its neighboring bands. The sections between bands are interbands.

Barr body—see X chromatin.

Base pair (bp)—in DNA, two bases—one a purine, the other a pyrimidine—lying opposite each other and joined by hydrogen bonds. Normal base pairs are A with T and C with G in a DNA double helix. Other pairs can form in ribosomal RNA.

B cells—B lymphocytes.

Bimodal distribution—refers to a frequency distribution curve with two peaks. If the frequency distribution curve of a population trait is bimodal, it is frequently evidence of two different phenotypes distinguished on a quantitative basis.

Bivalent (Haecker, 1892)—pairing configuration of two homologous chromosomes during the first meiotic division. As a rule the number of bivalents corresponds to half the normal number of chromosomes in diploid somatic cells. Bivalents are the cytogenetic prerequisite for crossing-over of nonsister chromatids. During meiosis, a trisomic cell forms a trivalent of the trisomic chromosomes.

Breakage-fusion-bridge cycle refers to a broken chromatid that fuses to its sister, forming a bridge.

Breakpoint—site of a break in a chromosomal alteration, e.g., translocation, inversion, or deletion.

Cadherins—dimeric cell adhesion molecules.

Carcinogen—a chemical substance that can induce cancer.

Caspase—a member of the family of specialized cysteine-containing aspartate proteases involved in apoptosis (programmed cell death).

CAT (or CAAT) box—a regulatory DNA sequence in the 5′ region of eukaryotic genes; transcription factors bind to this sequence.

Catenate—a link between molecules.

C-bands—specific staining of the centromeres of metaphase chromosomes.

CD region—common docking, a region involved in binding to a target protein.

cDNA—complementary DNA synthesized by the enzyme reverse transcriptase from RNA as the template.

Cell cycle (Howard & Pelc, 1953)—life cycle of an individual cell. In dividing cells, the following four phases can be distinguished: G1 (interphase), S (DNA synthesis), G2, and mitosis (M). Cells that do not divide are said to be in the G_0 phase.

Cell hybrid—a somatic cell generated by fusion of two cells in a cell culture. It contains the complete or incomplete chromosome complements of the parental cells. Cell hybrids are an important tool in gene mapping.

Centimorgan—a unit of length on a linkage map (100 centimorgans, cM = 1 Morgan). The distance between two gene loci in centimorgans corresponds to their recombination frequency expressed as percentage, i.e., one cM corresponds to one percent recombination frequency. Named after Thomas H. Morgan (1866–1945), who initiated the classic genetic experiments on *Drosophila* in 1910.

Centriole—small cylinder of microtubules.

Centromere (Waldeyer, 1903)—chromosomal region to which the spindle fibers attach during mitosis or meiosis. It appears as a constriction at metaphase. It contains chromosome-specific repetitive DNA sequences.

Chaperone—a protein needed to assemble or fold another protein correctly.

Chiasma (Janssens, 1909)—cytologically recognizable region of crossing-over in a bivalent. In some organisms the chiasmata move toward the end of the chromosomes (terminalization of the chiasmata) during late diplotene and diakinesis (see meiosis). The average number of chiasmata in autosomal bivalents is about 52 in human males, about 25–30 in females. The number of chiasmata in man was first determined in 1956 in a paper that confirmed the normal number of chromosomes in man (C. E. Ford and J.L. Hamerton, Nature 178: 1020, 1956).

Chimera (Winkler, 1907)—an individual or tissue that consists of cells of different genotypes of prezygotic origin.

Chylomicron—Lipoproteins secreted by intestinal epithelial cells; transport triglycerides and cholesterol frm the intestines to other tissues.

Chromatid (McClung, 1900)—longitudinal subunit of a chromosome resulting from chromosome replication; two chromatids are held together by the centromere and are visible during early prophase and metaphase of mitosis and between diplotene and the second metaphase of meiosis. Sister chromatids arise from the same chromosome; nonsister chromatids are the chromatids of homologous chromosomes.

After division of the centromere in anaphase, the sister chromatids are referred to as daughter chromosomes. A *chromatid break* or a chromosomal aberration of the chromatid type affects only one of the two sister chromatids. It arises after the DNA replication cycle in the S phase (see cell cycle). A break that occurs before the S phase affects both chromatids and is called an isolocus aberration (*isochromatid break*).

Chromatin (Flemming, 1882)—the stained material that can be observed in interphase nuclei. It is composed of DNA, basic chromosomal proteins (histones), nonhistone chromosomal proteins, and small amounts of RNA.

Chromatin remodeling—the energy-dependent displacement or reorganization of nucleosomes for transcription or replication.

Chromomere (Wilson, 1896)—each of the linearly arranged thickenings of the chromosome visible in meiotic and under some conditions also in mitotic prophase. Chromomeres are arranged in chromosome-specific patterns.

Chromosome (Waldeyer, 1888)—the gene-carrying structures, which are composed of chromatin and are visible during nuclear division as threadlike or rodlike bodies. *Polytene chromosomes* (Koltzhoff, 1934; Bauer, 1935) are a special form of chromosomes in the salivary glands of some diptera larvae (mosquitoes, flies).

Chromosome walking—sequential isolation of overlapping DNA sequences in order to find a gene on the chromosome studied.

Cis-acting—refers to a regulatory DNA sequence located on the same chromosome (cis), as opposed to trans-acting on a homologous chromosome.

Cis/trans (Haldane, 1941)—in analogy to chemical isomerism, refers to the position of genes of double heterozygotes (heterozygotes at two neighboring gene loci) on homologous chromosomes. When two certain alleles, e.g., mutants, lie next to each other on the same chromosome, they are in the *cis* position. If they lie opposite each other on different homologous chromosomes, they are in the *trans* position. The *cis/ trans* test (Lewis, 1951; Benzer, 1955) uses genetic methods (*genetic complementation*) to determine whether two mutant genes are in the *cis* or in the *trans* position. With reference to genetic linkage, the expressions *cis* and *trans* are analogous to the terms coupling and repulsion (q.v.).

Cistron (Benzer 1955)—a functional unit of gene effect as represented by the *cis/trans* test. If the phenotype is mutant with alleles in the *cis* position and the alleles do not complement each other (genetic complementation), they are considered alleles of the same cistron. If they complement each other they are considered to be nonallelic. This definition by Benzer was later expanded (Fincham, 1959): accordingly, a cistron now refers to a segment of DNA that encodes a unit of gene product. Within a cistron, mutations in the *trans* position do not complement each other. Functionally, the term cistron can be equated with the term gene.

Clade—a group of organisms evolved from a common ancestor.

Clathrin—a protein interacting with adaptor proteins to from the coat on vesicles that bud from the cytoplasm.

Clone (Webber, 1903)—a population of molecules, cells or organisms that have originated from a single cell or a single ancestor and are identical to it and to each other.

Clonal selection—selection of lymphocytes with a specific antigen receptor.

Cloning efficiency—a measure of the efficiency of cloning individual mammalian cells in culture.

Cloning vector—a plasmid, phage, or bacterial or yeast artificial chromosome (BAC, YAC) used to carry a foreign DNA fragment for the purpose of cloning (producing multiple copies of the fragment).

Coding strand of DNA—the strand of DNA that bears the same sequence as the RNA strand (mRNA) that is used as a template for translation (sense RNA). The other strand of DNA, which directs synthesis of the mRNA, is the template strand (see antisense RNA).

Codominant—expression of two dominant traits together, e.g., the AB blood group phenotype, see dominant.

Codon (Brenner, Crick, 1963)—a sequence of three nucleotides (a triplet) in DNA or RNA that codes for a certain amino acid or for the terminalization signal of an amino acid sequence.

Cohesin—proteins that hold sister chromatids together.

Colinear—the 1:1 representation of triplet nucleotides in DNA and the corresponding sequence of amino acids.

Complementation, genetic (Fincham, 1966)—complementary effect of (restoration of normal function by) double mutants at different gene loci. For example see genetic complementation groups for xeroderma pigmentosum (p. 92) or Fanconi anemia (p. 340).

Concatemer—an association of DNA molecules with complementary ends linked head to tail and repeated in tandem. Formed during replication of some viral and phage genomes.

Concordance—the occurrence of a trait or a disease in both members of a pair of twins (mono- or dizygotic).

Condensin—proteins involved in the preparation of chromosomes for cell division (chromosome condensation).

Conjugation (Hayes; Cavalli, Lederberg, Lederberg, 1953)—the transfer of DNA from one bacterium to another.

Consanguinity—blood relationship. Two or more individuals are referred to as consanguineous (related by blood) if they have one or more ancestors in common. A quantitative expression of consanguinity is the coefficient of inbreeding (q.v.).

Consensus sequence—a corresponding or identical DNA sequence in different genes or organisms.

Conserved in evolution—refers to genes or parts of chromosomes that have not changed in the course of evolution due to their importance for the organisms.

Contig—a series of overlapping DNA fragments (contiguous sequences).

Cosmid—a plasmid carrying the *cos* site (q.v.) of a phage in addition to sequences required for division. Serves as a cloning vector for DNA fragments up to 40 kb.

Cos site—a restriction site required of a small strand of DNA to be cleaved and packaged into the β phage head.

Coupling (Bateson, Saunders, Punnett, 1905)— *cis* configuration (q.v.) of double heterozygotes.

Covalent bond—a stable chemical bond holding molecules together by sharing one or more pairs of electrons (as opposed to a noncovalent hydrogen bond).

CpG island—a stretch of 1–2 kb in mammalian genomes that is rich in unmethylated CpG doublets; usually in the 5′ direction of a gene.

Crossing-over (Morgan and Cattell, 1912)—the exchange of genetic information between two homologous chromosomes by chiasma formation (q.v.) in the diplotene stage of meiosis I; leads to genetic recombination of neighboring (linked) gene loci. Unequal crossing-over (Sturtevant, 1925) results from mispairing of the homologous DNA segments at the recombination site. It results in structurally altered DNA segments or chromosomes, with a duplication in one and a deletion in the other. Crossing-over may also occur in somatic cells (Stern, 1936).

Cyclic AMP (cAMP)—cyclic adenine monophosphate, a second messenger produced in response to stimulation of G-protein-coupled receptors.

Cyclin—a protein involved in cell cycle regulation.

Cytokine—a small secreted molecule that can bind to cell surface receptors on certain cells to trigger their proliferation or differentiation.

Cytoplasmic inheritance—transmission of genetic information located in mitochondria. Since sperm cells do not contain mitochondria, the information transmitted is of maternal origin.

Cytoskeleton—network of stabilizing protein in the cytoplasm and cell membrane.

Dalton—a unit of atomic mass, approximately equal to the mass of a hydrogen atom $(1.66 \times 10^{24} \text{ g})$.

Deficiency (Bridges, 1917)—loss of a chromosomal segment resulting from faulty crossing-over, e.g., by unequal crossing-over or by crossing-over within an inversion (q.v.) or within a ring chromosome (q.v.). It arises at the same time as a complementary duplication (q.v.). This is referred to as duplication/deficiency.

Deletion (Painter & Muller, 1929)—loss of part of or a whole chromosome or loss of DNA nucleotide bases.

Denaturation—reversible separation of a double-stranded nucleic acid molecule into single strands. Rejoining of the complementary single strands is referred to as renaturation.

Diakinesis (Haecker, 1897)—a stage during late prophase I of meiosis.

Dicentric (Darlington, 1937)—refers to a structurally altered chromosome with two centromeres.

Dictyotene—a stage of fetal oocyte development during which meiotic prophase is interrupted. In human females, oocytes attain the stage of dictyotene about 4 weeks before birth; further development of the oocytes is arrested until ovulation, at which time meiosis is continued.

Differentiation—the process in which an unspecialized cell develops into a distinct, specialized cell.

D-loop—displacement loop; a DNA sequence that is formed in a part of opened DNA double helix, e.g., in mitochondrial DNA (p. 134) or in the telomere (p. 188).

Diploid (Strasburger, 1905)—cells or organisms that have two homologous sets of chromosomes, one from the father (paternal) and one from the mother (maternal).

Diplotene—a stage of prophase I of meiosis.

Direct repeat—repeated DNA sequences oriented in the same direction (see inverted repeat).

Discordance—the occurrence of a given trait or disease in only one member of a pair of twins (see concordance).

Disomy, uniparental (UPD)—presence of two chromosomes of a pair from only one of the parents. One distinguishes UPD due to isodisomy, in which the chromosomes are identical, and heterodisomy, in which they are homologous.

Dispermy—the penetration of a single ovum by two spermatozoa.

Dizygotic—twins derived from two different zygotes (fraternal twins), as opposed to monozygotic (identical) twins, derived from the same zygote.

D loop—displacement loop. A region in the DNA double helix that is opened and where a newly synthesized strand displaces a preexisting strand (see recombination and D loop in telomeres).

DNA (deoxyribonucleic acid)—the molecule containing the primary genetic information in the form of a linear sequence of nucleotides in groups of threes (triplets) (see codon).

Satellite DNA (sDNA) (Sueoka, 1961; Britten & Kohne, 1968)—contains tandem repeats of nucleotide sequences of different lengths. sDNA can be separated from the main DNA by density gradient centrifugation in cesium chloride, after which it appears as one or several bands (satellites) separated from that of the main body of DNA. In eukaryotes, light (AT-rich) and heavy (GC-rich) satellite DNA can be distinguished.

Microsatellites are small (2–10) tandem repeats of DNA nucleotides.

Minisatellites are tandem repeats of about 20–100 base pairs; classical satellite DNA consists of large repeats of 100–6500 bp (see p. 252).

DNAase—an enzyme that attacks bonds in DNA.

DNA library—a collection of cloned DNA molecules comprising the entire genome (genomic library) or of cDNA fragments obtained from mRNA produced by a particular cell type (cDNA library).

DNA microarray—a set of many thousands of different nucleotide sequences on a surface. Used to determine the gene expression pattern of thousands of genes simultaneously.

DNA polymerase—a DNA-synthesizing enzyme. To begin synthesis, it requires a primer of RNA or a complementary strand of DNA.

DNase (deoxyribonuclease)—an enzyme that digests DNA.

Domain—a distinctive functional region of the tertiary structure of a protein or a particular region of a chromosome.

Dominant (Mendel, 1865)—refers to a genetic trait that can be observed in the heterozygous state. The terms "dominant" and "recessive" refer to the effects of the alleles at a given gene locus. The effects observed depend in part on the accuracy of observation. When the effects of each of two different alleles at a (heterozygous) locus can be observed, the alleles are said to be codominant. At the DNA level, allelic genes at two homologous loci are codominant.

Dominant negative—a mutant allele that produces an undesirable effect resembling loss of function.

Dorsal—referes to the back of an animal (see ventral for the opposite)

Dosage compensation (H.J. Muller 1948) – refers to mechanisms that balances a difference in activity of alleles.

Downstream—the 3′ direction of a gene.

Drift, genetic (Wright, 1921)—random changes in gene frequency of a population. Especially relevant in small populations, where random differences in the reproductive frequency of a certain allele can change the frequency of the allele. Under some conditions an allele may disappear completely from a population (loss) or be present in all individuals of a population (fixation).

Duplication (Bridges, 1919)—addition of a chromosomal segment resulting from faulty crossing-over (see deficiency). It may also refer to additional DNA nucleotide base pairs. Duplication of genes (gene duplication) played an important role in the evolution of eukaryotes.

Ectoderm—one of the three primary cell layers of an embryo giving rise to epidermal tissues, the nervous system, and external sense organs (see endoderm and mesoderm).

Effector—a protein exerting a specific effect.

Electrophoresis (Tiselius, 1937)—separation of molecules by utilizing their different speeds of migration in an electrical field. As support medium, substances in gel form such as starch, agarose, acrylamide, etc. are used. Further molecular differences can be detected by modifications such as two-dimensional electrophoresis (electric field rotated 90° for the second migration) or cessation of migration at the isoelectric point (isoelectric focusing).

Elongation—addition of amino acids to a polypeptide chain.

Elongation factor (EF)—one of the proteins that associate with ribosomes while amino acids are added; EF in prokaryotes and eEF in eukaryotes.

Endocytosis—specific uptake of extracellular material at the cell surface. The material is surrounded by an invagination of the cell membrane, which fuses to form a membrane-bound vesicle containing the material.

Endoderm—the inner of the three primary cell layers of an embryo; gives rise to the gastrointestinal system and most of the respiratory tract (see ectoderm and mesoderm).

Endonuclease—a heterogeneous group of enzymes that cleave bonds between nucleotides of single- or double-stranded DNA or of RNA.

Endoplasmic reticulum—a complex system of membranes within the cytoplasm.

Endoreduplication (Levan & Hauschka, 1953)—chromosome replication during interphase without actual mitosis. Endoreduplicated chromosomes in metaphase consist of four chromatids lying next to each other, held together by two neighboring centromeres.

Enhancer (Baneriji 1981—a *cis*-acting regulatory DNA segment that contains binding sites for transcription factors. An enhancer is located at various distances from the promoter. It causes an increase in the rate of transcription (about tenfold).

Enzyme (E. Büchner, 1897)—a protein that catalyzes a biochemical reaction. Enzymes consist of a protein part (apoenzyme), responsible for the specificity, and a nonprotein part (coenzyme), needed for activity. Enzymes bind to their substrates, which become metaboli-cally altered or combined with other substances during the train of the reaction. Most of the enzymatically catalyzed chemical reactions can be classified into one of six groups:

(1) hydrolysis (cleavage with the addition of H_2O), by *hydrolases*;

(2) transfer of a molecular group from a donor to a receptor molecule, by *transferases*;

(3) oxidation and reduction, by *oxidases* and *reductases* (transfer of one or more electrons or hydrogen atoms from a molecule to be oxidized to another molecule that is to be reduced);

(4) isomerization, by *isomerases* (rearranging the position of an atom or functional group within a molecule);

(5) joining of two substrate molecules to form a new molecule, by *ligases* (*synthetases*);

(6) nonhydrolytic cleavage with formation of a double bond on one or both of the two molecules formed, by *lyases*.

Epigenetics—the study of genetic effects on the phenotype not caused by alteration of the DNA sequence.

Epigenetic—a heritable effect on gene or chromosomes function that is not accompanied by a change in the DNA sequence.

Episome (Jacob & Wollman, 1958)—a plasmid (q.v.) that can exist either independently in the cytoplasm or as an integrated part of the genome of its bacterial host.

Epistasis (Bateson, 1907)—nonreciprocal interaction of genes at the same gene locus (allelic) or at different gene loci (nonallelic) that alter the phenotypic expression of a gene.

Epitope—the part of an antigen molecule that binds to an antibody.

EST (expressed sequence tag)—a sequenced site from an expressed gene that "tags" a stretch of unsequenced cDNA next to it; used to map genes (see STS).

Eubacteria—a major class of prokaryotes (see first plate, Tree of Life, p. 24).

Euchromatin (Heitz, 1928)—chromosome or chromosomal segment that stains less intensely than heterochromatin (q.v.). Euchromatin corresponds to the genetically active part of chromatin that is not fully condensed in the interphase nucleus.

Eukaryote (E. Chatton, 1937)—cells in animals and plants that contain a nucleus and organelles in the cytoplasm (see eubacteria and prokaryote).

Euploid (Täckholm, 1922)—refers to cells, tissues, or individuals with the complete normal chromosomal complement characteristic of that species (cf. aneuploid, heteroploid, polyploid).

Excision repair—repair of bulk lesions in DNA in which a stretch of nucleotides (about 14 in prokaryotes and about 30 in eukaryotes) is excised from the affected strand and replaced by the normal sequence (resynthesis).

Exocytosis—specific process by which nondiffusable particles are transported through the cell membrane to be discharged into the cellular environment.

Exon (Gilbert, 1978)—a segment of DNA that is represented in the mature mRNA of eukaryotes (cf. intron).

Exonuclease—an enzyme that cleaves nucleotide chains at their terminal bonds only, at either the 5′ or 3′ end (cf. endonuclease).

Expression—the observable effects of an active gene.

Expression vector—a cloning vector containing DNA sequences that can be transcribed and translated.

Expressivity (Vogt, 1926)—refers to the degree of phenotypic expression of a gene or genotype. Absence of expressivity is also called nonpenetrance.

Fibroblast—type of connective tissue cell. Can be propagated in culture flasks containing suitable medium (*fibroblast cultures*).

Fingerprint, genetic—a characteristic pattern of small polymorphic fragments of DNA or proteins.

FISH—Fluorescence *in-situ* hybridization. Identification of a DNA stretch by a fluorescent marker.

Fitness, biological—refers to the probability (between 0.0 and 1.0) that a gene will be passed on to the next generation. For a given genotype and a given environment, the biological (or reproductive) fitness is determined by survival rate and fertility.

Fixation—a new allele that becomes permanently present in a population.

Founder effect—presence of a particular allele in a population due to a mutation in a single ancestor.

Gain-of-function—a mutation that causes a new type of function, usually undesirable.

Gamete (Strasburger, 1877)—a haploid germ cell, either a spermatozoon (male) or an ovum (female). In mammals, males are heterogametic (XY) and females homogametic (XX). In birds, females are heterogametic (ZW) and males homogametic (ZZ).

G-bands—a type of banding pattern of metaphase chromosomes used for their identification.

Gene (Johannsen, 1909)—a hereditary factor that constitutes a single unit of hereditary material. It corresponds to a segment of DNA that codes for the synthesis of a single polypeptide chain (cf. cistron).

Gene amplification (Brown & David, 1968)—selective production of multiple copies of a given gene without proportional increases of other genes.

Gene bank—a collection of cloned DNA fragments that together represent the genome they are derived from (gene library).

Gene cluster (Demerec & Hartman, 1959)—a group of two or more neighboring genes of similar function, e.g., the HLA system or the immunoglobulin genes.

Gene conversion (Winkler, 1930; Lindgren, 1953)—nonreciprocal transfer of genetic information. One gene serves as a sequence donor, remaining unaffected, while the other gene receives sequences and undergoes variation.

Gene dosage—refers to the quantitative degree of expression of a gene. Also used to refer to the number of copies of a gene in the genome.

Gene family—a set of evolutionarily related genes by virtue of identity or great similarity of some of their coding sequences.

Gene flow (Berdsell, 1950)—transfer of an allele from one population to another.

Gene frequency—the frequency of a given allele at a given locus in a population (allele frequency).

Gene locus (Morgan, Sturtevant, Muller, Bridges, 1915)—the position of a gene on a chromosome.

Gene map—the position of gene loci on chromosomes. A *physical map* refers to the absolute position of gene loci, their distance from each other being expressed by the number of base pairs between them. A *genetic map* expresses the distance of genetically linked loci by their frequency of recombination.

Gene product—the polypeptide or ribosomal RNA encoded by a gene (see protein).

Genetic code—the information contained in the triplets of DNA nucleotide bases used to incorporate a particular amino acid into a gene product.

Genetic marker—a polymorphic genetic property that can be used to distinguish the parental origin of alleles.

Genetics (Bateson, 1906)—the science of heredity and the hereditary basis of organisms; derived from Gk. *genesis* (origin).

Genome (Winkler, 1920)—all of the genetic material of a cell or of an individual.

Genome scan—a search with marker loci on all chromosomes for linkage with an unmapped locus.

Genomics—the scientific field dealing with the structure and function of entire genomes (see part II, Genomics).

Genotype (Johannsen, 1909)—all or a particular part of the genetic constitution of an individual or a cell (cf. phenotype).

Germ cell—a cell able to differentiate into gametes by meiosis (as opposed to somatic cells).

Germinal—refers to germ cells, as opposed to somatic cells.

Germline—the cell lineage giving rise to germ cells.

G protein—guanine nucleotide-binding proteins involved in signal transduction.

G6PD—glucose-6-phosphate dehydrogenase.

Growth factor – a protein, usually a small peptide acting as a ligand, that activates a receptor

Gyrase—a topoisomerase that unwinds DNA.

Haldane's rule—hybrid sterility or inviability preferentially affects the heterogametic sex.

Haploid (Strasburger, 1905)—refers to cells or individuals with a single chromosome complement; gametes are haploid.

Haploinsufficiency—refers to a diploid gene that does not exert normal function in the haploid state (e.g. following loss of the homologous gene locus).

Haplotype (Ceppellini et al., 1967)—a combination of alleles at two or more closely linked gene loci on the same chromosome, e.g., in the HLA system (q.v.).

Heavy strand—refers to differences in density of DNA that result from differences in A and G, and C and T bases, in contrast to the light strand. Occurs in mitochondrial DNA (p. 134).

Helicase—an enzyme that unwinds and separates the two strands of the DNA double helix by breaking the hydrogen bonds during transcription or repair.

Helix-loop-helix—a structural motif in DNA-binding proteins, such as some transcription factors (p. 220).

Hemizygous—refers to genes and gene loci that are present in only one copy in an individual, e.g., on the single X chromosome in male cells (XY) or because a homologous locus has been lost.

Heritability (Lush, 1950; Falconer, 1960)—the ratio of additive genetic variance to the total phenotypic variance. Phenotypic variance is the result of the interaction of genetic and non-genetic factors in a population.

Heterochromatin (Heitz, 1928)—a chromosome or chromosomal segment that remains darkly stained in interphase, early prophase, and late telophase because it remains condensed, as all chromosomal material is in metaphase. This contrasts with *euchromatin*, which becomes invisible during interphase. Heterochromatin corresponds to chromosomes or chromosome segments showing little or no genetic activity. *Constitutive* and *facultative* heterochromatin can be distinguished. An example of facultative heterochromatin is the heterochromatic X chromosome resulting from inactivation of one X chromosome in somatic cells of female mammals. An example of constitutive heterochromatin is the centric heterochromatin at centromeres that can be demonstrated as C bands.

Heterodisomy—presence of two homologous chromosomes from one parent only (cf. isodisomy and UPD).

Heteroduplex—refers to a region of a double-stranded DNA molecule with noncomplementary strands that originated from different duplex DNA molecules.

Heterogametic (Wilson, 1910)—producing two different types of gametes (q.v.), e.g., X and Y in (male) mammals or ZW in female birds.

Heterogeneity, genetic (Harris, 1953; Fraser, 1956)—an apparently uniform phenotype being caused by two or more different genotypes.

Heterokaryon (Ephrussi & Weiss, 1965; Harris & Watkins, 1965; Okada & Murayama, 1965)—a cell having two or more nuclei with different genomes.

Heteroploid (Winkler, 1916)—refers to cells or individuals with an abnormal number of chromosomes.

Heterosis (Shull, 1911)—increased reproductive fitness of heterozygous genotypes compared with the parental homozygous genotypes, in plants and animals.

Heterozygous (Bateson and Saunders, 1902)—having two different alleles at a given gene locus (cf. homozygous).

Hfr cell—a bacterium that possesses DNA sequences that lead to a high frequency of DNA transfer at conjugation.

HGPRT—hypoxanthine-guanine-phosphoribosyl transferase. An enzyme in purine metabolism that is inactive in Lesch-Nyhan syndrome.

Histocompatibility—tissue compatibility. Determined by the major histocompatibility complex MHC (see HLA).

Histone (Kossel, 1884)—chromosome-associated protein of the nucleosome. Histones H2A, H2B, H3, and H4 form a nucleosome (q.v.).

HLA (J. Dausset, P.I. Terasaki, 1954)—human leukocyte antigen system A. HLA is also said by some to refer to Los Angeles, where Terasaki made essential discoveries.

HMG proteins—Non-histone chromatin–associated proteins that exhibit high mobility during electrophoresis.

Hogness box—a nucleotide sequence that is part of the promoter in eukaryotic genes.

Holoprosencephaly—a developmental field defect resulting in a wide range of congenital malformations mainly involving the midline embryonic forebrain. The externally visible malformations are cyclopia without nose formation at the most severe end of the spectrum to a single incisor tooth and a flat face with closely set eyes (hypotelorism).

Homeobox—a highly conserved DNA segment in homeotic genes.

Homeosis—Transformation of one body part into another.

Homeotic gene—one of the developmental genes in *Drosophila* that can lead to the replacement of one body part by another by mutation.

Homologous—refers to a chromosome or gene locus of similar maternal or paternal origin.

Homozygosity mapping—mapping genes by identifying chromosomal regions that are homozygous as a result of identity by descent from a common ancestor in consanguineous matings (see identity by descent, IBD).

Homozygous (Bateson and Saunders, 1902)—having identical alleles at a given gene locus.

Hormone (W. Bayliss & E. Starling 1904)—an organic compound able to induce a specific response in target cells. Derived from Greek "to spur on").

Hox genes—clusters of mammalian genes containing homeobox sequences. They are important in embryonic development.

Hybridization—cross between two genotypically different plants or animals belonging to the same species. The term is often used in more narrow definitions: fusion of two single complementary DNA strands (DNA/DNA hybridization), fusion of complementary DNA and RNA strands (DNA/RNA hybridization), or the in vitro fusion of cultured cells of different species (cell hybridization).

Hybridoma—a clone of hybrid cells.

Hydrogen bond—a noncovalent weak chemical bond between an electronegative atom (usually oxygen or nitrogen) and a hydrogen atom; important in stabilizing the three-dimensional structure of proteins or base pairing in nucleic acids.

Identity by descent (IBD)—refers to homozygous alleles at one gene locus that are identical because they are inherited from a common ancestor (see consanguinity).

Immunoglobulin—an antigen-binding molecule.

Imprinting, genomic—different expression of an allele or chromosomal segment depending on the parental origin.

Inbreeding coefficient (Wright, 1929)—measure of the probability that two alleles at a gene locus of an individual are identical by descent, i.e., that they are copies of a single allele of an ancestor common to both parents (IBD, identity-by-descent). Also, the proportion of loci at which the individual is homozygous.

Incidence—the rate of occurrence of a disease in a population. In contrast, prevalence is the percentage of a population that is affected with a particular disease at a particular time.

Inducer—a molecule that induces the expression of a gene.

Initiation factor—a protein that associates with the small subunit of a ribosome when protein synthesis begins (IF in prokaryotes, eIF in eukaryotes).

Insertion—insertion of chromosomal material of nonhomologous origin into a chromosome without reciprocal translocation (q.v.).

Insertion sequence (IS)—a small bacterial transposon carrying genes for its own transposition (q.v.).

In silico—a process taking place within a computer; for example, analysis of biological data.

Intercalating agent—a chemical compound that can occupy a space between two adjacent base pairs in DNA.

In-situ **hybridization**—Hybridization of complementary single-stranded DNA or RNA in the original or natural position (see FISH).

Interphase—the period of the cell cycle between two cell divisions (see mitosis).

Intron (Gilbert, 1978)—a segment of noncoding DNA within a gene (cf. exon). It is transcribed, but removed from the primary RNA transcript before translation.

Inversion (Sturtevant, 1926)—structural alteration of a chromosome through a break at two sites with reversal of direction of the intermediate segment and reattachment. A *pericentric* inversion includes the centromere in the inverted segment. A *paracentric* inversion does not involve the centromere. An inversion per se does not cause clinical signs, but it represents a potential genetic risk because crossing-over may occur in the region of the inversion and lead to aneusomy in offspring (aneusomy by recombination). Chromosomal inversions played an important role in evolution.

Inverted repeat—two identical, oppositely oriented copies of the same DNA sequence. They are a characteristic feature of retroviruses.

In vitro—a biological process taking place outside a living organism or in an artificial environment in the laboratory.

In vivo—within a living organism.

Isochromatid break—a break in both chromatids at the same site.

Isochromosome (Darlington, 1940)—a chromosome composed of two identical arms connected by the centromere, e.g., two long or two short arms of an X chromosome. Implies duplication of the doubled arm and deficiency of the absent arm. An isochromosome may have one or two centromeres.

Isodisomy—presence of two identical chromosomes from one of the parents (cf. heterodisomy).

Isolate, genetic (Waklund, 1928)—a physically or socially isolated population that has not interbred with individuals outside of that population (no panmixis).

Isotype—closely related chains of immunoglobulins.

Isozyme or isoenzyme (Markert and Möller, 1959; Vessel 1959)—one of multiple distinguishable forms of enzymes of similar function in the same organism. Isoenzymes are a biochemical expression of genetic polymorphism.

Karyotype (Levitsky, 1924)—the chromosome complement of a cell, an individual, or a species.

Kilobase (kb)— 1000 base pairs.

Knockout (of a gene)—intentional inactivation of a gene in an experimental organism in order to obtain information about its function (same as targeted gene disruption).

Lagging strand of DNA—the new strand of DNA replicated from the 3′ to 5′ strand. It is synthesized in short fragments in the 5′ to 3′ direction (Okazaki fragments), which are subsequently joined together.

Lamins—Intermediate filament proteins forming a fibrous network, the nuclear lamina on the inner surface of the nuclear envelope.

Lampbrush chromosome (Rückert, 1892)—a special type of chromosome found in the primary oocytes of many vertebrates and invertebrates during the diplotene stage of meiotic division and in drosophila spermatocytes. The chromosomes show numerous lateral loops of DNA that are accompanied by RNA and protein synthesis.

Lariat—an intermediate form of RNA during splicing when a circular structure with a tail is formed by a 5′ to 3′ bond.

Leader sequence—a short N-terminal sequence of a protein that is required for directing the protein to its target.

Leaky mutant—a mutation causing only partial loss of function.

Leptotene—a stage of meiosis (q.v.).

Lethal equivalent (Morton, Crow, and Muller, 1956)—a gene or combination of genes that in the homozygous state is lethal to 100% of individuals. This may refer to a gene that is lethal in the homozygous state, to two different genes that each have 50% lethality, to three different genes each with 33% lethality, etc. It is assumed that each individual carries about 5–6 lethal equivalents.

Lethal factor (Bauer, 1908; Hadorn, 1959)—an abnormality of the genome that leads to death in utero, e.g., numerous chromosomal anomalies.

Leucine zipper—a specific DNA-binding protein that resembles a zipper and serves as a transcription factor.

Library—see DNA library.

Ligand—a molecule that can bind to a receptor and thereby induce a signal in the cell.

LINE (long interspersed nuclear element)—long interspersed repetitive DNA sequences.

Linkage disequilibrium (Kimura, 1956)—nonrandom association of alleles at closely linked gene loci that deviates from their individual frequencies as predicted by the Hardy-Weinberg equilibrium.

Linkage, genetic (Morgan, 1910)—localization of gene loci on the same chromosome close enough to cause deviation from independent segregation.

Linkage group—gene loci on the same chromosome that are so close together that they usually are inherited together without recombination.

Linker DNA—a synthetic DNA double strand that carries the recognition site for a restriction enzyme and that can bind two DNA fragments. Also, the stretch of DNA between two nucleosomes.

Locus—same as gene locus.

Long terminal repeat (LTR)—A repeat DNA sequence of up to 600 base pairs flanking the coding regions of retroviral DNA and viral transposons.

Lymphocyte—cell of the immune system, of one of two general types: B lymphocytes from the bone marrow and thymus-derived T lymphocytes.

Lysosome (de Duve 1955)—small cytoplasmic organelle containing hydrolytic enzymes.

Lysogeny—the ability of a phage to integrate into the bacterial chromosome.

Lytic infection—a phage infecting a bacterium and causing lysis.

Map distance—distance between gene loci, expressed in either physical terms (number of base pairs, e.g. kb 1000 bp or Mb million bp) or genetic terms (recombination frequency, expressed as cM, centimorgan. One cM corresponds to 1%).

Mapping—various methods to determine the position of a gene on a chromosome (physical map) or its relative distance to other gene loci and their order (genetic map).

Marker, genetic—an allele used to recognize a particular genotype.

Megabase (Mb)—1 million base pairs.

Meiosis (Strasburger 1884)—the special division of a germ cell nucleus that leads to reduction of the chromosome complement from the diploid to the haploid state. Prophase of the first meiotic division is especially important and consists of the following stages: leptotene, zygotene, pachytene, diplotene, diakinesis.

Mendelian inheritance (Castle, 1906)—inheritance according to the laws of Mendel as opposed to extrachromosomal inheritance, under the control of cytoplasmic hereditary factors (mitochondrial DNA).

Metabolic cooperation (Subak-Sharpe et al., 1969)—correction of a phenotype in cells in culture by contact with normal cells or cell products. An example of metabolic correction is the cross correction of cultured cells of different mucopolysaccharide storage diseases or correction of HGPRT-deficient cells by normal cells.

Metacentric—refers to chromosomes that are divided by the centromere into two arms of approximately the same length.

Metaphase (Strasburger, 1884)—stage of mitosis in which the contracted chromosomes are readily visible.

MHC (major histocompatibility complex) (Thorsby, 1974)—the principal histocompatibility system, consisting of class I and class III antigen genes of the HLA system including the class II genes.

microRNA (miRNA)—small RNA molecules of 21–23 nucelotides that associate with multiple proteins in an RNA-induced silencing complex (RISC, see RNAi).

Microfilament—Cytoskeletal fiber of ca. 7 nm diameter that results from polymerization of monoglobar (G) actin (actin filament).

Mismatch repair—a DNA repair mechanism to repair improperly paired DNA bases.

Missense mutation—a mutation that alters a codon to one for a different amino acid (see nonsense mutation).

Mitochondrion—plural: mitochondria. Cell organelles containing DNA.

Mitosis (Flemming, 1882)—nuclear division during the division of somatic cells, consisting of prophase, metaphase, anaphase, and telophase.

Mitosis index (Minot, 1908)—the proportion of cells present that are undergoing mitosis.

Mixoploidy (Nemec, 1910; Hamerton, 1971)—a tissue or individual having cells with different numbers of chromosomes (chromosomal mosaic).

MLC—*m*ixed *l*ymphocyte *c*ulture (Bach & Hirschhorn, Bach & Lowenstein, 1964). MLC is a test for differences in HLA-D phenotypes.

Mobile genetic element—a DNA sequence that an change its position (see transposon).

Modal number (White, 1945)—the number of chromosomes of an individual or a cell.

Monoclonal antibody—an antibody representing a single antigen specificity, produced from a single progenitor cell.

Monolayer (Abercrombie & Heaysman, 1957)—the single-layered sheet of cultured diploid cells on the bottom of a culture flask.

Monosomy (Blakeslee, 1921)—absence of a chromosome in an otherwise diploid chromosomal complement.

Monozygotic—twins with identical sets of nuclear genes (cf. dizygotic).

Morphogen—a protein present in embryonic tissues in a concentration gradient that induces a developmental process.

Mosaic—tissue or individuals made up of genetically different cells, as a rule of the same zygotic origin (cf. chimera).

mRNA (Brenner, Jacob, and Meselson, 1961; Jacob & Monod, 1961)—messenger RNA.

mtDNA—mitochondrial DNA.

Multigene family—a group of genes related by their common evolution.

Mutagen—a chemical or physical agent that can induce a mutation.

Mutation (de Vries, 1901)—permanent alteration of the genetic material. Different types include point mutations from exchange, loss, or insertion of base pairs within a gene and chromosomal mutation with alteration of the chromosome structure. A *missense* mutation is an alteration resulting in a gene product containing a substitution for a wrong amino acid. A *nonsense* mutation is an alteration that produces a stop codon in the midst of a genetic message so that a totally inadequate gene product is formed.

Mutation rate—the frequency of a mutation per locus per individual per generation.

Myosin—A class of motor proteins.

Necrosis—Cell death resulting from tissue damage

Neurotransmitter—Extracellular signaling molecules at the nerve-muscle junction.

N-linked oligosaccharides—Branched oligosaccharide chains attached to the amino group side-chain of asparagines in glycoproteins.

Noncovalent bond—a (weak) chemical bond between an electonegative atom (usually oxygen or nitrogen) and a hydrogen atom (see hydrogen bond), but not involving sharing of electrons.

Nondisjunction (Bridges, 1912)—faulty distribution of homologous chromosomes at meiosis. In mitotic nondisjunction, the distribution error occurs during mitosis.

Nonpolar—Molecules lacking a net electric charge or having an asymmetric distribution of positive and negative charges; generally insoluble in water.

Nonsense codon—a codon that does not have a normal tRNA molecule. Any of the three triplets (stop codons) that terminate translation (UAG, UAA, UGA).

Nonsense mutation—a mutation that results in lack of any genetic information, e.g., a stop codon (see missense mutation).

Northern blot—transfer of RNA molecules to a membrane by a procedure similar to that of a Southern blot (q.v.).

Nucleoside—compound of a purine or a pyrimidine base with a sugar (ribose or deoxyribose) (cf. nucleotide).

Nucleosome (Navashin, 1912; Kornberg, 1974)—a subunit of chromatin consisting of DNA wound around histone proteins in a defined spatial configuration.

Nucleotide—Single monomeric building block of a polynucleotide chain that makes up nucleic acid. A nucleotide is a phosphate ester consisting of a purine or a pyrimidine base, a sugar (ribose or deoxyribose as a pentose), and a phosphate group.

Nucleic acid—a molecule such as DNA and RNA that can store genetic information.

Ochre codon—the stop codon UAA (see amber codon, the stop codon UAG).

Okazaki fragment—a short nucleotide sequence that is synthesized on the lagging strand of DNA during replication (q.v.).

O-linked oligosaccharides—attachment to the hydoxyl groups of the side-chains of serine and threonin in glycoproteins (see N-linked).

Oncogene (Heubner & Todaro, 1969)—a DNA sequence of viral origin that can lead to malignant transformation of a eukaryotic cell after being integrated into the cell genome (see proto-oncogene).

Open reading frame (ORF)—a DNA sequence of variable length that does not contain stop codons and therefore is meaningful because it can be translated.

Operator (Jacob & Monod, 1959)—the recognition site of an operon at which the negative control of genetic transcription takes place by binding to a repressor.

Operon (Jacob et al., 1960)—in prokaryotes, a group of functionally and structurally related genes that are regulated together.

Origin of replication (ORI)—site of start of DNA replication.

Ortholog—a homologous DNA sequence or a gene that has evolved from a common ancestor between related species, e. g., the α- and β-globin genes (see paralog).

Pachytene (de Winiwarter, 1900)—stage of prophase meiosis I.

Palindrome (Wilson & Thomas, 1974)—a sequence of DNA nucleotides that read the same in the 5′ to 3′ direction as in the 3′ to 5′ direction. This occurs in recognition sequences of restriction enzymes, e.g. 5′AATG3′ in one strand and 3′GTAA5′ in the other.

Panmixis (Weismann, 1895)—pairing system with random partner selection, as opposed to assortative mating.

Paracrine—refers to signaling molecules acting on molecules nearby.

Paralog—a DNA sequence or a gene that has evolved from a common ancestor within a species, e.g., the two α-globin gene loci in humans (see ortholog).

Parasexual (Pontocorvo, 1954)—refers to genetic recombination by nonsexual means, e.g., by hybridization of cultured cells (see hybridization).

PCR (polymerase chain reaction) (Mullis 1985)—technique for in vitro propagation (amplification) of a given DNA sequence. It is a repetitive thermal cyclic process consisting of denaturation of genomic DNA of the sequence of interest, annealing the DNA to appropriate oligonucleotide primers, and replication of the DNA segment complementary to the primer.

Penetrance (Vogt, 1926)—the frequency or probability of expression of an allele (cf. expressivity).

Peptide—a compound of two or more amino acids joined by peptide bonds.

Phage—abbreviation for bacteriophage.

Phagocytosis—refers to cells that incorporate foreign cells , such as bacteria.

Phenocopy (Goldschmidt, 1935)—a nonhereditary phenotype that resembles a genetically determined phenotype.

Phenotype (Johannsen, 1909)—the observable effect of one or more genes on an individual or a cell.

Pheromone—a signaling molecule that can alter the behavior or gene expression of other individuals of the same species.

Phosphodiester bond—the chemical bond linking adjacent nucleotides of DNA or RNA.

Phytohemagglutinin (PHA)—a protein substance obtained from kidney beans (*Phaseolus vulgaris*). It is used to separate red from white blood cells. Nowell (1960) discovered its ability to induce blastic transformation (see Transformation) and cell division in lymphocytes. It is the basis of phytohemagglutinin-stimulated lymphocyte cultures for chromosomal analysis.

Plasmid (Lederberg, 1952)—autonomously replicating circular DNA structures found in bacteria. Although they are usually separate from the actual genome, they may become integrated into the host chromosome.

Plastids—any of several types of organelles found in plant cells, e.g., chloroplasts.

Pleiotropy (Plate, 1910)—expression of a gene with multiple, seemingly unrelated phenotypic features.

Point mutation—alteration of the genetic code within a single codon. The possible types are the exchange of a base: a pyrimidine for another pyrimidine (or a purine for another purine) as a *transition* (Frese, 1959), i.e., thymine for cytosine (or adenine for guanine); or the exchange of a pyrimidine by a purine or visa versa: *transversion*, i.e., thymine by adenine or visa versa (Frese, 1959). Besides the two types of exchange, a point mutation may be due to the insertion of a nucleotide base or the deletion of one or several base pairs.

Polar—refers to molecules with a net electric charge or a symmetric distribution of positive and negative charges.

Polar body (Robin, 1862)—an involutional cell arising during oogenesis that does not develop further as an oocyte.

Polyadenylation—the addition of multiple adenine residues at the 3′ end of eukaryotic mRNA after transcription.

Polycistronic messenger—mRNA including coding regions from more than one gene (in prokaryotes).

Polygenic (Plate, 1913; Mather, 1941)—refers to traits that are based on several or numerous genes whose effects cannot be individually determined. The term multigenic is sometimes used instead.

Polymerases—enzymes that catalyze the combining of nucleotides to form RNA or DNA (genetic transcription and DNA replication).

Polymorphism, genetic (Ford, 1940)—existence of more than one normal allele at a gene locus, with the rarest allele exceeding a frequency of 1%. A polymorphism may exist at several levels, i.e., variants in DNA sequence, amino acid sequence, chromosomal structure, or phenotypic traits.

Polypeptide—see peptide.

Polyploid (Strasburger, 1910)—refers to cells, tissues, or individuals having more than two copies of the haploid genome, e.g., three (triploid) or four (tetraploid). In man, triploidy and tetraploidy in a conceptus are usually lethal and as a rule lead to spontaneous abortion.

Polytene (Koltzoff, 1934; Bauer, 1935)—refers to a special type of chromosome resulting from repeated endoreduplication of a single chromosome. Giant chromosomes arise in this manner (cf. chromosome).

Population (Johannsen, 1903)—individuals of a species that interbreed and constitute a common gene pool (cf. race).

Position effect (Sturtevant 1925)—The phenomenon that the action of a gene is altered by its location (A. H. Sturtevant, Genetics 10: 117–147, 1925).

Premature chromosomal condensation (Johnson & Rao, 1970)—induction of chromosomal condensation in an interphase nucleus after fusion with a cell in mitosis. Condensed S-phase chromosomes appear pulverized (so-called chromosomal pulverization).

Prevalence—see incidence.

Pribnow box—part of a promoter (TATAAT sequence 10 bp upstream of the gene) in prokaryotes.

Primary transcript—the RNA transcribed from a eukaryotic gene before processing (splicing, addition of the cap, and polyadenylation).

Primer—a DNA or RNA oligonucleotide that after hybridization to an inversely complementary DNA has a 3′-OH end to which nucleotides can be added for synthesis of a new chain by DNA polymerase.

Prion—proteinaceous infectious particles that cause degenerative disorders of the central nervous system.

Proband—see Propositus.

Probe—a defined DNA or RNA fragment used to identify complementary sequences by specific hybridization.

Prokaryote—microorganisms without a cell nucleus or intracellular organelles, such as bacteria (see eukaryote and archea).

Promoter—a defined DNA region at the 5′ end of a gene that binds to transcription factors and RNA polymerase during the initiation of transcription. The –10 sequence is the consensus sequence TATAATG about 10 bp upstream of a prokaryotic gene (Pribnow box).

Prophage—a viral (phage) genome integrated into the bacterial (host) genome.

Prophase—an early stage of mitosis.

Propositus, proband—the individual in a pedigree that has brought a family to attention for genetic studies.

Protein—one or more polypeptides with a specific amino acid sequence and a specific three-dimensional structure. Proteins are biomolecules representing the key structural elements of living cells and participating in nearly all cellular and biochemical reactions (see gene product).

Proteome—the complete set of all protein-encoding genes or all the proteins produced by them.

Proto-oncogene (cellular oncogene)—a eukaryotic gene. It may be present in truncated form in a retrovirus, where it may behave as an oncogene.

Prototrophic—refers to cells or cell lines that do not require a special nutrient added to the culture medium (cf. auxotrophic).

Provirus—duplex DNA derived from an RNA retrovirus and incorporated into a eukaryotic genome.

Pseudogene—DNA sequences that closely resemble a gene but are without function due to an integral stop codon, deletion, or other structural change. A processed pseudogene consists of DNA sequences that resemble the mRNA copy of the parent gene, i.e., it does not contain introns.

Pseudohermaphroditism—a condition in which an individual has the gonads of one sex and phenotypic features of the opposite sex.

Quadriradial figure—the configuration assumed when homologous segments of chromosomes involved in a reciprocal translocation pair at meiosis. Rarely, such a figure may occur during mitosis.

Race—a population (q.v.) that differs from another population in the frequency of some of its gene alleles (L. C. Dunn: *Heredity and Evolution in Human Populations,* Harvard University Press, Cambridge, Mass., 1967). Accordingly, the concept of race is flexible and relative, defined in relation to the evolutionary process. The term race can be used to classify groups, whereas the classification of individuals is often uncertain and of biologically dubious value.

Reading frame—sequence of DNA nucleotides that can be read in triplets to code for a peptide (cf. open reading frame).

Receptor—a transmembrane or intracellular protein involved in transmission of a cell signal.

Recessive (Mendel, 1865)—refers to the genetic effect of an allele (q.v.) at a gene locus that is manifest as phenotype in the homozygous state only (q.v.).

Reciprocal translocation—mutual exchange of chromosome parts.

Recombinant DNA—A DNA molecule consisting of parts of different origin.

Recombination (Bridges and Morgan, 1923)—the formation of new combinations of genes as a result of crossing-over between homologous chromosomes during meiosis.

Recombination frequency—frequency of recombination between two or more gene loci. Expressed as the theta (Θ) value. A Θ of 0.01 (1% recombination frequency) corresponds to 1 centimorgan (cM).

Regulatory gene—a gene coding for a protein that regulates other genes.

Renaturation of DNA—combining of complementary single strands of DNA to form double-stranded DNA (cf. denaturation).

Repair (Muller, 1954)—correction of structural and functional DNA damage.

Replication—identical duplication of DNA.

Replication fork—the unwound region of the DNA double helix in which replication takes place.

Replicon (Huberman and Riggs, 1968)—an individual unit of discontinuous DNA replication in eukaryotic DNA.

Reporter gene—a gene used to analyze another gene, especially the regulatory region of the latter.

Repressor—a protein that suppresses gene function.

Repulsion (Bateson, Saunders, and Punnett, 1905)—term to indicate that the mutant alleles of neighboring heterozygous gene loci lie on opposite chromosomes, i.e., in trans configuration (see cis/trans).

Resistance factor—a plasmid gene causing antibiotic resistance.

Restriction enzyme, or restriction endonuclease (Meselson and Yuan, 1968)—endonuclease that cleaves DNA at a specific base sequence (restriction site or recognition sequence).

Restriction map—a segment of DNA characterized by a particular pattern of restriction sites.

Restriction site—a particular sequence of nucleotide bases in DNA that allow a particular restriction enzyme to cleave the DNA molecule at, or close to, that site (recognition site).

Retrotransposon—a mobile DNA sequence that can insert itself at a different position by using reverse transcriptase (see transposon).

Retrovirus—a virus with a genome consisting of RNA that multiplies in a eukaryotic cell by conversion into duplex DNA.

Reverse transcriptase—an enzyme complex that occurs in RNA viruses and that can synthesize DNA from an RNA template.

RFLP—restriction fragment length polymorphism. The production of DNA fragments of different lengths by a given restriction enzyme, due to inherited differences in a restriction site.

Rho factor—a protein involved in termination of transcription in *E. coli.*

Ribonuclease (RNAase)—an enzyme that can cleave RNA.

Ribosome (Roberts, 1958; Dintzis et al., 1958)—complex molecular structure in prokaryotic and eukaryotic cells consisting of specific proteins and ribosomal RNA in different subunits. The translation of genetic information occurs in ribosomes.

Ring chromosome—a circular chromosome. In prokaryotes the normal chromosome is ring-shaped. In mammals it represents a structural anomaly and implies that chromosomal material has been lost.

RNA (ribonucleic acid)—a polynucleotide with a structure similar to that of DNA except that the sugar is ribose instead of deoxyribose.

RNA editing—a change of RNA sequence following transcription; a mechanism of gene regulation.

RNAi (RNA interference)—inhibition of transcription by antisense RNA.

RNA polymerase—an enzyme that synthesizes RNA from a DNA template.

RNA silencing—the ability of double-stranded RNA to suppress a gene.

rRNA—ribosomal RNA. Any of many large RNA molecules in ribosomes.

RNA splicing—the processing of the primary transcript in mRNA (in eukaryotes).

RTK—Receptor tyrosine kinase; membrane bound proteins involved in signal transduction.

Satellite (Navashin, 1912)—small mass of chromosomal material attached to the short arm of an acrocentric chromosome (q.v.) by a constricted appendage or stalk. It is involved in the organization of the nucleolus. The stalk region can be stained by specific silver stain (NOR stain, nucleolus-organizing region). The size of the satellite, the length of its stalk, and the intensity of its fluorescence after staining with acridine are polymorphic cytogenetic markers.

Satellite DNA (sDNA) (Sueoka, 1961; Kit, 1961; Britten and Kohne, 1968)—DNA that is either heavier (GC-rich) or lighter (AT-rich) than the main DNA (see DNA). Not to be confused with the satellite regions of acrocentric chromosomes.

SCE (sister chromatid exchange) (Taylor, 1958)—an exchange between the two chromatids of a metaphase chromosome. After two replication cycles in a cell culture in the presence of a halogenated base analog (e.g., 5-bromodeoxyuridine), both DNA strands of one chromatid will be substituted with the halogenated base analog, whereas only one DNA strand of the other chromatid will be substituted. As a result, the two chromatids differ in staining intensity, and it is possible to determine where crossing-over of the two chromatids has occurred.

Segregation (Bateson and Saunders, 1902)—the separation of alleles at a gene locus at meiosis and their distribution to different gametes. Segregation accounts for the 1 : 1 distribution of allelic genes to different chromosomes.

Selection (Darwin, 1858)—preferential reproduction or survival of different genotypes under different environmental conditions.

Selection coefficient—quantitative expression (from 0 to 1) of the disadvantage that a genotype has (compared with a standard genotype) in transmitting genes to the next generation. The selection coefficient (s) is the numerical amount by which biological fitness (1-s) is decreased; i.e., a selection coefficient of 1 indicates a complete lack of biological fitness.

Selective medium—a medium that supports growth of cells in culture containing a particular gene.

Senescence—aging of cells in culture or referring to aging in general.

Semiconservative (Delbrück and Stent, 1957)—the normal type of DNA replication. One DNA strand is completely retained; the other is synthesized completely anew.

Serpentine—a seven-helix transmembrane protein.

Signal sequence—the N-terminal amino acid sequence of a secreted protein; required for transport of the protein to the right destination in the cell.

Silencer sequence—a eukaryotic DNA sequence that blocks the access of gene activity proteins required for transcription, by forming condensed heterochromatin in that particular area.

SINE (short interspersed nuclear element)—short repetitive DNA sequences (cf. LINE).

Smads—a class of cytosolic transcription factors activated by phosphorylation.

snRNA—small nuclear RNA. One of several small RNAs located in the nucleus; five are components of the spliceosome.

snRNPs (small nuclear ribonucleoprotein particles)—complexes of small nuclear RNA molecules and proteins.

Somatic—refers to cells and tissues of the body, as opposed to germinal (referring to germ cells).

Somatic cell—any cell of an organism that does not undergo meiosis and does not form gametes (as opposed to germ cells).

Somatic cell hybridization—formation of cell hybrids in culture.

Southern blot (Southern, 1975)—method of transferring DNA fragments from an agarose gel to a membrane after the fragments have been separated according to size by electrophoresis.

Speciation (Simpson, 1944)—formation of species during evolution. One of the first steps toward speciation is the establishment of a reproductive barrier against genetic exchange. A frequent mechanism is chromosomal inversion.

Species—a natural population in which there is interbreeding of the individuals, which share a common gene pool.

S phase (Howard and Pelc, 1953)—phase of DNA synthesis (DNA replication) between the G1 and the G2 phase of the eukaryotic cell cycle.

Spliceosome—an aggregation of different molecules that can splice RNA.

Splicing—a step in processing a primary RNA transcript in which introns are excised and exons are joined.

Splice junction—the sequences at the exon/intron boundaries.

Stem cell—an undifferentiated cell able to renew itself by division, retaining the potential for differentiation within a particular developmental pathway. One distinguishes omnipotent and pluripotent stem cells.

Steroid receptor—a transcription factor that responds to a steroid hormone.

Stop codon—a codon that terminates translation (UAG, UAA, UGA), originally called nonsense codons.

STS (sequence tagged site)—a short segment of DNA of known sequence.

Submetacentric—refers to a chromosome consisting of a short and a long arm because of the position of its centromere.

Superfamily—a set of genes or proteins related to each other by evolution.

Synapse—a region at the junction of a nerve and a muscle cell or between two nerve cells.

Synapsis (Moore, 1895)—the pairing of homologous chromosomes during meiotic prophase.

Synaptonemal complex (Moses, 1958)—parallel structures associated with chiasmata formation during meiosis, visible under the electron microscope.

Syndrome—within human genetics, a group of clinical and pathological characteristics that are etiologically related, regardless of whether the details of their relationship have yet to be identified.

Synteny (Renwick, 1971)—refers to gene loci that are located on the same chromosome, whether or not they are linked.

Tandem duplication—short identical DNA segments adjacent to each other.

TATA box—a conserved, noncoding DNA sequence about 25 bp in the 5′ region of most eukaryotic genes. It consists mainly of sequences of the TATAAAA motif. Also known as Hogness box (cf. Pribnow box, of prokaryotes).

T cells—T lymphocytes.

Telocentric (Darlington, 1939)—refers to chromosomes or chromatids with a terminally located centromere, without a short arm or satellite. They do not occur in man.

Telomerase—a ribonucleoprotein enzyme that adds nucleotide bases at the telomere.

Telomere (Muller, 1940)—the terminal areas of both ends of a chromosome containing specific consensus sequences.

Template—the molecule that determines the nucleotide sequence for the formation of another, similar (complementary) molecule (see DNA and RNA).

Teratogen (Ballantyne, 1894)—chemical or physical agent that leads to disturbances of embryological development and malformations.

Termination codon—one of the three triplets signaling the end of translation (UAG, UAA, UGA).

Terminator—a DNA sequence that signals the end of transcription.

Tetraploid (Nemec, 1910)—having a double diploid chromosome complement, i.e., four of each kind of chromosome are present (4 n instead of 2 n).

Topoisomerase—a class of enzymes that can control the three-dimensional structure of DNA by cutting one DNA strand, rotating it about the other, and resealing it (class I) or cutting and resealing both ends (class II). Used to unwind the DNA double helix at transcription.

Transcript—an RNA copy of a segment of the DNA of an active gene.

Transcription—the synthesis of messenger RNA (mRNA), the first step in relaying the information contained in DNA.

Transcription factor—any protein that regulates gene activity.

Transcription unit—all of the DNA sequences required to code for a given gene product (operationally corresponding to a gene). Includes promoter and coding and noncoding sequences.

Transduction (Zinder and Lederberg, 1952)—transfer of genes from one cell to another (usually bacteria) by special viruses, the bacteriophages.

Transfection—introduction of pure DNA into a living cell (cf. transformation).

tRNA—Transfer RNA, the intermediate between mRNA and protein synthesis carrying a specific amino acid to the growing polypeptide chain in the ribosome.

Transformation—this term has several different meanings in biology. In genetics, three main types of transformation are distinguished: 1) *malignant* transformation, the transition of a normal cell to a malignant state with loss of control of proliferation; 2) *genetic* transformation (Griffith, 1928; Avery et al. 1944), a change of genetic attributes of a cell by transfer of genetic information; and 3) *blastic* transformation, the reaction of lymphocytes to mitogenic substances (e.g., phytohemagglutinin or specific antigens) leading to cell division.

Transgene—a cloned gene introduced into a plant or animal and passed to subsequent generations.

Transgenic—refers to an animal or a plant into which a cloned gene has been introduced and stably incorporated. It reveals information about the biological function of the (trans)gene.

Transition—a mutation by replacement of a purine with another purine or a pyrimidine with another pyrimidine (see transversion).

Translation—the second step in the relay of genetic information. Here the sequence of triplets in mRNA are translated into a corresponding sequence of amino acids to form a polypeptide as the gene product.

Translocation—transfer of all or part of a chromosome to another chromosome. A translocation is usually *reciprocal*, leading to an exchange of nonhomologous chromosomal segments. A translocation between two acrocentric chromosomes that lose their short arms and fuse at their centromeres is called a *fusion* type translocation (Robertsonian translocation).

Transmembrane protein—a protein located in the cell membrane with domains inside and outside the cell.

Transposition—movement of a genetic element, a transposón, from one location in the genome to another.

Transposon—a DNA sequence with the ability to move and be inserted at a new location of the genome.

Transversion—a mutation with replacement of a purine with a pyrimidine or vice versa (see transition).

Triplet—a sequence of three nucleotides comprising a codon of a nucleic acid and representing the code for an amino acid (triplet code, see codon).

Trisomy (Blakeslee, 1922)—an extra chromosome in addition to a homologous pair of chromosomes.

Tumor suppressor gene—a gene that can suppress tumor development by one functional allele (see oncogene).

UPD—uniparental disomy. Refers to the presence of two chromosomes derived from the same parent, either as isodisomy when both chromosomes are identical, or heterodisomy when different.

Upstream—5' direction of a gene.

Variation—the differences among related individuals, e. g., parents and offspring, or among individuals in a population.

Variegation—different phenotypes within a tissue.

Vector—a molecule that can incorporate and transfer DNA.

Virion—a complete extracellular viral unit or particle.

Virus—DNA or RNA of defined size and sequence, enclosed in a protein coat encoded by its genes and able to replicate in a susceptible host cell only.

VNTR—variable number of tandem repeats; a type of DNA polymorphism.

Voltage-gated channel—an ion channel that is opened or closed by a gradient of electric current.

Western blot—technique to identify protein antigens, in principle similar to the Southern blot method (q.v.).

Wild-type—refers to the genotype or phenotype of an organism found in nature or under standard laboratory conditions, roughly meaning "normal."

X chromatin (formerly called Barr body or sex chromatin) (Barr and Bartram, 1949)—darkly staining condensation in the interphase cell nucleus representing an inactivated X chromosome

Xenogenic—refers to transplantation between individuals of different species.

X-inactivation (Lyon, 1961)—inactivation of one of the two X chromosomes in somatic cells of female mammals during the early embryonic period by formation of X chromatin.

X-linked—refers to genes on the X chromosome.

YAC (yeast artificial chromosome)—a yeast chromosome into which foreign DNA has been inserted for replication in dividing yeast cells. YACs can incorporate relatively large DNA fragments, up to about 1000 kb (see BAC).

Yeast two-hybrid system—a technique to identify genes or proteins that interact in function.

Y chromatin (F body; Pearson, Bobrow, Vosa, 1970)—the brightly fluorescent long arm of the Y chromosome visible in the interphase nucleus.

Z DNA—alternate conformation of DNA. Unlike normal B DNA, (Watson-Crick model), the helix is left-handed and angled (zigzag, thus Z DNA).

Zinc finger—A finger-shaped region found in many DNA-binding regulatory proteins, the "finger" being held together by a strategically placed zinc atom.

Zoo blot—a Southern blot containing conserved DNA sequences from related genes of different species. It is taken as evidence that the sequences are coding sequences from a gene (see p. 242).

Zygote (Bateson, 1902)—the new diploid cell formed by the fusion of the two haploid gametes, an ovum and a spermatozoon, at fertilization. The cell from which the embryo develops.

Zygotene (de Winiwarter, 1900)—a stage of prophase of meiosis I.

References to the Glossary

Bodmer WF, Cavalli-Sforza LL: Genetics and the Evolution of Man. W. H. Freeman & Co., San Francisco, 1976.

Brown TA: Genomes. 2nd ed. BIOS Scientific Publishers, Oxford, 2002.

Dorland's Illustrated Medical Dictionary, 28th ed. W.B. Saunders Co., Philadelphia, London, Toronto, Montreal, Sydney, Tokyo, 1994.

Griffiths AJF et al: An Introduction to Genetic Analysis, 7th ed. W.H. Freeman, New York, 2000.

Hellmuth L: A Genome Glossary. Science 291: 1197, 2001.

King RC, Stansfield WD: A Dictionary of Genetics, 6th ed. Oxford Univ. Press, Oxford, 2002.

Lewin, B.: Genes VIII. Pearson International, 2004.

Lodish H et al: Molecular Cell Biology, 5th ed. W.H. Freeman, New York, 2004.

Passarge, E.: Definition genetischer Begriffe (Glossar), pp. 311–323. In: Elemente der Klinischen Genetik. G. Fischer, Stuttgart, 1979.

Rieger R, Michaelis A, Green MM: Glossary of Genetics and Cytogenetics, 5th ed. Springer Verlag, Berlin, Heidelberg, New York, 1979.

Watson, J.D.: Molecular Biology of the Gene, 3rd ed. W.A. Benjamin, Menlo Park, California, 1976.

Whitehouse HLK: Towards the Understanding of the Mechanisms of Heredity, 3rd ed. Edward Arnold, London, 1973.

Web site

Glossary of Genetic Terms, National Institute of Human Genome Research
http://www.nhgri.nih.gov/DIR/VIP/Glossary/

Index

Note: page numbers in *italics* refer to figures.

dimethyl sulfate 62, *63*
disaccharides 32, *33*
discontinuous variation 170
disease 12–13
  chromosomal location 422, *423–427,*
      428–432, 446
  FGF receptors 437
  G proteins 438
  genomic instability 340, *341*
  geographical distribution 174, *175*
  hereditary muscle 388, *389*
  selective advantage of heterozygotes 174
  susceptibility 160
   see also mitochondrial diseases
disease locus 150, *151*
  linkage analysis 156, *157*
disheveled (Dsh) protein 274, *275*
disomy, uniparental (UPD) 198, *199,* 200, *201,*
      410, *411,* 452, 467
dispermy 198, 452
dissociation *(Ds)* 262, *263*
DMD gene 388, *389*
DMPK gene 406, *407*
DNA 2–3, 452
  A-form 48, *49*
  amplification 60, *61*
  B-form 48, *49*
  chloroplasts 132, 134, *135*
  chromosomes 186, *187*
  cleavage 66, *67*
  components 44, *45*
  copy number differences 206, *207*
  denaturation 46, *47,* 60
  depurination 82, *83*
  double helix 7–8, 44, *45,* 46, *47*
    physical dimensions 48, *49*
  double-strand breaks 84, *85*
  duplex loop
    formation 188, *189*
    recombination 84, *85*
  excision repair deficiency 92
  expressed 242, *243*
  footprinting 222, *223*
  G-rich tandem sequences 188, *189*
  hormone response element binding 220, *221*
  human genome 252, *253*
  intervening sequences 58, *59*
  long-range gene activation 218, *219*
  methylation 80, *81,* 228, *229*
  microarrays 452
  polymorphism 78, *79,* 244
  primer annealing 60
  recognition sequences 314, *315*

  regulatory protein binding 220, *221*
  renaturation 46, *47,* 463
  repair 4, 84, *85,* 90, *91*
  replication 46, *47,* 50, *51,* 463
  segment stability 76, *77*
  sequencing 62, *63,* 64, *65,* 240
  structure 7–8, 46, *47,* 48, *49*
  synthesis 60, *61,* 106, *107*
  transfer between cells 100, *101*
  transforming principle 7, 42, *43*
  transposons 252, *253*
  triple-stranded 48
  variability 78, *79*
  Z-form 48, *49,* 467
  zinc finger protein binding 220, *221*
   see also recombination; transcription;
      transduction; transfection; translation;
      transposition
DNA cloning 68, *69*
  cell-based 68, *69*
  overlapping clones 244, *245*
  vectors 68, *69,* 70, *71*
DNA libraries 68, 70, 72, *73*
DNA ligase 50, *51*
DNA polymerase 50, *51,* 188, *189,* 452
DNA-binding proteins 220, *221,* 222, *223*
DNA-mismatch repair genes 330, *331*
DNMT3B gene 228, *229*
domain shuffling 264
dominant negative effects 80, *81*
dominant traits 144, *145*
  independent distribution 142, *143*
  segregation 140, *141*
dot blot analysis 76, *77*
double-strand repair 90, *91*
double-stranded RNA (dsRNA) 224, *225*
Down syndrome 11, 200, *201,* 412, *413*
Drosophila melanogaster 3, 5, 138
  bicoid genes 298, *299*
  bithorax complex mutations 300, *301*
  chromosomes 178, *179*
  egg-polarity genes 298, *299,* 300, *301*
  embryonic development 298, *299*
  gap genes 298, *299,* 300, *301*
  genome *239*
  hedgehog gene 276, *277,* 298, *299*
  homeobox genes 300, *301*
  homeotic selector genes 298, *299*
  life cycle 298, *299*
  mutations 298, *299*
  nerve cell lateral inhibition 278, *279*
  nondisjunction 198, *199*
  notch mutant 278, *279*